GCSE
Biology

second edition

D.G. Mackean

JOHN MURRAY

ACKNOWLEDGEMENTS

I am very grateful to all the people who have provided photographs and would particularly like to thank Christine Hood, Dr C. J. Clegg and Dr Margaret Frayne, who read a substantial part of the manuscript and offered many constructive criticisms.

I also acknowledge the following Examining Groups who have kindly allowed me to reproduce questions from GCSE specimen papers and past examinations papers from 1992–94.

W = Welsh Joint Education Committee
N = Northern Examining Association
S = Southern Examining Group
L = University of London Examinations and Assessment Council
NI = Northern Ireland Council for the Curriculum, Examinations and Assessment
M = Midland Examining Group

First published in 1986
by John Murray (Publishers) Ltd
50 Albemarle Street
London W1X 4BD

Reprinted 1987, 1988, 1989 (with revisions), 1991, 1992, 1993 (twice)

Second edition 1995

Reprinted 1996 (twice, with revisions), 1997, 1998 (twice), 2000

Technical artwork by Art Construction
Natural history artwork by Chris Etheridge
The full-colour illustrations on pages 286-289 are by Pamela Haddon

Artwork labels and preparation by John Townson/Creation

Layouts by Fiona Webb

Typeset in 11/13.5pt Garamond Book Condensed by Wearset, Boldon, Tyne and Wear

Printed and bound in Great Britain by Butler & Tanner Limited, Frome and London

A CIP catalogue record for this book is available from the British Library

ISBN 0-7195-5302-4

Artwork figure acknowledgements

CHAPTER 2
Fig. 3 From *Principles of Plant Physiology* by Bonner and Galston. Copyright © 1952 by W.H. Freeman and Company. Used with permission.

CHAPTER 4
Fig. 8 Data of Brierley, *Plant Physiology*, 1954, reproduced by kind permission of the publishers, the Association for Science Education.

CHAPTER 5
Fig. 14 From Dr S. B. Verma and Professor N. J. Rosenburg, *Agriculture and the Atmospheric Carbon Dioxide Build-up* (*Span*, 22 February 1979).

CHAPTER 10
Fig. 7 Reproduced by permission of the Health Education Authority.

CHAPTER 12
Fig. 27 From *Smoking or Health: a report of the Royal College of Physicians*, 1977, Pitman Medical Publishing Co. Ltd.

CHAPTER 13
Fig. 10 From *Smoking and Health Now: a report of the Royal College of Physicians*, Pitman Medical Publishing Co. Ltd.

CHAPTER 16
Fig. 20 G.W. Corner, The Hormones in Human Reproduction. Copyright © 1942 by Princeton University Press. Reprinted/reproduced by permission of Princeton University Press.

CHAPTER 20
Fig. 2 From J. Besford, *Good Mouthkeeping*, 2nd Edn, Oxford University Press, 1984. Reproduced by permission of Oxford University Press.

CHAPTER 23
Fig. 13 After Grove and Newell, *Animal Biology,* 6th Edn, University Tutorial Press, 1961.

CHAPTER 24
Fig. 16 Reprinted with the permission of Simon & Schuster, Inc. from the Macmillan College text *Communities and Ecosystems 2/E* by Robert H. Whittaker. Copyright © 1975.

CHAPTER 25
Fig. 1 From P.W. Freeland, *School Science Review*, 1973, reproduced by kind permission of P.W. Freeland and the Association for Science Education.
Fig. 2 From J. Barker, *In the Balance*, Evans, 1975.
Fig. 15 From W. E. Shewell-Cooper, *The ABC of Soils*, English University Press.
Fig. 18 Fom Clive A. Edwards, *Soil Pollutants and Soil Animals*. Copyright © 1969 by Scientific American Inc. All rights reserved.
Fig. 25 First published in *New Scientist*, 27 May 1982.
Fig. 27 From J. E. Hansen and S. Lebedeff in *New Scientist*, 22 October 1985.

CHAPTER 26
Fig. 1 From J. Barker, *Breathing Space*, Evans, 1975.
Fig. 10 First appeared in *New Scientist*, 31 July 1986.

CHAPTER 27

Fig. 5 From P. S. Maitland, *The Impact of Farmed Atlantic Salmon on Wild Populations*, Nature Conservancy Council, Edinburgh, 1989. Reproduced with permission of Scottish Natural Heritage.

Fig. 10 From Ewer and Hall, *Ecological Biology*, Longman, 1978.

Fig. 11 From Bennet and Humphries, *Introduction to Field Biology*, Edward Arnold (Publishers) Limited, 1974. Reproduced by permission of Edward Arnold (Publishers) Limited.

CHAPTER 28

Fig. 1 From Lewis and Taylor, *Introduction to Experimental Ecology*, Academic Press, 1966.

Fig. 2 From Lewis and Taylor, *Introduction to Experimental Ecology*, Academic Press, 1966.

Fig. 4 From Lewis and Taylor, *Introduction to Experimental Ecology*, Academic Press, 1966.

Fig. 7 Reproduced with permission from James Bonner, *The world's people and the world's food supply*, 1980, Carolina Biology Readers Series © Carolina Biological Supply Company, Burlington, North Carolina, USA.

CHAPTER 33

Fig. 3 From N. Tinbergen, *The Study of Instinct*, Oxford University Press, 1951. Reproduced by permission of Oxford University Press.

CHAPTER 34

Fig. 6 After C. M. Jackson.

CHAPTER 35

Fig. 8 From James Gray, *How Animals Move*, Cambridge University Press, 1953.

CHAPTER 38

Fig. 1 After Brian Jones, *Introduction to Human and Social Biology*, 2nd Edn, 1985.

Fig. 6 After Brian Jones, *Introduction to Human and Social Biology*, 2nd Edn, 1985.

Photo acknowledgements

Thanks are due to the following for permission to reproduce copyright photographs:

Cover Manfred Danegger/NHPA; **Title page** Jane Burton/Bruce Coleman; **p.1** Weiss/Jerrican/Science Photo Library; **p.2** Fig. 1 Biophoto Associates; **p.3** Fig. 3 Biophoto Associates; **p.5** Fig. 10 Philip Harris Education; **p.9** Fig. 16 Biophoto Associates; **p.10** Fig. 17 Biophoto Associates; **p.13** Fig. 4 Biophoto Associates; **p.18** Fig. 11 D.G. Mackean; **p.32** Fig. 15 Biophoto Associates; **p.34** Fig. 18 D.G. Mackean; **p.35** Figs. 20a,b J.C. Revy/Science Photo Library; **p.40** Fig. 6 Heather Angel; **p.44** Fig. 11 Horticultural Research International, Littlehampton; **p.47** Fig. 16 Horticultural Research International, Littlehampton; **p.48** Fig. 19 Horticultural Research International, Littlehampton; **p.51** Hans Reinhard/Bruce Coleman; **p.53** Fig. 3 Sidney Moulds/Science Photo Library; **p.54** Fig. 4 Professor T. Mansfield; **p.55** Fig. 7b Gene Cox; **p.57** Fig. 10b Biophoto Associates; Fig. 10c*; **p.58** Fig. 12 Biophoto Associates; **p.59** Fig. 14 D.G. Mackean; **p.60** Fig. 18 Terry Heathcote/Oxford Scientific Films; **p.64** Fig. 4 John Mason/Ardea; **p.66** Fig. 5 Ronald Toms/Oxford Scientific Films; **p.72** Fig. 3 D.G. Mackean; **p.73** Fig. 7 Dr Michael Proctor; **p.74** Fig. 8 Biophoto Associates; Fig. 9 Heather Angel; **p.76** Figs. 12a,b,c, 15 D.G. Mackean; **p.78** Fig. 18 Kim Taylor/Bruce Coleman; **p.80** Figs. 23a,b D.G. Mackean; **p.81** Fig. 25 International Plant Laboratories Ltd.; **p.82** Fig. 26 Biophoto Associates; **p.85** Fig. 4 Jane Burton/Bruce Coleman; **p.86** Fig. 6 D.G. Mackean; **p.88** Fig. 9 D.G. Mackean; **p. 89** Fig. 12 D.G.

Mackean; **p.90** Figs. 13, 15 D.G. Mackean; **p.93** Gray Mortimore/AllSport; **p.97** Figs. 3a,b,c MMB/The Anthony Blake Photo Library, 3d Milk Marque/The Anthony Blake Photo Library; **p.99** Fig. 6 Last Resort; **p.101** Fig. 8 Last Resort; **p.110** Fig. 11 David Scharf/Science Photo Library; **p.117** Fig. 2 Biophoto Associates; **p.121** Fig. 15 Biophoto Associates; **p.125** Fig. 20 Biophoto Associates; **p.127** Fig. 23 Biophoto Associates; Fig. 24 Philip Harris Education; Fig. 25 Jerry Mason/Science Photo Library; **p.132** Fig. 3 Biophoto Associates; **p.134** Fig. 8 Philip Harris Education; **p. 136** Figs. 11a,b Biophoto Associates; Fig. 12 Popperfoto; **p.140** Fig. 4 Biophoto Associates; **p.143** Fig. 8 Biophoto Associates; **p.145** Fig. 1 Biophoto Associates; **p.146** Fig. 2 Biophoto Associates; **p.148** Fig. 6 Gerard Vandystadt/Agence Vandystadt/AllSport; **p.152** Fig. 7 John Walsh/Science Photo Library; **p.153** Fig. 9 Biophoto Associates; **p.154** Fig. 11 London Gynaecology and Fertility Centre Limited; **p.155** Fig. 15 Professor W.J. Hamilton*; **p.156** Fig. 16 reproduced with permission from Birth Atlas, Maternity Center Association, New York; **p.157** Fig. 17 Biophoto Associates; **p.158** Fig. 18 © Camilla Jessel; Fig. 19 D.G. Mackean; **p.164** Fig. 4 Rank Organisation*; **p.166** Fig. 8 Sporting Pictures (UK) Ltd.; **p.170** Fig. 13 Professor Crispian Scully, Eastman Dental Institute for Oral Healthcare Sciences, London; **p.171** Fig. 2 Biophoto Associates; **p.179** Fig. 1 Sporting Pictures (UK) Ltd; **p.183** Fig. 10 BBC Natural History Unit/Mike Benyon; **p.184** Fig. 11 Biophoto Associates; **p.185** Fig. 13 Biophoto Associates; **p.187** Fig. 15 Biophoto Associates; **p.190** Fig. 1 Sheila Terry/Science Photo Library; **p.192** Fig. 3 Mike Powell/AllSport USA; **p.197** Jane Burton/Bruce Coleman; **p.199** Fig. 2 Manfred Kage/Science Photo Library; **p.202** Fig. 8 Biophoto Associates; **p.204** Fig. 13 Dr. A. Lesk, Laboratory of Molecular Biology, Cambridge/Science Photo Library; **p.207** Fig. 17 Philippe Plailly/ Eurelios/Science Photo Library; Fig. 18 Cystic Fibrosis Trust (photograph: Camilla Jessel); **p.208** Figs. 19, 20 Zeneca Seeds; **p.213** Fig. 5 Philip Harris Education; **p.216** Fig. 9 Adam Hart-Davis/Science Photo Library; **p.218** Fig. 11 Sir Ralph Riley; **p.219** Fig. 1 Horticultural Research International, East Malling; **p.223** Fig. 6 Biophoto Associates; Fig. 7a Oxford Scientific Films; Figs. 7b,c,d Peter Parks/Oxford Scientific Films; Fig. 8 Biophoto Associates; **p.224** Fig. 10 Biophoto Associates; **p.231** David Woodfall/NHPA; **p.233** Fig. 3 Dr D.P. Wilson/FLPA Ltd.; Fig. 4 Manfred Danegger/NHPA; **p.235** Fig. 6a,b Colin Green; Fig. 7 N.G. Blake/Bruce Coleman; **p.238** Fig. 11 D.G. Mackean; **p.239** Fig. 13 Robert P. Carr/Bruce Coleman; **p.240** Fig. 17 Nigel Cattlin/Holt Studios; **p.243** Fig. 3 David Melville, WWF US; Fig. 4 Bob Gibbons/Ardea; **p.244** Fig. 5 Last Resort; Figs. 6, 7 Biophoto Associates; Fig. 8 Zeneca Agrochemicals; **p.246** Fig. 10 Heather Angel; Fig. 12 Alexandra Jones/Ecoscene; **p.247** Fig. 13 Biophoto Associates; **p.248** Fig. 14 Heather Angel; Fig. 16 ADAS Aerial Photography Unit - Crown Copyright; **p.250** Fig. 19 J. Swedberg/Ardea; **p.251** Fig. 20 Cuthbert/Ecoscene; Fig. 21 Norman Tomalin/Bruce Coleman; Fig. 22 Sally Morgan/Ecoscene; Fig. 23 Hans Reinhard/Bruce Coleman; **p.256** Fig. 2 Ford Motor Company Limited; Fig. 3 Peter Addis*; **p.257** Fig. 5 Thames Water Plc; Fig. 6 A. Pasking/Water Research Laboratory*; **p.258** Fig. 7 Thames Water Plc; Fig. 8 Ian Harwood/Ecoscene; **p.259** Fig. 9 Warren Spring Laboratory*; **p.260** Fig. 11 reproduced by kind permission of Cleanaway Limited; **p.261** Fig. 12 Alex Bartel/Science Photo Library; Fig. 13 London Waste Regulation Authority; **p.262** Fig. 14a Zeneca Agrochemicals; Fig. 14b B. Roth/Okapia/Oxford Scientific Films; Fig. 16 (top) Dr Eckart Pott/ Bruce Coleman; Fig. 16 (bottom) CSIRO Division of Entomolgy, Canberra; **p.263** Figs. 17a,b, 18 Nigel Cattlin/Holt Studios; Fig. 19 Agri Sense - BCS Ltd.; **p.266** Fig. 1 Ecosphere Associates Ltd., Tuscon, Arizona; Fig. 2 H. Schrempp/FLPA Ltd.; **p.267** Fig. 3 Last Resort; Fig. 4 P. Morris/Ardea; **p.268** Fig. 6 Doug Allan/Oxford Scientific Films; Fig. 7 Will McFadyen/The Wildfowl & Wetlands Trust; **p.269** Fig. 9 Topham Picture Point; **p.271** Fig. 15 Biophoto Associates; **p.272** Fig. 17 Julian Bateson/English Nature; **p.277** Fig. 3 Roger Wilmshurst; **p.280** Fig. 9 Mark Edwards/Still Pictures; **p.283** Heather Angel; **p.285** Fig. 2 Heather Angel; Fig. 3 National Institute of Agricultural Botany, Cambridge; **p.290** Fig. 9 Kim Taylor/Bruce Coleman; **p.294** Fig. 19 Hans Reinhard/Bruce Coleman; **p.295** Fig. 21 Heather Angel; Fig. 22 Dr Alan Beaumont; **p.296** Figs. 23, 26 Heather Angel; Figs. 24, 25 Biophoto Associates; **p.299** Fig. 2 Biophoto Associates; **p.302** Fig. 10 Heather Angel; **p.303** Fig. 11 Nigel Cattlin/Holt Studios; Fig. 12 Biophoto Associates; Fig. 13 Heather Angel; **p.304** Fig. 15 Biophoto Associates; **p.305** Fig. 17 Audio Visual Services, St. Mary's Medical School, London W2; Fig. 18 Dr David Hockley, National Institute for Biological Standards and Control; **p.306** Figs. 20, 21 Biophoto Associates; **p.310** Fig. 2 Biophoto Associates; Fig. 4 Heather Angel; **p.311** Fig. 5b Andy Callow/NHPA; Fig. 5c Andy Pucell/Bruce Coleman; **p.312** Fig. 7c D.G. Mackean; **p.313** Fig. 8 Dr Jeremy Burgess/Science Photo Library; **p.315** Fig. 3 David Thompson/Oxford Scientific Films; Fig. 4 Heather Angel; **p.317** Fig. 8 Dr K. Oates, University of Lancaster; **p.318** Fig. 2 Kim Taylor/Bruce Coleman; **p.319** Fig. 4 Kim Taylor/ Bruce Coleman; Fig. 5 Biophoto Associates; Fig. 6 Ardea; **p.320** Fig. 7 Heather Angel; Fig. 8 Owen Newman/Nature Photographers Ltd.; **p.321** Fig. 11 Biophoto Associates; Fig. 12 J.A.L. Cooke/Oxford Scientific Films; **p.322** Fig. 13 P. Morris/Ardea; Fig. 14 Heather Angel; **p.325** Fig. 5 London Scientific Films/Oxford Scientific Films; **p.329** Fig. 2 M.I. Walker/NHPA; **p.331** Fig. 7 Heather Angel; **p. 334** Fig. 14 Jane Burton/Bruce Coleman; Fig. 15 P. Morris/Ardea; **p.335** Figs. 1, 2 Heather Angel; Fig. 3 Hans Reinhard/Bruce Coleman; **p.336** Fig. 4a(inset) Biophoto Associates; Fig. 4b Heather Angel; **p. 339** CNRI/Science Photo Library; **p.341** Fig. 1 The Anthony Blake Photo Library; Fig. 2 Stuart Boreham/CEPHAS; **p.342** Fig. 3 The Anthony Blake Photo Library; **p.343** Fig. 4 SmithKline Beecham; **p.344** Fig. 6 Laurence White; **p.346** Fig. 8 Sinclair Stammers/Science Photo Library; **p.352** Fig. 5 Jonathan Watts/Science Photo Library; **p.353** Fig. 7 Tim Beddow/Science Photo Library; **p.354** Fig. 9 Thames Water Plc; **p.355** Figs. 10, 11 J. Sainsbury plc; **p.356** Fig. 12 Adam Hart-Davis/Science Photo Library; **Appendix 3** Fig. 1 Mary Evans Picture Library.

* Every effort has been made to contact these copyright holders, and the publishers apologise for any omissions, which they will be pleased to rectify at the earliest opportunity.

CONTENTS

GENETICS AND HEREDITY

ORGANISMS AND THEIR ENVIRONMENT

DIVERSITY OF ORGANISMS

MICRO-ORGANISMS AND HUMANS

TO THE STUDENT

This is a textbook to help you in studying biology for the GCSE. You will be following the GCSE syllabus of only one Examination Group, but this book contains the material needed by all the Groups. For this reason, amongst others, it is not expected that you will need to study or learn everything in the book.

Furthermore, the emphasis in GCSE is on the ability to understand and use biological information rather than on committing it all to memory. However, you will still need to use a book of this sort to find the facts and explanations before you can demonstrate your understanding or apply the biological principles.

The text is presented at two levels. *Core text* is in ordinary print and contains subject matter which occurs in nearly all the syllabuses. *Extension text* is indicated by a blue bar and covers topics that are beyond the basic syllabus requirements or are relevant to only one syllabus.

The questions included in a chapter are intended to test your understanding of the text you have just read. If you cannot answer the questions straightaway, read that section of text again with the question in mind.

There are check lists at the end of each chapter, summarising the important points covered.

The questions at the end of each section and those in Appendix 2 are selected from the examination papers published by the GCSE Examining Groups. In many cases, they are designed to test your ability to apply your biological knowledge. The question may provide certain facts and ask you to make interpretations or suggest explanations. In such cases, the factual information may not be covered in the text.

Looking up information   To find the information you need, use the index, the contents pages and the headings at the beginning of each chapter—where a blue bar indicates an extension topic. If the word you want does not appear in the index, try a related word. For example, information about 'sight' might be listed under 'vision', 'eyes' or 'senses'.

Practical work   Given standard laboratory equipment, it should be possible for you to attempt any of the practical work described in the book, though you will probably not have time to do it all.

For this reason, it has been necessary to give the expected results of the experiments so that you can appreciate the design and purpose of an experiment and consider alternative interpretations of the results even if you have not been able to do the experiment yourself.

Experimental Work in Biology, D. G. Mackean (Murray, 1983) describes about 140 tested experiments, with detailed instructions. Results are not given but you are asked questions to test your understanding of the experimental design and your ability to make critical interpretations of the results.

Second Edition   This edition reflects the requirements of the National Curriculum (Biology) as interpreted by the six UK Examining Groups.

Some Principles of Biology

1 CELLS AND TISSUES

Cell structure
How tissues are studied to see cells: taking sections. Cell components. Plant cells.

Cell division and specialization
Cell division and growth. Specialization of cells for different functions.

Tissues and organs
Definitions and examples of tissues, organs and systems.

Practical work
Preparing, observing and drawing plant and animal cells.

Cell structure

If a very thin slice of a plant stem is cut and studied under a microscope, it can be seen that the stem consists of thousands of tiny, box-like structures. These structures are called **cells**. Fig. 1 is a thin slice taken from the tip of a plant shoot and photographed through a microscope. Photographs like this are called **photomicrographs**. The one in Fig. 1 is 60 times larger than life, so a cell which appears to be 2 mm long in the picture, is only 0.03 mm long in life.

Figure 1 Longitudinal section through the tip of a plant shoot (x60). The slice is only one cell thick, so light can pass through it and allow the cells to be seen clearly.

Thin slices of this kind are called **sections**. If you cut *along the length* of the structure, you are taking a **longitudinal section**. Fig. 1 shows a longitudinal section, which passes through two small

developing leaves near the tip of the shoot, and two larger leaves below them. The leaves, buds and stem are all made up of cells. If you cut *across* the structure, you make a **transverse section** (Fig. 2).

(a) transverse section   **(b) longitudinal section**

Figure 2 Cutting sections of a plant stem

It is fairly easy to cut sections through plant structures just by using a razor blade. To make a microscopic study of animal structures is more difficult because they are mostly soft and flexible. Pieces of skin, muscle or liver, for example, first have to be soaked in melted wax. When the wax goes solid it is then possible to cut thin sections. The wax is dissolved away after making the section.

When sections of animal structures are examined under the microscope, they, too, are seen to be made up of cells but they are much smaller than plant cells and need to be magnified more. The photomicrograph of kidney tissue in Fig. 3 has been magnified 700 times to show the cells clearly. The sections are often treated with dyes, called 'stains', in order to show up the structures inside the cells more clearly.

Making sections is not the only way to study cells. Thin strips of plant tissue, only one cell thick, can be pulled off stems or leaves (Experiment 1, p. 9). Plant or animal tissue can be squashed or smeared on a microscope slide (Experiment 2, p. 10) or treated with chemicals to separate the cells before studying them.

Figure 3 Transverse section through a kidney tubule (×700). A section through a tube will look like a ring (see Fig. 12(b) on p. 6). In this case, each 'ring' consists of about 10 cells.

There is no such thing as a typical plant or animal cell because cells vary a great deal in their size and shape depending on their function. Nevertheless, it is possible to make a 'generalized' drawing like Fig. 4 to show features which are present in most cells. *All cells* have a **cell membrane** which is a thin boundary enclosing the **cytoplasm**. Most cells have a **nucleus**.

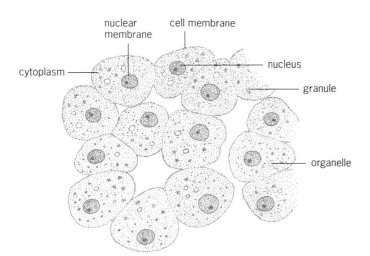

Figure 4 A group of animal cells, e.g. cells from the lining of the cheek

Cytoplasm

Under the ordinary microscope (light microscope), cytoplasm looks like a thick liquid with particles in it. In plant cells it may be seen to be flowing about. The particles may be food reserves such as oil droplets or granules of starch. Other particles are structures which have particular functions in the cytoplasm. These structures are the **organelles**. Examples are the **ribosomes** which build up the cell's proteins (see p. 11) and the **mitochondria** which generate energy for the cell's living processes (see p. 20).

When studied at much higher magnifications with the **electron microscope**, the cytoplasm no longer looks like a structureless jelly but appears to be organized into a complex system of membranes and vacuoles.

In the cytoplasm, a great many chemical reactions are taking place which keep the cell alive by providing energy and making substances that the cell needs (see pp. 11 and 20).

The liquid part of cytoplasm is about 90 per cent water with molecules of salts and sugars dissolved in it. Suspended in this solution there are larger molecules of fats (lipids) and proteins (see pp. 11–12). Lipids and proteins may be used to build up the cell structures, e.g. the membranes. Some of the proteins are **enzymes** (p. 13). Enzymes control the rate and type of chemical reactions which take place in the cells. Some enzymes are attached to the membrane systems of the cell, others float freely in the liquid part of the cytoplasm.

Cell membrane

This is a thin layer of cytoplasm round the outside of the cell. It stops the cell contents from escaping and also controls the substances which are allowed to enter and leave the cell. In general, oxygen, food and water are allowed to enter; waste products are allowed to leave and harmful substances are kept out. In this way the cell membrane maintains the structure and chemical reactions of the cytoplasm.

Nucleus (plural = nuclei)

Most cells contain one nucleus, usually seen as a rounded structure enclosed in a membrane and embedded in the cytoplasm. In drawings of cells, the nucleus may be shown darker than the cytoplasm because, in prepared sections, it takes up certain stains more strongly than the cytoplasm. The function of the nucleus is to control the type and quantity of enzymes produced by the cytoplasm. In this way it regulates the chemical changes which take place in the cell. As a result, the nucleus determines what the cell will be, e.g. a blood cell, a liver cell, a muscle cell or a nerve cell.

The nucleus also controls cell division as shown in Fig. 8. A cell without a nucleus cannot reproduce. Inside the nucleus are thread-like structures called **chromosomes** which can be seen most easily at the time when the cell is dividing. (See p. 198 for a fuller account of chromosomes.)

The term **protoplasm** is sometimes used to describe the cytoplasm, nucleus and cell membrane together.

Mitochondria

The mitochondria are tiny organelles present in plant and animal cells. They may be spherical, rod-like or elongated. They are most numerous in regions of rapid chemical activity and are responsible for producing energy from food substances (see 'Respiration', p. 20).

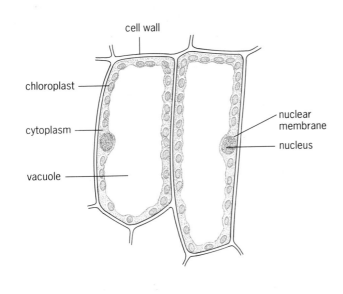

cell wall

chloroplast

cytoplasm

vacuole

nuclear membrane

nucleus

chloroplast

cell membrane

vacuole

cytoplasm

cell wall

nucleus

(b) transverse section

(a) longitudinal section

Figure 6 Structure of a palisade cell. It is important to remember that, although cells look flat in sections or in thin strips of tissue, they are in fact three-dimensional and may seem to have different shapes according to the direction in which the section is cut. If the cell is cut across it will look like (b); if cut longitudinally it will look like (a).

Plant cells

A few generalized animal cells are represented by Fig. 4, while Fig. 5 is a drawing of two palisade cells from a plant leaf (see p. 55).

Plant cells differ from animal cells in several ways:

1 Outside the cell membrane they all have a **cell wall** which contains cellulose and other compounds. It is non-living and allows water and dissolved substances to pass through. The cell wall is not selective like the cell membrane. (Note that plant cells *do* have a cell membrane but it is not easy to see or draw because it is pressed against the inside of the cell wall. See Fig. 6.)

Under the microscope, plant cells are quite distinct and easy to see because of their cell walls. In Fig. 1 it is only the cell walls (and in some cases the nuclei) which can be seen. Each plant cell has its own cell wall but the boundary between two cells side by side does not usually show up clearly. Cells next to each other therefore appear to be sharing the same cell wall.

2 Most mature plant cells have a large, fluid-filled space called a **vacuole**. The vacuole contains **cell sap**, a watery solution of sugars, salts and sometimes pigments. This large, central vacuole pushes the cytoplasm aside so that it forms just a thin lining inside the cell wall. It is the outward pressure of the vacuole on the cytoplasm and cell wall which makes plant cells and their tissues firm (see p.32). Animal cells may sometimes have small vacuoles in their cyotplasm but they are usually produced to do a particular job and are not permanent.

3 In the cytoplasm of plant cells are many organelles called **plastids** which are not present in animal cells. If they contain the green substance **chlorophyll**, the organelles are called **chloroplasts** (see p. 40). Colourless plastids usually contain starch which is used as a food store.

The shape of a cell when seen in a transverse section may be quite different when the same cell is seen in a longitudinal section and Fig. 6 shows why this is so. Figs 10(b) and (c) on p. 57 show the appearance of cells in a stem vein as seen in transverse and longitudinal section.

QUESTIONS

1 **a** What structures are usually present in all cells, whether they are from an animal or from a plant?
   b What structures are present in plant cells but not in animal cells?
2 What cell structure is largely responsible for controlling the entry and exit of substances into or out of the cell?
3 In what way does the red blood cell shown in Fig. 1 on p.116 differ from most other animal cells?
4 How does a cell membrane differ from a cell wall?
5 Why does the cell shown in Fig. 6(b) appear to have no nucleus?
6 **a** In order to see cells clearly in a section of plant tissue, would you have to magnify the tissue **i** ×5, **ii** ×10, **iii** ×100 or **iv** ×1000?
   b What is the approximate width (in mm) of one of the largest cells in Fig. 3?
7 In Fig. 3, the cell membranes are not always clear. Why is it still possible to decide roughly how many cells there are in each tubule section?

Cell division and cell specialization

Cell division

When plants and animals grow, their cells increase in numbers by dividing. Typical growing regions are the ends of bones, layers of cells in the skin, root tips and buds (Fig. 10). Each cell divides to produce two daughter cells. Both daughter cells may divide again, but usually one of the cells grows and changes its shape and structure and becomes adapted to do one particular job—in other words, it becomes **specialized** (Fig. 7). At the same time it loses its ability to divide any more. The other cell is still able to divide and so continue the growth of the tissue. **Growth** is, therefore, the result of cell division, followed by cell enlargement and, in many cases, cell specialization.

The process of cell division in an animal cell is shown in Fig. 8. The events in a plant cell are shown in Figs 9 and 10. Because of the cell wall, the cytoplasm cannot simply pinch off in the middle, and a new wall has to be laid down between the two daughter cells. Also a new vacuole has to form.

Organelles such as mitochondria and chloroplasts are able to divide and are shared more or less equally between the daughter cells at cell division.

Figure 10 Cell division in an onion root tip (x750). The nuclei are stained blue. Most of the cells have just completed cell division.

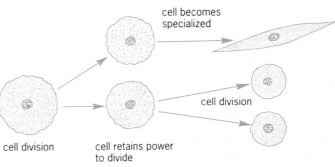

Figure 7 Cell division and specialization

(a) Animal cell about to divide.

(b) The nucleus divides first.

(c) The daughter nuclei separate and the cytoplasm pinches off between the nuclei.

(d) Two cells are formed — one may keep the ability to divide, and the other may become specialized.

Figure 8 Cell division in an animal cell

(a) A plant cell about to divide has a large nucleus and no vacuole.

(b) The nucleus divides first. A new cell wall develops and separates the two cells.

(c) The cytoplasm adds layers of cellulose on each side of the new cell wall. Vacuoles form in the cytoplasm of one cell.

(d) The vacuoles join up to form one vacuole. This takes in water and makes the cell bigger. The other cell will divide again.

Figure 9 Cell division in a plant cell

Specialization of cells

Most cells, when they have finished dividing and growing, become specialized. This means that:

1 They do one particular job.
2 They develop a distinct shape.
3 Special kinds of chemical change take place in their cytoplasm. The changes in shape and chemical reactions enable the cell to carry out its special function. Nerve cells and guard cells are examples of specialized cells.

Nerve cells (Fig. 11(e)):

1 Conduct electrical impulses to and from the brain.
2 Some of them are very long and connect distant parts of the body to the spinal cord and brain.
3 Their chemical reactions cause the impulses to travel along the fibre.

Guard cells of a stoma (Fig. 11(d)):

1 Open or close the stomatal pore in a leaf.
2 Their curved shape and unevenly thickened cell walls cause them to change shape when the pressure in the vacuole alters.
3 Changes in carbon dioxide concentration affect the chemical reactions in their cytoplasm and cell sap making the latter more or less concentrated.

The specialization of cells to carry out particular functions in an organism is sometimes referred to as **'division of labour'** within the organism. Similarly, the special functions of mitochondria, ribosomes and other cell organelles may be termed 'division of labour' within the cell.

QUESTIONS

8 Select from the following events and put them in the correct order for cell division in **i** animal cells,
ii plant cells:
a cytoplasm divides,
b vacuole forms in one cell,
c new cell wall separates cells,
d nucleus divides.
9 Look at Fig. 4 on p. 146. When a Malpighian cell of the skin divides, which daughter cell becomes specialized and which one keeps the ability to divide again?
10 Look at Fig. 2 on p. 53.
a Whereabouts in a leaf are the food-carrying cells?
b What other specialized cells are there in the leaf?

Tissues and organs

There are some microscopic organisms that consist of one cell only and can carry out all the processes necessary for their survival (see p. 284). The cells of the larger plants and animals cannot survive on their own. A muscle cell could not obtain its own food and oxygen. Other specialized cells have to provide the food and oxygen needed for the muscle cell to live. Unless these cells are grouped together in large numbers and made to work together, they cannot exist for long.

Tissue

A tissue such as bone, nerve or muscle in animals, and epidermis, phloem or pith (p. 56) in plants, is made up of many hundreds of cells of a few types. The cells of each type have similar structures and functions so that the tissue itself can be said to have a particular function, e.g. nerves conduct impulses, phloem carries food in plants. Fig. 12 shows how some cells are arranged to form simple tissues.

(a) cells forming an epithelium
A thin layer of tissue, e.g. the lining of the mouth cavity. Different types of epithelium form the internal lining of the windpipe, air passages, food canal, etc., and protect these organs from physical or chemical damage.

(b) cells forming a small tube
e.g. a kidney tubule (see p. 140). Tubules such as this carry liquids from one part of an organ to another.

(c) one kind of muscle cell
Forms a sheet of muscle tissue. Blood vessels, nerve fibres and connective tissues will also be present. Contractions of this kind of muscle help to move food along the food canal or to close down small blood vessels.

(d) cells forming part of a gland
The cells make chemicals which are released into the central space and carried away by a tubule such as shown in **b)**. Hundreds of cell groups like this would form a gland like the salivary gland.

Figure 12 How cells form tissues

Organs

Organs consist of several tissues grouped together to make a structure with a special function. For example, the stomach is an organ which contains tissues made from epithelial cells, gland cells and muscle cells. These cells are supplied with food and oxygen brought by blood vessels. The stomach also has a nerve supply. The heart, lungs, intestines, brain and eyes are further examples of organs in animals. In flowering plants, the root, stem and leaves are the organs. The tissues of the leaf are epidermis, palisade tissue, spongy tissue, xylem and phloem (see pp. 41 and 52–5).

(a) ciliated cells
These form the lining of the nose and windpipe, and the tiny cytoplasmic 'hairs', called cilia, are in a continually flicking movement keeping up a stream of fluid (mucus) that carries dust and bacteria away from the lungs.

(b) white blood cell
Occurs in the bloodstream and is specialized for engulfing harmful bacteria. It is able to change its shape and move about, even through the walls of blood vessels into the surrounding tissues.

(d) guard cells of a stoma
These two curved cells form an opening in the epidermis of a leaf and allow oxygen and carbon dioxide to pass in and out. They can change their shape and thus close the pore.

(e) nerve cell
Specialized for conducting impulses of an electrical nature along the fibre. A nerve consists of hundreds of fibres bound together. The fibres may be very long, e.g. from the foot to the spinal column.

(c) food-conducting cell in a plant (phloem cell)
Long cells, joined end to end, and where they meet, perforations occur in the walls. Through these holes the cytoplasm of one cell communicates with the next. Dissolved food is thought to pass through the holes during its transport through the stem.

Figure 11 Specialized cells (not to scale)

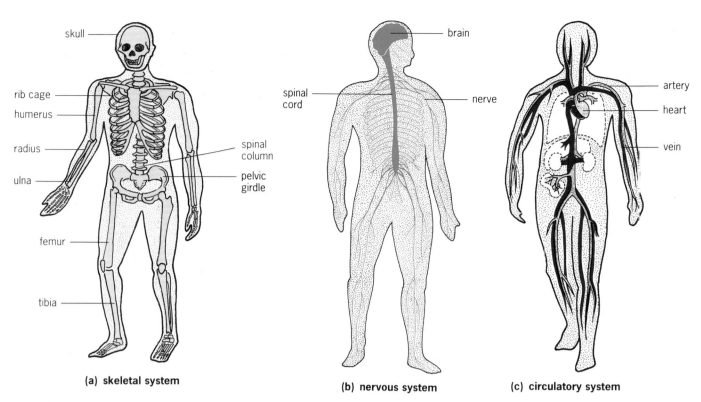

Figure 13 Three examples of systems in the human body

System

A system usually refers to a group of organs whose functions are closely related. For example, the heart and blood vessels make up the **circulatory system**; the brain, spinal cord and nerves make up the **nervous system** (Fig. 13). In a flowering plant, the stem, leaves and buds make up a system called the shoot (p. 52).

Organism

An organism is formed by the organs and systems working together to produce an independent plant or animal.

An example in the human body of how cells, tissues and organs are related is shown in Fig. 14.

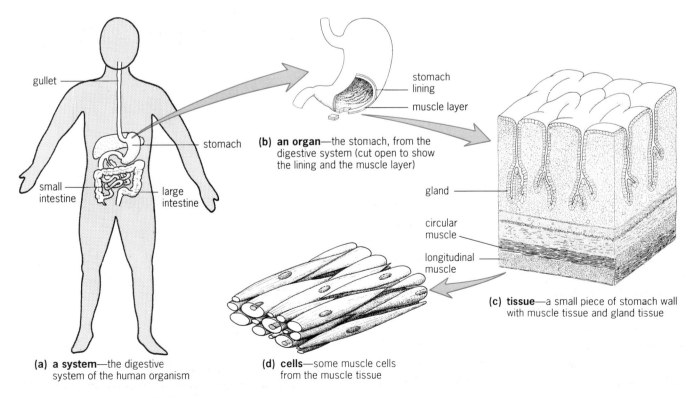

Figure 14 An example of how cells, tissues and organs are related

Tissue culture

It is possible to take samples of developing animal tissues, separate the cells by means of enzymes and make the cells grow and divide in shallow dishes containing a nutrient solution. This technique is called **tissue culture**. The cells do not become specialized but may move about, establish contact with each other and eventually cover the floor of the culture dish with a layer, one cell thick. At this point, they stop dividing unless they are separated and transferred to fresh culture vessels. Even so, most mammal cells cease to reproduce after about 20 divisions.

These tissue cultures are used to study cells and cell division, to test out new drugs and vaccines, to culture viruses or to test the effect of possible harmful chemicals. For some tests and experiments, tissue cultures are able to take the place of laboratory animals.

Large-scale tissue cultures are being developed in order to obtain cell chemicals which may be useful in medical treatments.

(a) Peel a strip of red epidermis from a piece of rhubarb stalk.

(b) Alternatively, peel the epidermis from the inside of an onion scale.

QUESTIONS

11 Say whether you think the following are cells, tissues, organs or organisms: lungs (p. 131), skin (p. 145), root hair (p. 67), mesophyll (p. 55), multi-polar neurone (p. 180).

12 What tissues are shown in the following drawings: Fig. 9 on p. 166; Fig. 9 on p. 182?

13 Look at Fig. 5 on p. 107.
   a Which of the structures do you think are organs?
   b What system is represented by the drawing?

(c) Place the epidermis in a drop of water or weak iodine solution on a slide and carefully lower a cover slip over it.

Figure 15 Looking at plant cells

PRACTICAL WORK

1 Plant cells

The outer layer of cells (epidermis) from a stem or an onion scale can be stripped off as shown in Fig. 15(a) and (b). A piece of onion or a rhubarb stalk is particularly suitable for this. A small piece of this epidermis is placed in a drop of weak iodine solution on a microscope slide and covered with a cover-slip (Fig. 15(c)). The tissue is then studied under the microscope. The iodine will stain the cell nuclei pale yellow and the starch grains will stain blue. If red epidermis from rhubarb stalk is used, you will see the red cell sap in the vacuoles.

To see chloroplasts, use fine forceps to pull a leaf from a moss plant, mount it in water on a slide as before and examine it with the high power objective of the microscope (Fig. 16). At the base of the leaf, the chloroplasts are less densely packed and, therefore, easier to see.

Figure 16 Cells in a moss leaf (x500). The vacuole occupies most of the space in each cell. The chloroplasts are confined to the layer of cytoplasm lining the cell wall.

2 Animal cells

Note The Department of Education and Science recommends that schools no longer use the technique which involves studying the epithelial cells which appear in a smear taken from the inside of the cheek, because of the very small risk of transmitting the AIDS virus. Some Local Education Authorities may therefore forbid the use of this technique in their schools, but the Institute of Biology suggests that if the following procedure is adopted the risk is negligible (*Biologist*, 35 (4) p. 211, September 1988).

Cotton buds from a freshly opened pack are rubbed lightly on the inside of the cheek and gums. The buds are rubbed on to clean slides and then dropped into a container of absolute alcohol. The smear on the slide is covered with a few drops of methylene blue solution before being examined under the microscope (Fig. 17). The slides are placed in laboratory disinfectant before washing.

Various alternatives to cheek epithelial smears have been put forward, e.g.

(a) Some Sellotape is pressed on to a 'well-washed' wrist. When the tape is removed and studied under the microscope, cells with nuclei can be seen. A few drops of methylene blue solution will stain the cells and make the nuclei more distinct.

(b) Microscope slides are pressed against the corneas of fresh or refrigerated bullock's eyes. Conjunctival cells stick to the slides and can be stained with methylene blue.

Figure 17 Cells from the lining epithelium of the cheek (x1500)

Checklist

- Nearly all plants and animals are made up of thousands or millions of microscopic cells.

- All cells contain cytoplasm enclosed in a cell membrane.

- Most cells have a nucleus.

- Cytoplasm contains organelles such as mitochondria, chloroplasts and ribosomes.

- Many chemical reactions take place in the cytoplasm to keep the cell alive.

- The nucleus directs the chemical reactions in the cell and also controls cell division.

- Plant cells have a cellulose cell wall and a large central vacuole.

- Cells are often specialized in their shapes and activities to carry out particular jobs.

- Large numbers of similar cells packed together form a tissue.

- Different tissues arranged together form organs.

- A group of related organs makes up a system.

2 THE CHEMICALS OF LIVING CELLS

Cell physiology

The term 'physiology' refers to all the normal functions that take place in a living organism. Digestion of food, circulation of blood and contraction of muscles are some aspects of human physiology. Absorption of water by roots, production of food in the leaves, and growth of shoots towards light are examples of plant physiology. The next three chapters are concerned with physiological events in individual cells.

The physiology of a whole organism is, in some ways, the sum of the physiology of its component cells. If all cells need a supply of oxygen then the whole organism must take in oxygen. Cells need chemical substances to make new cytoplasm and to produce energy. Therefore the organism must take in food to supply the cells with these substances. Of course, it is not quite as simple as this; most cells have specialized functions (p. 6) and so have differing needs. However, all cells need water, oxygen, salts and food substances and all cells consist of water, proteins, lipids, carbohydrates, salts and vitamins or their derivatives.

The chemical components of a cell

Water

Most cells contain about 75 per cent of water and will die if their water content falls much below this. Water is a good solvent and many substances move about the cell in a watery solution. Water molecules take part in a great many vital chemical reactions (e.g. photosynthesis, p. 37). In plants, it is the water pressure in the vacuoles which keeps the cells turgid (firm).

The physical and chemical properties of water differ from those of most other liquids but make it uniquely effective in supporting living activities. For example, water has a high capacity for heat (high thermal capacity). This means that it can absorb a lot of heat without its temperature rising to levels which damage the proteins in the protoplasm (see p. 12). However, because water freezes at 0 °C most cells are damaged if their temperature falls below this and ice crystals form in the protoplasm. (Oddly enough, rapid

freezing of cells in liquid nitrogen at below −196 °C does not harm them.)

Proteins

Some proteins contribute to the structures of the cell, e.g. to the cell membranes, the mitochondria, ribosomes and chromosomes. These proteins are called **structural proteins**.

There is another group of proteins called **enzymes**. Enzymes are present in the membrane systems, in the mitochondria, in special vacuoles and in the fluid part of the cytoplasm. Enzymes control the chemical reactions which keep the cell alive (see p. 13).

Although there are many different types of protein, they all contain carbon, hydrogen, oxygen, nitrogen and sulphur, and their molecules are made up of long chains of simpler chemicals called **amino acids**.

There are about 20 different amino acids in animal proteins, including alanine, leucine, valine, glutamine, cysteine, glycine and lysine. A small protein molecule might be made up from a chain consisting of a hundred or so amino acids, e.g. glycine–valine–valine–cysteine–leucine–glutamine–, etc.

Each different protein has the amino acids arranged in its own particular order.

The chain of amino acids in a protein takes up a particular shape as a result of cross-linkages. Cross-linkages form between amino acids which are not neighbours, as shown in Fig. 1. The shape of a protein molecule has a very important effect on its reactions with substances as explained in 'Enzymes' on p. 14.

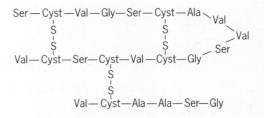

Figure 1 A small imaginary protein made from only five different kinds of amino acid. Note that cross-linkage occurs between cysteine molecules with the aid of sulphur atoms.

When a protein is heated to temperatures over 50 °C, the cross-linkages in its molecules break down; the protein molecules lose their shape and will not usually regain it even when cooled. The protein is said to have been **denatured**. Because the shape of the molecules has been altered, the protein will have lost its original properties.

Egg-white is a protein. When it is heated, its molecules change shape and the egg-white goes from a clear, runny liquid to a white solid and cannot be changed back again. The egg-white protein, albumen, has been denatured by heat.

Proteins form enzymes and many of the structures in the cell, so if they are denatured the enzymes and the cell structures will stop working and the cell will die. Whole organisms may survive for a time above 50 °C depending on the temperature, the period of exposure and the proportion of the cells which are damaged.

Lipids

Lipids are oils or fats and substances related to or derived from them. Fats are formed from carbon, hydrogen and oxygen only. A molecule of fat is made up of three molecules of an organic acid, called a **fatty acid**, combined with one molecule of glycerol.

$$\text{glycerol} \begin{cases} H_2 - C - O - \text{ stearic acid} \\ H - C - O - \text{ oleic acid} \\ H_2 - C - O - \text{ palmitic acid} \end{cases} \text{fatty acids}$$

Lipids form part of the cell membrane and the internal membranes of the cell such as the nuclear membrane. Droplets of fat or oil form a source of energy when stored in the cytoplasm.

Carbohydrates

These may be simple, soluble sugars or complex materials like starch and cellulose, but all carbohydrates contain carbon, hydrogen and oxygen only. A commonly occurring simple sugar is **glucose**, whose chemical formula is $C_6H_{12}O_6$.

The glucose molecule is often in the form of a ring, represented as

or more simply as

Two molecules of glucose can be combined to form a molecule of maltose $C_{12}H_{22}O_{11}$.

Sugars with a single carbon ring are called **monosaccharides**, e.g. glucose and fructose. Those sugars with two carbon rings in their molecules are called **disaccharides**, e.g. maltose and sucrose. Mono- and disaccharides are readily soluble in water.

When many glucose molecules are joined together, the carbohydrate is called a **polysaccharide**. **Glycogen** (Fig. 2) is a polysaccharide which forms a food storage substance in many animal cells. The **starch** molecule is made up of hundreds of glucose molecules joined together to form long chains. Starch is an important storage substance in the plastids of plant cells. **Cellulose** consists of even longer chains of glucose molecules. The chain molecules are grouped together to form microscopic fibres, which are laid down in layers to form the cell wall in plant cells (Figs 3 and 4).

Polysaccharides are not readily soluble in water.

one glucose unit

Figure 2 Part of a glycogen molecule

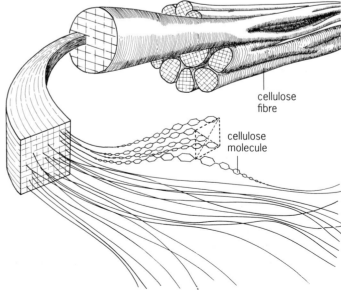

cellulose fibre

cellulose molecule

Figure 3 Cellulose. Plant cell walls are composed of long, interwoven and interconnected cellulose fibres which are large enough to be seen with the electron microscope. Each fibre is made up of many long-chain cellulose molecules.

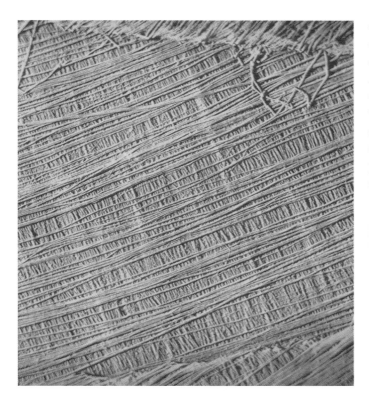

Figure 4 Electron micrograph of a plant cell wall (x18,000) showing the cellulose fibres

Salts

Salts are present in cells in the form of their ions, e.g. sodium chloride in a cell will exist as sodium ions (Na^+) and chlorine ions (Cl^-). The ions may be free to move about in the water of the cell or attached to other molecules such as the proteins or lipids.

Ions take part in and influence many chemical reactions in the cell, e.g. phosphate ions (PO_4^{3-}) are essential for energy transfer reactions (see 'ATP', p. 21). Ions are also involved in determining how much water enters or leaves the cell (see 'Osmosis', p. 31). Calcium, potassium and sodium ions are particularly important in chemical changes related to electrical activities of a cell, e.g. responding to stimuli or conducting nerve impulses. A shortage or excess of ions in the cells upsets their physiology and affects the normal functioning of the whole organism.

Vitamins

This is a category of substances which, in their chemical structure at least, have little in common. Plant cells can make their own vitamins. Animal cells have to be supplied with vitamins ready-made. Vitamins, or substances derived from them, play a part in chemical reactions in the cell, for example those which involve a transfer of energy from one compound to another. If cells are not supplied with vitamins or the substances needed to make them, the cell physiology is thrown out of order and the whole organism suffers.

Interconversion of substances in cells

Cells are able to build up and break down their proteins, lipids and carbohydrates or change one to the other. For example, animal cells build up glycogen from glucose; plant cells make starch and cellulose from glucose. All cells can make proteins from amino acids and they can build up fats from glycerol and fatty acids. Animal cells can change carbohydrates to lipids, and lipids to carbohydrates; they can also change proteins to carbohydrates but they cannot make proteins unless they are supplied with amino acids. Plant cells, on the other hand, can make their own amino acids starting from sugar and salts. The cells in the green parts of plants can even make glucose starting from only carbon dioxide and water (see pp. 37–41).

QUESTIONS

1 Why do you think it is essential for us to include proteins in our diet?
2 Which substances are particularly important in
   a forming the structures of the cell and b storing food in the cell?
3 Which of the following are a carbohydrates, b salts, c lipids: sugar, butter, iron sulphate, starch, sodium chloride, olive oil, cellulose?
4 What additional elements are needed to convert a carbohydrate into a protein? Where do you think plants get these elements from?
5 What chemical changes can occur in a plant cell that do not take place in animal cells?

Enzymes

Enzymes are proteins that act as **catalysts**. They are made in the cells. A catalyst is a chemical substance which speeds up a reaction but does not get used up during the reaction. One enzyme can be used many times over (Fig. 5).

How an enzyme molecule might work to join two other molecules together and so form a more complicated substance is shown by the diagram in Fig. 6.

An example of an enzyme-controlled reaction such as this is the joining up of two glucose molecules to form a molecule of maltose.

$C_6H_{12}O_6$ glucose $C_6H_{12}O_6$ glucose $C_{12}H_{22}O_{11}$ maltose H_2O water

In a similar way, hundreds of glucose molecules might be joined together, end to end, to form a long molecule of starch to be stored in the plastid of a plant cell. The glucose molecules might also be built up into a molecule of cellulose to be added to the cell wall. Protein molecules are built up by enzymes which join together tens or hundreds of amino acid molecules. These proteins are added to the cell membrane, to the cytoplasm or to the nucleus of the cell. They may also become the proteins which act as enzymes.

Reactions in which large molecules are built up from smaller molecules are called **anabolic** reactions (Fig. 6(a)).

13

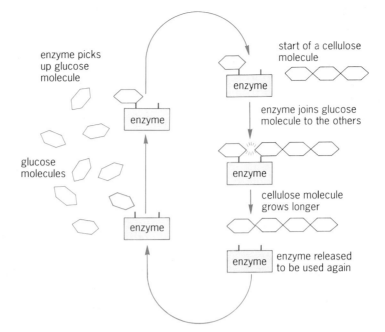

Figure 5 Building up a cellulose molecule

After the new substance has been formed, the enzyme is set free to start another reaction. Molecules of the two substances might have combined without the enzyme being present but they would have done so very slowly. By bringing the substances close together, the enzyme molecule makes the reaction take place much more rapidly. A chemical reaction which would take hours or days to happen on its own takes only a few seconds when the right enzyme is present.

Fig. 6(b) shows an enzyme speeding up a chemical change but this time it is a reaction in which the molecule of a substance is split into smaller molecules. If starch is mixed with water it will break down very slowly to sugar, taking several years. In your saliva there is an enzyme called **amylase** which can break down starch to sugar in minutes or seconds. In cells, many of the 'breaking-down' enzymes are helping to break down glucose to carbon dioxide and water in order to produce energy (see p. 20).

Reactions which split large molecules into smaller ones are called **catabolic** reactions.

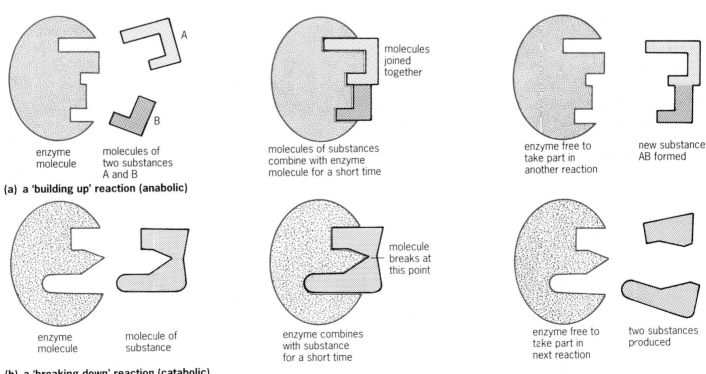

(a) a 'building up' reaction (anabolic)

(b) a 'breaking down' reaction (catabolic)

Figure 6 Possible explanation of enzyme action (the 'lock and key' model)

Enzymes and temperature

A rise in temperature increases the rate of most chemical reactions; a fall in temperature slows them down. In many cases a rise of 10 °C will double the rate of reaction in a cell. This is equally true for enzyme-controlled reactions, but above 50 °C the enzymes, being proteins, are denatured and stop working.

Fig. 6 shows how the shape of an enzyme molecule could be very important if it has to fit the substances on which it acts. Above 50 °C the shapes of enzymes are changed and the enzymes can no longer combine with the substances.

This is one of the reasons why organisms may be killed by prolonged exposure to high temperatures. The enzymes in their cells are denatured and the chemical reactions proceed too slowly to maintain life.

One way to test whether a substance is an enzyme is to heat it to boiling point. If it can still carry out its reactions after this, it cannot be an enzyme. This technique is used as a 'control' (see p. 25) in enzyme experiments.

Enzymes and pH

Acid or alkaline conditions alter the chemical properties of proteins, including enzymes. Most enzymes work best at a particular level of acidity or alkalinity (pH). The protein-digesting enzyme in your stomach, for example, works well at an acidity of pH 2. At this pH, the enzyme amylase, from your saliva, cannot work at all. Inside the cells, most enzymes will work best in neutral conditions (pH 7). The pH or temperature at which an enzyme works best is often called its **optimum** pH or temperature.

Although changes in pH affect the activity of enzymes, these effects are usually reversible, i.e. an enzyme which is inactivated by a low pH will resume its normal activity when its optimum pH is restored. Extremes of pH, however, may denature some enzymes irreversibly.

Enzymes are specific

This means simply that an enzyme which normally acts on one substance will not act on a different one. Fig. 6(a) shows how the shape of an enzyme could decide what substances it combines with. The enzyme in Fig. 6(a) has a shape which exactly fits the substances on which it acts, but would not fit the substance in 6(b). Thus, an enzyme which breaks down starch to maltose will not also break down proteins to amino acids. Also, if a reaction takes place in stages, e.g.

$$\text{starch} \rightarrow \text{maltose (stage 1)}$$
$$\text{maltose} \rightarrow \text{glucose (stage 2)}$$

a different enzyme is needed for each stage.

The names of enzymes usually end with **-ase** and they are named according to the substance on which they act, or the reaction which they promote. For example, an enzyme which acts on proteins may be called a **protease**; one which removes hydrogen from a substance is a **dehydrogenase**.

The substance on which an enzyme acts is called its **substrate**.

Thus, the enzyme **sucrase** acts on the substrate **sucrose** to produce the monosaccharides, glucose and fructose.

Rates of enzyme reactions

As explained above, the rate of an enzyme-controlled reaction depends on the temperature and pH. It also depends on the concentrations of the enzyme and its substrate. The more enzyme molecules produced by a cell, the faster the reaction will proceed, provided there are enough substrate molecules available. Similarly, an increase in the substrate concentration will speed up the reaction if there are enough enzyme molecules to cope with the additional substrate.

Intra- and extracellular enzymes

All enzymes are made inside cells. Most of them remain inside the cell to speed up reactions in the cytoplasm and nucleus. These are called **intracellular enzymes** ('intra' means 'inside'). In a few cases, the enzymes made in the cells are let out of the cell to do their work outside. These are **extracellular enzymes** ('extra' means 'outside'). Fungi (p. 303) and bacteria (p. 299) release extracellular enzymes in order to digest their food. A mould growing on a piece of bread releases starch-digesting enzymes into the bread and absorbs the soluble sugars which the enzyme produces from the bread. In the digestive systems of animals (p. 105), extra-cellular enzymes are released into the stomach and intestines in order to digest the food.

PRACTICAL WORK

Tests for proteins, fats and carbohydrates are described on pp. 102–3. Experiments on the digestive enzymes amylase and pepsin are described on pp. 113–4.

Further experiments In *Experimental Work in Biology*, 17 experiments on food tests and 11 experiments on enzymes are described. (See p. vi.)

1 Extracting and testing an enzyme from living cells

In this experiment, the enzyme to be extracted and tested is **catalase**, and the substrate is hydrogen peroxide (H_2O_2). Certain reactions in the cell produce hydrogen peroxide, which is poisonous. Catalase renders the hydrogen peroxide harmless by breaking it down to water and oxygen.

$$2H_2O_2 \xrightarrow{\text{catalase}} 2H_2O + O_2$$

Grind a small piece of liver with about 20 cm^3 water and a little sand, in a mortar. This will break open the liver cells and release their contents. Filter the mixture and share it between two test-tubes, A and B (Fig. 7). The filtrate will contain a great variety of substances dissolved out from the cytoplasm of the liver cells, including many enzymes. Because enzymes are specific, however, only one of these, catalase, will act on hydrogen peroxide. Add some drops of the filtrate from test-tube A to a few cm³ hydrogen peroxide in a test-tube. You will see a vigorous reaction as the hydrogen peroxide breaks down to produce oxygen. (The oxygen can be tested with a glowing splint.)

Now boil the filtrate in tube B for about 30 seconds. Add a few drops of the boiled filtrate to a fresh lot of hydrogen peroxide. There will be no reaction because boiling has denatured the catalase.

Next, shake a little manganese(IV) oxide powder in a test-tube with some water and pour this into some hydrogen peroxide. There will be a vigorous reaction similar to the one with the liver extract. If you now boil some manganese(IV) oxide with water and add this to hydrogen peroxide, the reaction will still occur. Manganese(IV) oxide is a catalyst but it is not an enzyme because heating has not altered its catalytic properties.

The experiment can be repeated with a piece of potato to compare its catalase content with that of the liver. The piece of potato should be about the same size as the liver sample.

2 Building starch from glucose

Grind a small piece of potato in a mortar with about 10 cm³ distilled water and a little clean sand. Filter the mixture into a clean test-tube. On a cavity tile place 2 rows of 4 drops of a 5 per cent solution of glucose phosphate (Fig. 8). Pick up a little of the filtered potato extract in a bulb pipette and place a drop in a spare cavity on the tile. Test this drop with iodine solution. If it goes blue it contains starch and the extract must be filtered again to remove all starch grains.

If the extract is free from starch, place 1 drop of it on each of the 4 drops of glucose phosphate in the top row on the tile. Now boil the rest of the potato extract, let it cool and then place a drop on each of the glucose phosphate drops in the bottom row on the tile.

After 5 minutes, test the first drop in each row with iodine solution. Repeat this test at 5-minute intervals with the other 3 pairs of drops.

Finally, use a spare cavity on the tile to test a drop of the glucose phosphate solution with iodine.

Result The last 2 glucose phosphate drops in the top row should turn blue. None of the drops in the bottom row should go blue.

Interpretation A blue colour with iodine shows that starch is present. Neither the glucose phosphate solution nor the potato extract contained starch, so the starch must have formed during the course of the experiment. Presumably, an enzyme in the potato extract had joined the glucose phosphate molecules together to form starch. This is supported by the fact that the boiled extract did not produce starch, as you would expect if the reaction was brought about by an enzyme. The enzyme in question is called **starch phosphorylase**.

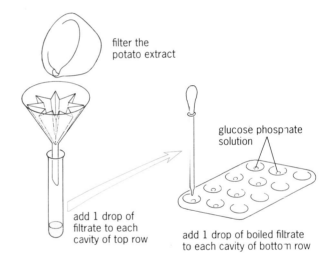

Figure 8 Building starch from glucose

Figure 7 Extracting and testing living cells for catalase

3 The effect of temperature on an enzyme reaction

Amylase is an enzyme which breaks down starch to a sugar (maltose).

Draw up 5 cm³ of 5 per cent amylase solution in a plastic syringe (or graduated pipette) and place 1 cm³ in each of 3 test-tubes labelled A, B and C. Rinse the syringe and use it to place 5 cm³ of a 1 per cent starch solution in each of 3 test-tubes labelled 1, 2 and 3.

To each of tubes 1 to 3, add 6 drops only of dilute iodine solution using a dropping pipette.

Prepare three water baths by half filling beakers or jars with the following:

1 ice and water, adding ice during the experiment to keep the temperature at about 10 °C;
2 water from the cold tap at about 20 °C;
3 warm water at about 35 °C by mixing hot and cold water from the tap.

Place tubes 1 and A in the cold water bath, tubes 2 and B in the water at room temperature, and tubes 3 and C in the warm water. Leave them for 5 minutes to reach the temperature of the water. (Fig. 9.)

leave all three for 5 minutes

note the time and add the amylase to the starch solution

Figure 9 The effect of temperature on an enzyme reaction

After 5 minutes, take the temperature of each water bath, then pour the amylase from tube A into the starch solution in tube 1 and return tube 1 to the water bath. Repeat this with tubes 2 and B, and 3 and C.

As the amylase breaks down the starch, it will cause the blue colour to disappear. Make a note of how long this takes in each case.

Answer the following questions:

(a) At what temperature did the amylase break down starch most rapidly?
(b) What do you think would have been the result if a fourth water bath at 90 °C had been used?

4 The effect of pH on an enzyme reaction

Label 5 test-tubes 1 to 5 and use a plastic syringe (or graduated pipette) to place 5 cm³ of a 1 per cent starch solution in each tube. Add acid or alkali to each tube as indicated in the table below. Rinse the syringe when changing from sodium carbonate to acid.

Tube	Chemical	Approximate pH	
1	1 cm³ sodium carbonate solution (M/20)	9	(alkaline)
2	0.5 cm³ sodium carbonate solution (M/20)	7–8	(slightly alkaline)
3	nothing	6–7	(neutral)
4	2 cm³ ethanoic (acetic) acid (M/10)	6	(slightly acid)
5	4 cm³ ethanoic (acetic) acid (M/10)	3	(acid)

Place several rows of iodine solution drops in a cavity tile.

Draw up 5 cm³ of 5 per cent amylase solution in a clean syringe and place 1 cm³ in each tube. Shake the tubes and note the time (Fig. 10).

Use a clean dropping pipette to remove a small sample from each tube in turn and let one drop fall on to one of the iodine drops in the cavity tile. Rinse the pipette in a beaker of water between each sample. Keep on sampling in this way.

When any of the samples fails to give a blue colour, this means that the starch in that tube has been completely broken down to sugar by the amylase. Note the time when this happens for each tube and stop taking samples from that tube. Do not continue sampling for more than about 15 minutes, but put a drop from each tube on to a piece of pH paper and compare the colour produced with a colour chart of pH values.

Answer the following questions:

(a) At what pH did the enzyme, amylase, work most rapidly?
(b) Is this its optimum pH?
(c) Explain why you might have expected the result which you got.
(d) Your stomach pH is about 2. Would you expect starch digestion to take place in the stomach?

Figure 10 The effect of pH on an enzyme reaction

QUESTIONS

6 Which of the following statements apply both to enzymes and to any other catalysts:
   a Their activity is stopped by high temperature.
   b They speed up chemical reactions.
   c They are proteins.
   d They are not used up during the reaction.
7 How would you expect the rate of an enzyme-controlled reaction to change if the temperature was raised **a** from 20 °C to 30 °C, **b** from 35 °C to 55 °C? Explain your answers.
8 There are cells in your salivary glands which can make an extracellular enzyme, amylase. Would you expect these cells to make intracellular enzymes as well? Explain your answer.
9 Apple cells contain an enzyme which turns the tissues brown when an apple is peeled and left for a time. Boiled apple does not go brown (Fig. 11). Explain why the boiled apple behaves differently.
10 Suppose that the enzyme in Fig. 6(a) is joining a glycine molecule (p. 11) to a valine molecule. Suggest why the same enzyme will not join glycine to serine.
11 If you tried Experiment 2, suggest what might be the value in a potato tuber of an enzyme which changes glucose to starch.

Figure 11 Enzyme activity in an apple. Slice A has been freshly cut. B and C were cut 2 days earlier but C was dipped immediately in boiling water for one minute.

Enzymes in industry

Brewing, baking and cheese-making are examples of industries which, for many years, have made use of enzymes on a large scale. In the case of brewing and baking it is the sugar-fermenting enzymes in living yeast cells which are exploited. These enzymes convert sugar to alcohol and carbon dioxide. It is the bubbles of carbon dioxide which make the dough 'rise' before baking and so give the bread a 'light' texture. In brewing, it is mainly the alcohol which is wanted but the carbon dioxide gives a 'sparkle' to beers and sparkling wine.

Brewing and baking exploit enzymes in living cells but for cheese-making an enzyme was extracted from calves' stomachs (it is now produced by genetically engineered bacteria, see p. 206). The enzyme is called **rennin** (the genetically engineered product is **chymosin**); it clots milk in the first stages of making cheese.

Of about 2000 known enzymes, about 50 are used commercially in large quantities. The advantages of using enzymes are:

a they work at low temperatures, so saving energy;

b they are not corrosive, as are the strong acids and alkalis that would otherwise be needed to carry out the chemical changes;

c they are specific, i.e. they act on only one substrate, so the process can be carefully controlled, and unwanted by-products avoided.

Two-thirds of all commercial enzyme production is taken up by protein-digesting enzymes (proteases) for detergents, and starch-digesting enzymes (amylases) for the food and textile industries. The proteases in detergents help to remove organic stains such as blood, gravy or fruit juice. The amylases are used to convert starch to glucose and fructose for sweeteners, or to destarch fabrics in the course of processing.

The commercial production and use of enzymes is described more fully on p. 344.

Checklist

- Living matter is made up of water, proteins, lipids, carbohydrates, salts and vitamins.

- Proteins are built up from amino acids joined together by chemical bonds.

- In different proteins the 20 or so amino acids are in different proportions and arranged in different sequences.

- Proteins are denatured by heat and some chemicals.

- Lipids include fats, fatty acids and oils.

- Fats are made from fatty acids and glycerol.

- Proteins and lipids form the membranes outside and inside the cell.

- Enzymes are proteins which catalyse chemical reactions in the cell.

- Enzymes are affected by pH and temperature and are denatured above 50 °C.

- Different enzymes may accelerate reactions which build up or break down molecules.

- Each enzyme acts on only one substance (breaking down), or a pair of substances (building up).

- The substance on which an enzyme acts is called the substrate.

3 ENERGY FROM RESPIRATION

Most of the processes taking place in cells need energy to make them happen. Building up proteins from amino acids or making starch from glucose needs energy. When muscle cells contract, or nerve cells conduct electrical impulses or plant cells form cell walls, they use energy. This energy comes from the food which cells take in. The food mainly used for energy in cells is glucose.

The process by which energy is released from food is called **respiration**.

Respiration is a chemical process which takes place in cells. It must not be confused with the process of breathing which is also sometimes called 'respiration'. To make the difference quite clear, the chemical process in cells is sometimes called **cellular respiration**, **internal respiration** or **tissue respiration**. The use of the word 'respiration' for breathing is best avoided altogether.

Aerobic respiration

The word **aerobic** means that oxygen is needed for this chemical reaction. The food molecules are combined with oxygen. The process is called **oxidation** and the food is said to be **oxidized**. All food molecules contain carbon, hydrogen and oxygen atoms. The process of oxidation converts the carbon to carbon dioxide (CO_2) and the hydrogen to water (H_2O) and, at the same time, sets free energy which the cell can use to drive other reactions.

Aerobic respiration can be summed up by the equation

$$\underset{\text{glucose}}{C_6H_{12}O_6} + \underset{\text{oxygen}}{6O_2} \xrightarrow{\text{enzymes}} \underset{\substack{\text{carbon} \\ \text{dioxide}}}{6CO_2} + \underset{\text{water}}{6H_2O} + \underset{\text{energy}}{2830 \text{ kJ}}$$

The 2830 kilojoules (kJ) is the amount of energy you would get by completely oxidizing 180 grams of glucose to carbon dioxide and water. In the cells, the energy is not released all at once. The oxidation takes place in a series of small steps and not in one jump as the equation suggests. Each small step needs its own enzyme and at each stage a little energy is released (Fig. 1).

(a) **molecule of glucose (H and O atoms not all shown)**

(b) **the enzyme attacks and breaks the glucose molecule into two 3-carbon molecules**

(c) **this breakdown sets free energy**

(d) **each 3-carbon molecule is broken down to carbon dioxide**

(e) **more energy is released and CO₂ is produced**

(f) **the glucose has been completely oxidized to carbon dioxide (and water), and all the energy released**

Figure 1 Aerobic respiration

Although the energy is used for the processes mentioned above, some of it always appears as heat. In 'warm-blooded' animals (birds and mammals) some of this heat is retained to keep up their body temperature. In 'cold-blooded' animals (e.g. reptiles and fish) the heat may build up for a time and allow the animal to move about faster. In plants the heat is lost to the surroundings (by conduction, convection and evaporation) as fast as it is produced.

Mitochondria

It is in the mitochondria that the chemistry of aerobic respiration takes place. The mitochondria generate ATP which is used by the cell as the source of energy for driving other chemical reactions in the cytoplasm and nucleus.

Anaerobic respiration

The word **anaerobic** means 'in the absence of oxygen'. In this process, energy is still released from food by breaking it down chemically but the reactions do not use oxygen though they do often produce carbon dioxide. A common example is the action of yeast on sugar solution to produce alcohol. The sugar is not completely oxidized to carbon dioxide and water but converted to carbon dioxide and alcohol. This process is called **fermentation** and is shown by the following equation:

$$\underset{\text{glucose}}{C_6H_{12}O_6} \xrightarrow{\text{enzymes}} \underset{\text{alcohol}}{2C_2H_5OH} + \underset{\substack{\text{carbon} \\ \text{dioxide}}}{2CO_2} + \underset{\text{energy}}{118 \text{ kJ}}$$

As with aerobic respiration, the reaction takes place in small steps and needs several different enzymes. The yeast uses the energy for its growth and living activities, but you can see from the equation that less energy is produced by anaerobic respiration than in aerobic respiration. This is because the alcohol still contains a great deal of energy which the yeast is unable to use.

In animals, the first stages of respiration in muscle cells are anaerobic and produce **pyruvic acid** (the equivalent of the yeast's alcohol). Only later on is the pyruvic acid completely oxidized to carbon dioxide and water.

$$\underset{\text{ANAEROBIC STAGE}}{\text{glucose} \xrightarrow{\text{enzymes}} \text{pyruvic acid}} \xrightarrow[\text{AEROBIC STAGE}]{\substack{\text{enzymes} \\ \text{and oxygen}}} CO_2 + H_2O$$

During exercise pyruvic acid may build up in a muscle faster than it can be oxidized. In this case it is turned into **lactic acid** and removed in the bloodstream. On reaching the liver, some of the lactic acid is oxidized to carbon dioxide and water, using up oxygen in the process. So, even after exercise has stopped, a high level of oxygen consumption may persist for a time as the excess of lactic acid is oxidized. This build up of lactic acid which is oxidized later is said to create an **oxygen debt**.

Accumulation of lactic acid in the muscles may be one of the causes of muscular fatigue. It is sometimes claimed that lactic acid also causes cramp but this is not established.

ATP and energy transfer

The energy released when glucose is broken down is not used directly in the cell. Instead it is transferred to other chemicals which act as a store of readily available energy. One of these chemicals is **adenosine triphosphate** (ATP) (Fig. 2). Adenosine is an organic base, adenine, combined with a 5-carbon sugar, ribose. It can further combine with one, two or three phosphate groups.

At those stages in a glucose molecule's breakdown when energy is released, the energy is used to combine a phosphate ion (PO_3^{2-}) with a molecule of adenosine diphosphate (ADP) to make a molecule of adenosine triphosphate (Fig. 3).

Figure 3 Energy transfer to ATP

From Fig. 3 it appears that four molecules of ATP are made for every molecule of glucose broken down. However, many of the intermediate steps have been omitted and, in fact, 38 molecules of ATP can be built up when one molecule of glucose is completely oxidized to carbon dioxide and water.

Figure 2 The adenosine phosphates

In the presence of an appropriate enzyme, ATP readily breaks down to ADP, releasing energy and a phosphate ion. The energy can be used to drive other chemical reactions such as those producing muscle contraction.

ATP is thus a kind of energy store in the cell and can be used directly for driving almost any reaction which needs energy. The breakdown of glucose to carbon dioxide and water is a comparatively slow process with many intermediate steps and enzymes. ATP breakdown is rapid and needs only a suitable ATP-ase enzyme to bring it about. This is true of all cells, plant, animal or micro-organism.

The breakdown of glucose might be compared to burning coal, in order to generate steam, which drives a turbine and so charges a battery; a fairly complex and long-winded business. The charged battery, however, is like the ATP. It can be used at a second's notice to do a wide variety of things such as light a lamp, ring a bell, drive a motor, etc.

Metabolism

All the chemical changes taking place inside a cell or a living organism are called its **metabolism**. The minimum turnover of energy needed simply to keep an organism alive, without movement or growth, is called the **basal metabolism**. Our basal metabolism maintains vital processes such as breathing, heart beat, digestion and excretion.

The processes which break substances down are sometimes called **catabolism**. Respiration is an example of catabolism in which carbohydrates are broken down to carbon dioxide and water. Chemical reactions which build up substances are called **anabolism**. Building up a protein from amino acids is an example of anabolism. The energy released by the **catabolic** process of respiration is used to drive the **anabolic** reactions which build up proteins.

You may have heard of anabolic steroids, in connection with drug-taking by athletes. These chemicals do reduce the rate of protein breakdown and may enhance the build up of certain proteins. However, their effects are complicated and not fully understood, they have undesirable side effects and their use contravenes athletics codes.

QUESTIONS

1 a If, in one word, you had to say what respiration was about, which word would you choose from this list: breathing, energy, oxygen, cells, food?
b In which parts of a living organism does respiration take place?
2 What are the two main differences between aerobic and anaerobic respiration?
3 What chemical substances **a** from outside the cell, **b** from inside the cell, must be provided for aerobic respiration to take place? **c** What are the products of aerobic respiration?
4 Victims of drowning who have stopped breathing are sometimes revived by a process called 'artificial respiration'. Why would a biologist object to the use of this expression? ('Resuscitation' is a better word to use.)
5 Which cell organelle would you expect to produce most ATP?
6 By comparing the equations for aerobic and anaerobic respiration on pp. 20 and 21, how many molecules of ATP might you expect to be produced by the anaerobic fermentation of one molecule of glucose?

PRACTICAL WORK

Experiments on respiration

If you look below at the chemical equation which represents aerobic respiration you will see that a tissue or an organism which is respiring should be (a) using up food, (b) using up oxygen, (c) giving off carbon dioxide, (d) giving off water and (e) releasing energy which can be used for other processes.

If we wish to test whether aerobic respiration is taking place:

1 '(d) giving out water' is not a good test because non-living material will give off water vapour if it is wet to start with.

2 '(a) using up food' can be tested by seeing if an organism loses weight. This is not so easy as it seems because most organisms lose weight as a result of evaporation of water and this may have nothing to do with respiration. It is the decrease in 'dry weight' which must be measured (p. 326).

3 (b), (c) and (e) are fairly easy to demonstrate either with whole organisms or pieces of living tissue as in Experiments 1 to 5.

Seeds are often used as the living organisms because when they start to grow (germinate) there is a high level of chemical activity in the cells. The seeds are easy to obtain and to handle and they fit into small-scale apparatus. In some cases blowfly maggots can be used as animal material.

1 Using up oxygen during respiration

The apparatus in Fig. 4 is a **respirometer** (a 'respire meter'), which can measure the rate of respiration by seeing how quickly oxygen is taken up. Germinating seeds or blowfly larvae are placed in the test-tube and, as they use up the oxygen for respiration, the level of liquid in the delivery tubing will go up.

Figure 4 To see if oxygen is taken up in respiration

There are two drawbacks to this. One is that the organisms usually give out as much carbon dioxide as they take in oxygen. So there may be no change in the total amount of air in the test-tube and the liquid level will not move. This drawback is overcome by placing **soda-lime** in the test-tube. Soda-lime will absorb carbon dioxide as fast as the organisms give it out. So only the uptake of oxygen will affect the amount of air in the tube. The second drawback is that quite small changes in temperature will make the air in the test-tube expand or contract and so cause the liquid to rise or fall whether or not respiration is taking place. To overcome this, the test-tube is kept in a beaker of water (a water bath). The temperature of water changes far more slowly than that of air, so there will not be much change during a 30-minute experiment.

Control To show that it is a living process which uses up oxygen, a similar respirometer is prepared but containing an equal quantity of germinating seeds which have been killed by boiling. (If blowfly larvae are used, the control can consist of an equivalent volume of glass beads. This is not a very good control but is probably more acceptable than killing an equivalent number of larvae.)

The apparatus is finally set up as shown in Fig. 4 and left for 30 minutes (10 minutes if blowfly larvae are used).

The capillary tube and reservoir of liquid are called a **manometer**.

Result The level of liquid in the experiment goes up more than in the control. The level in the control may not move at all.

Interpretation The rise of liquid in the delivery tubing shows that the living seedlings have taken up part of the air. It does not prove that it is oxygen which has been taken up. Oxygen seems the most likely gas, however, because (1) there is only 0.03 per cent carbon dioxide in the air to start with and (2) the other gas, nitrogen, is known to be less active than oxygen.

If the experiment is allowed to run for a long time, the uptake of oxygen could be checked at the end by placing a lighted splint in each test-tube in turn. If some of the oxygen has been removed by the living seedlings, the flame should go out more quickly than it does in the tube with dead seedlings.

2 Aerobic respiration in yeast

Methylene blue is a blue dye which goes colourless in the absence of oxygen.

Label two test-tubes A and B and place about 2 cm^3 1 per cent glucose in each. To tube A add about 30 mm (depth) of boiled yeast suspension. To tube B add an equal amount of unboiled yeast suspension. Place both tubes in a beaker of warm water at about 40 °C (Fig. 5). After 5 minutes add 2 cm^3 methylene blue solution, shake both tubes thoroughly and return them to the water bath. Do not disturb the tubes but watch to see how long it takes for the blue colour to disappear. (You can ignore a thin film of blue colour at the surface of the liquid. Oxygen from the air is being absorbed here.)

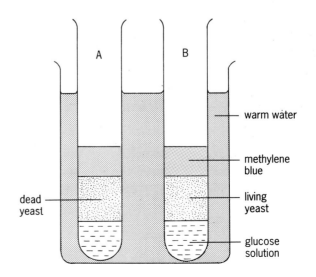

Figure 5 Aerobic respiration in yeast

Result The blue colour will disappear from tube B in a few minutes. The suspension of boiled yeast will remain blue.

If the tubes are shaken, oxygen from the air is introduced and the blue colour returns to tube B.

Interpretation The decoloration of methylene blue indicates that oxygen is being removed from the system. This happened only in the tube with living yeast and therefore some process in the live yeast cells is using up oxygen. We assume that the process is aerobic respiration.

3 Carbon dioxide from germinating seeds

Put some germinating wheat grains in a large test-tube. Cover the mouth of the tube with aluminium foil. After 15–20 minutes, take a sample of the air from the test-tube. Do this by pushing a glass tube attached to a 10 cm^3 plastic syringe through the foil and into the test-tube (Fig. 6(a)). Withdraw the syringe plunger enough to fill the syringe with air from the test-tube. Now slowly bubble this air sample through a little clear lime water in a small test-tube (Fig. 6(b)). Cover the mouth of the small test-tube and shake the lime water up.

Result The lime water will go milky.

(a) taking the air sample

(b) testing the air sample

Figure 6 Production of carbon dioxide by germinating seeds

Interpretation Lime water turning milky is evidence of carbon dioxide but it could be argued that the carbon dioxide came from the air or that the seeds give off carbon dioxide whether or not they are respiring. The only way to disprove these arguments is to do a **control experiment** (see p. 25).

Control Boil some of the germinating wheat grains before starting the experiment. When you set up the experiment, put an equal amount of boiled wheat grains in a large test-tube and cover the mouth of the tube with aluminium foil exactly as you did for the living seeds. When you test the air from the living seeds, also test the air from the dead seeds. It should not turn the lime water milky. This means that the carbon dioxide did not come from the air, nor was it given off by the dead seeds. It must be a living process in the seeds which produces carbon dioxide and this process is likely to be respiration. However, since you stopped all living processes by boiling the seeds in the control experiment, you have not been able to prove that it was respiration rather than some other chemical change which produced the carbon dioxide.

An experiment to show carbon dioxide production by plants, using hydrogencarbonate indicator, is described on p. 42.

4 Releasing energy in respiration

Fill a small vacuum flask with wheat grains which have been soaked for 24 hours and rinsed in 1 per cent formalin (or domestic bleach diluted 1 + 4) for 5 minutes. These solutions will kill any bacteria or fungi on the surface of the grains. Kill an equal quantity of soaked grains by boiling them for 5 minutes. Cool the boiled seeds in cold tap-water, rinse them in bleach or formalin for 5 minutes as before and then put them in a vacuum flask of the same size as the first one. This flask is the control.

Place a thermometer in each flask so that its bulb is in the middle of the seeds (Fig. 7). Plug the mouth of each flask with cotton wool and leave both flasks for 2 days, noting the thermometer readings whenever possible.

Result The temperature in the flask with the living seeds will be 5–10 °C higher than that of the dead seeds.

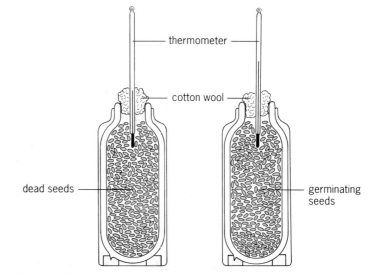

Figure 7 Energy release in germinating seeds

Interpretation Provided that there are no signs of the living seeds going mouldy, the heat produced must have come from living processes in the seeds, because the dead seeds in the control did not give out any heat. There is no evidence that this process is respiration rather than any other chemical change but the result is what you would expect if respiration does produce energy.

For an experiment comparing the energy value of different food substances, see p. 103.

5 Anaerobic respiration in yeast

Boil some water to expel all the dissolved oxygen. When cool, use the boiled water to make up a 5 per cent solution of glucose and a 10 per cent suspension of dried yeast. Place 5 cm^3 of the glucose solution and 1 cm^3 of the yeast suspension in a test-tube and cover the mixture with a thin layer of liquid paraffin to exclude atmospheric oxygen. Fit a delivery tube as shown in Fig. 8 and allow it to dip into clear lime water.

screw clip (closed when experiment starts)

liquid paraffin

yeast and glucose solution

lime water

Figure 8 Anaerobic respiration in yeast

Result After 10–15 minutes, with gentle warming if necessary, there should be signs of fermentation in the yeast-glucose mixture and the bubbles of gas escaping through the lime water should turn it milky.

Interpretation The fact that the lime water goes milky shows that the yeast-glucose mixture is producing carbon dioxide. If we assume that the production of carbon dioxide is evidence of respiration, then it looks as if the yeast is respiring. In setting up the experiment, you took care to see that oxygen was removed from the glucose solution and the yeast suspension, and the liquid paraffin excluded air (including oxygen) from the mixture. Any respiration taking place must, therefore, be anaerobic (i.e. without oxygen).

Control It might be suggested that the carbon dioxide came from a chemical reaction between yeast and glucose (as between chalk and acid) which had nothing to do with respiration or any other living process. A control should, therefore, be set up using the same procedure as before but with yeast which has been killed by boiling. The failure, in this case, to produce carbon dioxide supports the claim that it was a living process in the yeast in the first experiment which produced the carbon dioxide.

For an experiment on yeast in bread dough, see p. 342.

Experimental Work in Biology contains 16 experiments on respiration. (See p. vi.)

Controlled experiments

In most biological experiments, a second experiment called a **control** is set up. This is to make sure that the results of the first experiment are due to the conditions being studied and not to some other cause which has been overlooked.

In the experiment in Fig. 4, the liquid rising up the tube could have been the result of the test-tube cooling down, so making the air inside it contract. The identical experiment with dead seeds—the control—showed that the result was not due to a temperature change, because the level of liquid in the control did not move.

The term 'controlled experiment' refers to the fact that the experimenter (1) sets up a control and (2) controls the conditions in the experiment. In the experiment shown in Fig. 4 the seeds are enclosed in a test-tube, and soda-lime is added. This makes sure that any uptake or output of oxygen will make the liquid go up or down, and that the output of carbon dioxide will not affect the results. The experimenter had controlled both the amount and the composition of the air available to the germinating seeds.

If you did an experiment to compare the growth of plants in the house and in a greenhouse, you could not be sure whether it was the extra light or the high temperature of the greenhouse which caused better growth. This would not, therefore, be a properly controlled experiment. You must alter only one condition (*variable*) at a time, either the light or the temperature, and then you can compare the results with the control experiment.

A properly controlled experiment, therefore, alters only one variable at a time and includes a control which shows that it is this condition and nothing else which gave the result.

Dependent and independent variables

The **independent variable** is the one which is manipulated by the person doing the experiment. The **dependent variable** is the result of the experiment. In Experiment 5, the independent variable was the presence or absence of live yeast cells. The dependent variable was the production or non-production of carbon dioxide as revealed by lime water. The outcome of the experiment *depended* on whether or not we used living yeast.

In controlled experiments, it is important to alter only one independent variable.

25

QUESTIONS

7 Which of the following statements are true? If an organism is respiring you would expect it to be
a giving out carbon dioxide, **b** losing heat, **c** breaking down food, **d** using up oxygen, **e** gaining weight, **f** moving about?

8 What was the purpose of the soda-lime in Experiment 1 on p. 23, and the lime water in Experiments 3 and 5 on pp. 24 and 25?

9 In an experiment like the one shown in Fig. 4 on p. 23, the growing seeds took in 5 cm³ oxygen and gave out 7 cm³ carbon dioxide. What change in volume will take place
a if no soda-lime is present,
b if soda-lime is present?

10 The germinating seeds in Fig. 7 on p. 24 will release the same amount of heat whether they are in a beaker or a vacuum flask. Why then is it necessary to use a vacuum flask for this experiment?

11 What is the purpose of the control in the experiment to show carbon dioxide production (Fig. 6, p. 24)?

12 In Experiment 1, what was
a the independent variable,
b the dependent variable?

Hypothesis testing

You will have noticed that none of the experiments described above claim to have *proved* that respiration is taking place. The most we can claim is that they have not disproved the proposal that energy is produced from respiration. There are many reactions taking place in living organisms and, for all we know at this stage, some of them may be using oxygen or giving out carbon dioxide without releasing energy, i.e. they would not fit our definition of respiration.

This inability to 'prove' that a particular proposal is 'true' is not restricted to experiments on respiration. It is a feature of many scientific experiments. One way that science makes progress is by putting forward a **hypothesis**, making predictions from the hypothesis, and then testing these predictions by experiments. A hypothesis is an attempt to explain some event or observation using the information currently available. If the results of an experiment do not confirm the predictions, the hypothesis has to be abandoned or altered.

For example, biologists observing that living organisms take up oxygen might put forward the hypothesis that 'oxygen is used to convert food to carbon dioxide, so producing energy for movement, growth, reproduction, etc.'. This hypothesis can be tested by predicting that, '*if* the oxygen is used to oxidize food *then* an organism that takes up oxygen will also produce carbon dioxide'. Experiments 1 and 3 on pp. 23 and 24 test this and fulfil this prediction and, therefore, support the hypothesis. Looking at the equation for respiration, we might also predict that an organism which is respiring will produce carbon dioxide

and take up oxygen. Experiment 5 with yeast, however, does not fulfil this prediction and so does not support the hypothesis as it stands, because here is an organism producing carbon dioxide without taking up oxygen. The hypothesis will have to be modified, e.g. 'energy is released from food by breaking it down to carbon dioxide. Some organisms use oxygen for this process, others do not'.

There are still plenty of tests which we have not done. For example, we have not attempted to see whether it is food that is the source of energy and carbon dioxide. One way of doing this is to provide the organism with food, e.g. glucose, in which the carbon atoms are radioactive. Carbon-14 (^{14}C) is a radioactive form of carbon and can be detected by using a Geiger counter. If the organism produces radioactive carbon dioxide, it is reasonable to suppose that the carbon dioxide comes from the glucose.

$$C_6H_{12}O_6 + 6O_2 \longrightarrow 6CO_2 + 6H_2O + energy$$

This is **direct evidence** in support of the hypothesis. All the previous experiments have provided only **indirect evidence**.

Criteria for a good hypothesis
A good hypothesis must:

a explain **all** aspects of the observation;
b be the simplest possible explanation;
c be expressed in such a way that predictions can be made from it;
d be testable by experiment.

QUESTIONS

13 The experiment with yeast on p. 25 supported the claim that anaerobic respiration was taking place. The experiment was repeated using unboiled water and without the liquid paraffin. Fermentation still took place and carbon dioxide was produced. Does this mean that the design or the interpretation of the first experiment was wrong? Explain your answer.

14 Twenty seeds are placed on soaked cotton wool in a closed glass dish and after 5 days in the light 15 of the seeds had germinated. If the experiment is intended to see if light is needed for germination, which of the following would be a suitable control:
a exactly the same set-up but with dead seeds;
b the same set-up but with 50 seeds;
c an identical experiment but with 20 seeds of a different species;
d an identical experiment but left in darkness for 5 days?

15 Certain bacteria which live in sulphurous springs in areas of volcanic activity take up hydrogen sulphide (H_2S) and produce sulphates ($-SO_4$). Put forward a hypothesis to account for this chemical activity. Suggest one way of testing your hypothesis.

16 The table below shows the energy used up each day either as kilojoules per kilogram of body mass or as kilojoules per square metre of body surface.

	mass/kg	kJ per day	
		per kg body mass	per m² body surface
Pig	128.0	80	4510
Man	64.3	134	4360
Dog	15.2	216	4347
Mouse	0.018	2736	4971

From *Textbook of Physiology*, Emslie-Smith, Paterson, Scratcherd and Read (Churchill Livingstone, 1988)

a According to the table, what is the total amount of energy used each day by **i** a man, **ii** a mouse?
b Which of these two shows a greater rate of respiration in its body cells?
c Why, do you think, is there so little difference in the energy expenditure per square metre of body surface?

Checklist

- Respiration is the process in cells which releases energy from food.

- Aerobic respiration needs oxygen, anaerobic respiration does not.

- The oxidation of food produces carbon dioxide as well as releasing energy.

- The energy from oxidation of food is transferred to ATP.

- ATP acts as a store of readily available energy for driving other chemical reactions or producing movement.

- Experiments to investigate respiration try to detect uptake of oxygen, production of carbon dioxide, release of energy as heat or a reduction in dry weight.

- In a controlled experiment, the scientist tries to alter only one condition at a time, and sets up a control to check this.

- A control is a second experiment, identical to the first experiment except for the one condition being investigated.

- The control is designed to show that only the condition under investigation is responsible for the results.

- Experiments are designed to test predictions made from hypotheses; they cannot 'prove' a hypothesis.

4 HOW SUBSTANCES GET IN AND OUT OF CELLS

Cells need food materials which they can oxidize for energy or which they can use to build up their cell structures. They also need salts and water which play a part in chemical reactions in the cell. On the other hand they need to get rid of substances, such as carbon dioxide, which, if they accumulated in the cell, would upset some of the chemical reactions and even poison the cell.

Substances may pass through the cell membrane either passively by diffusion, or actively by some form of active transport.

Diffusion

The molecules of a gas such as oxygen are moving about all the time. So are the molecules of a liquid, or a substance such as sugar dissolved in water. As a result of this movement, the molecules spread themselves out evenly to fill all the available space (Fig. 1).

molecules moving about   become evenly distributed

Figure 1 Diffusion

This process is called **diffusion**. One effect of diffusion is that the molecules of a gas, a liquid or a dissolved substance will move from a region where there are a lot of them (i.e. concentrated) to regions where there are few of them (i.e. less concentrated) until the concentration everywhere is the same. Fig. 2(a) is a diagram of a cell with a high concentration of molecules (e.g. oxygen) outside and a low concentration inside. The effect of this difference in concentration is to make the molecules diffuse into the cell until the concentration inside and outside is the same (Fig. 2(b)).

Whether this will happen or not depends on whether the cell membrane will let the molecules through. Small molecules such as water (H_2O), carbon dioxide (CO_2) and oxygen (O_2) can pass through the cell membrane fairly easily. So diffusion tends to

(a) greater concentration outside cell

(b) concentrations equal on both sides of the cell membrane

Figure 2 Molecules entering a cell by diffusion

equalize the concentration of these molecules inside and outside the cell all the time.

When a cell uses up oxygen for its aerobic respiration, the concentration of oxygen inside the cell falls and so oxygen molecules diffuse into the cell until the concentration is raised again. During tissue respiration, carbon dioxide is produced and so its concentration inside the cell goes up. Once again diffusion takes place, but this time the molecules move out of the cell. In this way, diffusion can explain how a cell takes in its oxygen and gets rid of its carbon dioxide.

Rates of diffusion

The speed with which a substance diffuses through a cell wall or cell membrane will depend on temperature, pressure and many other conditions including (1) the distance it has to diffuse, (2) its concentration inside and outside the cell and (3) the size of its molecules or ions.

1 Cell membranes are all about the same thickness (about 0.007 μm) but plant cell walls vary in their thickness and permeability. Generally speaking, the thicker the wall, the slower the rate of diffusion.

2 The bigger the difference in concentration of a substance on either side of a membrane, the faster it will tend to diffuse. The difference is called a **concentration gradient** or **diffusion gradient** (Fig. 3). If a substance on one side of a membrane is steadily removed, the diffusion gradient is maintained. When oxygen molecules enter a red cell they combine with a chemical (haemoglobin) which takes them out of solution. Thus the concentration of free oxygen molecules inside the cell is kept very low and the diffusion gradient for oxygen is maintained.

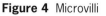

molecules will move from
the densely packed areas

Figure 3 Diffusion gradient

3 In general, the larger the molecules or ions, the slower they diffuse. However, many ions and molecules in solution attract water molecules around them (see p. 31) and so their effective size is greatly increased. It may not be possible to predict the rate of diffusion from the molecular size alone.

Controlled diffusion

Although for any one substance, the rate of diffusion through a cell membrane depends partly on the concentration gradient, the rate is often faster or slower than expected. Water diffuses more slowly and amino acids diffuse more rapidly through a membrane than might be expected. In some cases this is thought to happen because the ions or molecules can pass through the membrane only by means of special pores. These pores may be few in number or they may be open or closed in different conditions.

In other cases, the movement of a substance may be speeded up by an enzyme working in the cell membrane. So it seems that 'simple passive' diffusion, even of water molecules, may not be so simple or so passive after all, where cell membranes are concerned.

When a molecule gets inside a cell there are a great many structures and processes which may move it from where it enters to where it is needed. Simple diffusion is unlikely to play a very significant part in this movement.

Surface area

If 100 molecules diffuse through 1 mm^2 of a membrane in 1 minute, it is reasonable to suppose that an area of 2 mm^2 will allow twice as many through in the same time. Thus the rate of diffusion into a cell will depend on the cell's surface area. The greater the surface area, the faster is the total diffusion. Cells which are involved in rapid absorption, e.g. in the kidney or the intestine, often have their 'free' surface membrane formed into hundreds of tiny projections called **microvilli** (Fig. 4) which increase the absorbing surface.

The shape of a cell will also affect the surface area. For example, the cell in Fig. 5(a) has a greater surface area than that in Fig. 5(b), even though they each have the same volume.

Figure 4 Microvilli

Figure 5 Surface area. The cells have the same volume but (a) has a much greater surface area.

Endo- and exocytosis

Some cells can take in (**endocytosis**) or expel (**exocytosis**) solid particles or drops of fluid through the cell membrane. Endocytosis occurs in single-celled 'animals' such as *Paramecium* (p. 284) when they feed, or in certain white blood cells (phagocytes, p. 116) when they engulf bacteria, a process called **phagocytosis**. Exocytosis takes place in the cells of some glands. A secretion, e.g. a digestive enzyme, forms vacuoles or granules in the cytoplasm and these are expelled through the cell membrane to do their work outside the cell (Figs 6 and 7).

Figure 6 Endocytosis (phagocytosis) in a white blood cell

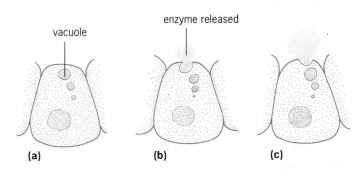

Figure 7 Exocytosis in a gland cell

Active transport

If diffusion were the only method by which a cell could take in substances, it would have no control over what went in or out. Anything that was more concentrated outside would diffuse into the cell whether it was harmful or not. Substances which the cell needed would diffuse out as soon as their concentration inside rose above that outside the cell. The cell membrane, however, has a great deal of control over the substances which enter and leave the cell.

In some cases, substances are taken into or expelled from the cell against the concentration gradient. For example, sodium ions may continue to pass out of a cell even though the concentration outside is greater than inside. The processes by which such reverse concentrations are produced are not fully understood and may be quite different for different substances but are all generally described as **active transport**.

Anything which interferes with respiration, e.g. lack of oxygen or glucose, prevents active transport taking place (Fig. 8). Also, during active transport, ATP is broken down to ADP (see p. 21). Thus it seems that active transport needs a supply of energy from respiration.

In some cases, a combination of active transport and controlled diffusion seems to occur. For example, sodium ions are thought to get into a cell by diffusion through special pores in the membrane and are expelled by a form of active transport. The reversed diffusion gradient for sodium ions created in this way is very important in the conduction of nerve impulses in nerve cells.

Plants need to absorb mineral salts from the soil, but these salts are in very dilute solution. Active transport enables the cells of the plant roots to take up salts from this dilute solution against the concentration gradient.

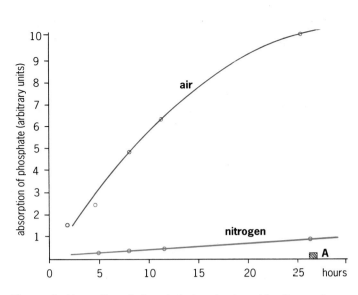

Figure 9 Absorption of phosphate ions in air and in nitrogen by roots of beech. (See question 4.) **A** represents concentration of phosphate in external solution.

QUESTIONS

1 Look at Fig. 3 on p. 117. If the symbol O_2 represents an oxygen molecule, explain why oxygen is entering the cells drawn on the left but leaving the cells on the right.
2 Look at Fig. 9 on p. 134 representing one of the small air pockets (an alveolus) which form the lung.
   a Suggest a reason why the oxygen and carbon dioxide are diffusing in opposite directions.
   b What might happen to the rate of diffusion if the blood flow were to speed up?
3 List the ways in which a cell membrane might regulate the flow of substances into the cell.
4 What is your interpretation of the results shown by the graph in Fig. 9?

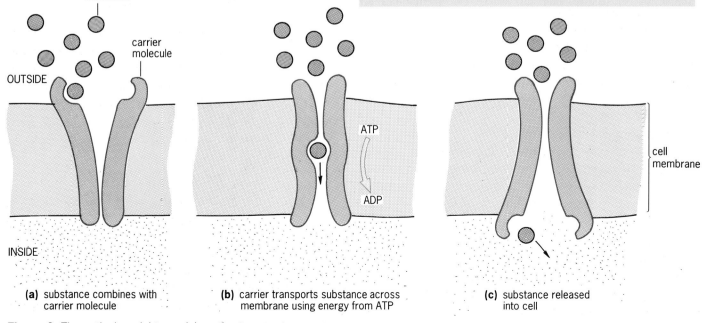

(a) substance combines with carrier molecule

(b) carrier transports substance across membrane using energy from ATP

(c) substance released into cell

Figure 8 Theoretical model to explain active transport

Osmosis

If a dilute solution is separated from a concentrated solution by a **partially permeable** membrane, water diffuses across the membrane from the dilute to the concentrated solution. This is known as **osmosis**.

A partially permeable membrane is porous but allows water to pass through it more rapidly than dissolved substances.

Figure 10 Osmosis. Water will diffuse from the dilute to the concentrated solution through the partially permeable membrane. As a result, the liquid level will rise on the left and fall on the right.

Figure 11 The diffusion gradient for water. There are more free water molecules on the left, so more will diffuse from left to right than in the other direction. Sugar molecules will diffuse more slowly from right to left.

Since a dilute solution contains, in effect, more water molecules than a concentrated solution, there is a diffusion gradient which favours the passage of water from the dilute to the concentrated solution.

In living cells, the cell membrane is partially permeable and the cytoplasm and vacuole (in plant cells) contain dissolved substances. As a consequence, water tends to diffuse into cells by osmosis if they are surrounded by a weak solution, e.g. fresh water. If the cells are surrounded by a stronger solution, e.g. sea water, the cells may lose water by osmosis.

These effects are described more fully on p. 32.

Explanation of osmosis

When a substance such as sugar dissolves in water, the sugar molecules attract some of the water molecules and stop them moving freely. This, in effect, reduces the concentration of water molecules. In Fig. 11 the sugar molecules on the right have 'captured' half the water molecules. There are more free water molecules on the left of the membrane than on the right, so water will diffuse more rapidly from left to right across the membrane than from right to left.

The partially permeable membrane does not act like a sieve, in this case. The sugar molecules can diffuse from right to left but, because they are bigger and surrounded by a cloud of water molecules, they diffuse more slowly than the water.

Artificial partially permeable membranes are made from cellulose acetate in sheets or tubes and used for **dialysis** (p. 36) rather than for osmosis. The pore size can be adjusted during manufacture so that large molecules cannot get through at all.

The cell membrane behaves like a partially permeable membrane. The partial permeability may depend on pores in the cell membrane but the processes involved are far more

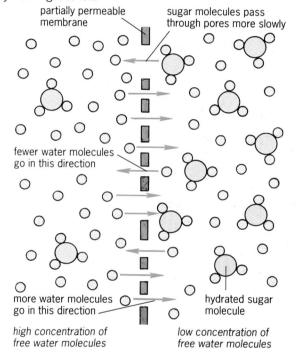

Figure 12 The diffusion theory of osmosis

complicated than in an artificial membrane and depend on the structure of the membrane and on living processes in the cytoplasm (see p. 29). The cell membrane contains lipids and proteins. Anything which denatures proteins, e.g. heat, also destroys the structure and the partially permeable properties of a cell membrane. If this happens, the cell will die as essential substances diffuse out of the cell and harmful chemicals diffuse in.

Water potential

The water potential of a solution is a measure of whether it is likely to lose or gain water molecules from another solution. A dilute solution, with its high proportion of free water molecules, is said to have a higher water potential than a concentrated solution, because water will flow from the dilute to the concentrated solution (from a high potential to a low potential). Pure water has the highest possible water potential because water molecules will flow from it to any other aqueous solution, no matter how dilute.

Animal cells

In Fig. 13 an animal cell is shown very simply. The coloured circles represent molecules in the cytoplasm. They may be sugar, salt or protein molecules. The grey circles represent water molecules.

The cell is shown surrounded by pure water. Nothing is dissolved in the water; it has 100 per cent concentration of water molecules. So the concentration of free water molecules outside the cell is greater than that inside and, therefore, water will diffuse into the cell by osmosis.

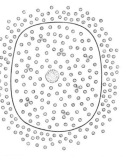

(a) There is a higher concentration of free water molecules outside the cell than inside, so water diffuses into the cell.

(b) The extra water makes the cell swell up.
(Note that molecules are really far too small to be seen at this magnification.)

Figure 13 Osmosis in an animal cell

The membrane allows water to go through either way. So in our example, water can move into or out of the cell.

The cell membrane is partially permeable to most of the substances dissolved in the cytoplasm. So although the concentration of these substances inside may be high, they cannot diffuse freely out of the cell.

The water molecules move into and out of the cell, but because there are more of them on the outside, they will move in faster than they move out. The liquid outside the cell does not have to be 100 per cent pure water. As long as the concentration of water outside is higher than that inside, water will diffuse in by osmosis.

Water entering the cell will make it swell up, and unless the extra water is expelled in some way the cell will burst.

Conversely, if the cells are surrounded by a solution which is more concentrated than the cytoplasm, water will pass out of the cell by osmosis and the cell will shrink. Excessive uptake or loss of water by osmosis may damage cells.

For this reason, it is very important that the cells in an animal's body are surrounded by a liquid which has the same concentration as the liquid inside the cells. The outside liquid is called 'tissue fluid' (see p. 121) and its concentration depends on the concentration of the blood. In vertebrate animals the blood's concentration is monitored by the brain and adjusted by the kidneys, as described on p. 141.

By keeping the blood concentration within narrow limits, the concentration of tissue fluid remains more or less constant (see pp. 124 and 143) and the cells are not bloated by taking in too much water, or dehydrated by losing too much.

Plant cells

The cytoplasm of a plant cell and the cell sap in its vacuole contain salts, sugars and proteins which effectively reduce the concentration of free water molecules inside the cell. The cell wall is freely permeable to water and dissolved substances but the cell membrane of the cytoplasm is partially permeable. If a plant cell is surrounded by water or a solution more dilute than its contents, water will pass into the vacuole by osmosis. The vacuole will expand and press outwards on the cytoplasm and cell wall. The cell wall of a mature plant cell cannot be stretched, so there comes a time when the inflow of water is resisted by the unstretchable cell wall (Fig. 14).

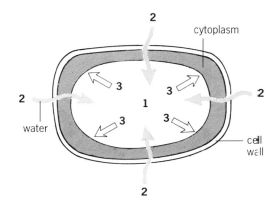

1 Since there is effectively a lower concentration of water in the cell sap
2 water diffuses into the vacuole
3 and makes it push out against the cell wall.

Figure 14 Osmosis in a plant cell

This has a similar effect to inflating a soft bicycle tyre. The tyre represents the firm cell wall, the floppy inner tube is like the cytoplasm and the air inside corresponds to the vacuole. If enough air is pumped in, it pushes the inner tube against the tyre and makes the tyre hard. A plant cell with the vacuole pushing out on the cell wall is said to be **turgid** and the vacuole is exerting **turgor pressure** on the cell wall.

If all the cells in a leaf and stem are turgid, the stem will be firm and upright and the leaves held out straight. If the vacuoles lose water for any reason, the cells will lose their turgor and become **flaccid** (Experiment 3, p. 34). A leaf with flaccid cells will be limp and the stem will droop. A plant which loses water to this extent is said to be 'wilting' (Fig. 15).

(a) plant wilting

(b) plant recovered after watering

Figure 15 Wilting

QUESTIONS

5 A 10 per cent solution of copper sulphate is separated by a partially permeable membrane from a 5 per cent solution of copper sulphate. Will water diffuse from the 10 per cent to the 5 per cent solution, or from the 5 per cent to the 10 per cent solution?

6 If a fresh beetroot is cut up, the pieces washed in water and then left for an hour in a beaker of water, little or no red pigment escapes from the cells into the water. If the beetroot is boiled first, the pigment does escape into the water. Bearing in mind the properties of a living cell membrane, offer an explanation for this difference.

7 When doing experiments with animal tissues they are usually bathed in Ringer's solution which has a concentration similar to that of blood or tissue fluid. Why do you think this is necessary?

8 Why does a dissolved substance reduce the number of 'free' water molecules in a solution?

9 When a plant leaf is in daylight, its cells make sugar from carbon dioxide and water (see p. 37). The sugar is at once turned into starch and deposited in plastids. What is the osmotic advantage of doing this? (Sugar is soluble in water, starch is not. See p. 12.)

PRACTICAL WORK

Experiments on osmosis and dialysis

Experiments 1, 2 and 5 use 'Visking' dialysis tubing. It is made from cellulose and is partially permeable, allowing water molecules to diffuse through freely, but restricting the passage of dissolved substances to varying extents. It is used in kidney machines because it lets the small molecules of harmful waste products (e.g. urea, p. 142) out of the blood but retains the blood cells and large protein molecules.

1 Osmosis and turgor

Take a 20 cm length of dialysis tubing which has been soaked in water and tie a knot tightly at one end. Place 3 cm³ of a strong sugar solution in the tubing using a plastic syringe (Fig. 16(a)) and then knot the open end of the tube (Fig. 16(b)). The partly filled tube should be quite floppy (Fig. 16(c)). Place the tubing in a test-tube of water for 30–45 minutes. After this time, remove the dialysis tubing from the water and look for any changes in how it looks or feels.

Result   The tubing will now be firm, distended by the solution inside.

Interpretation   The dialysis tubing is partially permeable and the solution inside has fewer free water molecules than outside.

(a) Place 3 cm³ syrup in the dialysis tube.

(b) Knot tightly, after expelling the air bubbles.

(c) The partly filled tube should be flexible enough to bend.

water

dialysis tube containing syrup

Figure 16 Experiment to illustrate turgor in a plant cell

Water has, therefore, diffused in and increased the volume and the pressure of the solution inside.

This is a crude model of what is thought to happen to a plant cell when it becomes turgid. The sugar solution represents the cell sap and the dialysis tubing represents the cell membrane and cell wall combined.

2 Osmosis and water flow

Tie a knot in one end of a piece of soaked dialysis tubing and fill it with sugar solution as in the previous experiment. Then fit it over the end of a length of capillary tubing and hold it in place with an elastic band. Push the capillary tubing into the dialysis tubing until the sugar solution enters the capillary. Now clamp the capillary tubing so that the dialysis tubing is totally immersed in a beaker of water as shown in Fig. 17. Watch the level of liquid in the capillary tubing over the next 10 or 15 minutes.

capillary tube

first level

elastic band

cellulose tube containing sugar solution (with red dye)

water

Figure 17 Demonstration of osmosis

Result   The level of liquid in the capillary tube will be seen to rise.

Interpretation   Water must be passing into the sugar solution from the beaker. This is what you would expect when a concentrated solution is separated from water by a partially permeable membrane. The results are similar to those in Experiment 1 but instead of the expanding solution distending the dialysis tube, it escapes up the capillary.

A process similar to this might be partially responsible for moving water from the roots to the stem of a plant.

3 Plasmolysis

Peel a small piece of epidermis (outer layer of cells) from a red area of a rhubarb stalk (see Fig. 15(a), p. 9). Place the epidermis on a slide with a drop of water and cover with a cover-slip (see Fig. 15(c), p. 9). Place a 30 per cent solution of sugar at one edge of the cover-slip with a pipette and then draw the solution under the cover-slip by placing a piece of blotting-paper on the opposite side (Fig. 18). While you are doing this, study a small group of cells under the microscope and watch for any changes in their appearance (Fig. 20).

Figure 18 Changing the water for sugar solution

Result   The red cell sap will appear to shrink and get darker and pull the cytoplasm away from the cell wall leaving clear spaces. (It will not be possible to see the cytoplasm but its presence can be inferred from the fact that the red cell sap seems to have a distinct outer boundary in those places where it has separated from the cell wall.)

Interpretation   The interpretation in terms of osmosis is outlined in Fig. 19. The cells are said to be **plasmolysed**.

The plasmolysis can be reversed by drawing water under the cover-slip in the same way that you drew the sugar solution under. It may need two or three lots of water to flush out all the sugar. If you watch a group of cells, you should see their vacuoles expanding to fill the cells once again.

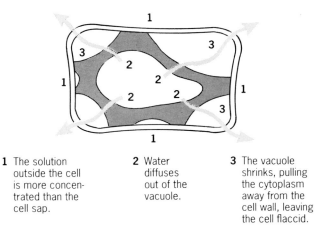

1 The solution outside the cell is more concentrated than the cell sap.

2 Water diffuses out of the vacuole.

3 The vacuole shrinks, pulling the cytoplasm away from the cell wall, leaving the cell flaccid.

Figure 19 Plasmolysis

Figure 20 (a) Turgid cells (x 100). The cells are in a strip of epidermis from a red onion scale. The cytoplasm is pressed against the inside of the cell wall by the vacuole.

(b) Plasmolysed cells (x100). The same cells as they appear after treatment with salt solution. The vacuole has lost water by osmosis, shrunk and pulled the cytoplasm away from the cell wall.

Rhubarb is used for this experiment because the coloured cell sap shows up. If rhubarb is not available the epidermis from a red onion scale can be used with results similar to those in Fig. 20.

4 Turgor in potato tissue

Push a No. 4 or No. 5 cork borer into a large potato.
Caution Do not hold the potato in your hand but use a board as in Fig. 21(a).
Push the potato tissue out of the cork borer using a pencil as in Fig. 21(b). Prepare a number of potato cylinders in this way and choose the two longest (at least 50 mm). Cut these two accurately to the same length, e.g. 50, 60 or 70 mm. Measure carefully.

Label two test-tubes A and B and place a potato cylinder in each. Cover the potato tissue in tube A with water; cover the tissue in B with a 20 per cent sugar solution. Leave the tubes for a day.

After this time, remove the cylinder from tube A and measure its length. Notice also whether it is firm or flabby. Repeat this for the potato in tube B, but rinse it in water before measuring it.

Result The cylinder from tube A should have gained a millimetre or two and feel firm. The cylinder from tube B should be a millimetre or two shorter and feel flabby.

Interpretation If the potato cells were not fully turgid at the beginning of the experiment, they would take up water by osmosis (tube A), and cause an increase in length.

In tube B, the sugar solution is stronger than the cell sap of the potato cells, so these cells will lose water by osmosis. The cells will lose their turgor and the potato cylinder will become flabby and shorter.

An alternative to measuring the potato cores is to weigh them before and after the 24 hours immersion in water or sugar solution. The core in tube A should gain weight and that in tube B should lose weight. It is important to blot the cores dry with a paper towel before each weighing.

Whichever method is used, it is a good idea to pool the results of the whole class since the changes may be quite small. A gain in length of 1 or 2 mm might be due to an error in measurement, but if most of the class record an increase in length, then experimental error is unlikely to be the cause.

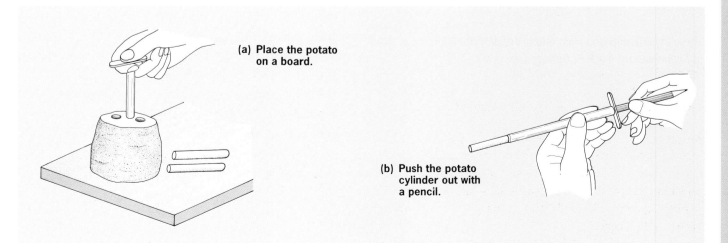

(a) Place the potato on a board.

(b) Push the potato cylinder out with a pencil.

Figure 21 Obtaining cylinders of potato tissue

5 Partial permeability (dialysis)

Take a 15 cm length of dialysis tubing which has been soaked in water and tie a knot tightly at one end. Use a dropping pipette to partly fill the tubing with 1 per cent starch solution. Put the tubing in a test-tube and hold it in place with an elastic band as shown in Fig. 22. Rinse the tubing and test-tube under the tap to remove all traces of starch solution from the outside of the dialysis tube.

Fill the test-tube with water and add a few drops of iodine solution to colour the water yellow. Leave for 10–15 minutes.

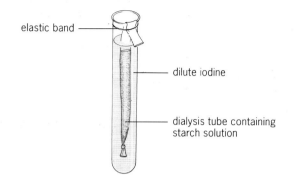

elastic band

dilute iodine

dialysis tube containing starch solution

Figure 22 Dialysis

Result   The starch inside the dialysis tubing goes blue but the iodine solution outside stays yellow.

Interpretation   The blue colour is characteristic of the reaction which takes place between starch and iodine, and is used as a test for starch (see p. 102). The results show that iodine molecules have passed through the dialysis tubing into the starch but the starch molecules have not moved out into the iodine. This is what we would expect if the dialysis tubing is partially permeable on the basis of its pore size. Starch molecules are very large (see p. 12) and probably cannot get through the pores. Iodine molecules are much smaller and can, therefore, get through.

Note   This experiment illustrates the process of dialysis rather than osmosis. The movement of water is not necessarily involved and the pore size of the membrane makes it genuinely partially permeable with respect to iodine and starch.

These experiments and eight others are described in greater detail in *Experimental Work in Biology* (see p. vi).

QUESTIONS

10 In Experiment 2 (Fig. 17), what do you think would happen
 a if a much stronger sugar solution was placed in the cellulose tube?
 b if the beaker contained a weak sugar solution instead of water?
 c if the sugar solution was in the beaker and the water was in the cellulose tube?

11 In Experiment 1 (Fig. 16), what might happen if the cellulose tube filled with sugar solution was left in the water for several hours?

12 In Experiment 3, Fig. 19 explains why the vacuole shrinks. Give a brief explanation of why it swells up again when the cell is surrounded by water.

13 An alternative interpretation of the results of Experiment 5 might be that the dialysis tubing allowed molecules (of any size) to pass in but not out. Describe an experiment to test this possibility and say what results you would expect   **a** if it were correct and   **b** if it were false.

14 In Experiment 2 on p. 34, the column of liquid accumulating in the capillary tube exerts an ever-increasing pressure on the solution in the dialysis tube. Bearing this in mind and assuming a very long capillary, at what stage would you expect the net flow of water from the beaker into the dialysis tubing to cease?

Checklist

- Diffusion is the result of molecules of liquid, gas or dissolved solid moving about.

- The molecules of a substance diffuse from a region where they are very concentrated to a region where they are less concentrated.

- Substances may enter cells by simple diffusion, controlled diffusion, active transport or endocytosis.

- Osmosis is the diffusion of water through a partially permeable membrane.

- Water diffuses from a dilute solution of salt or sugar to a concentrated solution because the concentrated solution contains fewer free water molecules.

- Cell membranes are partially permeable and cytoplasm and cell sap contain many substances in solution.

- Cells take up water from dilute solutions but lose water to concentrated solutions because of osmosis.

- Osmosis maintains turgor in plant cells.

5 PHOTOSYNTHESIS AND NUTRITION IN PLANTS

Photosynthesis
Introductory explanation; definition and equation.

Practical work
Experiments to test photosynthesis: principles of experimental design; destarching and testing a leaf for starch; need for chlorophyll, light and carbon dioxide; production of oxygen.

Process of photosynthesis
The process in leaf cells.

Gaseous exchange
Uptake of carbon dioxide and output of oxygen in photosynthesis. Uptake of oxygen and output of carbon dioxide in respiration.
Compensation point.

Adaptation of leaves for photosynthesis
Shape and internal structure.

Effects of external factors on rate of photosynthesis
Light intensity; carbon dioxide concentration; temperature. Limiting factors.

Use of photosynthetic products in plants
Storage, energy and growth.

Sources of mineral elements
Nitrates, sulphates and fertilizers.

Water cultures
Experiment to demonstrate the importance of certain minerals.

All living organisms need food. They need it as a source of raw materials to build new cells and tissues as they grow. They also need food as a source of energy. Food is a kind of 'fuel' which drives essential living processes and brings about chemical changes (see pp. 20 and 94). Animals take in food, digest it, and use the digested products to build their tissues or to produce energy.

Plants also need energy and raw materials but, apart from a few insect-eating species, plants do not appear to take in food. The most likely source of their raw materials would appear to be the soil. However, experiments show that the weight gained by a growing plant is far greater than the weight lost by the soil it is growing in. So there must be additional sources of raw materials.

These additional sources can only be water and air. A hypothesis to explain the source of food in a plant is that it **makes it** from air, water and soil salts. Carbohydrates (p. 12) contain the elements carbon, hydrogen and oxygen, as in glucose ($C_6H_{12}O_6$). The carbon and oxygen could be supplied by carbon dioxide (CO_2) from the air, and the hydrogen could come from the water (H_2O) in the soil. The nitrogen and sulphur needed for making proteins (p. 11) could come from nitrates and sulphates in the soil.

This building-up of complex food molecules from simpler substances is called a **synthesis** and it needs enzymes and energy to make it happen. The enzymes are present in the plant's cells and the energy for the first stages in the synthesis comes from sunlight. The process is, therefore, called **photosynthesis** ('photo' means 'light'). There is evidence to suggest that the green substance **chlorophyll**, in the chloroplasts of plant cells, plays a part in photosynthesis. Chlorophyll absorbs sunlight and makes the energy from sunlight available for chemical reactions. Thus, in effect, the function of chlorophyll is to convert light energy to chemical energy.

Our working hypothesis for photosynthesis is, therefore, the building-up of food compounds from carbon dioxide and water by green plants using energy from sunlight which is absorbed by chlorophyll.

A chemical equation for photosynthesis would be

$$6CO_2 + 6H_2O \xrightarrow{\text{light energy}} C_6H_{12}O_6 + 6O_2$$

carbon   water   glucose   oxygen
dioxide

In order to keep the equation simple, glucose is shown as the food compound produced. This does not imply that it is the only substance synthesized by photosynthesis.

PRACTICAL WORK

Experiments to test photosynthesis

The design of biological experiments was discussed on p. 25 and this should be revised before studying the next section. It would also be helpful to read the section on 'Hypothesis testing' (p. 26) if you have not already done so.

A **hypothesis** is an attempt to explain certain observations. In this case the hypothesis is that plants make their food by photosynthesis. The equation given below is one way of stating the hypothesis and is used here to show how it might be tested.

$$6CO_2 + 6H_2O \xrightarrow[\text{chlorophyll}]{\text{sunlight}} C_6H_{12}O_6 + 6O_2$$

  ↑   ↑   ↓   ↓
uptake uptake   production   release
of carbon of water   of   of oxygen
dioxide   sugar
                         (or starch)

If photosynthesis is going on in a plant, then the leaves should be producing sugars. In many leaves, as fast as sugar is produced, it is turned into starch. Since it is easier to test for starch than for sugar, we regard the production of starch in a leaf as evidence that photosynthesis has taken place.

The first three experiments described below are designed to see if the leaf can make starch without chlorophyll, or sunlight, or carbon dioxide, in turn. If the photosynthesis story is sound, then the lack of any one of these three conditions should stop photosynthesis, and so stop the production of starch. If a leaf without a supply of carbon dioxide can still produce starch, then the hypothesis is no good and must be altered or rejected.

In designing the experiments, it is very important to make sure that only *one* independent variable is altered. If, for example, the method of keeping light from a leaf also cuts off its carbon dioxide supply, it would be impossible to decide whether it was the lack of light or lack of carbon dioxide which stopped the production of starch. To make sure that the experimental design has not altered more than one variable, a **control** is set up in each case. This is an identical situation, except that the condition missing from the experiment, e.g. light, carbon dioxide or chlorophyll, is present in the control (see p. 25).

Destarching a plant

If the production of starch is your evidence that photosynthesis is taking place, then you must make sure that the leaf does not contain any starch at the beginning of the experiment. This is done by **destarching** the leaves. It is not possible to remove the starch chemically, without damaging the leaves, so a plant is destarched simply by leaving it in darkness for 2 or 3 days. Potted plants are destarched by leaving them in a dark cupboard for a few days. In the darkness, any starch in the leaves will be changed to sugar and carried away from the leaves to other parts of the plant. For plants in the open, the experiment is set up on the day before the test. During the night, most of the starch will be removed from the leaves. Better still, wrap the leaves in aluminium foil for 2 days while they are still on the plant. Then test one of the leaves to see that no starch is present.

Testing a leaf for starch

Iodine solution (yellow) and starch (white) form a deep blue colour when they mix. The test for starch, therefore, is to add iodine solution to a leaf to see if it goes blue. First, however, the leaf has to be treated as follows:

1 Heat some water to boiling point in a beaker and then **turn out the Bunsen flame**.
2 Use forceps to dip a leaf in the hot water for about 30 seconds. This kills the cytoplasm, denatures the enzymes and makes the leaf more permeable to iodine solution.
3 Push the leaf to the bottom of a test-tube and cover it with alcohol (ethanol). Place the tube in the hot water (Fig. 1). The alcohol will boil and dissolve out most of the chlorophyll. This makes colour changes with iodine easier to see.

Figure 1 To remove chlorophyll from a leaf

4 Pour the green alcohol into a spare beaker, remove the leaf and dip it once more into the hot water to soften it.
5 Spread the decolorized leaf flat on a white tile and drop iodine solution on to it. The parts containing starch will turn blue; parts without starch will stain brown or yellow with iodine.

1 Is chlorophyll necessary for photosynthesis?

It is not possible to remove chlorophyll from a leaf without killing it, and so a **variegated** leaf, which has chlorophyll only in patches, is used. A leaf of this kind is shown in Fig. 2(a). The white part of the leaf serves as the experiment, because it lacks chlorophyll, while the green part with chlorophyll is the control. After being destarched, the leaf—still on the plant—is exposed to daylight for a few hours. Remove a leaf from the plant; draw it carefully to show where the chlorophyll is (i.e. the green parts), and test it for starch as described above.

Result Only the parts that were previously green turn blue with iodine. The parts that were white stain brown (Fig. 1(b)).

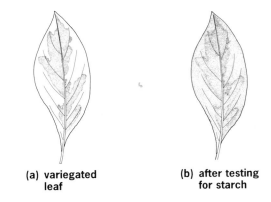

(a) variegated leaf

(b) after testing for starch

Figure 2 To show that chlorophyll is necessary

Interpretation Since starch is present only in the parts which originally contained chlorophyll, it seems reasonable to suppose that chlorophyll is needed for photosynthesis.

It must be remembered, however, that there are other possible interpretations which this experiment has not ruled out; for example, starch could be made in the green parts and sugar in the white parts. Such alternative explanations could be tested by further experiments.

2 Is light necessary for photosynthesis?

Cut a simple shape from a piece of aluminium foil to make a stencil and attach it to a destarched leaf (Fig. 3(a)). After 4 to 6 hours of daylight, remove the leaf and test it for starch.

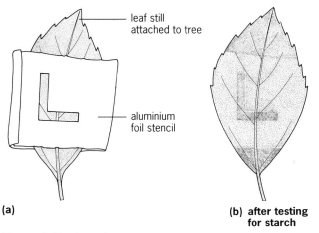

(a)

(b) **after testing for starch**

Figure 3 To show that light is necessary

Result Only the areas which had received light go blue with iodine.

Interpretation As starch has not formed in the areas which received no light, it seems that light is needed for starch formation and thus for photosynthesis.

You could argue that the aluminium foil had stopped carbon dioxide from entering the leaf and that it was shortage of carbon dioxide rather than absence of light which prevented photosynthesis taking place. A further control could be designed, using transparent material instead of aluminium foil for the stencil.

3 Is carbon dioxide needed for photosynthesis?

Water two destarched potted plants and enclose their shoots in polythene bags. In one pot place a dish of soda-lime to absorb the carbon dioxide from the air (the experiment). In the other place a dish of sodium hydrogencarbonate to produce carbon dioxide (the control), as shown in Fig. 4. Place both plants in the light for several hours and then test a leaf from each for starch.

Result The leaf which had no carbon dioxide does not turn blue. The one from the polythene bag containing carbon dioxide does turn blue.

plastic bag

either soda-lime or sodium hydrogencarbonate solution

Figure 4 To show that carbon dioxide is necessary

Interpretation The fact that starch was made in the leaves which had carbon dioxide, but not in the leaves which had no carbon dioxide, suggests that this gas must be necessary for photosynthesis. The control rules out the possibility that high humidity or high temperature in the plastic bag prevents normal photosynthesis.

4 Is oxygen produced during photosynthesis?

Place a short-stemmed funnel over some Canadian pondweed in a beaker of water. Fill a test-tube with water and place it upside-down over the funnel stem (Fig. 5). (The funnel is raised above the bottom of the beaker to allow the water to circulate.) Place the apparatus in sunlight. Bubbles of gas should appear from the cut stems and collect in the test-tube. Set up a control in a similar way but place it in a dark cupboard. When sufficient gas has collected from the plant in the light, remove the test-tube and insert a glowing splint.

gas collecting

sunlight

Canadian pondweed

support to keep funnel off bottom

Figure 5 To show that oxygen is produced

Result   The glowing splint bursts into flames.

Interpretation   The relighting of a glowing splint does not prove that the gas collected in the test-tube is *pure* oxygen, but it does show that it contains extra oxygen and this must have come from the plant. The oxygen is given off only in the light.

> Note that water contains dissolved oxygen, carbon dioxide and nitrogen. These gases may diffuse in or out of the bubbles as they pass through the water and collect in the test-tube. The composition of the gas in the test-tube may not be the same as that in the bubbles leaving the plant.

Controls

When setting up an experiment and a control, which of the two procedures constitutes the 'control' depends on the way the prediction is worded. For example, if the prediction is that 'in the absence of light, the pondweed will not produce oxygen', then the 'control' is the plant in the light. If the prediction is that 'the pondweed in the light will produce oxygen', then the 'control' is the plant in darkness. As far as the results and interpretation are concerned, it does not matter which is the 'control' and which is the 'experiment'.

The results of these four experiments support the hypothesis of photosynthesis as stated on p. 37 and as represented by the equation. Starch formation (our evidence for photosynthesis) does not take place in the absence of light, chlorophyll or carbon dioxide, and oxygen production occurs only in the light.

If starch or oxygen production had occurred in the absence of any one of these conditions, we should have to change our hypothesis about the way plants obtain their food. Bear in mind, however, that although our results support the photosynthesis theory, they do not prove it. For example, it is now known that many stages in the production of sugar and starch from carbon dioxide do not need light (the 'dark' reaction).

Figure 6   All the reactions involved in producing food take place in the leaves. Notice how little the leaves overlap.

QUESTIONS

1 Which of the following are needed for starch production in a leaf: **a** carbon dioxide, **b** oxygen, **c** nitrates, **d** water, **e** chlorophyll, **f** soil, **g** light?
2 In Experiment 1 (concerning the need for chlorophyll), why was it not necessary to set up a separate control experiment?
3 What is meant by 'destarching' a leaf? Why is it necessary to destarch leaves before setting up some of the photosynthesis experiments?
4 In Experiment 3 (concerning the need for carbon dioxide), what were the functions of **a** the soda-lime, **b** the sodium hydrogencarbonate, **c** the polythene bag?
5 Why do you think a pondweed, rather than a land plant, is used for Experiment 4 (concerning production of oxygen)? In what way might this choice make the results less useful?
6 A green plant makes sugar from carbon dioxide and water. Why do we not try the experiment of depriving a plant of water to see if that stops photosynthesis?
7 Does the method of destarching a plant take for granted the results of Experiment 2? Explain your answer.
8 In Experiments 1 to 3, state in each case which was **a** the independent variable and **b** the dependent variable.

The process of photosynthesis

Although the details of photosynthesis vary in different plants, the hypothesis as stated in this chapter has stood up to many years of experimental testing and is universally accepted. The next section describes how photosynthesis takes place in a plant.

The process takes place mainly in the cells of the leaves and is summarized in Fig. 7. In land plants water is absorbed from the soil by the roots and carried in the water vessels of the veins, up the stem to the leaf. Carbon dioxide is absorbed from the air through the stomata (pores in the leaf, see p. 54). In the leaf cells, the carbon dioxide and water are combined to make sugar; the energy for this reaction comes from sunlight which has been absorbed by the green pigment **chlorophyll**. The chlorophyll is present in the **chloroplasts** of the leaf cells and it is inside the chloroplasts that the reaction takes place. Chloroplasts (Fig. 7(d)) are small, green structures present in the cytoplasm of the leaf cells. Chlorophyll is the substance which gives leaves and stems their green colour. It is able to absorb energy from light and use it to split water molecules into hydrogen and oxygen (the 'light' reaction). The oxygen escapes from the leaf and the hydrogen molecules are added to carbon dioxide molecules to form sugar (the 'dark' reaction).

There are four types of chlorophyll which may be present in various proportions in different species. There are also a number of photosynthetic pigments, other than chlorophyll, which may mask the colour of chlorophyll even when it is present, e.g. the brown and red pigments which occur in certain seaweeds.

QUESTIONS

9 What substances must a plant take in, in order to carry on photosynthesis? Where does it get each of these substances from?

10 Look at Fig. 7(a) on p. 55. Identify the palisade cells, the spongy mesophyll cells and the cells of the epidermis. In which of these would you expect photosynthesis to occur **a** most rapidly, **b** least rapidly, **c** not at all? Explain your answer.

11 **a** What provides a plant with energy for photosynthesis?
b What chemical process provides a plant with energy to carry on all other living activities?

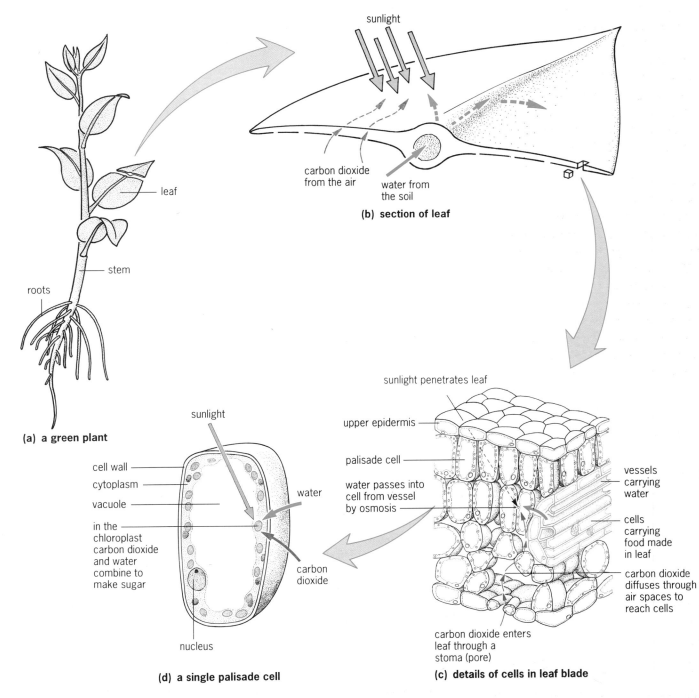

(b) section of leaf

(a) a green plant

(d) a single palisade cell

(c) details of cells in leaf blade

Figure 7 Photosynthesis in a leaf

Gaseous exchange in plants

Air contains the gases nitrogen, oxygen, carbon dioxide and water vapour. Plants and animals take in or give out these last three gases and this process is called **gaseous exchange**.

You can see from the equation for photosynthesis (p. 37) that one of its products is oxygen. Therefore, in daylight, when photosynthesis is going on in green plants, they will be taking in carbon dioxide and giving out oxygen. This exchange of gases is the opposite of that resulting from respiration (p. 20) but it must not be thought that green plants do not respire. The energy they need for all their living processes—apart from photosynthesis—comes from respiration and this is going on all the time, using up oxygen and producing carbon dioxide.

During the daylight hours, plants are photosynthesizing as well as respiring, so that all the carbon dioxide produced by respiration is used up by photosynthesis. At the same time, all the oxygen needed by respiration is provided by photosynthesis. Only when the rate of photosynthesis is faster than the rate of respiration will carbon dioxide be taken in and the excess oxygen be given out (Fig. 8).

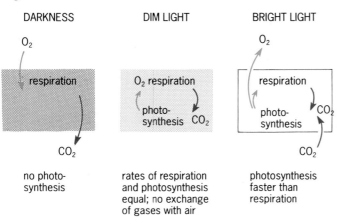

DARKNESS — O_2 → respiration → CO_2 — no photosynthesis

DIM LIGHT — O_2 respiration, photosynthesis CO_2 — rates of respiration and photosynthesis equal; no exchange of gases with air

BRIGHT LIGHT — O_2 respiration, photosynthesis CO_2 — photosynthesis faster than respiration

Figure 8 Respiration and photosynthesis

Compensation point

As the light intensity increases during the morning and fades during the evening, there will be a time when the rate of photosynthesis exactly matches the rate of respiration. At this point, there will be no net intake or output of carbon dioxide or oxygen. This is the compensation point. The sugar produced by photosynthesis exactly compensates for the sugar broken down by respiration.

PRACTICAL WORK

5 Gaseous exchange during photosynthesis

Wash three test-tubes first with tap-water, then with distilled water and finally with hydrogencarbonate indicator. Then place 2 cm³ hydrogencarbonate indicator (see Appendix 1) in each tube.

Place a green leaf in tubes 1 and 2 so that it is held against the walls of the tube and does not touch the indicator (Fig 9). Close the three tubes with bungs. Cover tube 1 with aluminium foil and place all three in a rack in direct sunlight, or a few centimetres from a bench lamp, for about 40 minutes.

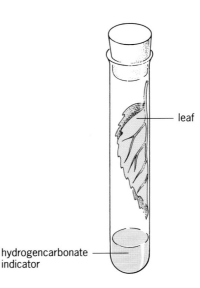

leaf

hydrogencarbonate indicator

Figure 9 Gaseous exchange during photosynthesis and respiration

Result The indicator (which was originally orange) should not change colour in tube 3, the control; that in tube 1, with the leaf in darkness, should turn yellow; and in tube 2 with the illuminated leaf, the indicator should be scarlet or purple.

Interpretation Hydrogencarbonate indicator is a mixture of dilute sodium hydrogencarbonate solution with the dyes cresol red and thymol blue. It is a pH indicator in equilibrium with the atmospheric carbon dioxide, i.e. its original colour represents the acidity produced by the carbon dioxide in the air. Increase in atmospheric carbon dioxide makes it more acid and it changes colour from orange to yellow. Decrease in atmospheric carbon dioxide makes it less acid and causes a colour change to red or purple.

The results, therefore, provide evidence that in darkness (tube 1) leaves produce carbon dioxide (from respiration), while in light (tube 2) they use up more carbon dioxide in photosynthesis than they produce in respiration. Tube 3 is the control, showing that it is the presence of the leaf which causes a change in the atmosphere in the test-tube.

The experiment can be criticized on the grounds that the hydrogencarbonate indicator is not a specific test for carbon dioxide but will respond to any change in acidity or alkalinity. In tube 2 there would be the same change in colour if the leaf produced an alkaline gas such as ammonia, and in tube 1, any acid gas produced by the leaf would turn the indicator yellow. However, a knowledge of the metabolism of the leaf suggests that these are less likely events than changes in the carbon dioxide concentration.

12 What gases would you expect a leaf to be **i** taking in **ii** giving out, **a** in bright sunlight **b** in darkness?

13 Measurements on a leaf show that it is giving out carbon dioxide and taking in oxygen. Does this prove that photosynthesis is *not* going on in the leaf? Explain your answer.

14 How could you adapt the experiment with hydrogencarbonate indicator to find the light intensity which corresponded to the compensation point?

How would you expect the compensation points to differ between plants growing in a wood and those growing in a field?

Adaptation of leaves for photosynthesis

When biologists say that something is **adapted**, they mean that its structure is well suited to its function. The detailed structure of the leaf is described on p. 52 to 55, and although there are wide variations in leaf shape the following general statements apply to a great many leaves, and are illustrated in Figs 7(b) and (c).

1 Their broad, flat shape offers a large surface area for absorption of sunlight and carbon dioxide.

2 Most leaves are thin and the carbon dioxide has to diffuse across only short distances to reach the inner cells.

3 The large spaces between cells inside the leaf provide an easy passage through which carbon dioxide can diffuse.

4 There are many stomata (pores) in the lower surface of the leaf. These allow the exchange of carbon dioxide and oxygen with the air outside.

5 There are more chloroplasts in the upper (palisade) cells than in the lower (spongy mesophyll) cells. The palisade cells, being on the upper surface, will receive most sunlight and this will reach the chloroplasts without being absorbed by too many cell walls.

6 The branching network of veins provides a good water supply to the photosynthesizing cells. No cell is very far from a water-conducting vessel in one of these veins.

Although photosynthesis takes place mainly in the leaves, any part of the plant which contains chlorophyll will photosynthesize. Many plants have green stems in which photosynthesis takes place.

The rate of photosynthesis

The rate of the light reaction will depend on the light intensity. The brighter the light, the faster will water molecules be split in the chloroplasts. The 'dark' reaction will be affected by temperature. A rise in temperature will increase the rate at which carbon dioxide is combined with hydrogen to make carbohydrate.

Limiting factors

If you look at Fig. 10(a), you will see that an increase in light intensity does indeed speed up photosynthesis, but only up to a point. Beyond that point, any further increase in light intensity has only a small effect. This limit on the rate of increase could be because all available chloroplasts are fully occupied in light absorption. So, no matter how much the light intensity increases, no more light can be absorbed and used. Alternatively, the limit could be imposed by the fact that there is not enough carbon dioxide in the air to cope with the increased supply of hydrogen atoms produced by the light reaction. Or, it may be that low temperature is restricting the rate of the 'dark' reaction.

Fig. 10(b) shows that, if the temperature of a plant is raised, then the effect of increased illumination is not limited so much. Thus, in Fig. 10(a), it seems likely that the increase in the rate of photosynthesis could have been limited by the temperature. Any one of the external factors, temperature, light intensity or carbon dioxide concentration, may limit the effects of the other two. A temperature rise may cause photosynthesis to speed up, but only to the point where the light intensity limits further increase. In such conditions, the external factor which restricts the effect of the others is called the **limiting factor**.

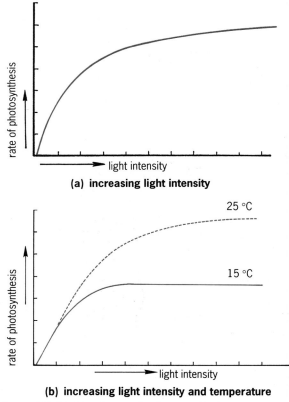

Figure 10 Limiting factors in photosynthesis

Since there is only 0.03 per cent of carbon dioxide in the air, it might seem that shortage of carbon dioxide could be an important limiting factor. Indeed, experiments do show that an increase in carbon dioxide concentration does allow a faster rate of photosynthesis. However, recent work in plant physiology has shown that the extra carbon dioxide affects reactions other than photosynthesis.

The main effect of extra carbon dioxide is to slow down the rate of oxidation of sugar by a process called **photorespiration** and this produces the same effect as an increase in photosynthesis.

Although carbon dioxide concentration limits photosynthesis only indirectly, artificially high levels of carbon dioxide in greenhouses do effectively increase yields of crops (Fig. 11).

Figure 11 Lettuce plants grown in increasing concentrations of carbon dioxide. Top left 0.02%, rising to 0.2% bottom right. Atmospheric concentration of carbon dioxide is about 0.03%.

Greenhouses also maintain a higher temperature and so reduce the effect of low temperature as a limiting factor.

The concept of limiting factors does not apply only to photosynthesis. Adding fertilizer to the soil, for example, may increase crop yields, but only up to the point where the roots can take up all the nutrients and the plant can build them into proteins, etc. The uptake of mineral ions is limited by the absorbing area of the roots, rates of respiration, aeration of the soil and availability of carbohydrates from photosynthesis.

Currently there is debate about whether athletic performance is limited by the ability of the heart and lungs to supply oxygenated blood to muscles, or by the ability of the muscles to take up and use the oxygen.

The role of the stomata

The stomata (p. 54) in a leaf may affect the rate of photosynthesis according to whether they are open or closed. When photosynthesis is taking place, carbon dioxide in the leaf is being used up and its concentration falls. At low concentrations of carbon dioxide, the stomata will open. Thus, when photosynthesis is most rapid, the stomata are likely to be open, allowing carbon dioxide to diffuse into the leaf. When the light intensity falls, photosynthesis will slow down and the build-up of carbon dioxide from respiration will make the stomata close. In this way, the stomata are normally regulated by the rate of photosynthesis rather than photosynthesis being limited by the stomata. However, if the stomata close during the daytime as a result of excessive water loss from the leaf, their closure will restrict photosynthesis by preventing the inward diffusion of atmospheric carbon dioxide.

Normally the stomata are open in the daytime and closed at night. Their closure at night, when intake of carbon dioxide is not necessary, reduces the loss of water vapour from the leaf (p. 65).

PRACTICAL WORK

6 The effect of light intensity on the rate of photosynthesis

Fill a beaker or jar with tap-water and add about 5 cm³ saturated sodium hydrogencarbonate solution. (This is to maintain a good supply of carbon dioxide.)

Select a pondweed shoot about 5–10 cm long. Partly prise open a small paper-clip and slide it over the tip of the shoot and drop the shoot into the jar of water. The paper-clip should hold the shoot under water as shown in Fig. 12.

bubbles appear from the cut end of the stem

paper clip holds pond weed upside down

Figure 12 Light intensity and oxygen production

Switch on a bench lamp and bring it close to the jar. After a minute or two, bubbles should appear from the cut end of the stem. If they do not, try a different piece of pondweed.

When the bubbles are appearing steadily, switch off the bench lamp and observe any change in the production of bubbles.

Now place the lamp about 25 cm from the jar. Switch on the lamp and try to count the number of bubbles coming off in a minute. Move the lamp to about 10 cm from the jar and count the bubbles again.

If the bubbles appear too rapidly to count, try tapping a pen or pencil on a sheet of paper at the same rate as the bubbles appear and get your partner to slide the paper slowly along for 15 seconds. Then count the dots. (Fig. 13.)

pull the paper along steadily

after 15 seconds count the dots

tap rhythmically in time with the bubbles appearing

Figure 13 Estimating the rate of bubble production

Result When the light is switched off, the bubbling should stop. The rate of bubbling should be faster when the lamp is closer to the plant.

Interpretation Assuming that the bubbles contain oxygen produced by photosynthesis, it seems that an increase in light intensity produces an increase in the rate of photosynthesis. We are assuming also that the bubbles do not change in size during the experiment. A fast stream of small bubbles might represent the same volume of gas as a slow stream of large bubbles.

The plant's use of photosynthetic products

The glucose molecules produced by photosynthesis are quickly built up into starch molecules and added to the growing starch granules in the chloroplast. If the glucose concentration was allowed to increase in the mesophyll cells of the leaf, it could disturb the osmotic balance between the cells (p. 32). Starch is a relatively insoluble compound and so does not alter the osmotic potential of the cell contents.

The starch, however, is steadily broken down to sucrose (p. 12) and this soluble sugar is transported out of the cell into the food-carrying cells (see p. 57) of the leaf veins. These veins will distribute the sucrose to all parts of the plant which do not photosynthesize, e.g. the growing buds, the ripening fruits, the roots and the underground storage organs.

The cells in these regions will use the sucrose in a variety of ways (Fig. 14).

Respiration

The sugar can be used to provide energy. It is oxidized by respiration (p. 20) to carbon dioxide and water, and the energy released is used to drive other chemical reactions such as the building-up of proteins described below.

Storage

Sugar which is not needed for respiration is turned into starch and stored. Some plants store it as starch grains in the cells of their stems or roots. Other plants such as the potato or parsnip have special storage organs (tubers) for holding the reserves of starch (p. 80). Sugar may be stored in the fruits of some plants; grapes, for example, contain a large amount of glucose.

Synthesis of other substances

As well as sugars for energy and starch for storage, the plant needs cellulose for its cell walls, lipids for its cell membranes, proteins for its cytoplasm and pigments for its flower petals, etc. All these substances are built up (synthesized) from the sugar molecules and other molecules produced in photosynthesis.

By joining hundreds of glucose molecules together, the long-chain molecules of cellulose (Fig. 3, p. 12) are built up and added to the cell walls.

Amino acids (see p. 11) are made by combining **nitrogen** with sugar molecules or smaller carbohydrate molecules. These amino acids are then joined together to make the proteins which form the enzymes and the cytoplasm of the cell. The nitrogen for this synthesis comes from **nitrates** which are absorbed from the soil by the roots.

Proteins also need **sulphur** molecules and these are absorbed from the soil in the form of **sulphates** (SO_4). **Phosphorus** is needed for DNA (p. 202) and for energy transfer reactions (p. 21). It is taken up as **phosphates** (PO_4).

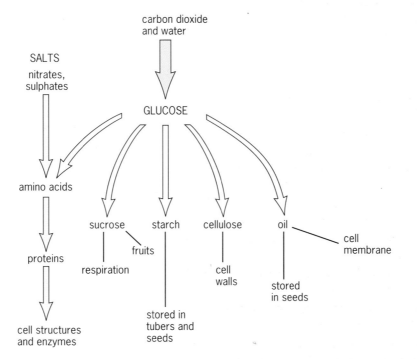

Figure 14 Green plants can make all the materials they need from carbon dioxide, water and salts

The chlorophyll molecule needs **magnesium** (Mg). This metallic element is also obtained in salts from the soil (see the salts listed under 'Water cultures', below).

Many other elements, e.g. iron, manganese, boron, in very small quantities, are also needed for healthy growth. These are often referred to as **trace elements**.

The metallic and non-metallic elements are all taken up in the form of their ions (p. 13).

All these chemical processes, such as the uptake of salts and the building-up of proteins, need energy from respiration to make them happen.

QUESTIONS

15 What substances does a green plant need to take in, to make **a** sugar, **b** proteins? What must be present in the cells to make reactions **a** and **b** work?

16 A molecule of carbon dioxide enters a leaf cell at 4 p.m. and leaves the same cell at 6 p.m. What is likely to have happened to the carbon dioxide molecule during the 2 hours it was in the leaf cell?

17 In a partially controlled environment such as a greenhouse, **a** how could you alter the external factors to obtain maximum photosynthesis, **b** which of these alterations might not be cost effective?

18 Fig. 15 is a graph showing the average daily change in the carbon dioxide concentration, 1 metre above an agricultural crop in July. From what you have learned about photosynthesis and respiration, try to explain the changes in the carbon dioxide concentration.

Figure 15 Daily changes in concentration of carbon dioxide one metre above a plant crop

The sources of mineral elements

The mineral elements needed by plants are absorbed from the soil in the form of salts. For example, a plant's needs for potassium (K) and nitrogen (N) might be met by absorbing the ions of the salt **potassium nitrate** (KNO_3). Salts like this come originally from rocks which have been broken down to form the soil. They are continually being taken up from the soil by plants or washed out of the soil by rain. They are replaced partly from the dead remains of plants and animals. When these organisms die and their bodies decay, the salts they contain are released back into the soil. This process is explained in some detail, for nitrates, on p. 237.

In arable farming, the ground is ploughed and whatever is grown is removed. There are no dead plants left to decay and replace the mineral salts. The farmer must replace them by spreading animal manure, sewage sludge or artificial fertilizers in measured quantities over the land.

Three manufactured fertilizers in common use are ammonium nitrate, superphosphate and compound NPK.

Ammonium nitrate (NH_4NO_3)

The formula shows that ammonium nitrate is a rich source of nitrogen but no other plant nutrients. It is sometimes mixed with calcium carbonate to form a compound fertilizer such as 'Nitro-chalk'.

Superphosphates

These fertilizers are mixtures of minerals. They all contain calcium and phosphate and some have sulphate as well.

Compound NPK fertilizer

'N' is the chemical symbol for nitrogen, 'P' for phosphorus and 'K' for potassium. NPK fertilizers are made by mixing ammonium sulphate, ammonium phosphate and potassium chloride in varying proportions. They provide the ions of nitrate, phosphate and potassium which are the ones most likely to be below the optimum level in an agricultural soil.

Water cultures

It is possible to demonstrate the importance of the various mineral elements by growing plants in water cultures. A full water culture is a solution containing the salts which provide all the necessary elements for healthy growth, e.g.

- potassium nitrate for potassium and nitrogen;
- magnesium sulphate for magnesium and sulphur;
- potassium phosphate for potassium and phosphorus;
- calcium nitrate for calcium and nitrogen.

From these elements, plus the carbon dioxide, water and sunlight needed for photosynthesis, a green plant can make all the substances it needs for a healthy existence.

Some branches of horticulture, e.g. growing of glasshouse crops, make use of water cultures on a large scale. Tomatoes may be grown with their roots in flat polythene tubes. The appropriate water culture solution is pumped along these tubes (Fig. 16). This method has the advantage that the yield is increased and the need to sterilize the soil each year, to destroy pests, is eliminated. This kind of technique is sometimes described as hydroponics or soil-less culture.

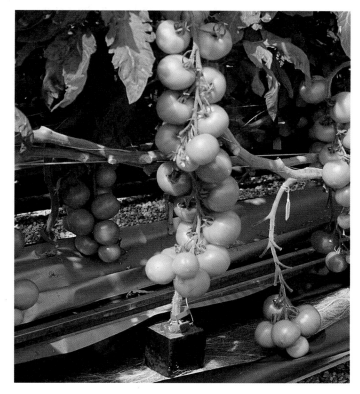

Figure 16 Soil-less culture. The tomato plants are growing in a nutrient solution circulated through troughs of polythene. The network of roots can be seen in the polythene.

Figure 17 To set up a water culture

Figure 18 Result of water culture experiment

PRACTICAL WORK

7 The importance of different mineral elements

Place wheat seedlings in test-tubes containing water cultures as shown in Fig. 17. Cover the tubes with aluminium foil to keep out light and so stop green algae from growing in the solution (p. 295). Some of the solutions have one of the elements missing. For example, in the list of chemicals on p. 46, magnesium chloride is used instead of magnesium sulphate and so the solution will lack sulphur. In a similar way, solutions lacking nitrogen, potassium and phosphorus can be prepared.

Leave the seedlings to grow in these solutions for a few weeks, keeping the tubes topped up with distilled water.

Result The kind of results which might be expected from wheat seedlings is shown in Fig. 18. Generally, the plants in a complete culture will be tall and sturdy, with large, dark green leaves. The plants lacking nitrogen will usually be stunted and have small, pale leaves. In the absence of magnesium, chlorophyll cannot be made, and these plants will be small with yellow leaves. Fig. 19 shows the effect of a phosphate deficiency.

Interpretation The healthy plant in the full culture is the control and shows that this method of raising plants does not affect them. The other, less healthy plants show that a full range of mineral elements is necessary for normal growth.

Quantitative results
Although the effects of mineral deficiency can usually be seen simply by looking at the wheat seedlings, it is better if actual measurements are made.

The height of the shoot, or the total length of all the leaves on one plant can be measured. The total root length can also be measured, though this is difficult if root growth is profuse.

Alternatively, the dry weight (see p. 326) of the shoots and roots can be measured. In this case, it is best to pool the results of several experiments. All the shoots from the complete culture are placed in a labelled container; all those from the 'no nitrate'

47

culture solution are placed in another container, and so on for all the plants from the different solutions. The shoots are then dried at 110 °C for 24 hours and weighed. The same procedure can be carried out for the roots.

You would expect the roots and shoots from the complete culture to weigh more than those from the nutrient-deficient cultures. Full instructions for the Practical Work in this chapter and ten other experiments, are contained in *Experimental Work in Biology* (see p. vi).

Figure 19 Effect of mineral salts on growth. The chrysanthemum plant on the left was planted at the same time as that on the right but has been deprived of phosphate.

QUESTIONS

19 What salts would you put in a water culture which is to contain *no* nitrogen?

20 What mineral elements do you think are provided by **a** bone meal (p. 162), **b** dried blood (p. 116)?

21 How can a floating pond plant, such as duckweed, survive without having its roots in soil?

22 In the water culture experiment, why should **a** lack of nitrate, **b** lack of phosphate cause reduced growth?

23 Fig. 20 shows the increased yield of winter wheat in response to adding more nitrogenous fertilizer.
a If the applied nitrogen is doubled from 50 to 100 kg per hectare, how much extra wheat does the farmer get?
b If the applied nitrogen is doubled from 100 to 200 kg per hectare, how much extra wheat is obtained?
c What sort of calculations will a farmer need to make before deciding to increase the applied nitrogen from 150 to 200 kg per hectare?

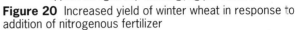

Figure 20 Increased yield of winter wheat in response to addition of nitrogenous fertilizer

Checklist

- Photosynthesis is the way plants make their food.

- They combine carbon dioxide and water to make sugar.

- To do this, they need energy from sunlight, which is absorbed by chlorophyll.

- Chlorophyll converts light energy to chemical energy.

- The equation to represent photosynthesis is

$$6CO_2 + 6H_2O \xrightarrow[\text{absorbed by chlorophyll}]{\text{energy from sunlight}} C_6H_{12}O_6 + 6O_2$$

- Plant leaves are adapted for the process of photosynthesis by being broad and thin, with many chloroplasts in their cells.

- From the sugar made by photosynthesis, a plant can make all the other substances it needs, provided it has a supply of mineral salts like nitrate, phosphate and potassium.

- In daylight, respiration and photosynthesis will be taking place in a leaf; in darkness, only respiration will be taking place.

- In daylight a plant will be taking in carbon dioxide and giving out oxygen.

- In darkness a plant will be taking in oxygen and giving out carbon dioxide.

- Experiments to test photosynthesis are designed to exclude light, or carbon dioxide, or chlorophyll, to see if the plant can still produce starch.

- The rate of photosynthesis may be limited by light intensity and temperature.

EXAMINATION QUESTIONS

Section 1 Some Principles of Biology

Do not write on this page. Where necessary copy drawings, tables or sentences.

1 Which one of the following is found in the cells of both animals and plants?

A Cell wall   **C** Cell membrane
B Cellulose   **D** Chloroplast

[N]

2 (a) The Figure shows two cells A and B.

(i) Which cell is the plant cell? (1)
(ii) State the reasons for your choice in (i). (2)
(iii) Name the two cells. (2)
(b) Describe an investigation you could carry out to find if a leaf contains starch. (5)

[M]

3 The diagram below shows a palisade cell.

(a) Name, on the diagram, the structures labelled A,B,C and D. (4)
(b) Which TWO of the structures labelled A,B,C and D would you not find in a human cheek cell? (1)

[L]

4 Organisms are made up of cells, organs, systems and tissues. In the table, list these four terms in order of increasing size. Write down an example of each term next to its name. The first one has been done for you. (5)

[L]

Term	Example
1 Cell	Cheek cell
2	
3	
4	

5 The diagram below shows a specialised plant cell.

(a) Name the parts labelled A,B,C and D. (4)
(b) What is the function of the chloroplast? (1)
(c) Why are chloroplasts found in stems and leaves and not in the cells of the root? (1)

[NI]

6 (a) Complete the table below by stating 'yes' if the process occurs, or 'no' if it does not occur for each organism. (4)

	Processes	
Organism	*Respiration*	*Photosynthesis*
A green plant at night		
A fungus		
A green plant during the day		
An animal		

(b) Name **two** substances made during photosynthesis which are needed by humans. (2)

[M]

7 The Figure shows an experiment to investigate gaseous exchange in aquatic plants and animals.
The tubes were left in darkness for 24 hours.

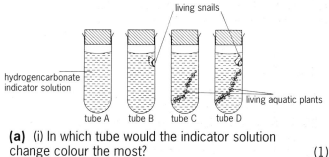

(a) (i) In which tube would the indicator solution change colour the most? (1)
(ii) Explain why this tube would show the greatest colour change. (1)
(b) (i) In which tube would there be no colour change? (1)
(ii) Explain why this tube was used in the experiment. (2)

[M]

8 Plants absorb substances such as nitrate ions by active uptake from the soil.
(a) What is meant by the term active uptake? (2)
(b) Why do plants need to take in nitrates? (1)
(c) Describe the use of another named ion taken in by plants. (2)
(d) Describe a practical investigation to show that plants need light for photosynthesis. (5)

[M]

49

9 The diagram below summarises photosynthesis in a chloroplast.

(a) Name A, B, C and D. (2)
(b) Name an insoluble storage material which the leaf can make from D. (1)
(c) Explain the effect on photosynthesis of greasing the underside of a leaf. (3)
[M]

10 Both photosynthesis and respiration occur in green plants. Compare these two processes using the following headings:
(a) Raw materials, products, energy conversions. (10)
(b) Where and when each process occurs. (4)
(c) The main function of each process. (2)
(d) Factors which affect the rate of each process. (4)
[M]

11 Give an account of the structure and position of leaves of a flowering plant. Explain how these features are linked to their functions. (12)
[L]

12 200 years ago a scientist breathed into a jar as part of his experiment. He wrote "A sprig of a nettle plant was put in a jar full of air fouled by breathing so as to extinguish a candle. The jar was put in the sunshine, and in the space of two hours the air was found to be nearly as good as common air." Which process in the plant produced "air nearly as good as common air"?
A Transpiration   **C** Osmosis
B Photosynthesis   **D** Respiration
[N]

13 The graph shows the amount of sugar contained in the leaves of a group of plants in a glasshouse over a period of a week.

(a) (i) Fill in the spaces in the word equation to show how sugar is produced.

$$............ + ... \xrightarrow[\text{chlorophyll}]{\text{sunlight}} \text{sugar} + \text{oxygen} \quad (2)$$

(ii) Name the process shown by the equation. (1)
(b) (i) Explain the rise and fall in the sugar level each day. (2)
(ii) Suggest a reason for the overall increase in the amount of sugar present each day. (1)
(iii) Suggest a reason for the lower peak on day 5. (1)
(c) Suggest TWO reasons why putting a gas fire in the glasshouse might increase the amount of sugar produced. (2)
[L]

14 A pupil wanted to see if starch was present in the leaves of a potted plant during the hours of daylight and darkness. The pupil removed small discs from a large leaf of the plant. The discs were removed at 3 pm and 3 am the following day. The discs were each tested for starch. The appearance of the discs is shown in the diagram.

(a) (i) At which time did the discs contain starch? (1)
(ii) Give a reason for the presence of starch at this time. (1)
(iii) Suggest a reason for the lack of starch at the other time. (1)
(b) (i) Which chemical solution is used to test for starch? (1)
(ii) Why is it necessary to boil the discs in alcohol or acetone before using the chemical named in (b) (i)? (1)
[L]

15 Which one of the following foods will give the highest yield of energy per gram when fully oxidised?
A Protein   **C** Starch
B Fat   **D** Vitamin   *[N]*

16 Lack of magnesium in plants
A Causes young leaves to wilt
B Causes leaves to turn dark green
C Increases starch production
D Reduces chlorophyll formation

Flowering Plants

6 PLANT STRUCTURE AND FUNCTION

A young sycamore plant is shown in Fig. 1. It is typical of many flowering plants in having a **root system** below the ground and a **shoot** above ground. The shoot consists of an upright stem, with leaves and buds. The buds on the side of the stem are called **lateral buds**. When they grow, they will produce branches. The bud at the tip of the shoot is the **terminal bud** and when it grows, it will continue the upward growth of the stem. The lateral buds and the terminal buds may also produce flowers.

The region of stem from which leaves and buds arise is called a **node**. The region of stem between two nodes is the **internode**.

The leaves make food by photosynthesis (p. 37) and pass it back to the stem.

The stem carries this food to all parts of the plant which need it and also carries water and dissolved salts from the roots to the leaves and flowers.

In addition, the stem supports and spaces out the leaves so that they can receive sunlight and absorb carbon dioxide which they need for photosynthesis.

An upright stem also holds the flowers above the ground, helping the pollination by insects or the wind (p. 73). A tall stem may help in seed dispersal later on (p. 77).

The roots anchor the plant in the soil and prevent it falling over or being blown over by the wind. They also absorb the water and salts which the plant needs for making food in the leaves.

The structure and functions of the plant organs will be considered in more detail in this chapter.

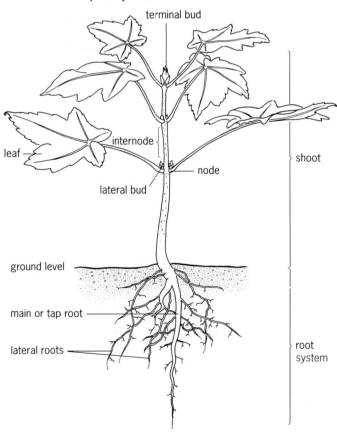

Figure 1 Structure of a typical flowering plant

Leaf

A typical leaf of a broad-leaved plant is shown in Fig. 2(a) (Fig. 2(b) shows a transverse section through the leaf). It is attached to the stem by a **leaf stalk** which continues into the leaf as a **midrib**. Branching from the midrib is a network of veins which deliver water and salts to the leaf cells and carry away the food made by them.

As well as carrying food and water, the network of veins forms a kind of skeleton which supports the softer tissues of the leaf blade.

The **leaf blade** (or **lamina**) is broad and a vertical section through a small part of a leaf blade is shown in Fig. 2(c) and Fig. 3 is a photograph of a leaf section under the microscope.

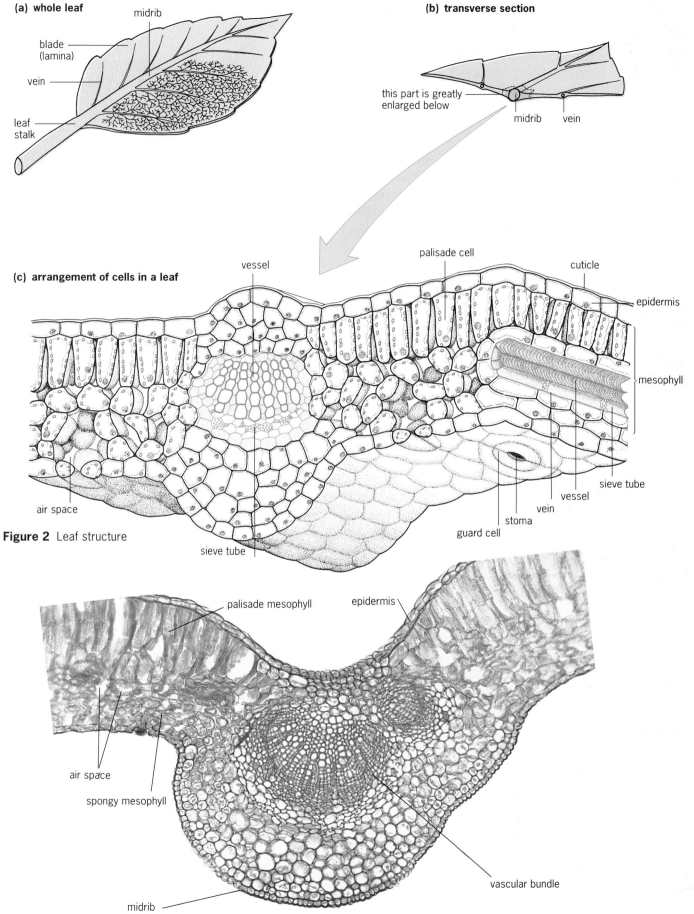

(a) whole leaf

midrib

blade (lamina)

vein

leaf stalk

(b) transverse section

this part is greatly enlarged below

midrib

vein

(c) arrangement of cells in a leaf

vessel

palisade cell

cuticle

epidermis

mesophyll

sieve tube

vessel

vein

stoma

guard cell

air space

sieve tube

Figure 2 Leaf structure

palisade mesophyll

epidermis

air space

spongy mesophyll

vascular bundle

midrib

Figure 3 Transverse section through a leaf (x30)

Epidermis

The **epidermis** is a single layer of cells on the upper and lower surface of the leaf. The epidermis helps to keep the leaf's shape. The closely fitting cells (Fig. 2(c)) reduce evaporation from the leaf and prevent bacteria and fungi from getting in. There is a thin waxy layer called the **cuticle** over the epidermis which helps to reduce water loss.

Stomata

In the leaf epidermis there are structures called **stomata** (singular = stoma). A stoma consists of a pair of **guard cells** (Fig. 4) surrounding an opening, or stomatal pore. Changes in the turgor (p. 32) and shape of the guard cells can open or close the stomatal pore. In most dicotyledons (i.e. the broad-leaved plants; see p. 294), the stomata occur only in the lower epidermis. In monocotyledons (i.e. narrow-leaved plants such as grasses) the stomata are equally distributed on both sides of the leaf.

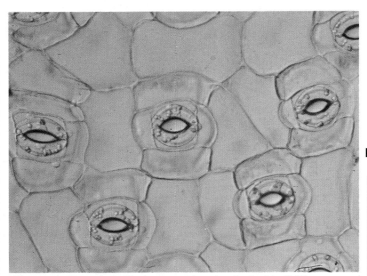

Figure 4 Stomata in the lower epidermis of a leaf (x 350)

In very general terms, stomata are open during the hours of daylight but closed during the evening and most of the night (Fig. 5). This pattern, however, varies greatly with the plant species. A satisfactory explanation of stomatal rhythm has not been worked out, but when the stomata are open (i.e. mostly during daylight), they allow carbon dioxide to diffuse into the leaf where it is used for photosynthesis.

If the stomata close, the carbon dioxide supply to the leaf cells is virtually cut off and photosynthesis stops. However, in many species, the stomata are closed during the hours of darkness, when photosynthesis is not taking place anyway.

It seems, therefore, that stomata allow carbon dioxide into the leaf when photosynthesis is taking place and prevent excessive loss of water vapour (see pp. 64 and 65) when photosynthesis stops, but the story is likely to be more complicated than this.

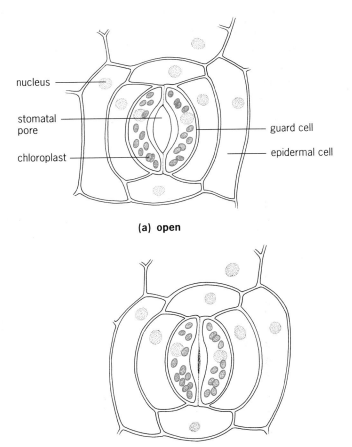

(a) open

(b) closed

Figure 5 Stoma

The detailed mechanism by which stomata open and close is not fully understood, but it is known that, in the light, the potassium concentration in the guard cell vacuoles increases. This lowers the water potential of the cell sap and water enters the guard cells by osmosis from their neighbouring epidermal cells. This inflow of water raises the turgor pressure inside the guard cells.

The cell wall next to the stomatal pore is thicker than elsewhere in the cell and is less able to stretch (Fig. 6). So, although the increased turgor tends to expand the whole guard cell, the thick inner wall cannot expand. This causes the guard cells to curve in such a way that the stomatal pore between them is opened.

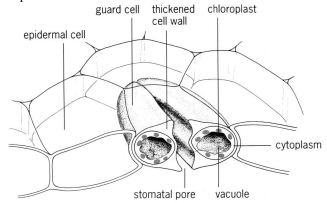

Figure 6 Structure of guard cells

When potassium ions leave the guard cell, the water potential rises, water passes out of the cells by osmosis, the turgor pressure falls and the guard cells straighten up and close the stoma.

Where the potassium ions come from and what triggers their movement into or out of the guard cells is still under active investigation.

You will notice from Figs 5 and 6 that the guard cells are the only epidermal cells containing chloroplasts. At one time it was thought that the chloroplasts built up sugar by photosynthesis during daylight, that the sugars made the cell sap more concentrated and so caused the increase in turgor. In fact, little or no photosynthesis takes place in these chloroplasts and their function has not been explained, though it is known that starch accumulates in them during the hours of darkness. In some species of plants, the guard cells have no chloroplasts.

Mesophyll

The tissue between the upper and lower epidermis is called **mesophyll** (Fig. 2(c)). It consists of two zones: the upper, **palisade mesophyll** and the lower, **spongy mesophyll** (Fig. 7). The palisade cells are usually long and contain many chloroplasts. The spongy mesophyll cells vary in shape and fit loosely together, leaving many air spaces between them (see Fig. 7, p. 41).

The function of the palisade cells and—to a lesser extent—of the spongy mesophyll cells is to make food by photosynthesis. Their chloroplasts absorb sunlight and use its energy to join carbon dioxide and water molecules to make sugar molecules as described on p. 37.

In daylight, when photosynthesis is rapid, the mesophyll cells are using up carbon dioxide. As a result, the concentration of carbon dioxide in the air spaces falls to a low level and more carbon dioxide diffuses in (p. 28) from the outside air, through the stomata (Fig. 7). This diffusion continues through the air spaces, up to the cells which are using carbon dioxide. These cells are also producing oxygen as a by-product of photosynthesis. When the concentration of oxygen in the air spaces rises, it diffuses out through the stomata.

Veins (vascular bundles)

The water needed for making sugar by photosynthesis is brought to the mesophyll cells by the **veins**. The mesophyll cells take in the water by osmosis (p. 31) because the concentration of free water molecules in a leaf cell, which contains sugars, will be less than the concentration of water in the water vessels of a vein. The branching network of leaf veins means that no cell is very far from a water supply.

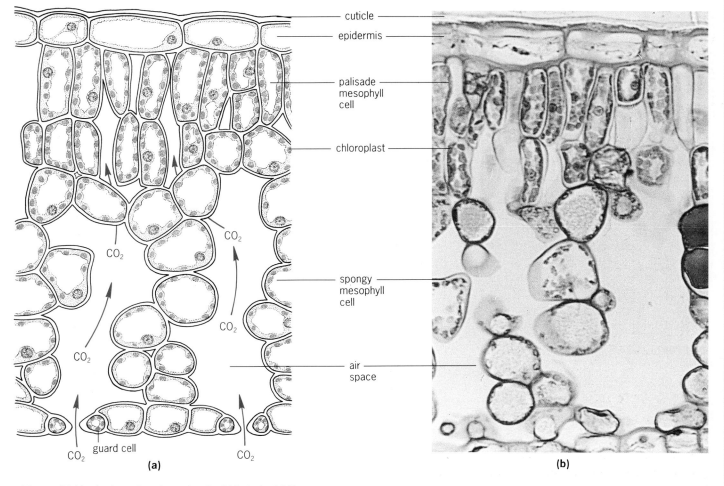

Figure 7 Vertical section through a leaf blade (×300)

The sugars made in the mesophyll cells are passed to the phloem cells (see below) of the veins, and these cells carry the sugars away from the leaf into the stem.

The ways in which a leaf is thought to be well adapted to its function of photosynthesis are listed on p. 43.

QUESTIONS

1 What are the functions of **a** the epidermis, **b** the mesophyll of a leaf?
2 Look at Fig. 7. Why do you think that photosynthesis does not take place in the cells of the epidermis?
3 During bright sunlight, what gases are **a** passing out of the leaf through the stomata, **b** entering the leaf through the stomata?
4 What types of leaves do you know which do not have any midrib?
5 In some plants, the stomata close for a period at about midday. Suggest some possible advantages and disadvantages of this to the plant.

Stem

In Fig. 8 a stem is shown cut across (transversely) and down its length (longitudinally) to show its internal structure.

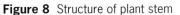

Figure 8 Structure of plant stem

Epidermis

Like the leaf epidermis, this is a single layer of cells which helps to keep the shape of the stem and cuts down the loss of water vapour. There are stomata in the epidermis which allow the tissues inside to take up oxygen and get rid of carbon dioxide. In woody stems, the epidermis is replaced by bark which consists of many layers of dead cells.

Vascular bundles

These are made up of groups of specialized cells which conduct water, dissolved salts and food up or down the stem (Fig. 10). The vascular bundles in the roots, stem, leaf stalks and leaf veins all connect up to form a transport system throughout the entire plant (Fig. 9). The two main tissues in the vascular bundles are called **xylem** and **phloem**. Food substances travel in the phloem; water and salts travel mainly in the xylem. The cells in each tissue form elongated tubes called **vessels** (in the xylem) or **sieve tubes** (in the phloem) and they are surrounded and supported by other cells.

Figure 9 Distribution of veins from root to leaf

Vessels

The cells in the xylem which carry water become vessels. A vessel is made up of a series of long cells joined end to end (Fig. 11(a)). Once a region of the plant has ceased growing, the end walls of these cells are digested away to form a continuous, fine tube (Fig. 10(c)). At the same time, the cell walls are thickened and impregnated with a substance called **lignin**, which makes the cell wall very strong and impermeable. Since these lignified cell walls prevent the free passage of water and nutrients, the cytoplasm dies. This does not affect the passage of water in the vessels. Xylem also contains many elongated, lignified supporting cells called **fibres**.

epidermis

cortex

phloem

xylem

pith

(b) transverse section through sunflower stem (x40)

vessel   pith   epidermis   cortex

vessel   sieve tube
(phloem)

(a) diagram showing cells

starch
grain   water vessels   phloem   cortex

(c) longitudinal section through sunflower stem (x200)

Figure 10 Structure of plant stem

Sieve tubes

The conducting cells in the phloem remain alive and form sieve tubes. Like vessels, they are formed by vertical columns of cells (Fig. 11(b)). Perforations appear in the end walls, allowing substances to pass from cell to cell, but the cell walls are not lignified and the cell contents do not die, although they do lose their nuclei. The perforated end walls are called **sieve plates**.

Phloem contains supporting cells as well as sieve tubes.

Functions of vascular bundles

In general, water travels up the stem in the xylem from the roots to the leaves. Food may travel either up or down the stem in the phloem, from the leaves where it is made, to any part of the plant which is using or storing it.

Vascular bundles have a supporting function as well as a transport function, because they contain vessels, fibres and other thick-walled, lignified, elongated cells. In many stems, the vascular bundles are arranged in a cylinder, a little way in from the epidermis. This pattern of distribution helps the stem to resist the sideways bending forces caused by the wind. In a root, the vascular bundles are in the centre (Fig. 12) where they resist the pulling forces which the root is likely to experience when the shoot is being blown about by the wind.

The network of veins in many leaves supports the soft mesophyll tissues and resists stresses which could lead to tearing.

The methods by which water, salts and food are moved through the vessels and sieve tubes are discussed in Chapter 7 (p. 63).

(a) cells forming a vessel   (b) cells forming a sieve tube

Figure 11 Conducting structures in a plant

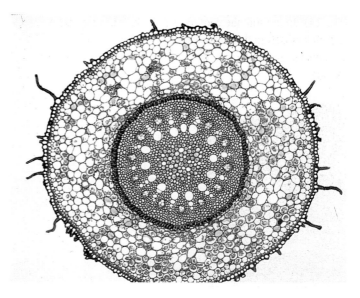

Figure 12 Transverse section through a root (x40). Notice that the vascular tissue is in the centre. Some root hairs can be seen in the outer layer of cells.

Cortex and pith

The tissue between the vascular bundles and the epidermis is called the **cortex**. Its cells often store starch. In green stems, the outer cortex cells contain chloroplasts and make food by photosynthesis. The central tissue of the stem is called **pith**. The cells of the pith and cortex act as packing tissues and help to support the stem in the same way that a lot of blown-up balloons packed tightly into a plastic bag would form quite a rigid structure.

QUESTIONS

6 Make a list of the types of cells or tissues you would expect to find in a vascular bundle.

7 What structures help to keep the stem's shape and upright position?

8 What are the differences between xylem and phloem
  a in structure   **b** in function?

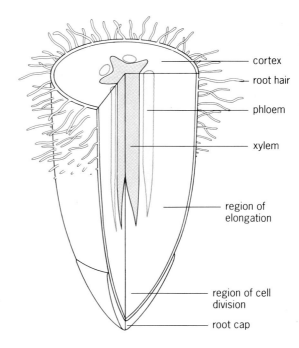

Figure 13 Root structure

Root

The internal structure of a typical root is shown in Fig. 13. The vascular bundle is in the centre of the root (Fig. 12), unlike the stem where the bundles form a cylinder in the cortex.

The xylem carries water and salts from the root to the stem. The phloem will bring food from the stem to the root, to provide the root cells with substances for their energy and growth.

Outer layer and root hairs

There is no distinct epidermis in a root. At the root tip are several layers of cells forming the **root cap**. These cells are continually replaced as fast as they are worn away when the root tip is pushed through the soil.

In a region above the root tip, where the root has just stopped growing, the cells of the outer layer produce tiny, tube-like outgrowths called **root hairs** (Fig. 7, p. 67). These can just be seen as a downy layer on the roots of seedlings grown in moist air

(Fig. 14). In the soil, the root hairs grow between the soil particles and stick closely to them. The root hairs take up water from the soil by osmosis, and absorb mineral salts (as ions) by active transport (p. 30).

Figure 14 Root hairs (x5) as they appear on a root grown in moist air

The large number of tiny root hairs greatly increases the absorbing surface of a root system. The surface area of the root system of a mature rye plant has been estimated at about 200 m^2. The additional surface provided by the root hairs was calculated to be 400 m^2.

Root hairs remain alive for only a short time. The region of root just below a root hair zone is producing new root hairs, while the root hairs at the top of the zone are shrivelling (Fig. 15). Above the root hair zone, the cell walls of the outer layer become less permeable. This means that water cannot get in so easily.

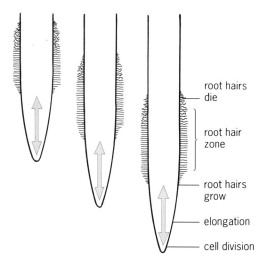

root hairs die

root hair zone

root hairs grow

elongation

cell division

Figure 15 The root hair zone changes as the root grows

Tap root (Fig. 16(a))

When a seed germinates, a single root grows vertically down into the soil. Later, lateral roots grow from this at an acute angle outwards and downwards, and from these laterals other branches may arise. Where a main root is recognizable the arrangement is called a tap-root system.

Fibrous root (Fig. 16(b))

When a seed of the grass and cereal group germinates, several roots grow out at the same time and laterals grow from them. There is no distinguishable main root, and it is called a fibrous system.

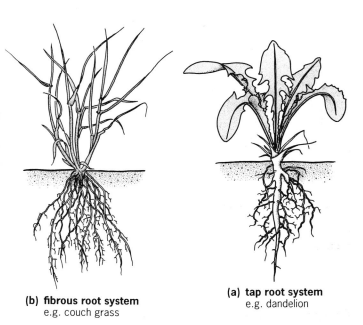

(b) fibrous root system
e.g. couch grass

(a) tap root system
e.g. dandelion

Figure 16 Types of root system

Adventitious root

Where roots grow, not from a main root, but directly from the stem as they do in bulbs, corms, rhizomes or ivy, they are called adventitious roots, but such a system may also be described as a fibrous rooting system.

QUESTIONS

9 State briefly the functions of the following: xylem, palisade cell, root hair, root cap, stoma, epidermis.

10 If you were given a cylindrical structure cut from part of a plant, how could you tell whether it was a piece of stem or a piece of root **a** with the naked eye, **b** with the aid of a microscope or hand lens?

11 Describe the path taken by **a** a carbon dioxide molecule from the air and **b** a water molecule from the soil, until they reach a mesophyll cell of a leaf to be made into sugar.

12 Why do you think that root hairs are produced only on the parts of the root system that have stopped growing?

13 Discuss whether you would expect to find a vascular bundle in a flower petal.

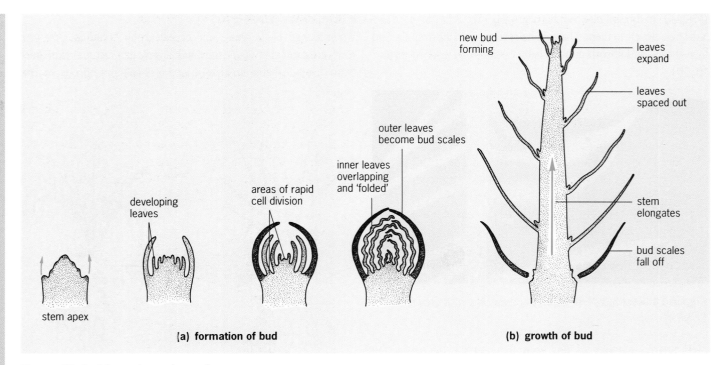

Figure 17 Bud formation and growth

Buds

In woody plants, cell division at the tip of the shoot produces tiny, closely packed leaves on a short stem, so forming a **bud** (Fig. 17). The outermost leaves of the bud may become thick and tough, forming **bud scales**, which protect the more delicate inner leaves from frost and from attack by insects and fungi. The new leaves and stem pass the winter in the bud. In the spring, the bud's stem elongates very rapidly by cell expansion, pushing the bud scales apart and spacing out the new leaves (Fig. 18). The bud scales fall off but the new leaves expand and grow to full size.

All this growth takes place in a few weeks, after which the stem and leaves do not grow any more that year. New **terminal buds** are formed at the shoot tips; **lateral buds** develop at the junction of the leaf stalk and stem. In the following season, the growth of the terminal bud will increase the length of the shoot. The lateral buds, when they grow, will form branches. Terminal or lateral buds may produce flowers as well as leaves. Bud formation allows deciduous trees and shrubs to come into full leaf very quickly, as soon as conditions become favourable.

QUESTIONS

14 Which type of bud might **a** form a branch, **b** produce a flower, **c** continue growth in length of the stem?

15 Assuming that all the buds in Fig. 1 develop next season, make a sketch to show the appearance of the shoot. Indicate by colour which regions are last year's and which are this year's growth.

16 Cutting a hedge removes all the terminal buds. Suggest why it is that the hedge will **a** continue to grow in height, **b** become more bushy.

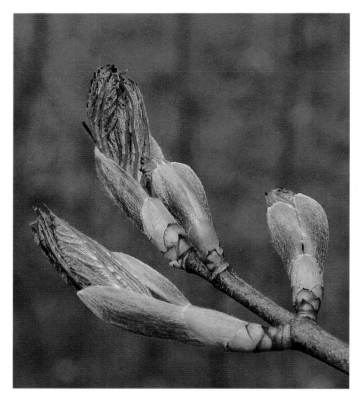

Figure 18 Bud burst in a sycamore. The bud scales have opened out to release the green foliage leaves.

Structure and function

It is always tempting when studying an organism, to ascribe some function to the structures which are being observed. In some cases this is easy. It is obvious, for example, that the function of a mammal's hind limb is locomotion because the limb can be seen in action.

The functions of internal organs are not so obvious and guesses about their function may be quite inaccurate. At one time it was thought that the arteries carried air because the arteries seen in the dissection of dead animals often contained no blood. The chloroplasts in the guard cells (p. 54) were assumed for a long time to be the site of photosynthesis and it was only after conducting experiments that this assumption was shown to be false.

The functions of tissues described in this chapter have been stated as if they were certainly known and without offering any evidence for these functions. In Chapter 7 experiments will be described which do provide some evidence for the statements.

In general, although function can be guessed at from studying anatomy of dead organisms, it cannot be confirmed without experiments to test the guesses (hypotheses). It is usually most unwise to assume a particular function from simply studying anatomy.

PRACTICAL WORK

1 Stomata

Strip off a piece of the lower epidermis of a rhubarb leaf and place it on a slide, in a little water, under the microscope (Fig. 19). Draw a drop of 10 per cent salt solution under the cover-slip as shown in Fig. 17 on p. 34. This solution withdraws water by osmosis from the guard cells which lose their turgor. As a result the stomata close. Now flush out the salt solution by drawing water under the cover-slip, the guard cells take up water and become turgid.

Figure 19 Lower the cover-slip carefully to exclude air bubbles

2 To show the tension in stems

Partly remove a strip of epidermis from a rhubarb stalk as shown in Fig. 20(a). When you replace the epidermis in position it will have become too short, showing the shrinking stress that exists in the epidermis.

Push a cork borer into the pith and withdraw it without removing any tissue. The cylinder of pith so formed, freed from the constraint of the epidermis, expands and protrudes slightly. This shows the elongating tendency of the inner tissues (Fig. 20(b)).

These two opposing stresses in the stem help to give it rigidity when the cells are turgid (p. 32).

(a)

(b)

Figure 20 To show opposite tensions in pith and epidermis

3 Cells and vessels

Some celery or rhubarb stalks which have been cut up into pieces about 2 cm long, are left for a few days in a macerating fluid of 10 per cent chromic and nitric acids. This fluid breaks down the material between the cells, so that the cells can be teased apart with mounted needles. The macerating acids will have been washed off before you use the material.

Place a little of the material on a slide with a little water and pull it to pieces with a pair of mounted needles. When you study it under the microscope, you should see individual cells and vessels.

Checklist

- The shoot of a plant consists of the stem, leaves, buds and flowers.

- The leaf makes food by photosynthesis in its mesophyll cells.

- The water for photosynthesis is carried in the leaf's veins.

- The carbon dioxide for photosynthesis enters the leaf through the stomata and diffuses through the air spaces in the leaf.

- Closure of the stomata stops the entry of carbon dioxide into a leaf but also reduces water loss.

- Sunlight is absorbed by the chloroplasts in the mesophyll cells.

- The food made in the leaf is carried away in the phloem cells.

- The stem supports the leaves and flowers.

- The stem contains vascular bundles (veins).

- The water vessels in the veins carry water up the stem to the leaves.

- The phloem in the veins carries food up or down the stem to wherever it is needed.

- The position of vascular bundles helps the stem to withstand sideways bending and the root to resist pulling forces.

- The roots hold the plant in the soil and absorb the water and mineral salts needed by the plant for making sugars and proteins.

- The root hairs make very close contact with soil particles and are the main route by which water and mineral salts enter the plant.

- Buds are condensed shoots; their immature foliage leaves are protected inside thick bud scales.

- Buds grow by rapid elongation of the stem and expansion of the leaves.

7 TRANSPORT IN PLANTS

Transport of water
Transpiration: its function and control. Rates of transpiration.

Transport of salts

Transport of food
Movement of solutes in the phloem.

Uptake of water and salts
Osmosis in the roots. Possible mechanisms for salt uptake.

Transport of gases
Diffusion through stomata and intercellular spaces.

Practical work
Experiments on transport in plants: the potometer; rates of transpiration; role of vascular bundles; evaporation from a leaf.

Transport of water

Transpiration

The main force which draws water from the soil and through the plant is caused by a process called **transpiration**. Water evaporates from the leaves and causes a kind of 'suction' which pulls water up the stem (Fig. 1). The water travels up the vessels in the vascular bundles (see Fig. 9, p. 56) and this flow of water is called the **transpiration stream**. The cells in part of a leaf blade are shown in Fig. 2. As explained on p. 32, the cell sap in each cell is exerting a turgor pressure outwards on the cell wall. This pressure forces some water out of the cell wall and into the air

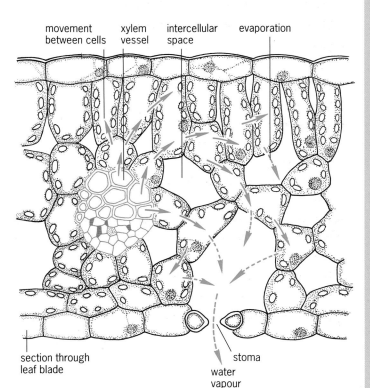

Figure 2 Movement of water through a leaf

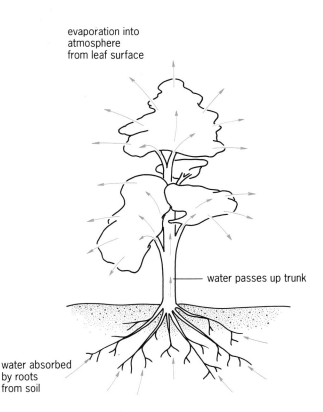

Figure 1 The transpiration stream

space between the cells. Here the water evaporates and the water vapour passes by diffusion through the air spaces in the mesophyll and out of the stomata. It is this loss of water vapour from the leaves which is called 'transpiration'.

The cell walls which are losing water in this way replace it by drawing water from the nearest vein. Most of this water travels along the cell walls without actually going inside the cells (Fig. 3). Thousands of leaf cells are evaporating water like this and drawing water to replace it from the xylem vessels in the veins. As a result, water is pulled through the xylem vessels and up the stem from the roots. This transpiration pull is strong enough to draw up water 50 metres or more in trees (Fig. 4).

63

In addition to the water passing along the cell walls, a small amount will pass right through the cells. When leaf cell A (Fig. 3) loses water, its turgor pressure will fall. This fall in pressure allows the water in the cell wall to enter the vacuole and so restore the turgor pressure. In conditions of water shortage, cell A may be able to get water by osmosis from cell B more easily than B can get it from the xylem vessels. In this case, all the mesophyll cells will be losing water faster than they can absorb it from the vessels, and the leaf will wilt (see p. 32).

Figure 3 Probable pathway of water through leaf cells

Importance of transpiration

A tree, on a hot day, may draw up hundreds of litres of water from the soil. Most of this water evaporates from the leaves; only a tiny fraction is retained for photosynthesis and to maintain the turgor of the cells. The advantage to the plant of this excessive evaporation is not clear. A rapid water flow may be needed to obtain sufficient mineral salts, which are in very dilute solution in the soil. Evaporation may also help to cool the leaf when it is exposed to intense sunlight.

Against the first possibility, it has to be pointed out that, in some cases, an increased transpiration rate does not increase the uptake of minerals.

The second possibility, the cooling effect, might be very important. A leaf exposed to direct sunlight will absorb heat and its temperature may rise to a level which could kill the cytoplasm. Water evaporating from a leaf absorbs its latent heat and cools the leaf down. This is probably one value of transpiration. However, there are plants whose stomata close at around midday, greatly reducing transpiration. How do these plants avoid overheating?

Many biologists regard transpiration as an inevitable consequence of photosynthesis. In order to photosynthesize, a leaf has to take in carbon dioxide from the air. The pathway which allows carbon dioxide in will also let water vapour out whether the

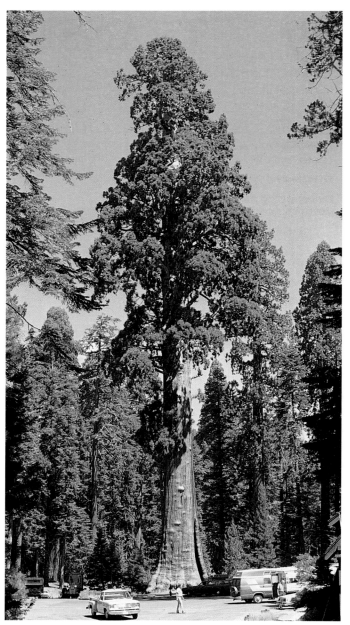

Figure 4 Californian redwoods. Some of these trees are over 100 metres tall. Transpiration from their leaves pulls water up the trunk.

plant needs to lose water or not. In all probability, plants have to maintain a careful balance between the optimum intake of carbon dioxide and a damaging loss of water. Plants achieve this balance in different ways, some of which are described below.

QUESTION

1 Describe the pathway followed by a water molecule from the time it enters a plant root to the time it escapes into the atmosphere from a leaf.

The role of stomata

The opening and closing of stomata can be triggered by a variety of factors, principally light intensity, carbon dioxide concentration and humidity. These factors interact with each other. For example, a rise in light intensity will increase the rate of photosynthesis and so lower the carbon dioxide concentration in the leaf. These are the conditions you would expect to influence stomatal aperture if the stomata are to control the balance between loss of water vapour and uptake of carbon dioxide.

The stomata also respond if the leaf is under water stress, i.e. losing water by transpiration faster than it is being taken up by the roots. Before wilting sets in, the stomata start to close. Although they do not prevent wilting, the stomata do seem to delay its onset.

Rate of transpiration

Transpiration is the evaporation of water from the leaves, so any change which increases or reduces evaporation will have the same effect on transpiration.

Light intensity

Light itself does not affect evaporation, but in daylight the stomata (p. 54) of the leaves are open. This allows the water vapour in the leaves to diffuse out into the atmosphere. At night, when the stomata close, transpiration is greatly reduced.

Generally speaking, then, transpiration speeds up when light intensity increases because the stomata respond to changes in light intensity.

Sunlight may also warm up the leaves and increase evaporation (see below).

Humidity

If the air is very humid, i.e. contains a great deal of water vapour, it can accept very little more from the plants and so transpiration slows down. In dry air, the diffusion of water vapour from the leaf to the atmosphere will be rapid.

Air movements

In still air, the region round a transpiring leaf will become saturated with water vapour so that no more can escape from the leaf. In these conditions, transpiration would slow down. In moving air, the water vapour will be swept away from the leaf as fast as it diffuses out. This will speed up transpiration.

Temperature

Warm air can hold more water vapour than cold air. Thus evaporation or transpiration will take place more rapidly into warm air.

Furthermore, when the sun shines on the leaves, they will absorb heat as well as light. This warms them up and increases the rate of evaporation of water.

Experiments which investigate the effect of some of these conditions on the rate of transpiration are described on p. 68 (Experiment 1).

QUESTIONS

1 What kind of climate and weather conditions do you think will cause a high rate of transpiration?
2 What would happen to the leaves of a plant which was losing water by transpiration faster than it was taking it up from the roots?
3 In what two ways does sunlight increase the rate of transpiration?
4 Apart from drawing water through the plant, what else may be drawn up by the transpiration stream?
5 Transpiration has been described in this chapter as if it took place only in leaves. What other parts of a plant might transpire?

Water movement in the xylem

You may have learned that you cannot draw water up by 'suction' to a height of more than about 10 metres. Many trees are taller than this yet they can draw up water effectively. The explanation offered is that, in long vertical columns of water in very thin tubes, the attractive forces between the water molecules are greater than the forces trying to separate them. So, in effect, the transpiration stream is pulling up thin threads of water which resist the tendency to break.

There are still problems, however. It is likely that the water columns in some of the vessels do have air breaks in them and yet the total water flow is not affected.

Evidence for the pathway of water

Experiment 3 on p. 70 uses a dye to show that, in a cut stem, the dye, and, therefore, presumably the water, travels in the vascular bundles. Closer examination with a microscope would show that it travels in the xylem vessels.

Removal of a ring of bark (which includes the phloem) does not affect the passage of water along a branch (Experiment 4). Killing parts of a branch by heat or poisons does not interrupt the flow of water, but anything which blocks the vessels does stop the flow.

The evidence all points to the non-living xylem vessels as the main route by which water passes from the soil to the leaves.

Adaptations to arid conditions

In both hot and cold climates, plants may suffer from water shortage. High temperatures accelerate evaporation from leaves. At very low temperatures the soil water becomes frozen and therefore unavailable to the roots of plants.

It is thought that the autumn leaf-fall of deciduous trees and shrubs is an essential adaptation to winter 'drought'. Loss of leaves removes virtually all evaporating surfaces at a time when water may become unavailable. Without leaves, however, the plants cannot make food by photosynthesis and so they enter a dormant condition in which metabolic activity is at a low level.

Evergreen trees, e.g. pine and spruce (p. 296), which survive in cold climates, have small, compact needle-like leaves. The small surface area of such leaves offers little resistance to high winds and can reduce the amount of water lost in transpiration. However, photosynthesis can continue whenever water is available.

Other adaptations to drought include leaves with thick waxy cuticles, leaves with few stomata often sunken below the level of the epidermis, hairy leaves or rolled-up leaves. These last four adaptations help to retain a static layer of air, saturated with water vapour, close to the leaf. This effectively reduces the diffusion gradient. Waxy cuticles reduce evaporation.

Cacti are adapted to hot, dry conditions in several ways. Often they have no leaves. Photosynthesis is carried on by a thick green stem which offers only a small surface area for evaporation. Cacti are succulent, i.e. they store water in their fleshy tissues and draw on this store for photosynthesis (Fig. 5).

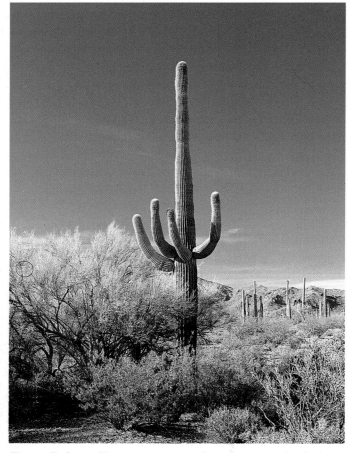

Figure 5 Organ Pipe cactus; a succulent cactus growing in the Arizona desert

The stomata of many cacti are closed during the day, when temperatures are high, and open at night when evaporation is at a minimum. This strategy requires a slightly different form of photosynthesis. At night, carbon dioxide diffuses in through the open stomata and is 'fixed' (i.e. incorporated) into an organic acid. Little water vapour is lost at night. In the daytime the stomata are closed but the organic acid breaks down to yield carbon dioxide which is then built into sugars by photosynthesis. Closure of the stomata in daytime greatly reduces water loss.

Transport of salts

The liquid which travels in the xylem is not, in fact, pure water. It is a very dilute solution, containing from 0.1 to 1.0 per cent dissolved solids, mostly amino acids, other organic acids and mineral salts. The organic acids are made in the roots; the mineral salts come from the soil. The faster the flow in the transpiration stream, the more dilute is the xylem sap.

Experimental evidence suggests that salts are carried from the soil to the leaves mainly in the xylem vessels.

Transport of food

The xylem sap is always a very dilute solution, but the phloem sap may contain up to 25 per cent of dissolved solids, the bulk of which consists of sucrose and amino acids. There is a good deal of evidence to support the view that sucrose, amino acids and many other substances are transported in the phloem. This is called **translocation**.

The movement of water and salts in the xylem is always upwards, from soil to leaf, but in the phloem, the solutes may be travelling up or down the stem. The carbohydrates made in the leaf during photosynthesis are converted to sucrose and carried out of the leaf to the stem. From here, the sucrose may pass upwards to growing buds and fruits or downwards to the roots and storage organs. All parts of a plant which cannot photosynthesize will need a supply of nutrients brought by the phloem. It is quite possible for substances to be travelling upwards and downwards at the same time in the phloem.

There is no doubt that substances travel in the sieve tubes (p. 57) of the phloem but the mechanism by which they are moved is not fully understood. We do know that translocation depends on living processes because anything which inhibits cell metabolism, e.g. poisons or high temperatures, also arrests translocation.

QUESTIONS

6 How do sieve tubes and vessels differ **a** in the substances they transport, **b** in the directions these substances are carried?

7 A complete ring of bark cut from round the circumference of a tree-trunk causes the tree to die. The xylem continues to carry water and salts to the leaves which can make all the substances needed by the tree. So why does the tree die?

8 Make a list of all the non-photosynthetic parts of a plant which need a supply of sucrose and amino acids.

Uptake of water and salts

Uptake of water

The water tension developed in the vessels by a rapidly transpiring plant is thought to be sufficient to draw water through the root from the soil. The precise pathway taken by the water is the subject of some debate, but the path of least resistance seems to be in or between the cell walls rather than through the cells.

When transpiration is slow, e.g. at night-time or just before bud burst in a deciduous tree, then osmosis may play a more important part in the uptake of water. In Fig. 6, showing a root hair in the soil, the cytoplasm of the root hair is partially permeable to water. The soil water is more dilute than the cell sap and so water passes by osmosis (p. 31) from the soil into the cell sap of the root hair cell. This flow of water into the root hair cell raises the cell's turgor pressure (p. 32). So water is forced out through the cell wall into the next cell and so on, right through the cortex of the root to the xylem vessels (Fig. 7).

One problem for this explanation is that it has not been possible to demonstrate that there is an osmotic gradient across the root cortex which could produce this flow of water from cell to cell. Nevertheless, root pressure developed probably by osmosis does force water up the root system and into the stem.

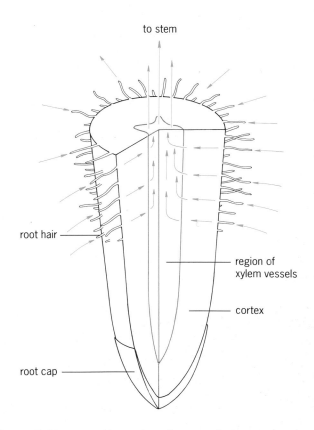

Figure 7 Diagrammatic section of root to show passage of water from the soil

It may be that diffusion from a relatively high concentration in the soil to a lower concentration in the root cells accounts for uptake of some individual salts, but it has been shown (a) that salts can be taken from the soil even when their concentration is below that in the roots and (b) that anything which interferes with respiration impairs the uptake of salts. This suggests that 'active transport' (p. 30) plays an important part in the uptake of salts.

The growing region of the root and the root hair zone seem to be most active in taking up salts. Most of the salts appear to be carried at first in the xylem vessels, though they soon appear in the phloem as well.

The salts are used by the plant's cells to build up essential molecules. Nitrates, for example, are combined with carbohydrates to make amino acids in the roots. These amino acids are used later to make proteins.

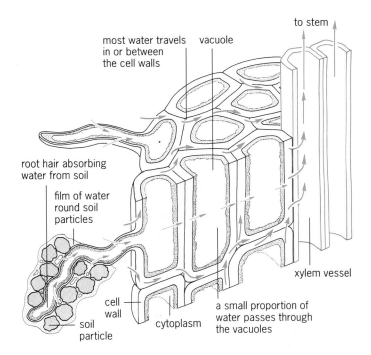

Figure 6 The probable pathways of water through a root

Uptake of salts

The methods by which roots take up salts from the soil are not fully understood. Some salts may be carried in with the water drawn up by transpiration and pass mainly along the cell walls in the root cortex and into the xylem.

QUESTIONS

9 If root hairs take up water from the soil by osmosis, what would you expect to happen if so much nitrate fertilizer was put on the soil that the soil water became a stronger solution than the cell sap of the root hairs?

10 A plant's roots may take up water and salts less efficiently from a waterlogged soil than from a fairly dry soil. Revise 'Active transport' (p. 30) and suggest reasons for this.

11 Why do you think that, in a deciduous tree in spring, transpiration is negligible before bud burst?

Transport of gases

The process of diffusion described on p. 28 accounts for the movement of gases in and out of a plant. During respiration, oxygen is taken in and carbon dioxide given out. When photosynthesis is faster than respiration, carbon dioxide diffuses in and oxygen diffuses out (Fig. 8, p. 42). In leaves and green stems, the gases enter and leave through the stomata (p. 54). Then they diffuse through the air spaces between the cells to reach all parts of the plant shoot. In woody plants, the stems have no stomata and the gases have to pass through small openings in the bark called **lenticels**.

Roots obtain their oxygen from the air spaces in the soil. Much of this oxygen will be dissolved in the soil water which enters the root through the growing region and the root hairs.

Absorption by leaves

Leaves are able to absorb certain substances if these are sprayed on to them. Mineral ions in solution can be absorbed through the cuticle or stomata and, for some crops, this is a method of applying 'fertilizer'. (The process is called **foliar feeding**.)

Some insecticide and fungicide sprays are absorbed through the leaves and translocated through the plant. Such pesticides are called **systemic** because they enter the plant's system. A caterpillar which ate part of a leaf treated with a systemic insecticide would be poisoned by the chemical in the cells of the leaf.

It is important, of course, that systemic pesticides are broken down to harmless compounds by the plant, long before its leaves or fruits are used for human consumption.

PRACTICAL WORK

Experiments on transport in plants

1 Rates of water uptake in different conditions

The apparatus shown in Fig. 8 is called a **potometer**. It is designed to measure the rate of uptake of water in a cut shoot.

Fill the syringe with water and attach it to the side arm of the 3-way tap. Turn the tap downwards (i) and press the syringe until water comes out of the rubber tubing at the top.

Collect a leafy shoot and push its stem into the rubber tubing as far as possible. Set up the apparatus in a part of the laboratory that is not receiving direct sunlight.

Turn the tap up (ii) and press the syringe till water comes out of the bottom of the capillary tube. Turn the tap horizontally (iii).

As the shoot transpires, it will draw water from the capillary tube and the level can be seen to rise. Record the distance moved by the water column in 30 seconds or a minute.

Figure 8 A potometer

Turn the tap up and send the water column back to the bottom of the capillary. Turn the tap horizontally and make another measurement of the rate of uptake. In this way obtain the average of three readings.

The conditions can now be changed in one of the following ways:

1 Move the apparatus into sunlight or under a fluorescent lamp.
2 Blow air past the shoot with an electric fan or merely fan it with an exercise book.
3 Cover the shoot with a plastic bag.

After each change of conditions, take three more readings of the rate of uptake and notice whether they represent an increase or a decrease in the rate of transpiration.

You might expect results as follows:

1 An increase in light intensity should make the stomata open and allow more rapid transpiration.
2 Moving air should increase the rate of evaporation and, therefore, the rate of uptake.
3 The plastic bag will cause a rise in humidity round the leaves and suppress transpiration.

Ideally, you should change only one condition at a time. If you took the experiment outside, you would be changing the light intensity, the temperature and the air movement. When the rate of uptake increased, you would not know which of these three changes was mainly responsible.

To obtain reliable results, you should really keep taking readings until three of them are nearly the same. A change in conditions may take 10 or 15 minutes before it produces a new, steady rate of uptake. In practice, you may not have time to do this, but even your first three readings should indicate a trend towards increased or decreased uptake.

Limitations of the potometer

Although we use the potometer to compare rates of transpiration, it is really the rates of uptake that we are observing. Not all the water taken up will be transpired; some will be used in photosynthesis; some may be absorbed by cells to increase their turgor. However, these quantities are very small compared with the volume of water transpired and they can be disregarded.

The rate of uptake of a cut shoot may not reflect the rate in the intact plant. If the root system were present, it might offer resistance to the flow of water or it could be helping the flow by means of its root pressure.

2 To find which surface of a leaf loses more water vapour

Cut four leaves of about the same size from a plant (do not use an evergreen plant). Protect the bench with newspaper and then treat each leaf as follows:

(a) smear a thin layer of Vaseline (petroleum jelly) on the lower surface;
(b) smear Vaseline on the upper surface;
(c) smear Vaseline on both surfaces;
(d) no Vaseline on either surface.

Place a little Vaseline on the cut end of the leaf stalk and then suspend the four leaves from a retort stand with cotton threads for several days.

Result All the leaves will have shrivelled and curled up to some extent but the ones which lost most water will be the most shrivelled (Fig. 9).

Interpretation The Vaseline prevents evaporation. The untreated leaf and the leaf with its upper surface sealed show the greatest degree of shrivelling, so it is from the lower surface that leaves evaporate most water.

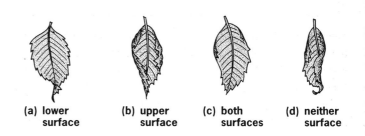

Figure 9 The results of evaporation from leaves subjected to different treatments

(a) lower surface (b) upper surface (c) both surfaces (d) neither surface

More accurate results may be obtained by weighing the leaves at the start and the end of the experiment. It is best to group the leaves from the whole class into their respective batches and weigh each batch. Ideally, the weight loss should be expressed as a percentage of the initial weight.

More rapid results can be obtained by sticking small squares of blue cobalt chloride paper to the upper and lower surface of the same leaf using transparent adhesive tape (Fig. 10). Cobalt chloride paper changes from blue to pink as it takes up moisture. By comparing the time taken for each square to go pink, the relative rates of evaporation from each surface can be compared.

The results of either experiment can be correlated with the numbers of stomata on the upper and lower epidermis. This can be done by painting clear nail varnish over each surface and allowing it to dry. The varnish is then peeled off and examined under the microscope. The outlines of the guard cells can be seen and counted.

'Sellotape'

cobalt chloride paper

Figure 10 To find which surface of a leaf loses more water vapour

3 Transport in the vascular bundles

Place the shoots of several leafy plants in a solution of 1 per cent methylene blue. 'Busy Lizzie' (*Impatiens*) or celery stalks with leaves are usually effective. Leave the shoots in the light for 30 minutes or more.

Result   In some cases, after this time, the blue dye will appear in the leaf veins. If some of the stems are cut across, the dye will be seen in the vascular bundles (see Fig. 2, p. 2).

Interpretation   These results show that the dye and, therefore, probably also the water, travel up the stem in the vascular bundles. Closer study would show that they travel in the xylem vessels.

4 Transport of water in the xylem

Cut three leafy shoots from a deciduous tree or shrub. Each shoot should have about the same number of leaves.

On one twig remove a ring of bark about 5 mm wide, about 100 mm up from the cut base. With the second shoot, smear a layer of Vaseline over the cut base so that it blocks the vessels. The third twig is a control.

Place all three twigs in a jar with a little water. The water level must be below the region from which you removed the ring of bark. Leave the twigs where they can receive direct sunlight.

Result   After an hour or two, you will probably find that the twig with blocked vessels shows signs of wilting. The other two twigs should still have turgid leaves.

Interpretation   Removal of the bark (including the phloem) has not prevented water from reaching the leaves, but blocking the xylem vessels has. The vessels of the xylem, therefore, offer the most likely route for water passing up the stem.

Full details of these experiments and eleven others can be found in *Experimental Work in Biology* (see p. vi).

QUESTIONS

12 A leafy shoot, plus the beaker of water in which it is placed, weighs 275 grams. Two hours later, it weighs 260 grams. An identical beaker of water, with no plant, loses 3 grams over the same period of time. What is the rate of transpiration of the shoot?
13 In Experiment 2, suggest **a** why forceps should be used to handle the cobalt chloride paper squares, **b** why it was water vapour from the leaf and not water vapour from the air which made the cobalt chloride paper change from blue to pink.
14 An interesting experiment is to use a syringe, attached to the cut end of a woody shoot, to force air through the xylem vessels. If the shoot is held under water while you press the syringe, would you expect to see air escaping from the stomata? If you cut away the top half of each leaf, what would you expect to see when the syringe is pressed? Explain your reasoning.
15 In Experiment 1, identify the independent and dependent variables for each condition.

Checklist

- Transpiration is the evaporation of water from the leaves of a plant.
- Transpiration produces the force which draws water up the stem.
- The water travelling in the transpiration stream will contain dissolved salts.
- Closure of stomata and shedding of leaves may help to regulate the transpiration rate.
- The rate of transpiration is increased by sunlight, high temperature, low humidity and air movements.
- Root pressure forces water up the stem as a result of osmosis in the roots.
- Salts are taken up from the soil by roots, and are carried in the xylem vessels.
- Water and salts move up the stem from the roots to the leaves.
- Food made in the leaves moves up or down the stem in the phloem.
- Oxygen and carbon dioxide move in or out of the leaf by diffusion through the stomata.

8 REPRODUCTION IN FLOWERING PLANTS

Sexual reproduction
Flower structure; wallflower and grasses.

Pollination
Cross- and self-pollination. Wallflower and grass. Adaptation to wind and insect pollination.

Fertilization and fruit formation
Fertilization: the pollen tube. Fruit and seed formation; the ovary forms the fruit.

Dispersal of fruit and seeds
Wind, animal and explosive methods.

Practical work
Growth of pollen tubes.

Vegetative propagation
Stolons; rhizomes; strawberry runners; bulbs and corms. Food storage. Advantages of vegetative reproduction; clones.

Artificial propagation
Cuttings; tissue culture.

Sexual reproduction

Flowers are reproductive structures; they contain the reproductive organs of the plant. The male organs are the **stamens** which produce pollen. The female organs are the **carpels**. After fertilization, part of the carpel becomes the fruit of the plant and contains the seeds. In the flowers of most plants there are both stamens and carpels. These flowers are, therefore, male and female at the same time, a condition known as **bisexual** or **hermaphrodite**.

Some species of plants have unisexual flowers, i.e. any one flower will contain either stamens or carpels but not both. Sometimes both male and female flowers are present on the same plant, e.g. the hazel, which has male and female catkins on the same tree. In the willow tree, on the other hand, the male and female catkins are on different trees.

Flower structure

The basic structure of a flower is shown in Figs 1 and 2.

Petals

These are usually brightly coloured and sometimes scented. They are arranged in a circle (Fig. 1) or a cylinder. Most flowers have from four to ten petals. Sometimes they are joined together to form a tube (Fig. 3) and the individual petals can no longer be distinguished. The colour and scent of the petals attract insects to the flower; the insects may bring about pollination (p. 73).

The flowers of grasses and many trees do not have petals but small, leaf-like structures which enclose the reproductive organs.

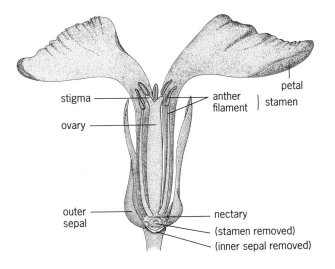

Figure 1 Wallflower; structure of flower (1 sepal, 2 petals and stamen removed)

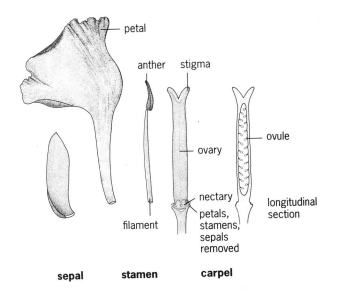

Figure 2 Floral parts of wallflower

Figure 3 Daffodil flower cut in half. The inner petals form a tube. Three stamens are visible round the long style and the ovary contains many ovules.

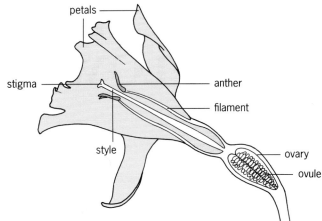

Figure 4 Daffodil flower. Outline drawing of Fig. 3. In daffodils, lilies, tulips, etc. (monocots) there is no distinction between sepals and petals.

Sepals

Outside the petals is a ring of sepals. They are often green and much smaller than the petals. They may protect the flower when it is in the bud.

Stamens

The stamens are the male reproductive organs of a flower. Each stamen has a stalk called the **filament**, with an **anther** on the end. Flowers such as the buttercup and blackberry have many stamens; others such as the tulip have a small number, often the same as, or double, the number of petals or sepals. Each anther consists of four **pollen sacs** in which the pollen grains are produced by cell division. When the anthers are ripe, the pollen sacs split open and release their pollen (see Fig. 6).

Carpels

These are the female reproductive organs. Flowers such as the buttercup and blackberry have a large number of carpels while others, such as the lupin, have a single carpel. Each carpel consists of an **ovary**, bearing a **style** and a **stigma**.

Inside the ovary there are one or more **ovules**. Each blackberry ovary contains one ovule but the wallflower ovary contains several. The ovule will become a **seed**, and the whole ovary will become a **fruit**. (In biology, a fruit is the fertilized ovary of a flower, not necessarily something to eat.)

The style and stigma project from the top of the ovary. The stigma has a sticky surface and pollen grains will stick to it during pollination. The style may be quite short (e.g. wallflower, Fig. 1) or very long (e.g. daffodil, Fig. 4).

Receptacle

The flower structures just described are all attached to the expanded end of a flower stalk. This is called the **receptacle** and, in a few cases after fertilization, it becomes fleshy and edible, e.g. apple and pear.

Wallflower

The wallflower (Fig. 1) has four sepals (two outer and two inner), and four petals. The sepals and the bases of the petals are not joined up but form a kind of tube which encloses the stamens and ovary.

You might expect to find four or eight stamens but, in fact, there are six; two outer and an inner ring of four.

At the base of each outer stamen there is a swelling which forms a **nectary**. This produces a sugary solution called **nectar** which collects in the cup-like base of the outer sepals. Insects visit the flower to pick up this nectar.

The ovary consists of two carpels joined to make a fairly long pod-like ovary with a short style and a two-lobed stigma at the top. There are two rows of about ten ovules in each carpel.

In order to show the structure of a flower, it is usual to draw a 'half flower'. This shows the flower as it would appear if cut vertically down the middle. In the wallflower, the cut would have to pass between the petals or the sepals. So, Fig. 1 is not a true vertical section or half flower but a flower from which one sepal, two petals and one outer stamen have been removed to show the structure.

Grasses

The flowers of the grasses are tiny, inconspicuous and in dense inflorescences (groups of flowers). There are no petals or sepals in the usual sense, but the reproductive organs are enclosed in two green, leaf-like structures called **bracts**. The ovary contains one ovule and bears two styles with feathery stigmas. There are three stamens, the anthers of which, when ripe, hang outside the bracts.

The cereals wheat, oats, barley, maize, etc. are grasses especially bred and cultivated by humans for the sake of the food stored in the fruits or seeds of their flowers.

Rye grass

The inflorescence and flower of this grass are illustrated in Fig. 5.

QUESTIONS

1 Working from outside to inside, list the parts of a bisexual flower.
2 What features of flowers might attract insects?
3 Which part of a flower becomes
   a the seed, **b** the fruit?

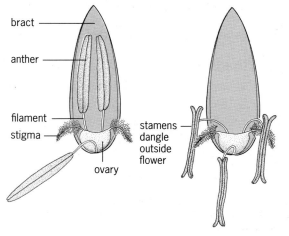

(a) flowering stem **(b) single spikelet** **(c) immature flower opened** **(d) mature flower opened**

Figure 5 Ryegrass; flower structure

Pollination

The transfer of pollen from the anthers to the stigma is called 'pollination'. The anthers split open, exposing the microscopic pollen grains (Fig. 6). The pollen grains are then carried away on the bodies of insects, or simply blown by the wind, and may land on the stigma of another flower. In **self-pollinating** plants, the pollen which reaches the stigma comes from the same flower or another flower on the same plant. In **cross-pollination**, the pollen is carried from the anthers of one flower to the stigma in a flower of another plant of the same species.

If a bee carried pollen from one of the younger flowers near the middle of a wallflower plant to an older flower near the bottom, this would be self-pollination. If, however, the bee flew off to visit a separate wallflower plant and pollinated one of its flowers, this would be cross-pollination.

> The term 'cross-pollination', strictly speaking, should be applied only if there are genetic differences between the two plants involved. The flowers on a single plant all have the same genetic constitution. The flowers on plants growing from the same rhizome or rootstock (p. 78) will also have the same genetic constitution. Pollination between such flowers is little different from self-pollination in the same flower.

Pollination of the wallflower

Nectar is produced by the nectaries at the base of the outer stamens and collects in the cup-like base of the two outer sepals. The sepals and bases of the petals form a narrow tube which is partially blocked by the stamens and ovary. Thus only insects with long probosces ('tongues') can reach the nectar.

Long-tongued bees still have to push some way into the petal tube (Fig. 7) and, in so doing, become dusted with pollen from the ripe anthers. When the same bee visits an older flower, some of the pollen adhering to its body will be picked up by the sticky surface of the stigma.

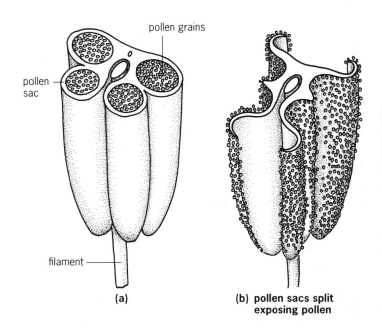

(a) **(b) pollen sacs split exposing pollen**

Figure 6 Structure of anther (top cut off)

Figure 7 Pollination of the wallflower. The bee pushes past the petals to reach the nectary and pollen is deposited on its body.

Butterflies and moths have longer proboscises than bees do. Even so, when a butterfly collects nectar from the flower, pollen is likely to adhere to its proboscis and be transferred to another flower.

Wind pollination

Grasses, cereals and many trees are pollinated, not by insects but by wind currents. The flowers are often quite small with inconspicuous, green, leaf-like bracts, rather than petals. They produce no nectar. The anthers and stigma are not enclosed by the bracts but are exposed to the air. The pollen grains being light and smooth, may be carried long distances by the moving air and some of them will be trapped on the stigmas of other flowers.

In the grasses, at first, the feathery stigmas protrude from the flower, and pollen grains floating in the air are trapped by them. Later, the anthers hang outside the flower (Figs 5(b) and 8), the pollen sacs split, and the wind blows the pollen away. This sequence varies with species.

Figure 8 Grass flowers. The stigmas are the fluffy white structures. Notice that the anthers and stigmas protrude from the bracts.

If the branches of a birch or hazel tree with ripe male catkins, or the flowers of the ornamental pampas grass, are shaken, a shower of pollen can easily be seen (Fig. 9).

Adaptation

Insect-pollinated flowers are considered to be adapted in various ways to their method of pollination. The term **'adaptation'** implies that, in the course of evolution, the structure and physiology of a flower have been modified in ways which improve the chances of successful pollination by insects.

Most insect-pollinated flowers have brightly coloured petals and scent which attract a variety of insects. Some flowers produce nectar which is also attractive to many insects. The dark lines ('honey guides') on petals are believed to help direct the insects to the nectar source and thus bring them into contact with the stamens and stigma.

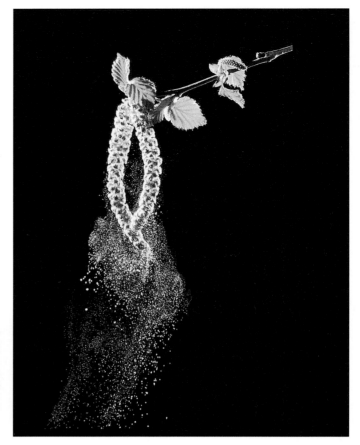

Figure 9 Birch catkins shedding pollen

These features are adaptations to insect pollination in general, but are not necessarily associated with any particular insect species. The various petal colours and the nectaries of the wallflower attract a variety of insects. Many flowers, however, have modifications which adapt them to pollination by only one type or species of insect. Flowers such as the honeysuckle, with narrow, deep petal tubes are likely to be pollinated only by moths or butterflies whose long 'tongues' can reach down the tube to the nectar.

Tube-like flowers such as foxgloves need to be visited by fairly large insects to effect pollination. The petal tube is often lined with dense hairs which impede small insects that would take the nectar without pollinating the flower. A large bumble-bee, however, pushing into the petal tube, is forced to rub against the anthers and stigma.

Some orchid flowers resemble certain species of bee so closely in appearance and odour that male bees try to copulate with the flower and, in doing so, encounter the anthers and stigma.

Many tropical and sub-tropical flowers are adapted to pollination by birds or even by mammals, e.g. bats and mice.

Wind-pollinated flowers are adapted to their method of pollination by producing large quantities of light pollen, and having anthers and stigmas which project outside the flower. Many grasses have anthers which are not rigidly attached to the filaments and can be shaken by the wind. The stigmas of grasses are feathery and act as a net which traps passing pollen grains.

4 Put the following events in the correct order for pollination in a wallflower plant: **a** bee gets dusted with pollen, **b** pollen is deposited on stigma, **c** bee visits older flower, **d** bee visits young flower, **e** anthers split open.

5 Which of the following trees would you expect to be pollinated by insects: apple, hazel, oak, cherry, horse-chestnut, sycamore?

6 In what ways do you think **a** an antirrhinum flower, **b** a nasturtium flower are adapted to insect pollination?

7 In the course of evolution, some flowers may have become adapted to pollination by certain insect species. Discuss whether the insects are likely to have become adapted to the flowers. What sort of adaptation might you expect?

8 Draw up a table to contrast the features of a typical insect-pollinated flower with those of a typical wind-pollinated flower. Include features such as petals (or their equivalent), structure and position of anthers and stigmas, pollen and nectar.

Fertilization and fruit formation

Pollination is complete when pollen from an anther has landed on a stigma. If the flower is to produce seeds, pollination has to be followed by a process called **fertilization**. In all living organisms, fertilization happens when a male sex cell and a female sex cell meet and join together (they are said to **fuse** together). The cell which is formed by this fusion is called a **zygote** and develops into an embryo of an animal or a plant (Fig. 10). The sex cells of all living organisms are called **gametes**.

In animals, the male gamete is the sperm and the female gamete is the egg or ovum (p. 150).

In flowering plants, the male gamete is in the pollen grain; the female gamete, called the **egg cell**, is in the ovule. For fertilization to occur, the nucleus of the male cell from the pollen grain has to reach the female nucleus of the egg cell in the ovule, and fuse with it. The following account explains how this happens.

Fertilization

The pollen grain absorbs liquid from the stigma and a microscopic **pollen tube** grows out of the grain. This tube grows down the style and into the ovary where it enters a small hole, the **micropyle**, in an ovule (Fig. 11). The nucleus of the pollen grain travels down the pollen tube and enters the ovule. Here it combines with the nucleus of the egg cell. Each ovule in an ovary needs to be fertilized by a separate pollen grain.

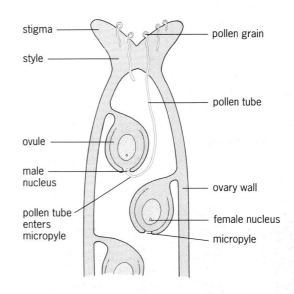

Figure 11 Diagram of fertilization, e.g. wallflower

Although pollination must occur before the ovule can be fertilized, pollination does not necessarily result in fertilization. A bee may visit many flowers on a Bramley apple tree, transferring pollen from one flower to the other. The Bramley, however, is 'self-sterile'; pollination with its own pollen will not result in fertilization. Pollination with pollen from a different variety of apple tree, for example a Worcester, can result in successful fertilization and fruit formation.

Fruit and seed formation

After the pollen nucleus has fused with the egg nucleus, the egg cell divides many times and produces a miniature plant called an **embryo**. The embryo consists of a tiny root and shoot with two special leaves called **cotyledons**. In dicot plants (p. 294) food made in the leaves of the parent plant is carried in the phloem to

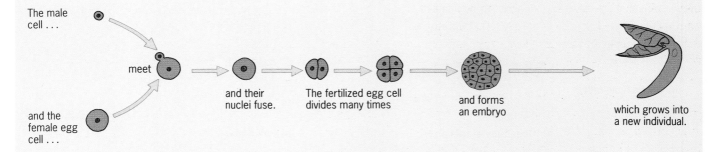

Figure 10 Fertilization. (a) The male cell and the female egg cell meet and their nuclei fuse (b). (c) The fertilized egg cell divides many times and forms an embryo which grows into a new individual (d).

the cotyledons. The cotyledons grow so large with this stored food that they completely enclose the embryo (Fig. 1, p. 83). In monocot plants (p. 294) the food store is laid down in a special tissue called endosperm (p. 85) which is outside the cotyledons. In both cases the outer wall of the ovule becomes thicker and harder, and forms the seed coat or **testa** (see p. 83).

As the seeds grow, the ovary also becomes much larger and the petals and stamens shrivel and fall off (Figs 12(b) and 13). The ovary is now called a **fruit**. The biological definition of a fruit is a fertilized ovary; it is not necessarily edible. The wallflower ovary forms a dry capsule.

(a) tomato flowers
The petals of the older flowers are shrivelling.

(b) after fertilisation
The petals have dropped and the ovary is growing.

(c) ripe fruit
The ovary has grown and ripened. The green sepals remain and the dried stigma is still attached.

Figure 12 Tomato; fruit formation

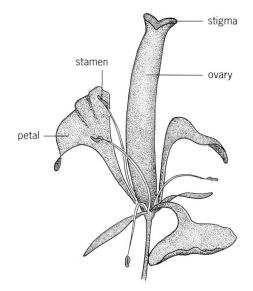

Figure 13 Wallflower. As the ovary grows, the petals, sepals and stamens shrivel and fall off.

Figure 15 Blackberry flowers and fruits. Most of the flowers have been pollinated and fertilized. The petals have dropped and the carpels are growing to form fruits.

(a) petals have fallen

(b) longitudinal section

Figure 14 Blackberry; fruit formation

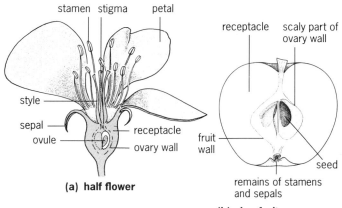

(a) half flower

(b) ripe fruit
(*longitudinal section*)

Figure 16 Apple; fruit formation

A plum is a good example of a fleshy, edible fruit. Tomatoes (Fig. 12) and cucumbers are also fruits although they are classed as vegetables in the shops. Blackberries (Figs 14 and 15) and raspberries are formed by many small fruits clustered together. In the apple (Fig. 16) and pear the edible part consists of the swollen receptacle surrounding and fused to the ovary wall.

QUESTIONS

9 Which structures in a flower produce **a** the male gametes, **b** the female gametes?

10 In not more than two sentences, show that you understand the difference between pollination and fertilization.

11 In flowering plants, **a** can pollination occur without fertilization, **b** can fertilization occur without pollination?

12 Which parts of a tomato flower **a** grow to form the fruit, **b** fall off after fertilization, **c** remain attached to the fruit?

13 Which of the following edible plant products do you think are, biologically, **a** fruits, **b** seeds, **c** neither: runner beans, peas, grapes, baked beans, marrow, rhubarb, tomatoes?

Dispersal of fruits and seeds

When the seeds are mature, the whole fruit or the individual seeds fall from the parent plant to the ground and the seeds may then germinate (p. 84). In many plants, the fruits or seeds are adapted in such a way that they are carried a long distance from the parent plant. This reduces competition for light and water between members of the same species. It may also result in plants growing in new places. The main adaptations are for dispersal by the wind and by animals but some plants have 'explosive' pods that scatter the seeds.

In the wallflower, the fruit splits open from the bottom (Fig. 17(b)), exposing the seeds, which remain attached to a central septum. When the wind shakes the plant, the seeds become detached from the septum and are scattered, though not very far, away from the parent plant.

Wind dispersal

'Parachute' fruits and seeds

Clematis, thistles, willow herb and dandelion have seeds or fruits of this kind. Feathery hairs project from the fruit or seed and so increase its surface area. As a result, the seed 'floats' over long distances before sinking to the ground. It is, therefore, likely to be carried a long way from the parent plant by slight air currents.

'Winged' fruits

Fruits of the sycamore (Fig. 17(e)) and ash trees have wing-like outgrowths from the ovary wall, or leaf-like structures on the flower stalk. These 'wings' cause the fruit to spin as it falls from the tree and so slow down its fall. This delay increases the chances of the fruit being carried away in air currents.

'Explosive' fruits

The pods of flowers in the pea family, e.g. gorse, broom, lupin and vetches, dry in the sun and shrivel. The tough fibres in the fruit wall shrink and set up a tension. When the fruit splits in half down two lines of weakness, the two halves curl back suddenly and flick out the seeds (Fig. 17(c)).

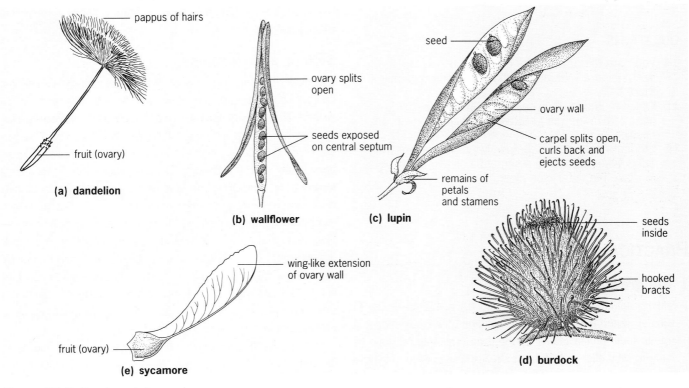

(a) dandelion — pappus of hairs, fruit (ovary)

(b) wallflower — ovary splits open, seeds exposed on central septum

(c) lupin — seed, ovary wall, carpel splits open, curls back and ejects seeds, remains of petals and stamens

(d) burdock — seeds inside, hooked bracts

(e) sycamore — wing-like extension of ovary wall, fruit (ovary)

Figure 17 Fruit and seed dispersal

Animal dispersal

Mammals and hooked fruits

The inflorescence of the burdock is surrounded by bracts which form hooks (Fig. 17(d)). These hooks catch in the fur of passing mammals and the seeds fall out as the mammal moves about.

In other hooked fruits, e.g. agrimony, herb bennet and goosegrass, hooks develop from different parts of the flower.

Birds, mammals and succulent fruits

Fruits such as the blackberry and elderberry are eaten by birds and mammals (Fig. 18). The hard pips containing the seed are not digested and so pass out with the droppings of the bird away from the parent plant. The soft texture and, in some cases, the bright colour of these fruits may be regarded as an adaptation to this method of dispersal.

Figure 18 Animal dispersal. Thrush eating a red currant.

QUESTIONS

14 What are the advantages of a plant dispersing its seeds over a wide area? What disadvantages might there be?

15 Which methods of dispersal are likely to result in seeds travelling the greatest distances? Explain your answer.

16 Students sometimes confuse 'wind pollination' and 'wind dispersal'. Write a sentence or two about each to make the difference clear.

PRACTICAL WORK

1 The growth of pollen tubes

1 Make a solution of 15 g sucrose and 0.1 g sodium borate in 100 cm^3 water. Put a drop of this solution on a cavity slide and scatter some pollen grains on the drop. This can be done by scraping an anther (which must already have opened to expose the pollen) with a mounted needle, or simply by touching the anther on the liquid drop.

Cover the drop with a cover-slip and examine the slide under the microscope at intervals of about 15 minutes. In some cases, pollen tubes may be seen growing from the grains.

Suitable plants include lily, narcissus, tulip, bluebell, lupin, wallflower, sweet pea or deadnettle, but a 15 per cent sucrose solution may not be equally suitable for all of them. It may be necessary to experiment with solutions ranging from 5 to 20 per cent.

2 Cut the stigma from a mature flower, e.g. honeysuckle, crocus, evening primrose or chickweed and place it on a slide in a drop of 0.5 per cent methylene blue. Squash the stigma under a cover-slip (if the stigma is large, it may be safer to squash it between two slides), and leave it for 5 minutes.

Put a drop of water on one side of the slide, just touching the edge of the cover-slip and draw it under the cover-slip by holding a piece of filter paper against the opposite edge (see Fig. 17, p. 34). This will remove excess stain.

If the squash preparation is now examined under the microscope, pollen tubes may be seen growing between the spread-out cells of the stigma.

Vegetative propagation

In addition to reproducing by seeds, many plants are able to produce new individuals asexually, i.e. without gametes, pollination and fertilization. This form of asexual reproduction is also called **vegetative propagation**. When vegetative propagation takes place naturally, it usually results from the growth of a bud on a stem which is close to, or under the soil. Instead of just making a branch, the bud produces a complete plant with roots, stem and leaves. When the old stem dies, the new plant is independent of the parent which produced it.

Stolons and rhizomes

The flowering shoots of plants such as the strawberry and the creeping buttercup are very short and, for the most part, below ground. The stems of shoots such as these are called **rootstocks**. The rootstocks bear leaves and flowers. After the main shoot has flowered, the lateral buds produce long shoots which grow horizontally over the ground (Fig. 19). These shoots are called **stolons** (or 'runners'), and have only small scale-leaves at their nodes and very long internodes. At each node there is a bud which can produce not only a shoot, but roots as well. Thus a complete plant may develop and take root at the node, nourished for a time by food sent from the parent plant through the stolon. Eventually, the stolon dries up and withers, leaving an independent daughter plant growing a short distance away from the parent. In this way a strawberry plant can produce many daughter plants by vegetative propagation in addition to producing seeds.

In many plants, horizontal shoots arise from lateral buds near the stem base, and grow under the ground. Such underground horizontal stems are called **rhizomes**. At the nodes of the rhizome are buds which may develop to produce shoots above the ground.

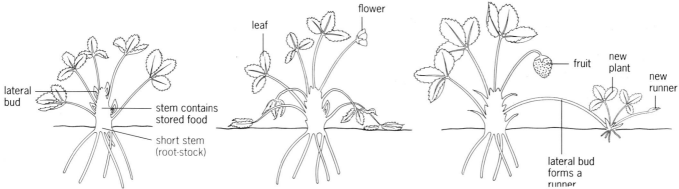

Figure 19 Strawberry runner developing from rootstock

The shoots become independent plants when the connecting rhizome dies.

Many grasses propagate by rhizomes; the couch grass (Fig. 20) is a good example. Even a small piece of rhizome, provided it has a bud, can produce a new plant.

In the bracken, the entire stem is horizontal and below ground. The bracken fronds you see in summer are produced from lateral buds on a rhizome many centimetres below the soil.

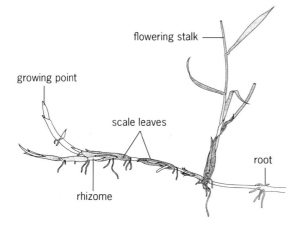

Figure 20 Couch grass rhizome

Bulbs and corms

Bulbs such as those of the daffodil and snowdrop are very short shoots. The stem is only a few millimetres long and the leaves, which encircle the stem, are thick and fleshy with stored food.

In spring, the stored food is used by a rapidly growing terminal bud which produces a flowering stalk and a small number of leaves. During the growing season, food made in the leaves is sent to the leaf bases and stored. The leaf bases swell and form a new bulb ready for growth in the following year.

Vegetative reproduction occurs when some of the food is sent to a lateral bud as well as to the leaf bases. The lateral bud grows inside the parent bulb and, next year, will produce an independent plant (Fig. 21).

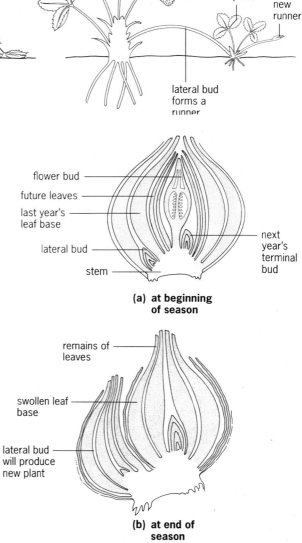

Figure 21 Daffodil bulb; vegetative reproduction

The corms of crocuses and anemones have life cycles similar to those of bulbs but it is the stem, rather than the leaf bases, which swells with stored food. Vegetative reproduction takes place when a lateral bud on the short, fat stem grows into an independent plant.

QUESTIONS

17 Plants can often be propagated from stems but rarely from roots. What features of shoots account for this difference?

18 The plants which survive a heath fire are often those which have a rhizome (e.g. bracken). Suggest a reason why this is so.

19 Suggest why pulling up couch grass is unlikely to be effective in eliminating this weed from the garden.

Food storage

In many cases the organs associated with asexual reproduction also serve as food stores.

Food in the storage organs enables very rapid growth in the spring. A great many of the spring and early summer plants have bulbs, corms, rhizomes or tubers: e.g. daffodil, snowdrop and bluebell, crocus and cuckoo pint, iris and lily-of-the-valley and lesser celandine.

Potatoes are **stem tubers**. Lateral buds at the base of the potato shoot produce underground shoots (rhizomes). These rhizomes swell up with stored starch and form tubers (Fig. 22(a)). Because the tubers are stems, they carry buds. If the tubers are left in the ground or transplanted, the buds will produce shoots, using food stored in the tuber (Fig. 22(b)). In this way, the potato plant can propagate vegetatively.

Carrots, parsnips and beetroot are examples of roots which are modified as storage organs. In the first year of growth, the leaves send food down to the tap root which swells up to form a tuber. In the second year, the food store is used to promote rapid growth and the production of flowers and seeds. The parent plant then dies back.

This is an example of a **biennial** plant. Plants which continue to grow and reproduce asexually each year are called **perennials**. Plants which flower, produce seeds and then die are called **annuals**.

The food reserve in plant storage organs is often starch but in some cases it is glucose (as in the onion) or sucrose (as in sugarbeet).

In agriculture, humans have exploited many of these types of plant and bred them for bigger and more nutritious storage organs for their own consumption, e.g. onion, leek, potato, beetroot.

Artificial propagation

Agriculture and horticulture exploit vegetative reproduction in order to produce fresh stocks of plants. This can be done naturally, e.g. by planting potatoes, dividing up rootstocks or pegging down stolons at their nodes to make them take root. There are also methods which would not occur naturally in the plant's life cycle. Two methods of artificial propagation are by taking cuttings and by tissue culture.

Cuttings

It is possible to produce new individuals from certain plants by putting the cut end of a shoot into water or moist earth. Roots (Fig. 23) grow from the base of the stem into the soil while the shoot continues to grow and produce leaves.

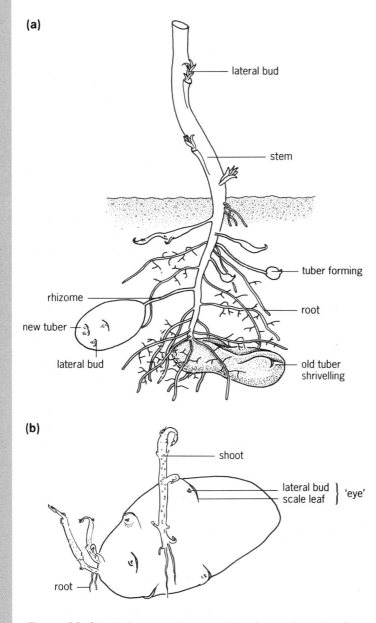

(a)

(b)

Figure 22 Stem tubers growing on potato plant and potato tuber sprouting

(a) roots developing from Busy Lizzie stem

(b) roots growing from coleus cutting

Figure 23 Rooted cuttings

In practice, the cut end of the stem may be treated with a rooting 'hormone' (see p. 337) to promote root growth, and evaporation from the shoot is reduced by covering it with polythene or a glass jar. Carnations, geraniums and chrysanthemums are commonly propagated from cuttings.

Tissue culture

Once a cell has become part of a tissue it usually loses the ability to reproduce. However, the nucleus of any cell in a plant still holds all the 'instructions' (genes, p. 199) for making a complete plant and in certain circumstances they can be brought back into action.

In laboratory conditions, single plant cells can be induced to divide and grow into complete plants. One technique is to take small pieces of plant tissue from a root or stem and treat it with enzymes to separate it into individual cells. The cells are then provided with particular plant 'hormones' which induce cell division and, eventually, the formation of roots, stems and leaves.

An alternative method is to start with a small piece of tissue and place it on a nutrient jelly (agar, p. 305). Cells in the tissue start to divide and produce many cells forming a shapeless mass called a **callus**. If the callus is then provided with the appropriate hormones it develops into a complete plant (Fig. 24).

Figure 25 Tissue culture. Plants grown from small amounts of unspecialized tissue on an agar culture medium.

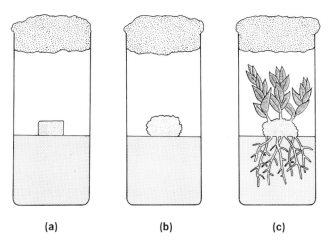

(a)   (b)   (c)

Figure 24 Propagation by tissue culture using nutrient jelly

Using the technique of tissue culture, large numbers of plants can be produced from small amounts of tissue (Fig. 25) and they have the advantage of being free from fungal or bacterial infections. The plants produced in this way form clones, because they have been produced from a single parent plant.

Sexual reproduction and vegetative propagation compared

Variation

In sexual reproduction, the offspring are the result of the fusion of male and female gametes. These gametes may come from the same plant or from different plants of the same species. In either case, the production and subsequent fusion of gametes produces a good deal of variation among the offspring (p. 219). This may result from new combinations of characteristics, e.g. petal colour of one parent combined with fruit size of the other. It may also be the result of spontaneous changes in the gametes when they are produced (mutations, p. 205).

In vegetative propagation no gametes are involved and all the new plants are produced by cell division (mitosis, p. 198) from only one parent. Consequently they are genetically identical; there is no variation. A population of genetically identical individuals produced from a single parent is called a **clone** (p. 206).

A clone has the advantage that the good characteristics of the parent are passed on to all the offspring. This is a very useful feature in horticulture where the grower wants to preserve the quality of the produce. In natural conditions it might be a disadvantage, in the long term, if the climate or other conditions change. If a clone of vegetatively produced plants has no resistance to a particular disease, the whole population could be wiped out. In a population of plants which have been produced sexually, there is a chance that at least some of the offspring will have resistance to the disease. These plants will survive and produce further offspring with disease-resistance.

In fact, plants which reproduce asexually by vegetative propagation also reproduce sexually by flowers and seeds; so they enjoy the benefits of both methods.

Dispersal

The seeds produced as a result of sexual reproduction will be scattered over a relatively wide range. Some will land in inhospitable environments, perhaps lacking light or water. These seeds will fail to germinate. Nevertheless, most methods of seed dispersal result in some of the seeds establishing populations in new habitats.

A plant which reproduces vegetatively will already be growing in a favourable situation, so all the offspring will find themselves in a suitable environment. However, there is no vegetative dispersal mechanism and the plants will grow in dense colonies (Fig. 26), competing with each other for water and minerals. The dense colonies, on the other hand, leave little room for competitors of other species.

As mentioned before, most plants which reproduce vegetatively also produce flowers and seeds. In this way they are able to colonize more distant habitats.

Food storage

The store of food in tubers, tap roots, bulbs, etc. enables the plants to grow rapidly as soon as conditions become favourable. Early growth enables the plant to flower and produce seeds before competition with other plants (for water, mineral salts and light) reaches its maximum. This must be particularly important in woods where, in summer, the leaf canopy prevents much light from reaching the ground and the tree roots tend to drain the soil of moisture over a wide area.

The seeds produced by sexual reproduction all contain some stored food but it is quickly used up during germination which produces only a miniature plant. It takes a long time for a seedling to become established and eventually produce seeds of its own.

Figure 26 Vegetative reproduction. Each clump of daffodils is probably derived from one or two parent bulbs.

Checklist

- Flowers contain the reproductive organs of plants.
- The stamens are the male organs. They produce pollen grains which contain the male gamete.
- The carpels are the female organs. They produce ovules which contain the female gamete and will form the seeds.
- The flowers of most plant species contain male and female organs. A few species have unisexual flowers.
- Brightly coloured petals attract insects, which pollinate the flower.
- Pollination is the transfer of pollen from the anthers of one flower to the stigma of the same or another.
- Pollination may be done by insects or by the wind.
- Flowers which are pollinated by insects are usually brightly coloured and have nectar. The shape of the petals may adapt the flower to pollination by one type of insect.
- Flowers which are pollinated by the wind are usually small and green. Their stigmas and anthers hang outside the flower where they are exposed to air movements.
- Fertilization occurs when a pollen tube grows from a pollen grain into the ovary and up to an ovule. The pollen nucleus passes down the tube and fuses with the ovule nucleus.
- After fertilization, the ovary grows rapidly to become a fruit and the ovules become seeds.
- Seeds and fruits may be dispersed by the wind, by animals, or by an 'explosive' method.
- Dispersal scatters the seeds so that the plants growing from them are less likely to compete with each other and with their parent plant.
- Asexual reproduction occurs without gametes or fertilization.
- Plants reproduce asexually when some of their buds grow into new plants.
- Asexual reproduction in plants is called vegetative propagation.
- The stolon of the strawberry plant is a horizontal stem which grows above the ground, takes root at the nodes and produces new plants.
- The couch grass rhizome is a horizontal stem which grows below the ground and sends up shoots from its nodes.
- Bulbs are condensed shoots with circular fleshy leaves. Bulb-forming plants reproduce asexually from lateral buds.
- Rhizomes, corms, bulbs and tap roots may store food which is used to accelerate early growth.
- Vegetative propagation produces (genetically) identical individuals.
- A clone is a population of organisms produced asexually from a single parent.
- Artificial propagation from cuttings or grafts preserves the desirable characteristics of a crop plant.
- Whole plants can be produced from single cells or small pieces of tissue.

9 GERMINATION AND TROPISMS

The previous chapter described how a seed is formed from the ovule of a flower as a result of fertilization, and is then dispersed from the parent plant. If the seed lands in a suitable place it will germinate, i.e. grow into a mature plant.

Flowering plants can be divided into two major groups, the **monocotyledons** and the **dicotyledons**. A **cotyledon** is a modified leaf in a seed. The cotyledon plays a part in supplying food to the growing plant embryo. Monocotyledons (usually abbreviated to 'monocots') have only one of these seed leaves in the seed. Dicotyledons ('dicots') have two cotyledons in the seed and these two cotyledons store food.

Other characteristics and examples of monocots and dicots are discussed on p. 294.

To follow the process of germination, the structure of a dicot seed, in this case the broad bean, will first be described.

Seed structure and germination

Seed structure (broad bean)

The seed (Figs 1 and 2) contains a miniature plant, the **embryo**, which consists of a root or **radicle**, and a shoot or **plumule**. The embryo is attached to two leaves called the **cotyledons**, which are swollen with stored food. This stored food, mainly starch, is used by the embryo when it starts to grow. The embryo and cotyledons are enclosed in a tough seed coat or **testa**. The **micropyle**, through which the pollen tube entered (p. 75), remains as a small hole in the testa and is an important route for the entry of water in some seeds. The **hilum** is the scar left where the seed was attached to the pod.

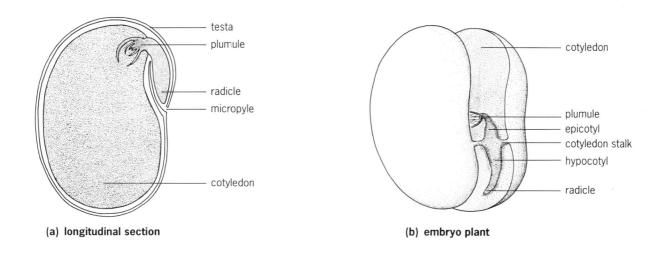

(a) longitudinal section

(b) embryo plant

Figure 1 Seed structure of a dicotyledon

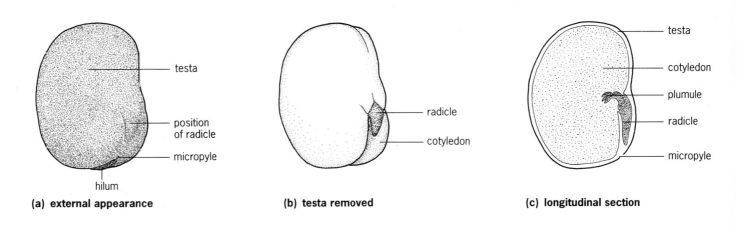

(a) external appearance

(b) testa removed

(c) longitudinal section

Figure 2 Broad bean seed

Germination

The stages of germination of a broad bean are shown in Fig. 3.

A seed just shed from its parent plant contains only 5–20 per cent water, compared with 80–90 per cent in mature plant tissues. Once in the soil, some seeds will absorb water and swell up, but will not necessarily start to germinate until other conditions are suitable (see pp. 85–6).

The radicle grows first and bursts through the testa. The radicle continues to grow down into the soil, pushing its way between soil particles and small stones. Its tip is protected by the root cap (see p. 58). Branches, called **lateral roots**, grow out from the side of the main root and help to anchor it firmly in the soil. On the main root and the lateral roots, microscopic **root hairs** grow out. These are fine outgrowths from some of the outer cells. They make close contact with the soil particles and absorb water from the spaces between them (p. 67).

Once the radicle is securely anchored, the **epicotyl** (Fig. 1(b)) starts to grow and elongates so rapidly that it arches out from between the cotyledons and 'elbows' up through the soil. In this way the plumule is dragged out backwards from the cotyledons and up through the soil. In this position the plumule leaves are least likely to be broken off by their passage through the soil. Once the plumule leaves are above the soil, the epicotyl straightens.

Up to this point, all the food needed for making new cells and producing energy has come from the cotyledons.

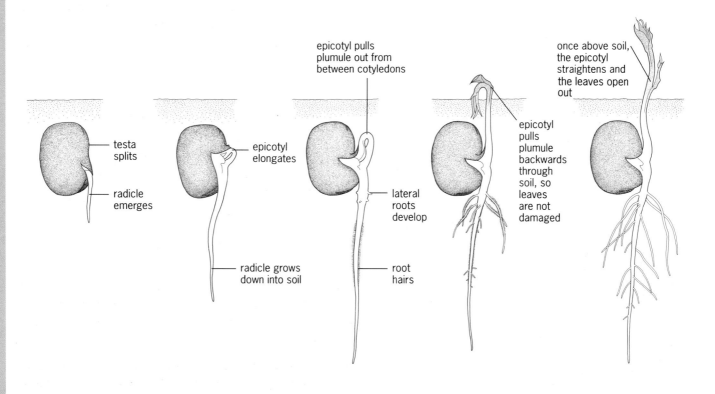

Figure 3 Germination of broad bean

The main type of food stored in the cotyledons is starch. Before this can be used by the growing shoot and root, the starch has to be turned into a soluble sugar. In this form, it can be transported by the phloem cells. The change from starch to sugar in the cotyledons is brought about by enzymes which become active as soon as the seed starts to germinate. The cotyledons shrivel as their food reserve is used up.

By now the plumule leaves have grown much larger, turned green and started to absorb sunlight and make their own food by photosynthesis (p. 37). Between the plumule leaves is a growing point which continues the upward growth of the stem and the production of new leaves. The embryo has now become an independent plant, absorbing water and mineral salts from the soil, carbon dioxide from the air and making food in its leaves.

Other forms of germination

Not all seeds germinate in the same way as the broad bean. In many dicot seeds, it is the **hypocotyl** (Fig. 1(b)) which elongates once the radicle is anchored firmly in the soil. This elongating hypocotyl pulls the cotyledons, sometimes still enclosed in the testa, out of the soil (Fig. 4). The cotyledons swell and push the testa off. They also become green and start to photosynthesize. The plumule, which was protected by the cotyledons during its passage through the soil, grows and forms the first two foliage leaves. The cotyledons then fall off.

In grasses and cereals, the food reserve is not in the cotyledon but in a region called the **endosperm**. There is only one cotyledon (hence the classification as monocots), and this remains in the seed, digesting, absorbing and transferring the stored food in the endosperm. During germination, the shoot grows straight up through the soil, protected at first by a sheath-like **coleoptile**. Once above ground, the coleoptile stops growing and the first leaf bursts through.

QUESTIONS

1 What are the functions of **a** the radicle, **b** the plumule and **c** the cotyledons of a seed?
2 Rewrite the following in the correct order for germination in a broad bean seed: **a** epicotyl straightens, **b** root hairs develop, **c** radicle grows down into the soil, **d** epicotyl starts to elongate, **e** plumule pulled out of the soil, **f** radicle emerges from the testa, **g** testa splits.
3 During germination of the broad bean, how are **a** the plumule, **b** the radicle protected from damage as they are forced through the soil?
4 List all the possible purposes for which a growing seedling might use the food stored in its cotyledons.
5 At what stage of development is a seedling able to stop depending on the cotyledons for its food?
6 What do you think are the advantages to a germinating seed of having its radicle growing some time before the shoot starts to grow?

Figure 4 Germinating seeds. The brown testa is shed, the cotyledons turn green and the plumule expands.

PRACTICAL WORK

Experiments on the conditions for germination

The environmental conditions which might be expected to affect germination are temperature, light intensity and the availability of water and air. The relative importance of some of these conditions can be tested by the experiments which follow. The effects of these conditions are discussed on p. 87.

1 The need for water

Label three containers A, B and C and put dry cotton wool in the bottom of each. Place a number of soaked seeds in B and C, and an equal number of dry seeds in A. Leave A quite dry; add water to B to make the cotton wool moist; add water to C until all the seeds are completely covered (Fig. 5). Put lids on the containers and leave them all at room temperature for a week.

A
soaked peas, dry cotton wool

B
soaked peas, wet cotton wool

C
soaked peas covered with water

Figure 5 The need for water in germination

Result The seeds in B will germinate normally. Those in A will not germinate. The seeds in C may have started to germinate but will probably not be as advanced as those in B and may have died and started to decay.

Interpretation Although water is necessary for germination, too much of it may prevent germination by cutting down the oxygen supply to the seed.

2 The need for oxygen

Set up the experiment as shown in Fig. 7. (**CARE:** Pyrogallic acid and sodium hydroxide is a caustic mixture. Use eye shields, handle the liquids with care and report any spillage at once.)

If the moist cotton wool is rolled in some cress seeds, they will stick to it (Fig. 6). The bungs must make an airtight seal in the flask and the cotton wool must not touch the solution. Pyrogallic acid and sodium hydroxide absorb oxygen from the air, so the cress seeds in flask A are deprived of oxygen. Flask B is the control (see p. 25). This is to show that germination can take place in these experimental conditions provided oxygen is present. Leave the flasks for a week at room temperature.

Figure 6 Roll the moist cotton wool in a dish containing cress seeds

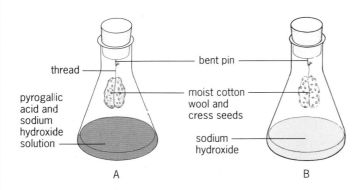

Figure 7 The need for oxygen

Result The seeds in flask B will germinate but there will be little or no germination in flask A.

Interpretation The main difference between flasks A and B is that A lacks oxygen. Since the seeds in this flask have not germinated, it looks as if oxygen is needed for germination.

To show that the chemicals in flask A had not killed the seeds, the cotton wool can be swapped from A to B. The seeds from A will now germinate.

Note Sodium hydroxide absorbs carbon dioxide from the air. The mixture (sodium hydroxide + pyrogallic acid) in flask A, therefore, absorbs both carbon dioxide and oxygen from the air in this flask. In the control flask B, the sodium hydroxide absorbs carbon dioxide but not oxygen. When the seeds in B germinate, it can then be argued that lack of carbon dioxide did not affect them, whereas lack of oxygen did.

3 Temperature and germination

Soak some maize grains for a day and then roll them up in three strips of moist blotting paper as shown in Fig. 8. Put the rolls into plastic bags. Place one in a refrigerator (about 4 °C), leave one upright in the room (about 20 °C) and put the third in a warm place such as over a radiator or, better, in an incubator set to 30 °C.

Because the seeds in the refrigerator will be in darkness, the other seeds must also be enclosed in a box or a cupboard, to exclude light. Otherwise it could be objected that it was lack of light rather than low temperature which affected germination.

After a week, examine the seedlings and measure the length of the roots and shoots.

Figure 8 Temperature and germination. Roll the seeds in moist blotting-paper, and stand the rolls upright in plastic bags.

Result The seedlings kept at 30 °C will be more advanced than those at room temperature. The grains in the refrigerator may not have started to germinate at all.

Interpretation Seeds will not germinate below a certain temperature. The higher the temperature, the faster the germination, at least up to 35–40 °C.

QUESTIONS

7 List the external conditions necessary for germination.
8 Do any of the results of Experiments 1–3 suggest whether or not light is necessary for germination? Explain.
9 **a** In Experiment 3 it could be argued that the low temperature of the refrigerator had killed the seeds and that this explains why they did not germinate. How could you check on this possibility?
   b How could you modify Experiment 3, at least in theory, to find the minimum temperature at which maize seeds would germinate?
10 In Experiment 3, identify **a** the independent variable, **b** the controlled variables and **c** the dependent variable.

Controlling the variables

These experiments on germination illustrate one of the problems of designing biological experiments. You have to decide what conditions (the 'variables') could influence the results and then try to change only one condition at a time. The dangers are that (1) some of the variables might not be controllable, (2) controlling some of the variables might also affect the condition you want to investigate, and (3) there might be a number of important variables you have not thought of.

1 In your germination experiments, you were unable to control the quality of the seeds, but had to assume that the differences between them would be small. If some of the seeds were dead or diseased, they would not germinate in any conditions and this could distort the results. This is one reason for using as large a sample as possible in the experiments.
2 You had to ensure that, when temperature was the variable, the exclusion of light from the seeds in the refrigerator was not an additional variable. This was done by putting all the seeds in darkness.
3 A variable you might not have considered could be the way the seeds were handled. Some seeds can be induced to germinate more successfully by scratching or chipping the testa.

The importance of water, oxygen and temperature

Use of water in the seedling

Most seeds, when first dispersed, contain very little water. In this dehydrated state, their metabolism is very slow and their food reserves are not used up. The dry seeds are also resistant to extremes of temperature and to desiccation. Before the metabolic changes necessary for germination can take place, the seeds must absorb water.

Water is absorbed firstly through the micropyle, in some species, and then through the testa as a whole. Once the radicle has emerged, it will absorb water from the soil, particularly through the root hairs. The water which reaches the embryo and cotyledons is used to:

1 activate the enzymes in the seed;
2 help the conversion of stored starch to sugar, and proteins to amino acids (p. 11);
3 transport the sugar in solution from the cotyledons to the growing regions;
4 expand the vacuoles of newly formed cells and so cause the root and shoot to grow and the leaves to expand (p. 325);
5 maintain the turgor (p. 32) of the cells and thus keep the shoot upright and the leaves expanded;
6 provide the water needed for photosynthesis once the plumule and young leaves are above ground;
7 transport salts from the soil to the shoot.

Uses of oxygen

In some seeds the testa is not very permeable to oxygen, and the early stages of germination are probably anaerobic (p. 21). The testa when soaked or split open allows oxygen to enter. The oxygen is used in aerobic respiration, which provides the energy for the many chemical changes involved in mobilizing the food reserves and making the new cytoplasm and cell walls of the growing seedling.

Importance of temperature

On p. 15 it was explained that a rise in temperature speeds up most chemical reactions, including those taking place in living organisms. Germination, therefore, occurs more rapidly at high temperatures, up to about 40 °C. Above 45 °C, the enzymes in the cells are denatured and the seedlings would be killed. Below certain temperatures (e.g. 0–4 °C) germination may not start at all in some seeds. However, there is considerable variation in the range of temperatures at which seeds of different species will germinate.

Germination and light

Since a great many cultivated plants are grown from seeds which are planted just below soil level, it seems obvious that light is not necessary for germination. Similarly, experiments in the laboratory, with seeds grown in glass dishes, show that light does not prevent germination. There are some species, however, in which the seeds need some exposure to light before they will germinate, e.g. foxgloves and some varieties of lettuce. In all seedlings, once the shoot is above ground, light is necessary for photosynthesis.

QUESTIONS

11 Describe the natural conditions in the soil that would be most favourable for germination. How could a gardener try to create these conditions?
12 Suppose you planted some bean seeds in the garden but no plants appeared after 3 weeks. List all the possible causes for this failure. How could you investigate some of these possibilities?

Dormancy

When plants shed their seeds in summer and autumn, there is usually no shortage of water, oxygen and warmth. Yet, in a great many species, the seeds do not germinate until the following spring. These seeds are said to be **dormant**, i.e. there is some internal control mechanism which prevents immediate germination even though the external conditions are suitable.

If the seeds did germinate in the autumn, the seedlings might be killed by exposure to frost, snow and freezing conditions. The advantage of a dormant period is that germination is delayed until these adverse conditions are past.

The controlling mechanisms are very varied and are still the subject of investigation and discussion. The factors known to influence dormancy are plant growth substances (p. 336), the testa, low temperature and light, or a combination of these.

Tropisms

Tropisms are growth movements related to directional stimuli, e.g. a shoot will grow towards a source of light but away from the direction of gravity. Growth movements of this kind are usually in response to the direction of light, or gravity. Responses to light are called **phototropisms**; responses to gravity are **geotropisms** (or **gravitropisms**).

If the plant organ responds by growing towards the stimulus, the response is said to be 'positive'. If the response is growth away from the stimulus, it is said to be 'negative'. For example, if a plant is placed horizontally, its stem will change its direction and grow upwards, away from gravity (Fig. 9). The shoot is **negatively geotropic**. The roots, however, will change their direction of growth to grow vertically downwards towards the pull of gravity (Experiment 4). Roots, therefore, are **positively geotropic**.

Phototropism and geotropism are best illustrated by some simple controlled experiments. Seedlings are good material for experiments on sensitivity because their growing roots and shoots respond readily to the stimuli of light and gravity.

Figure 9 Negative geotropism. The tomato plant has been left on its side for 24 hours.

PRACTICAL WORK

Experiments on tropisms

4 Geotropism in pea radicles

Soak about 20 peas in water for a day and then let them germinate in a roll of moist blotting-paper (Fig. 8, p. 86). After 3 days, choose 12 seedlings with straight radicles and pin six of these to the turntable of a clinostat so that the radicles are horizontal. Pin another six seedlings to a cork that will fit in a wide-mouthed jar (Fig. 10). Leave the jar on its side. A **clinostat** is a clockwork or electric turntable which rotates the seedlings slowly about four times an hour. Although gravity is pulling sideways on their roots, it will pull equally on all sides as they rotate.

Place the jar and the clinostat in the same conditions of lighting or leave them in darkness for 2 days.

Figure 10 Geotropism in roots

Result The radicles in the clinostat will continue to grow horizontally but those in the jar will have changed their direction of growth, to grow vertically downwards.

Interpretation The stationary radicles have responded to the stimulus of one-sided gravity by growing towards it. The radicles are positively geotropic.

The radicles in the clinostat are the controls. Rotation of the clinostat has allowed gravity to act on all sides equally and there is no one-sided stimulus, even though the radicles were horizontal.

5 Phototropism in shoots

Select two potted seedlings, e.g. sunflower or runner bean, of similar size and water them both. Place one of them under a cardboard box with a window cut in one side so that light reaches the shoot from one direction only (Fig. 11). Place the other plant in an identical situation but on a clinostat. This will rotate the plant about four times per hour and expose each side of the shoot equally to the source of light. This is the control.

Figure 11 Phototropism in a shoot

Result   After 1 or 2 days, the two plants are removed from the boxes and compared. It will be found that the stem of the plant with one-sided illumination has changed its direction of growth and is growing towards the light (Fig. 12). The control shoot has continued to grow vertically.

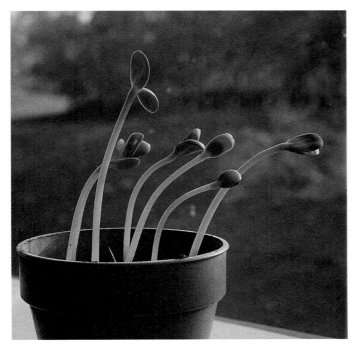

Figure 12 Positive phototropism. The sunflower seedlings have received one-sided lighting for a day.

Interpretation   The results suggest that the young shoot has responded to one-sided lighting by growing towards the light. The shoot is said to be positively phototropic because it grows towards the direction of the stimulus.

However, the results of an experiment with a single plant cannot be used to draw conclusions which apply to green plants as a whole. The experiment described here is more of an illustration than a critical investigation. To investigate phototropisms thoroughly, a large number of plants from a wide variety of species would have to be used.

Advantages of tropic responses

Positive phototropism of shoots
By growing towards the source of light, a shoot brings its leaves into the best situation for photosynthesis. Similarly, the flowers are brought into an exposed position where they are most likely to be seen and pollinated by flying insects.

Negative geotropism in shoots
Shoots which are negatively geotropic grow vertically. This lifts the leaves and flowers above the ground and helps the plant to compete for light and carbon dioxide. The flowers are brought into an advantageous position for insect or wind pollination. Seed dispersal may be more effective from fruits on a long, vertical stem. However, these advantages are a product of a tall shoot rather than negative geotropism.

Stems which form rhizomes (p. 78) are not negatively geotropic; they grow horizontally below the ground, though the shoots which grow up from them are negatively geotropic.

Branches from upright stems are not negatively geotropic; they grow at 90 degrees or, usually, at a more acute angle to the directional pull of gravity. The lower branches of a potato plant must be partially **positively** geotropic when they grow down into the soil and produce potato tubers (p. 80).

Positive geotropism in roots
By growing towards gravity, roots penetrate the soil which is their means of anchorage and their source of water and mineral salts. Lateral roots are not positively geotropic; they grow at right angles or slightly downwards from the main root. This **diageotropism** of lateral roots enables a large volume of soil to be exploited and helps to anchor the plants securely.

QUESTIONS

13 **a** To what directional stimuli do **i** roots,
   ii shoots respond?
   b Name the plant organs which are **i** positively phototropic, **ii** positively geotropic, **iii** negatively geotropic.
14 Why is it incorrect to say
   a 'Plants grow towards the light',
   b 'If a root is placed horizontally, it will bend towards gravity'?
15 Explain why a clinostat is used for the controls in tropism experiments.
16 Look at Fig. 9. What will the shoot look like in 24 hours after the pot has been stood upright again? (Just draw the outline of the stem.)
17 Why does a bean seed still germinate successfully even if you plant it 'upside-down'?
18 What do you think might happen if a potted plant were placed on its side and the shoot illuminated from below (i.e. light and gravity are acting from the same direction)?

---

6 Region of response

Grow pea seedlings as described for maize on p. 86 and select four with straight radicles about 25 mm long. Mark all the radicles with lines about 1 mm apart (Fig.13). Use four strips of moist cotton wool to wedge two seedlings in each of two Petri dishes (Fig. 14). Leave the dishes on their sides for two days, one (A) with the radicles vertical and the other (B) with the radicles horizontal.

Result   The ink marks will be more widely spaced in the region of greatest extension (Fig. 14). By comparing the seedlings in the two dishes, it can be seen that the region of curvature in the B seedlings corresponds to the region of extension in the A seedlings.

Interpretation The response to the stimulus of one-sided gravity takes place in the region of extension. It may be that this is also the region which detects the stimulus.

Further experiments can be found in *Experimental Work in Biology*, Combined Edition by DG Mackean (see p. vi).

Figure 13 Marking a root. A piece of cotton is held by the hairpin and dipped into black ink.

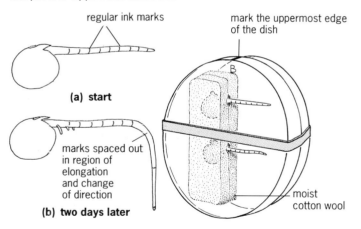

Figure 14 Region of response in radicles. Result of Experiment 6 on the B seedlings.

QUESTIONS

19 In Fig. 15 the two sets of pea seedlings were sown at the same time, but the pot on the left was kept under a light-proof box. From the evidence in the picture, what effects does light appear to have on growing seedlings?

20 It is suggested that it is the very tip of the radicle which detects the one-sided pull of gravity even though it is the region of extension which responds. How could you modify Experiment 6 to test this hypothesis? What snags might you encounter in interpreting your results? (See p. 336 for root growth.)

Figure 15 Effect of light on shoots

Checklist

- A dicot seed consists of an embryo with two cotyledons enclosed in a seed coat (testa).

- The embryo consists of a small root (radicle) and shoot (plumule).

- The cotyledons contain the food store that the embryo will use when it starts to grow.

- The food stored in the cotyledons has to be made soluble by enzymes and transported to the growing regions.

- When germination takes place, (a) the radicle of the embryo grows out of the testa and down into the soil and (b) the plumule is pulled backwards out of the soil.

- In many seeds, the cotyledons are also brought above the soil. They photosynthesize for a while before falling off.

- Germination is influenced by temperature and the amount of water and oxygen available.

- Some seeds have a dormant period after being shed. During this period they will not germinate.

- The roots or shoots of plants may respond to the stimuli of light or gravity.

- A response related to the direction of the stimulus is a tropism.

- Phototropism is a growth response to the direction of light.

- Geotropism is a growth response to the direction of gravity.

- Growth towards the direction of the stimulus is called 'positive'; growth away from the stimulus is called 'negative'.

- Tropic responses bring shoots and roots into the most favourable positions for their life-supporting functions.

EXAMINATION QUESTIONS

Section 2 Flowering Plants

Do not write on this page. Where necessary copy drawings, tables or sentences.

1 The diagram below shows a transverse (cross) section of an unthickened dicotyledonous stem showing the position of the xylem.

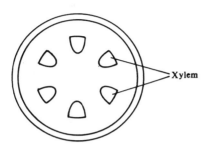

(a) Draw in the position of the phloem on the diagram. (1)
(b) Describe the function of (i) the xylem and (ii) the phloem. (4)
[L]

2 (a) (i) What is the main function of a leaf of a flowering plant such as grass?
(ii) What is found inside the cells of grass leaves to enable them to do this? (2)
(b) Give TWO functions of the roots of flowering plants. (2)
[M]

3 Give an account of the structure and position of roots and stems of a flowering plant. Explain how these features are linked to their functions. (12)
[L]

4 The diagrams below show two different plant cells.

(a) Name the parts labelled A and B. (2)
(b) How does the part labelled C help the root to carry out one of its functions? (2)
(c) How does the part labelled D help the leaf to carry out its main function? (2)
[L]

5 The following sets of external conditions affect transpiration. Which one would result in least loss of water vapour from a leafy shoot?
A Cold, moist, still   **C** Cold, dry, still
B Warm, dry, still   **D** Cold, moist, windy
[N]

6 Give an account of the processes involved in the movement of water through a plant as it passes from the soil to the air. (15)
[L]

7 Flowers in the bud stage are usually protected by:
A Sepals   **C** Petals
B Stigmas   **D** Stamens
[N]

8 The diagram below shows a section through the flower of a daffodil.
(a) In the spaces provided on the diagram, name the structures A and B. (2)

(b) Insects transfer pollen from structure B to structure A. State two ways, visible in the diagram, in which the daffodil flower is adapted for insect pollination. (2)
[S]

9 (a) Explain the functions of the parts of a named herbaceous dicotyledonous flower. (7)
(b) Give an account of the reproductive processes which bring about the formation of a fruit containing seeds. (8)
[L]

10 (a) Name a structure which forms each of the following in a flower: (i) pollen (ii) seed (iii) fruit.
(b) Explain the difference between self-pollination and cross-pollination. (5)
[L]

11 (a) Complete the drawing of a section through a wallflower by drawing **and labelling** the following: sepals, nectaries, filament, anther, stigmas, style, ovary, ovule. (4)

(b) Name a part of the flower where: (i) pollen is placed during pollination; (ii) reduction division (meiosis) takes place; (iii) the seeds develop. (3)
(c) Give two features shown in your completed drawing which tell you that the wallflower is pollinated by insects. (2)
[W]

12 The diagram shows a section through a flower.

(a) Label with a guideline and the appropriate letter: A the stigma; B a part where male gametes are produced; C a part where fertilisation occurs. (3)
(b) Give two ways in which you would expect a wind-pollinated flower to differ from this flower.
(c) For each of the following parts of a flower, name the part in a mammal which is most similar in function: stamen; pollen. (2)
(d) Describe how hormones control the menstrual cycle in a mammal. (2)
[N]

13 The diagram below shows a wind-pollinated flower.

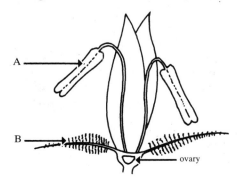

(a) Name the parts labelled A and B. (2)
(b) Give two features of wind-pollinated flowers shown in the diagram. (2)
(c) Explain how this flower is cross-pollinated by the wind. (2)
[NI]

14 Describe the events which take place in a flower following pollination and leading up to the formation of a fruit. (6)
[W]

15 Pollen grains were placed in different concentrations of sugar solutions. They were left for one hour. Samples of fifty pollen grains had their pollen tubes measured and their average length was recorded. The table shows the results.

Concentration of sugar (%)	0	2	4	6	8	10
Average length of pollen tubes (mm)	2	8	10	7	6	3

(a) (i) Plot these results as a graph. Plot the average length of the pollen tubes against the concentration of sugar. (3)
(ii) What is the average length of pollen tube in a 5% sugar solution? (1)
(b) Which part of the flower produces sugar to help the growth of the pollen tube? (1)
(c) State the function of the pollen tube. (1)
[W]

16 When flowering plants reproduce, pollen grains are moved from the male to the female parts of the flower.
(a) What is this process called?
(b) Where on the female part of the flower does the pollen grain land and germinate?
(c) The diagram shows a germinating pollen grain.
(i) Name the structure labelled Z. (ii) Name the process in which nuclei X and Y take part to produce a zygote. (4)
[L]

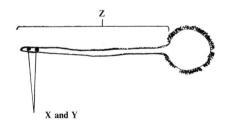

17 It is thought that seeds need oxygen, which is present in the air, to germinate. Describe how you could set up an experiment to test this. (4)
[L]

18 Describe how, after pollination, fertilisation occurs in flowering plants. (7)
[L]

Human Physiology

10 FOOD AND DIET

The need for food

All living organisms need food. An important difference between plants and animals is that the green plants can make food in their leaves but animals have to take it in 'ready-made' by eating plants or the bodies of other animals. In all plants and animals, food is used as follows:

For growth

It provides the substances needed for making new cells and tissues.

As a source of energy

Energy is required for the chemical reactions which take place in living organisms to keep them alive. When food is broken down during respiration (see p. 20), the energy from the food is used for chemical reactions such as building complex molecules (p. 11). In animals the energy is also used for activities such as movement, heart beat and nerve impulses. Mammals and birds use energy to maintain their body temperature.

For replacement of worn and damaged tissues.

The substances provided by food are needed to replace—for example—the millions of our red blood cells that break down each day, and to replace the skin which is worn away, and to repair wounds.

Classes of food

There are three classes of food: carbohydrates, proteins and fats. The chemical structure of these substances is described on p. 11–13.

Carbohydrates

Sugar and **starch** are important carbohydrates in our diet. Starch is abundant in potatoes, bread, maize, rice and other cereals. Sugar appears in our diet mainly as **sucrose** (table sugar) which is added to drinks and many prepared foods such as jam, biscuits and cakes.

Glucose and fructose are sugars which occur naturally in many fruits and some vegetables.

Although all foods provide us with energy, carbohydrates are the cheapest and most readily available source of energy. They contain the elements carbon, hydrogen and oxygen (e.g. glucose is $C_6H_{12}O_6$). When carbohydrates are oxidized to provide energy by respiration they are broken down to carbon dioxide and water (see p. 20). One gram of carbohydrate can provide, on average, 16 kilojoules (kJ) of energy.

If we eat more carbohydrates than we need for our energy requirements, the excess is converted in the liver to either glycogen (see p. 111) or fat. The glycogen is stored in the liver and muscles; the fat is stored in fat depots in the abdomen, round the kidneys or under the skin (Fig. 1).

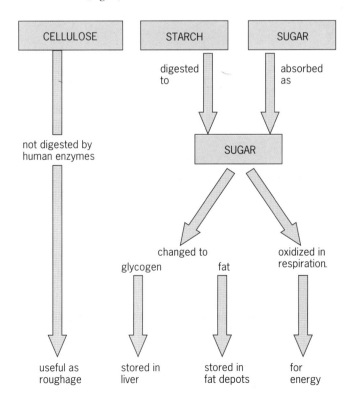

Figure 1 Digestion and use of carbohydrate

The **cellulose** in the cell walls of all plant tissues is a carbohydrate. We probably derive relatively little nourishment from cellulose but it is important in the diet as **fibre** (see p. 96) which helps to maintain a healthy digestive system.

Proteins

Lean meat, fish, eggs, milk and cheese are important sources of animal protein. All plants contain some protein, but beans or cereals like wheat and maize are the best sources.

Proteins, when digested, provide the chemical substances needed to build cells and tissues, e.g. skin, muscle, blood and bones. Neither carbohydrates nor fats can do this and so it is essential to include some proteins in the diet.

Protein molecules consist of long chains of **amino acids** (see p. 11). When proteins are digested, the molecules are broken up into the constituent amino acids. The amino acids are absorbed into the bloodstream and used to build up different proteins. These proteins form part of the cytoplasm and enzymes of cells and tissues. Such a rearrangement of amino acids is shown in Fig. 2.

The amino acids which are not used for making new tissues cannot be stored, but the liver removes their amino (—NH$_2$) groups and changes the residue to glycogen. The glycogen can be stored or oxidized to provide energy (p. 111). One gram of protein can provide 17 kJ of energy.

Chemically, proteins differ from both carbohydrates and fats because they contain nitrogen and sulphur as well as carbon, hydrogen and oxygen.

(a) part of a plant protein of 14 amino acids

(b) digestion breaks up protein into its amino acids

(c) our body builds up the same 14 amino acids but into a protein it needs

key ala = alanine, glu = glycine, leu = leucine,
cys = cysteine , glu = glutamine, lys =lysine,
val = valine, s = sulphur atom

Figure 2 A model for digestion and use of a protein molecule

Fats

Animal fats are found in meat, milk, cheese, butter and egg-yolk. Plant fats occur as oils in fruits (e.g. palm oil) and seeds (e.g. sunflower seed oil), and are used for cooking and making margarine. Fats and oils are sometimes collectively called **lipids**.

Lipids are used in the cells of the body to form part of the cell membrane and other membrane systems (p. 3). Lipids can also be oxidized in respiration, to carbon dioxide and water. When used to provide energy in this way, 1 g fat gives 37 kJ of energy. This is more than twice as much energy as can be obtained from the same weight of carbohydrate or protein.

Fats can be stored in the body, so providing a means of long-term storage of energy in fat depots. The fatty tissue, **adipose tissue**, under the skin forms a layer which, if its blood supply is restricted, can reduce heat losses from the body.

QUESTIONS

1 What sources of protein-rich foods are available to a vegetarian who **a** will eat animal products but not meat itself, **b** will eat only plants and their products?
2 Why must all diets contain some protein?
3 Could you survive on a diet which contained no carbohydrate? Justify your answer.
4 In what sense can the fats in your diet be said to contribute to 'keeping you warm'?
5 How do proteins differ from fats (lipids) in
  a their chemical composition (pp. 11 and 12)
  b their energy value
  c their role in the body?
6 Construct a flow chart for the digestion and use of proteins, similar to the one for carbohydrates in Fig. 1.

Diet

In addition to proteins, carbohydrates and fats, the diet must include salts, vitamins, water and vegetable fibre (roughage). These substances are present in a balanced diet and do not normally have to be taken in separately.

Salts

These are sometimes called 'mineral salts' or just 'minerals'. Proteins, carbohydrates and fats provide the body with carbon, hydrogen, oxygen, nitrogen, sulphur and phosphorus but there are several more elements which the body needs and which occur as salts in the food we eat.

Iron

The red blood cells contain the pigment haemoglobin (p. 116). Part of the haemoglobin molecule contains iron and this plays an important part in carrying oxygen round the body. Millions of red cells break down each day and their iron is stored by the liver and used to make more haemoglobin. However, some iron is lost and adults need to take in about 15 mg each day. Iron is needed also in the muscles and for enzyme systems in all the body cells.

Red meat, especially liver and kidney, is the richest source of iron in the diet, but eggs, groundnuts, bread, spinach and other green vegetables are also important sources.

If the diet is deficient in iron, a person may suffer from some form of anaemia. Insufficient haemoglobin is made and the oxygen-carrying capacity of the blood is reduced.

Calcium

Calcium, in the form of calcium phosphate, is deposited in the bones and the teeth and makes them hard. It is present in blood plasma and plays an essential part in normal blood clotting (p. 125). Calcium is also needed for the chemical changes which make muscles contract and for the transmission of nerve impulses.

The richest sources of calcium are milk (liquid, skimmed or dried). and cheese, but calcium is present in most foods in small quantities and also in 'hard' water.

> Many calcium salts are not soluble in water and may pass through the alimentary canal without being absorbed. Simply increasing the calcium in the diet may not have much effect unless the calcium is in the right form, the diet is balanced and the intestine is healthy. Vitamin D and bile salts (p. 108) are needed for efficient absorption.

Iodine

This is needed in only small quantities, but it forms an essential part of the molecule of **thyroxine**. Thyroxine is a hormone (p. 186) produced by the thyroid gland in the neck.

Specially rich sources of iodine are sea fish and shellfish but it is present in most vegetables, provided that the soil in which they grow is not deficient in the mineral. In some parts of the world, where soils have little iodine, potassium iodide may be added to table salt to bring the iodine in the diet to a satisfactory level.

Phosphorus

Phosphorus is needed for the calcium phosphate of bone, and also for ATP (p. 21) and DNA (p. 202). It is present in nearly all food but is particularly abundant in cheese, meat and fish.

Vitamins

All proteins are similar to each other in their chemical structure and so are all carbohydrates. Vitamins, on the other hand, are a group of organic substances quite unrelated to each other in their chemical structure. The features shared by them all are:

1 They are not digested or broken down for energy.
2 Mostly, they are not built into the body structures.
3 They are essential in very small quantities for normal health.
4 They are needed for chemical reactions in the cells, working in association with enzymes.

Plants can make these vitamins in their leaves, but animals have to take them in ready-made either from plants or from other animals.

If any one of the vitamins is missing from our diet, or at a low level, we will develop a vitamin-deficiency disease. These diseases can be cured, at least in the early stages, simply by adding the vitamin to the diet.

Fifteen or more vitamins have been identified and they are sometimes grouped into two classes: water-soluble or fat-soluble. The fat-soluble vitamins are found mostly in animal fats or vegetable oils, which is one reason why our diet should include some of these fats. The water-soluble vitamins are present in green leaves, fruits and cereal grains.

Table 1 describes vitamins, gives examples of foods which contain them and mentions the vitamin-deficiency diseases which result from their absence.

Dietary fibre (roughage)

When we eat vegetables and other fresh plant material, we take in a large quantity of plant cells. The cell walls of plants consist mainly of cellulose, but we do not have enzymes for digesting this substance. The result is that the plant cell walls reach the large intestine (colon) without being digested. This undigested part of the diet is called **fibre** or roughage. The colon contains many bacteria which can digest some of the substances in the plant cell walls to form fatty acids (see p. 12). Vegetable fibre, therefore, may supply some useful food material, but it has other important functions.

The fibre itself and the bacteria which multiply from feeding on it, add bulk to the contents of the colon and help it to retain water. This softens the faeces (p. 110) and reduces the time needed for the undigested residues to pass out of the body. Both of these effects help to prevent constipation and keep the colon in a healthy condition.

Most vegetables and whole cereal grains contain fibre, but white flour and white bread do not contain much. Good sources of dietary fibre are wholemeal bread and bran.

Water

About 70 per cent of most tissue consists of water; it is an essential part of cytoplasm. The body fluids, blood, lymph and tissue fluid (Chapter 12) are composed mainly of water.

Digested food, salts and vitamins are carried round the body as a watery solution in the blood (p. 124) and excretory products such as excess salt and urea are removed from the body in solution by the kidneys (p. 139). Water thus acts as a solvent and as a transport medium for these substances.

Digestion is a process which uses water in a chemical reaction to break down insoluble substances to soluble ones (p. 106). These products then pass, in solution, into the bloodstream. In all cells there are many reactions in which water plays an essential part as a reactant and a solvent.

Since we lose water by evaporation, sweating, urinating and breathing, we have to make good this loss by taking in water with the diet.

Table 1 Vitamins

Name and source of vitamin	Diseases and symptoms caused by lack of vitamin	Notes

Retinol (vitamin A; fat-soluble):
 	
Liver, cheese, butter, margarine, milk, eggs.
 	
Carotene (vitamin A precursor; water-soluble).
 	
Fresh green leaves and carrots.
	Reduced resistance to disease, particularly those which enter through the epithelium. Poor night vision. Cornea of eyes becomes dry and opaque leading to **keratomalacia** and blindness.	The yellow pigment, carotene, present in green leaves and carrots is turned into retinol by the body.
		    Retinol forms part of the light-sensitive pigment in the retina (p. 174).
		    Retinol is stored in the liver.

Ascorbic acid (vitamin C; water-soluble): Oranges, lemons, grapefruit, tomatoes, fresh green vegetables, potatoes.	Fibres in connective tissue of skin and blood vessels do not form properly, leading to bleeding under the skin, particularly at the joints, swollen, bleeding gums and poor healing of wounds. These are all symptoms of **scurvy**.	Possibly acts as a catalyst in cell respiration. Scurvy is only likely to occur when fresh food is not available. Cows' milk and milk powders contain little ascorbic acid so babies may need additional sources. Cannot be stored in the body; daily intake needed.

Calciferol (vitamin D; fat-soluble):
 	
Butter, milk, cheese, egg-yolk, liver, fish-liver oil.
	Calcium is not deposited properly in the bones, causing **rickets** in young children because the bones remain soft and are deformed by the child's weight.
	    Deficiency in adults causes **osteomalacia**; fractures are likely.
		Calciferol helps the absorption of calcium from the intestine and the deposition of calcium salts in the bones.
		    Natural fats in the skin are converted to a form of calciferol by sunlight.

The B vitamins There are ten or more water-soluble vitamins which occur together, particularly in whole cereals, peas and beans. A deficiency of any one of these vitamins is likely to occur only in communities living on restricted diets such as maize or milled rice. **Folic acid** is a B vitamin which, recently, has been shown to be effective in reducing the incidence of birth defects such as spina bifida. Women planning a pregnancy may be advised to take supplements of folic acid. It is present naturally in green vegetables, root vegetables and whole grain products.

There are several other substances classed as vitamins, e.g. **riboflavin** (B$_2$), **tocopherol** (E), **phylloquinone**, but these are either (1) unlikely to be missing from the diet, or (2) their functions are not fully understood.

Vitamin K plays a part in the blood clotting process (p. 125). It is widely available in green vegetables and is also made by the bacteria living normally in the intestine. Consequently it is unlikely to be lacking except in people whose intestinal bacteria have been reduced by heavy doses of antibiotics.

Figure 3 Examples of types of food in a balanced diet (see question 8)

QUESTIONS

7 Which tissues of the body need **a** iron, **b** glucose, **c** calcium, **d** protein, **e** iodine?

8 In Fig. 3 some examples of the food that would give a balanced diet are shown. Consider each picture in turn and say what class of food or item of diet is mainly present. For example, the meat is mainly protein but will also contain some iron.

9 What is the value of leafy vegetables, such as cabbage and lettuce, in the diet?

10 Why is a diet consisting mainly of one type of food, e.g. rice or potatoes, likely to be unsatisfactory even if it is sufficient to meet our energy needs?

11 A zoologist is trying to find out whether rabbits need vitamin C in their diet. Assuming that a sufficiently large number of rabbits is used and adequate controls are applied, the best design of experiment would be to give the rabbits:
  a an artificial diet of pure protein, carbohydrate, fats, minerals and vitamins but lacking vitamin C;
  b an artificial diet as above but with extra vitamin C;
  c a natural diet of grass, carrots etc. but with added vitamin C;
  d natural food but of one kind only, e.g. exclusively grass or exclusively carrots.
    Justify your choice and say why you exclude the other alternatives.

Balanced diets

A balanced diet must contain enough carbohydrates and fats to meet our energy needs. It must also contain enough protein of the right kind to provide the essential amino acids to make new cells and tissues for growth or repair. The diet must also contain vitamins and mineral salts, plant fibre and water. The composition of four food samples is shown in Fig. 4.

Figure 4 An analysis of four food samples
Note The percentage of water includes any salts and vitamins. There are wide variations in the composition of any given food sample according to its source and the method of preservation and cooking. 'White fish' (e.g. cod, haddock, plaice) contains only 0.5 per cent fat whereas herring and mackerel contain up to 14 per cent. White bread contains only 2–3 per cent fibre. Frying the food greatly adds to its fat content.

Energy requirements

Energy can be obtained from carbohydrates, fats and proteins. The cheapest energy-giving food is usually carbohydrate; the greatest amount of energy is available in fats; proteins give about the same energy as carbohydrates but are expensive. Whatever mixture of carbohydrate, fat and protein makes up the diet, the total energy must be sufficient (1) to keep our internal body processes working (e.g. heart beating, breathing action), (2) to keep up our body temperature and (3) to meet the needs of work and other activities.

The amount of energy that can be obtained from food is measured in calories or joules. One gram of carbohydrate or

protein can provide us with 16 or 17 kJ (kilojoules). A gram of fat can give 37 kJ. We need to obtain about 12 000 kJ of energy each day from our food. Table 2 shows how this figure is obtained. However, the figure will vary greatly according to our age, occupation and activity (Fig. 5). It is fairly obvious that a person who does hard manual work, such as digging, will use more energy than someone who sits in an office.

Table 2 Energy requirements in kJ

8 hours asleep	2400
8 hours awake; relatively inactive physically	3000
8 hours physically active	6600
Total	12 000

The 2400 kJ used during 8 hours sleep represents the energy needed for **basal metabolism** which maintains the circulation, breathing, body temperature, brain function and essential chemical processes in the liver and other organs.

If the diet includes more food than is needed to supply the energy demands of the body, the surplus food is stored either as glycogen in the liver (see p. 112) or as fat below the skin and in the abdomen.

In 1984, a national study recommended that, for a balanced diet, 55 per cent of our energy intake should be made up of carbohydrate, 35 per cent of fat and 10 per cent of protein (though not necessarily used for energy). This means that a growing adolescent should take in about 63 g protein, 97 g fat and 344 g carbohydrate each day.

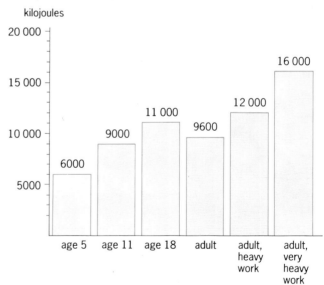

Figure 5 The changing energy requirements with age and occupation

Protein requirements

As explained on p. 95, proteins are an essential part of the diet because they supply the amino acids needed to build up our own body structures. Estimates of how much protein we need have changed over the last few years. A recent FAO/WHO report recommended that an average person needs 0.57 g protein for every kilogram of body weight. That is, a 70 kg person would need $70 \times 0.57 = 39.9$, i.e. about 40 g protein per day. This could be supplied by about 200 g (7 ounces) lean meat or 500 g bread but 2 kg potatoes would be needed to supply this much protein and even this will not contain all the essential amino acids.

Special needs

Pregnancy

A pregnant woman who is already receiving an adequate diet needs to take in no extra food. Her body's metabolism will adapt to the demands of the growing baby. If, however, her diet is deficient in protein, calcium, iron, vitamin D or folic acid, she will need to increase her intake of these substances to meet the needs of the baby.

The baby needs protein for making its tissues; calcium and vitamin D are needed for bone development, and iron is used to make the haemoglobin in its blood.

Lactation

'Lactation' means the production of breast milk for feeding the baby. The production of milk, rich in proteins and minerals, makes a large demand on the mother's resources. If her diet is already adequate, her metabolism will adjust to these demands. Otherwise, she may need to increase her intake of proteins, vitamins and calcium to produce milk of adequate quality and quantity.

Growing children

Most children up to the age of about 12 years need less food than adults, but they need more in proportion to their body weight. For example, an adult may need 0.57 g protein per kg body weight, but a 6–11 month baby needs 1.53 g per kg, and a 10-year-old needs 0.8 g per kg. The extra protein is needed for making new tissues as the child grows.

In addition to protein, children need extra calcium for their growing bones, iron for their red blood cells, vitamin D to help calcification of their bones and vitamin A for disease resistance.

Western diets

In the affluent societies of the USA and Europe for example, there is no general shortage of food and most people can afford a diet with an adequate energy and protein content. Consequently there are few people who suffer from malnutrition, but it seems very likely that many people eat too much food of the wrong kind. This causes illnesses in middle age and old age.

Too much sugar

White sugar (sucrose) is made by refining (purifying) the sugar from sugar-cane or sugar-beet. Refined sugar is used in sweets, jam, biscuits and soft drinks. It is also added to many other processed foods.

There is plenty of evidence to show that sugar is an important cause of tooth decay (p. 191), but refined sugar also affects us in many other ways. It is a very concentrated source of energy. You can absorb a lot of sugar from biscuits, ice-cream, sweets, soft drinks,

tinned fruits and sweet tea without ever feeling 'full up', so you tend to take in more sugar than your body needs. A high intake of refined sugar, therefore, causes some people to become overweight, and this leads to other forms of illness as described below.

Too much fat

The condition known as **atherosclerosis** results from the formation of fatty plaques in the lining of the arteries (p. 128). If these plaques form in the coronary arteries supplying the heart (p. 118) they may lead to a blockage and a coronary heart attack. The plaques contain lipids, including one called **cholesterol**. The more fat and cholesterol you have in your blood, the more likely you are to suffer a coronary heart attack. Many doctors and scientists think that if you eat too much fat, you raise the level of fats and cholesterol in the blood and so put yourself at risk. There is still some argument about this, but until more is known, it seems to be a good idea to keep a low level of fats in your diet.

There are reasons to believe that animal fats such as butter, cream, some kinds of cheese, egg-yolk and the fat present in meat might be more harmful than some vegetable oils. Animal fats are digested to give what are called **saturated** fatty acids (because of the structure of their molecules). Many of the fats and oils from plants, such as the oil from sunflower seeds, contain **unsaturated** fatty acids (Fig. 6). These unsaturated fatty acids are thought to be less likely to cause fatty deposits in the arteries. For this reason, it seems to be better to fry food in vegetable oil and to use margarine from certain vegetable oils rather than butter, though this is still a matter for controversy.

Figure 6 Margarine from sunflower oil. Notice that the saturated fatty acids are low and the unsaturated fatty acids (polyunsaturates) are high.

Too little fibre

We eat too many processed foods, such as sugar, which have been purified from their vegetable sources, and too much white bread, from which the bran has been removed. Unprocessed foods such as potatoes, vegetables and fruit contain a large amount of fibre. Fibre prevents constipation and probably other disorders and diseases of the intestine, possibly including cancer. Eating a diet with a lot of fibre makes you feel 'full up' and so stops you from over-eating. A 100 g portion of boiled potato provides only 340 kJ. (A potato about the size of an egg weighs 50–70 g.) You could feel quite full after eating 300 g of potatoes but would take in only 1000 kJ. A 100g portion of milk chocolate will give you 2400 kJ but it is not filling. So a high fibre diet helps to keep your weight down without leaving you feeling hungry all the time.

Too much of everything

If you eat more food than your body requires for its energy needs or for building tissues, you are likely to store the surplus as fat and so become overweight (**obese**) (Fig. 7). An overweight person is much more likely to suffer from high blood pressure, coronary heart disease and diabetes than a person whose weight is about right. Being fat also makes you less willing to take exercise because you have to carry the extra weight around.

Whether you put on weight or not depends to some extent on genetics. You may inherit the tendency to get fat. Some people seem able to 'burn off' their excess food as heat and never get fat, no matter how much they eat. You can't change your genetics but you can avoid putting on too much weight by controlling your diet. This does not necessarily mean eating less but simply eating differently. Avoid sugar and all processed food with a high sugar level, such as sweets, cakes and biscuits, and include more vegetables, fruit and wholemeal bread in your diet. Your teeth, waistline, intestines and health in general will benefit from such a change in diet.

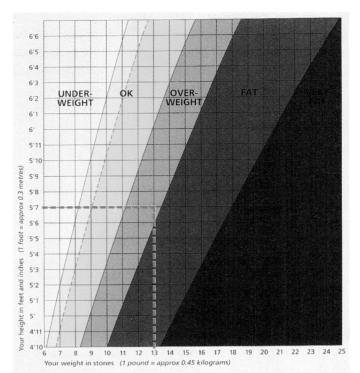

Figure 7 Height and weight. This table is intended for adults who have reached their full height. They would find their height on the vertical scale and look along the line until they reached their weight on the horizontal scale. The broken blue line shows how the chart would be read by a 5ft 7in person who weighs 13 stone.

QUESTIONS

12 Make a list of the food and other substances that are needed to make up a balanced diet.

13 Why is it better to take in regular, small amounts of protein rather than to eat a large amount of protein at one meal? (Revise pp. 94–95.)

14 Select one food class and one mineral salt which are particularly important in the diet of *all three* of the following: pregnant woman, woman breast-feeding a baby, growing child.

15 **a** If you feel 'peckish' between meals, why is it better to eat an apple than a bar of chocolate?
b If you are going to do a long-distance walk, why is it better to take chocolate bars than apples?

16 100 g of boiled potato will give you 240 kJ of energy, but 100 g of chips give you 900 kJ. Why do you think there is such a big difference?

17 Why should a 'high fibre' diet help to stop you putting on weight?

18 It is sometimes believed that a person who does hard, physical work needs to eat a lot of protein. Try to explain why this is not true.

19 How much protein would a 5 kg baby need each day?

20 From Table 2 on p. 98 work out the approximate minimum amount of energy needed each day to maintain your basal metabolism.

Preserving and processing

Food preservation

If food is kept for any length of time before it is eaten it may start to 'go off'. This may be because it is attacked by its own enzymes, oxidized by the air or, more important, decomposed by bacteria and fungi. All these processes make the food taste and smell unpleasant, but the greatest harm is likely to result from the fungi and bacteria.

Both these organisms may produce poisonous compounds (toxins) which make us ill if we eat them, e.g. *Salmonella* poisoning by bacteria. Cooking the food may kill the organisms but will not necessarily destroy the toxins they have already produced.

Methods of food preservation try to prevent the food's own enzymes from working and to stop the growth of fungi and bacteria.

Drying (dehydration)

Removal of water stops the enzymes from working and prevents the growth of bacteria and fungi. The food is usually dried under low air pressure. This makes the water evaporate without raising the food to a temperature at which it would start to cook. The dehydrated food is sealed in packets and kept at room temperature until it is reconstituted by adding water.

Refrigeration and freezing

Most refrigerators are kept at 4 °C. At this temperature bacteria reproduce only very slowly but they are not killed. The activity of any enzymes in the food is also slowed down. The temperature of −20 °C in the freezer stops the growth and reproduction of bacteria but it still does not kill them all.

Canning

Food is enclosed in cans which are sealed and heated to 120 °C to destroy enzymes and bacteria. Bacteria cannot enter the cans. So canned food can remain good to eat for years, provided the cans are not damaged or corroded.

Pasteurization

Milk is heated to 72 °C for 15 seconds (or to 60 °C for 30 minutes) and then cooled rapidly. This does not sterilize the milk, but does destroy the bacteria which might cause diseases like tuberculosis.

Ultra-high temperature (UHT)

Milk is heated under pressure to 132 °C for one second and sealed in cartons or plastic bottles. All the bacteria are killed and so the milk will keep for months. There is a slight change of flavour.

Salting

Impregnating food with salt, e.g. sodium chloride or sodium nitrate, lowers its water potential. Any bacteria that land on the food will be killed because the salt extracts water from them by osmosis (p. 31).

Syrup

Immersing food in concentrated sugar solutions has the same osmotic effect on bacteria as does salt.

Pickling

The ethanoic (acetic) acid in vinegar lowers the pH to the extent that bacteria cannot grow and enzymes are inhibited.

Irradiation

Food is sealed in plastic wrapping and exposed to low doses of X-rays or gamma-rays which kill any bacteria or parasites in the food. The sealed wrapping prevents any more bacteria getting at the food. The food does not become radioactive because the radiation dose is too low.

Irradiation can also improve the keeping quality of some fresh fruits and vegetables.

The disadvantage is that the flavour, texture and colour of some foods may be altered and the vitamin content may be reduced (though this can be said of most methods of preservation). Of greater concern is the fact that the technique might be used to sterilize food which is already contaminated. The bacteria would be killed but any toxins they had produced would not be destroyed.

Irradiated food is used for patients whose immune systems have been suppressed but its general sale is not yet permitted in Britain.

Food additives

About 3500 different chemicals may be used by the processed food industry. These chemicals have no food value but are added to food (1) to stop it going bad, (2) to 'improve' its colour or (3) to alter or enhance its flavour.

Preservatives

There is always a time lag between harvesting a perishable food, processing it, packaging it and sending it to the food shops. The food also spends time on the shelf in the shop and in the kitchen. Thus, except for dried or frozen food, some chemical method of preservation is needed to stop bacterial growth.

Two chemical preservatives are sodium nitrite (E250) and sulphur dioxide (E220). Sodium nitrite is added to cured meat (ham or bacon, for example). It prevents the growth of bacteria, particularly *Clostridium botulinum* which causes a deadly form of food poisoning called botulism. Sulphur dioxide is added to jams, fruit juices, beer and wine. It suppresses the growth of bacteria and fungi in these products.

Other additives

Of the 3500 additives, only about 1 per cent are preservatives. The rest are flavourings, colourings, stabilizers and bulking agents (Fig. 8).

SAFEWAY

8 POTATO WAFFLES

Ingredients:
Potato, Vegetable Oil, Potato Granules [Contain Emulsifier (Mono- and Di-glycerides of Fatty Acids), Preservative (Sodium Metabisulphite)], Salt, Stabiliser (Hydroxypropylmethylcellulose), Pepper

NUTRITION INFORMATION		
AVERAGE VALUES	Per 100g	Approx Per Waffle (Oven Baked)
ENERGY	906 kJ (217 kcal)	513 kJ (123 kcal)
PROTEIN	2.8g	1.6g
CARBOHYDRATE	21.1g	12.7g
of which sugars	0.8g	0.5g
starch	20.3g	12.3g
FAT	13.5g	7.3g
of which saturates	6.4g	3.4g
polyunsaturates	1.2g	0.7g
FIBRE	1.8g	1.0g
SODIUM	0.4g	0.3g

This pack provides approximately 8 servings
ACCEPTABLE FOR GLUTEN FREE, COW'S MILK FREE
AND VEGETARIAN DIETS

Figure 8 Food labelling. The manufacturer must list all the ingredients including additives. (The stabilizer maintains the pH.) The nutritional information is optional.

Flavouring

One widely used flavouring is monosodium glutamate, made from sugar-beet pulp and wheat protein. It is described as a 'flavour enhancer', and may achieve its effect by promoting the flow of saliva or stimulating the taste receptors on the tongue.

Sugar is a widely used flavouring agent. We expect to find it in cakes and biscuits. But it turns up in many unexpected places too—in instant baby cereal (34 per cent in certain brands), tomato sauce (14 per cent), frozen peas (7 per cent) and baked beans (5 per cent).

101

Colouring

Many substances (for example chlorophyll, saffron and turmeric) used for colouring food are from natural sources. This does not necessarily mean that they are harmless. Others are synthetic dyes. For example, tartrazine, sunset yellow and carmoisine are 'azo' dyes, and are known to cause allergic reactions in some people.

'Bulking agents'

Some substances, such as sodium polyphosphate (E450), are added to meat (ham and poultry particularly) as so-called 'tenderizers'. In fact they serve mainly to promote the absorption of water into the meat by osmosis, so that it weighs more.

Possible harmful effects

Only about 300 of the additives are regulated by law, but the food industry tests all the additives, usually by feeding them in large doses to animals. The results, however, are not necessarily applicable to humans. Some workers, for example, estimate that tests for cancer-causing properties may be only 37 per cent successful in animal trials.

There is little widely accepted evidence to show that food additives are harmful. Most allergic reactions to food are caused by naturally occurring substances in the food, though the yellow dye, tartrazine (E102), does cause an allergic reaction in a small number of people. Eczema in children, in some cases, has been relieved by eliminating artificial colour and flavour from their diets. The nitrates and nitrites used to 'cure' ham and bacon are sometimes suspected, without much evidence, as possible carcinogens (cancer-causers).

Most of us eat from 3 to 7 kg of additives each year with no obvious harm, but it is very difficult to know whether the additives have a long-term effect. The risks of cancer from eating cured meats are far lower than the risk of serious bacterial poisoning from eating unprocessed meat. Nevertheless, some people are uneasy that there are so many food additives, whose long-term effects are not known for sure, and which are not really essential for a safe and healthy diet.

World food

The world population is increasing at the rate of about 50 million people per year, and the rate of increase is accelerating. This introduces many problems, but one of the most urgent is the supply of food. Whether increase in food production can continue to match this increase in numbers is uncertain.

Theoretically, a 10- or 20-fold increase in world food production seems possible but although food production has increased by 2.7 per cent per year, the demand for food by an expanding population has risen by 3.9 per cent per year. In Africa and the Near East, food production has not kept up with population growth, and in Latin America it has barely kept pace. Clearly, a large proportion of the world population receives inadequate food, a fact made obvious by the droughts and famines in Africa in recent years.

Famine is caused by a number of interrelated events. Long-term climatic changes may gradually make agriculture impossible. For example, meteorological evidence suggests that the rainfall in Ethiopia, Sudan and Chad has been declining for the last 200 years. Attempts to continue cultivation in these arid conditions leads to the spread of the desert margins.

Overpopulation puts more pressure on poor quality land, leading to soil erosion and the spread of deserts (see p. 248). The land in these over-grazed, over-cultivated regions is also much more susceptible to the effects of drought. In some cases, the adverse changes in climate can be attributed to removal of forests and other plant cover to make way for agriculture. Some disastrous floods, which wash away topsoil, may be the result of deforestation of mountainous areas.

Economic pressures also cause food shortages. In the simplest case, a country may have an adequate food supply but the people are just too poor to buy it. Some developing countries use their best quality land for raising cash crops, such as coffee and tea. These have no food value but are exported to earn foreign currency.

In many developing countries, even if the people do get enough to eat in terms of energy content, their diets are unbalanced and lack necessary proteins and vitamins. These people will suffer from malnutrition even though they are not starving.

PRACTICAL WORK

1 Food tests

See Appendix 1 for the preparation of iodine and Benedict's solutions.

a Test for starch   Shake a little starch powder in a test-tube with some cold water and then boil it to make a clear solution. When the solution is cold, add 3 or 4 drops of **iodine solution**. A dark blue colour should be produced.

b Test for glucose   Heat a little glucose with some **Benedict's solution** in a test-tube. The heating is done by placing the test-tube in a beaker of boiling water (see Fig. 9). The solution will change from clear blue to cloudy green, then yellow and finally to a red precipitate (deposit) of copper(I) oxide.

c Test for protein   (Biuret test) To a 1 per cent solution of albumen (the protein of egg-white) add 5 cm^3 dilute sodium hydroxide (CARE: this solution is caustic), followed by 5 cm^3 1 per cent copper sulphate solution. A purple colour indicates protein.

d Test for fat Shake 2 drops of cooking oil with about 5 cm³ ethanol in a dry test-tube until the fat dissolves. Pour the alcoholic solution into a test-tube containing a few cm³ water. A cloudy white emulsion will form. This shows that the solution contained some fat or oil.

e Test for vitamin C Draw up 2 cm³ fresh lemon juice into a plastic syringe. Add this juice drop by drop to 2 cm³ of a 0.1 per cent solution of PIDCP (a blue dye) in a test-tube. The PIDCP will become colourless quite suddenly as the juice is added. The amount of juice added from the syringe should be noted down. Repeat the experiment but with orange juice in the syringe. If it takes more orange juice than lemon juice to decolourize the PIDCP, the orange juice must contain less vitamin C.

2 Application of the food tests

The tests can be used on samples of food such as milk, potato, raisins, onion, beans, egg-yolk, peanuts, to find what food materials are present (Fig. 9). The solid samples are crushed in a mortar and shaken with warm water to extract the soluble products. Separate samples of the watery mixture of crushed food are tested for starch, glucose or protein as described above. To test for fats, the food must first be crushed in ethanol, not water, and then filtered. The clear filtrate is poured into water to see if it goes cloudy, indicating the presence of fats.

3 Energy from food

Arrange the apparatus as shown in Fig. 10. Use a measuring cylinder to place 100 cm³ cold water in the metal can. With a thermometer, find the temperature of the water and make a note of it. In the nickel crucible or tin lid place 1 g sugar and heat it with the Bunsen flame until it begins to burn. As soon as it starts burning, slide the crucible under the can so that the flames heat the water. If the flame goes out, do not apply the Bunsen burner to the crucible while it is under the can, but return the crucible to the Bunsen flame to start the sugar burning again and replace the crucible beneath the can as soon as the sugar catches alight. When the sugar has finished burning and cannot be ignited again, gently stir the water in the can with the thermometer and record its new temperature. Calculate the rise in temperature by subtracting the first from the second temperature. Work out the quantity of energy transferred to the water from the burning sugar as follows:

Figure 9 Food tests

4.2 joules raise 1 g water 1 °C

100 cm³ cold water weighs 100 g

Let the rise in temperature be T °C

To raise 1 g water 1 °C needs 4.2 joules

∴ To raise 100 g water 1 °C needs 100 × 4.2 joules

∴ To raise 100 g water T °C needs T × 100 × 4.2 joules

∴ 1 g burning sugar produced 420 × T joules

The experiment may now be repeated using 1 g vegetable oil instead of sugar and replacing the warm water in the can with 100 cm³ cold water.

Although the experiment is very inaccurate, the sources of error are more or less the same for both substances. Consequently, the results may be used for purposes of comparison.

Try to point out some of the faults in the design of the experiment.

Figure 10 Energy from food

Checklist

- Our diets must contain proteins, carbohydrates, fats, minerals, vitamins, fibre and water.

- Fats, carbohydrates and proteins provide energy.

- Proteins provide amino acids for the growth and replacement of the tissues.

- Mineral salts like calcium and iron are needed in tissues such as bone and blood.

- Vegetable fibre helps to maintain a healthy intestine.

- Adolescents and adults need about 10–12 000 kJ of energy each day from their food.

- Vitamins are essential in small quantities for chemical reactions in cells.

- The fat-soluble vitamins A and D occur mainly in animal products.

- Most cereals contain vitamins of the B group.

- Vitamin C occurs in certain fruits and in green leaves.

- Lack of vitamin A can lead to blindness; shortage of vitamin C causes scurvy; inadequate vitamin D causes rickets.

- Growing children have special dietary needs.

- Western diets often contain too much sugar and fat and too little fibre.

- Methods of food preservation aim to stop enzymes working and to suppress growth of fungi and bacteria.

11 DIGESTION, ABSORPTION AND USE OF FOOD

The alimentary canal
Structure, function, peristalsis.

Digestion
Definition. Mouth, stomach, small intestine.

Absorption
Absorption in the small and large intestine.

Use and storage of digested food
Glucose, fats, amino acids.

The liver
Functions.

Practical work
Experiments on digestion. Salivary amylase, pepsin and lipase.

Feeding involves taking food into the mouth, chewing it and swallowing it down into the stomach. This satisfies our hunger, but for food to be of any use to the whole body it has first to be **digested**. This means that the solid food is dissolved. The soluble products then have to be **absorbed** into the bloodstream and carried by the blood all round the body. In this way, the blood delivers dissolved food to the living cells in all parts of the body such as the muscles, brain, heart and kidneys. This chapter describes how the food is digested and absorbed. Chapter 12 describes how the blood carries it round the body.

The alimentary canal

The **alimentary canal** is a tube, running through the body. Food is digested in the alimentary canal. The soluble products are absorbed and the indigestible residues expelled (egested). A simplified diagram of an alimentary canal is shown in Fig. 1.

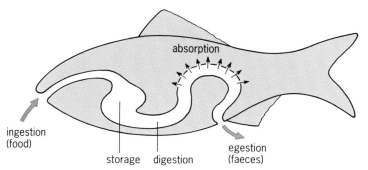

Figure 1 The alimentary canal (generalized)

The inside of the alimentary canal is lined with layers of cells forming what is called an **epithelium**. New cells in the epithelium are being produced all the time to replace the cells worn away by

Figure 2 The general structure of the alimentary canal

the movement of the food. There are also cells in the lining which produce **mucus**. Mucus is a slimy liquid that lubricates the lining of the canal and protects it from wear and tear. Mucus may also protect the lining from attack by the **digestive enzymes** which are released into the alimentary canal.

Some of the digestive enzymes are produced by cells in the lining of the alimentary canal, as in the stomach lining. Others are produced by **glands** which are outside the alimentary canal but pour their enzymes through tubes (called **ducts**) into the

105

alimentary canal. The **salivary glands** (Fig. 6) and the **pancreas** (Fig. 8) are examples of such digestive glands.

The alimentary canal has a great many blood vessels in its walls, close to the lining. These bring oxygen needed by the cells and take away the carbon dioxide they produce. They also absorb the digested food from the alimentary canal.

Peristalsis

The alimentary canal has layers of muscle in its walls (Fig. 2). The fibres of one layer of muscles run round the canal (**circular muscle**) and the others run along its length (**longitudinal muscle**). When the circular muscles in one region contract, they make the alimentary canal narrow in that region.

A contraction in one region of the alimentary canal is followed by another contraction just below it so that a wave of contraction passes along the canal pushing food in front of it. The wave of contraction, called **peristalsis**, is illustrated in Fig. 3.

muscular wall of gullet

circular muscle contracting

food

Figure 3 Diagram to illustrate peristalsis

Digestion

Digestion is mainly a chemical process and consists of breaking down large molecules to small molecules. The large molecules are usually not soluble in water, while the smaller ones are. The small molecules can pass through the epithelium of the alimentary canal, through the walls of the blood vessels and into the blood.

Some food can be absorbed without digestion. The glucose in fruit juice, for example, could pass through the walls of the alimentary canal and enter the blood vessels. Most food, however, is solid and cannot get into blood vessels. Digestion is the process by which solid food is dissolved to make a solution.

The chemicals which dissolve the food are **enzymes**, described on p. 13. A protein might take 50 years to dissolve if just placed in water but is completely digested by enzymes in a few hours. All the solid starch in foods such as bread and potatoes is digested to **glucose** which is soluble in water. The solid proteins in meat, egg and beans are digested to soluble substances called **amino acids**. Fats are digested to two soluble products called **glycerol** and **fatty acids** (see p. 12).

The chemical breakdown usually takes place in stages. For example, the starch molecule is made up of hundreds of carbon, hydrogen and oxygen atoms. The first stage of digestion breaks it down to a 12-carbon sugar molecule called **maltose**. The last stage of digestion breaks the maltose molecule into two 6-carbon sugar molecules called glucose (Fig. 4). Protein molecules are digested first to smaller molecules called **peptides** and finally into completely soluble molecules called amino acids.

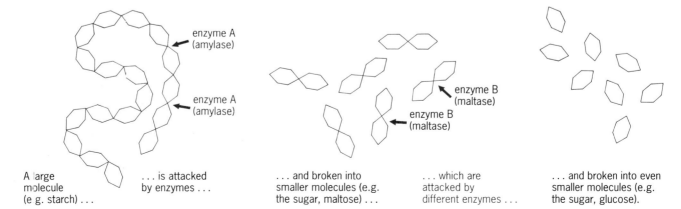

enzyme A (amylase)

enzyme A (amylase)

enzyme B (maltase)

enzyme B (maltase)

A large molecule (e.g. starch) . . .

. . . is attacked by enzymes . . .

. . . and broken into smaller molecules (e.g. the sugar, maltose) . . .

. . . which are attacked by different enzymes . . .

. . . and broken into even smaller molecules (e.g. the sugar, glucose).

Figure 4 Enzymes acting on starch. A large molecule (e.g. starch) is attacked by enzymes (A) and broken into smaller molecules (e.g. the sugar maltose) which are attacked by different enzymes (B) and broken into even smaller molecules (e.g. the sugar glucose).

$$\text{starch} \rightarrow \text{maltose} \rightarrow \text{glucose}$$
$$\text{protein} \rightarrow \text{peptide} \rightarrow \text{amino acid}$$

These stages take place in different parts of the alimentary canal. The progress of food through the canal and the stages of digestion will now be described (Figs 5 and 6).

The mouth

The act of taking food into the mouth is called **ingestion**. In the mouth, the food is chewed and mixed with **saliva**. The chewing breaks the food into pieces which can be swallowed and it also increases the surface area for the enzymes to work on later. Saliva is a digestive juice produced by three pairs of glands whose ducts lead into the mouth (Fig. 6). It helps to lubricate the food and make the small pieces stick together. Saliva contains one enzyme, **salivary amylase** (sometimes called **ptyalin**), which acts on cooked starch and begins to break it down into maltose.

Strictly speaking, the 'mouth' is the aperture between the lips. The space inside, containing the tongue and teeth, is called the **buccal cavity**. Beyond the buccal cavity is the 'throat' or **pharynx**.

Swallowing

By studying Fig. 6(a), it can be seen that for food to enter the gullet (oesophagus), it has to pass over the windpipe. All the complicated actions which occur during swallowing ensure that food does not enter the windpipe and cause choking.

1 The tongue presses upwards and back against the roof of the mouth, forcing a pellet of food, called a **bolus**, to the back of the mouth.
2 The soft palate closes the nasal cavity at the back.
3 The larynx cartilage round the top of the windpipe is pulled upwards so that the opening of the windpipe (the **glottis**) lies under the back of the tongue.
4 The glottis is also partly closed by the contraction of a ring of muscle.
5 The **epiglottis**, a flap of cartilage (gristle) helps to prevent the food from going down the windpipe instead of the gullet.

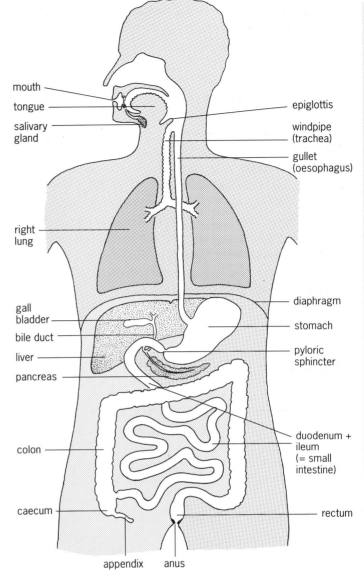

Figure 5 The alimentary canal

The beginning of the swallowing action is voluntary, but once the food reaches the back of the mouth, swallowing becomes an automatic or reflex action. The food is forced into and down the

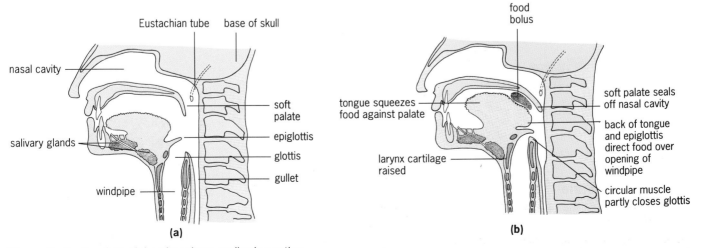

Figure 6 Section through head to show swallowing action

oesophagus, or gullet, by peristalsis. This takes about 6 seconds with relatively solid food and then the food is admitted to the stomach. Liquid travels more rapidly down the gullet.

The stomach

The stomach has elastic walls which stretch as the food collects in it. The **pyloric sphincter** is a circular band of muscle at the lower end of the stomach which stops solid pieces of food from passing through. The main function of the stomach is to store the food from a meal, turn it into a liquid and release it in small quantities at a time to the rest of the alimentary canal.

Glands in the lining of the stomach (Fig. 7) produce **gastric juice** containing the enzyme **pepsin**. Pepsin is a **protease** (or proteinase), i.e. it acts on proteins and breaks them down into soluble compounds called **peptides**. The stomach lining also produces hydrochloric acid which makes a weak solution in the gastric juice. This acid provides the best degree of acidity for pepsin to work in (see p. 15) and kills many of the bacteria taken in with the food.

The regular, peristaltic movements of the stomach, about once every 20 seconds, mix up the food and gastric juice into a creamy liquid. How long food remains in the stomach depends on its nature. Water may pass through in a few minutes; a meal of carbohydrate such as porridge may be held in the stomach for less than an hour, but a mixed meal containing protein and fat may be in the stomach for 1 or 2 hours.

The pyloric sphincter lets the liquid products of digestion pass, a little at a time, into the first part of the small intestine called the **duodenum**.

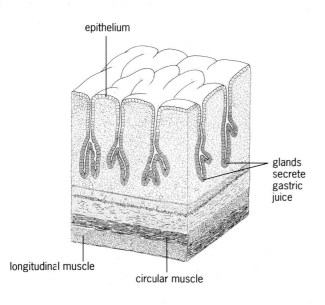

Figure 7 Diagram of section through stomach wall

The small intestine

A digestive juice from the pancreas (**pancreatic juice**) and bile from the liver are poured into the duodenum to act on food there. The pancreas is a digestive gland lying below the stomach (Fig. 8).

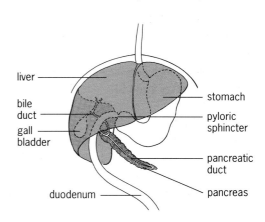

Figure 8 Relationship between stomach, liver and pancreas

It makes a number of enzymes, which act on all classes of food. There are several proteases which break down proteins to peptides and **amino acids**. **Pancreatic amylase** attacks starch and converts it to maltose. **Lipase** digests fats (lipids) to fatty acids and glycerol.

Pancreatic juice contains sodium hydrogencarbonate which partly neutralizes the acid liquid from the stomach. This is necessary because the enzymes of the pancreas do not work well in acid conditions.

Bile is a green, watery fluid made in the liver, stored in the gall-bladder and delivered to the duodenum by the bile duct (Fig. 8). It contains no enzymes, but its green colour is caused by bile pigments which are formed from the breakdown of haemoglobin in the liver. Bile also contains bile salts which act on fats rather like a detergent. The bile salts **emulsify** the fats. That is, they break them up into small drops which are more easily digested by lipase.

All the digestible material is thus changed to soluble compounds which can pass through the lining of the intestine and into the bloodstream. The final products of digestion are:

food	final products
starch	→ glucose
proteins	→ amino acids
fats (lipids)	→ fatty acids and glycerol

The small intestine itself does not appear to liberate digestive enzymes. The structure labelled 'crypt' in Fig. 10 is not a digestive gland, though some of its cells do produce mucus and other secretions. The main function of the crypts is to produce new epithelial cells (see 'Absorption' below) to replace those lost from the tips of the villi.

The epithelial cells of the villi contain, in their cell membranes, enzymes which complete the breakdown of sugars and peptides, before they pass through the cells on their way to the bloodstream. For example, the enzyme **maltase** converts the disaccharide maltose into the monosaccharide, glucose.

Prevention of self-digestion

The gland cells of the stomach and pancreas make protein-digesting enzymes (proteases) and yet the proteins of the cells which make these enzymes are not digested. One reason for this is that the proteases are secreted in an inactive form. Pepsin is produced as **pepsinogen** and does not become the active enzyme until it encounters the hydrochloric acid in the stomach. The lining of the stomach is protected from the action of pepsin probably by the layer of mucus.

Similarly, trypsin, one of the proteases from the pancreas, is secreted as the inactive **trypsinogen** and is activated by **enterokinase**, an enzyme secreted by the lining of the duodenum.

Control of secretion

The sight, smell and taste of food set off nerve impulses from the sense organs to the brain. The brain relays these impulses to the stomach and initiates gastric secretion. When the food reaches the stomach, it stimulates the stomach lining to produce a hormone (p. 185) called **gastrin**. This hormone circulates in the blood and when it returns to the stomach in the blood-stream, it stimulates the gastric glands to continue secretion. Thus, gastric secretion is maintained all the time food is present.

In a similar way, the pancreas is affected first by nervous impulses and then by the hormone **secretin**. Secretin is released into the blood from cells in the duodenum when they are stimulated by the acid contents of the stomach. When secretin reaches the pancreas, it stimulates it to produce pancreatic juice.

Caecum and appendix

In some grass-eating animals (herbivores) such as horses and rabbits, the caecum and appendix are quite large. It is in these organs that digestion of plant cell walls takes place, largely as a result of bacterial activity.

In humans, the caecum and appendix are small structures, possibly without digestive function. The appendix, however, contains lymphoid tissue (p. 123) and may have an immunological function (p. 125).

Absorption

The small intestine consists of the duodenum and the **ileum**. Nearly all the absorption of digested food takes place in the ileum which is efficient at this for the following reasons:

1 It is fairly long and presents a large absorbing surface to the digested food.
2 Its internal surface is greatly increased by circular folds (Fig. 9) bearing thousands of tiny projections called **villi** (singular = villus) (Figs 10 and 11). These villi are about 0.5 mm long and may be finger-like or flattened in shape.
3 The lining epithelium is very thin and the fluids can pass rapidly through it. The outer membrane of each epithelial cell has microvilli (p. 29) which increase by 20 times the exposed surface of the cell.
4 There is a dense network of blood capillaries (tiny blood vessels, see p. 121) in each villus (Fig. 10).

circular folds and villi

villus
capillary network
blood vessels supplying villus
epithelium
circular muscle
longitudinal muscle

Figure 9 The absorbing surface of the ileum

QUESTIONS

4 Why can you not breathe while you are swallowing?
5 Why is it necessary for our food to be digested? Why do plants not need a digestive system? (See p. 37.)
6 In which parts of the alimentary canal is **a** starch **b** protein digested?
7 Study the characteristics of enzymes on pp. 14 and 15.
   In what ways does pepsin show the characteristics of an enzyme?

The small molecules of the digested food, for example, glucose and amino acids, pass into the epithelial cells and then through the wall of the capillaries in the villus and into the bloodstream. They are then carried away in the capillaries which join up to form veins. These veins unite to form one large vein, the **hepatic portal vein** (see Fig. 11 on p. 120). This vein carries all the blood from the intestine to the liver, which may store or alter any of the digestion products. When these products are released from the liver, they enter the general blood circulation.

Some of the fatty acids and glycerol from the digestion of fats enter the blood capillaries of the villi. However, a large proportion of the fatty acids and glycerol may be combined to form fats again in the intestinal epithelium. These fats then pass into the **lacteals** (Fig. 10). The fluid in the lacteals flows into the **lymphatic system** which forms a network all over the body and eventually empties its contents into the bloodstream (see p. 123).

Absorption of the products of digestion and other dietary items is not just a matter of simple diffusion, except perhaps for alcohol and, sometimes, water. Although the mechanisms for transport across the intestinal epithelium have not been fully worked out, it seems likely that various forms of active transport (p. 30) are involved. Even water can cross the epithelium against an osmotic gradient (p. 31). Amino acids, sugars and salts are, almost certainly, taken up by active transport. Glucose, for example, crosses the epithelium faster than fructose although their rates of diffusion would be about the same.

Water-soluble vitamins may diffuse into the epithelium but fat-soluble vitamins are carried in the microscopic fat droplets that enter the cells. The ions of mineral salts are probably absorbed by active transport. Calcium ions need vitamin D for their effective absorption.

The epithelial cells of the villi are constantly being shed into the intestine. Rapid cell division in the epithelium of the crypts (Fig. 10), replaces these lost cells. In effect there is a steady procession of epithelial cells moving up from the crypts to the villi.

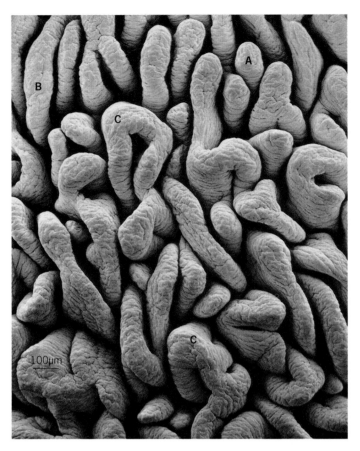

Figure 11 Scanning electron micrograph of the human intestinal lining (x60). The villi are about 0.5mm long. In the duodenum they are mostly leaf-like (C), but further towards the ileum they become narrower (B), and in the ileum they are mostly finger-like (A). This micrograph is of a region in the duodenum.

The large intestine (colon and rectum)

The material passing into the large intestine consists of water with undigested matter, largely cellulose and vegetable fibres (roughage), mucus and dead cells from the lining of the alimentary canal. The large intestine secretes no enzymes but the bacteria in the colon digest part of the fibre to form fatty acids which the colon can absorb. Bile salts are absorbed and returned to the liver by the blood circulation. The colon also absorbs much of the water from the undigested residues. About 7 litres of digestive juices are poured into the alimentary canal each day. If the water from these was not absorbed by the ileum and colon, the body would soon be dehydrated.

The semi-solid waste, the **faeces** or 'stool', is passed into the rectum by peristalsis and is expelled at intervals through the anus. The residues may spend from 12 to 24 hours in the intestine. The act of expelling the faeces is called **egestion** or **defecation**.

Use of digested food

The products of digestion are carried round the body in the blood. From the blood, cells absorb and use glucose, fats and amino acids. This uptake and use of food is called **assimilation**.

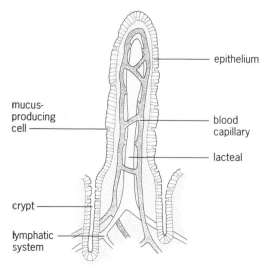

epithelium

mucus-producing cell

blood capillary

lacteal

crypt

lymphatic system

Figure 10 Structure of a single villus

Glucose

During respiration in the cells, glucose is oxidized to carbon dioxide and water (see p. 20). This reaction provides energy to drive the many chemical processes in the cells which result in, for example, the building-up of proteins, contraction of muscles or electrical changes in nerves.

Fats

These are built into cell membranes and other cell structures. Fats also form an important source of energy for cell metabolism. Fatty acids produced from stored fats or taken in with the food, are oxidized in the cells to carbon dioxide and water. This releases energy for processes such as muscle contraction. Fats can provide twice as much energy as sugars.

Amino acids

These are absorbed by the cells and built up, with the aid of enzymes, into proteins. Some of the proteins will become plasma proteins in the blood (p. 117). Others may form structures such as the cell membrane or they may become enzymes which control the chemical activity within the cell. Amino acids not needed for making cell proteins are converted by the liver into glycogen which can then be used for energy.

QUESTIONS

8 What are the products of digestion of **a** starch, **b** protein, **c** fats, which are absorbed by the ileum?

9 What characteristics of the small intestine enable it to absorb digested food efficiently?

10 State briefly what happens to a protein molecule in food, from the time it is swallowed, to the time its products are built up into the cytoplasm of a muscle cell.

11 List the chemical changes which a starch molecule undergoes from the time it reaches the duodenum to the time its carbon atoms become part of carbon dioxide molecules. Say where in the body these changes occur.

Storage of digested food

If more food is taken in than the body needs for energy or for building tissues, such as bone or muscle, it is stored in one of the following ways.

Glucose

The sugar not required immediately for energy in the cells is changed in the liver to **glycogen**. The glycogen molecule is built up by combining many glucose molecules into a long-chain molecule similar to that of starch (see p. 12). Some of this insoluble glycogen is stored in the liver and the rest in the muscles. When the blood sugar falls below a certain level, the liver changes its glycogen back to glucose and releases it into the circulation. The muscle glycogen is not returned to the circulation but is used by muscle cells as a source of energy during muscular activity.

The glycogen in the liver is a 'short-term' store, sufficient for only about 6 hours. Excess glucose not stored as glycogen is converted to fat and stored in the fat depots.

Fat

Unlike glycogen, there is no limit to the amount of fat stored and because of its high energy value (p. 95) it is an effective 'long-term' store. The fat is stored in adipose tissue in the abdomen, round the kidneys and under the skin. These are the **fat depots**. The adipose tissue of the skin is shown in Figs 1 to 3 on p. 146.

> Certain cells accumulate drops of fat in their cytoplasm. As these drops increase in size and number, they join together to form one large globule of fat in the middle of the cell, pushing the cytoplasm into a thin layer and the nucleus to one side (Fig. 12). Groups of fat cells form adipose tissue.

Figure 12 Accumulation of fat in a fat cell

Amino acids

Amino acids are not stored in the body. Those not used in protein formation are **deaminated** (see below). The protein of the liver and other tissues can act as a kind of protein store to maintain the protein level in the blood, but absence of protein in the diet soon leads to serious disorders.

All these conversions of one substance to another need specific enzymes to make them happen.

Body weight

The rate at which glucose is oxidized or changed into glycogen and fat is controlled by hormones (p. 185). When intake of carbohydrate and fat is more than enough to meet the energy requirements of the body, the surplus will be stored mainly as fat. Some people never seem to get fat no matter how much they eat, while others start to lay down fat when their intake only just exceeds their needs. Putting on weight is certainly the result of eating more food than the body needs, but it is not clear why people should differ so much in this respect. The explanation probably lies in the balance of hormones which, to some extent, is determined by heredity.

QUESTIONS

12 List the ways in which the body can store an excess of carbohydrates taken in with the diet.

13 If you were deprived of food for several days, how would your body meet the demands for energy by your heart and other organs?

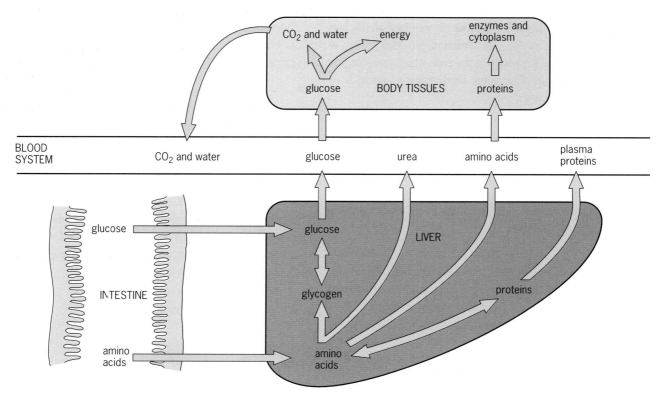

Figure 13 Some functions of the liver

The liver

The liver has been mentioned several times in connection with the digestion, use and storage of food. This is only one aspect of its many important functions, some of which are listed below. It is a large, reddish-brown organ which lies just beneath the diaphragm and partly overlaps the stomach. All the blood from the blood vessels of the alimentary canal passes through the liver, which adjusts the composition of the blood before releasing it into the general circulation (Fig. 13).

Regulation of blood sugar

After a meal, the liver removes excess glucose from the blood and stores it as glycogen. In the periods between meals, when the glucose concentration in the blood starts to fall, the liver converts some of its stored glycogen into glucose and releases it into the bloodstream. In this way, the concentration of sugar in the blood is kept at a fairly steady level.

The concentration of glucose in the blood of a person who has not eaten for 8 hours is usually between 90 and 100 mg/100 cm^3 blood. After a meal containing carbohydrate, the blood sugar level may rise to 140 mg/100 cm^3 but 2 hours later, the level returns to about 95 mg as the liver has converted the excess glucose to glycogen.

About 100 g glycogen is stored in the liver of a healthy man. If the concentration of glucose in the blood falls below about 80 mg/100 cm^3 blood, some of the glycogen stored in the liver is converted by enzyme action into glucose and it enters the circulation. If the blood sugar level rises above 160 mg/100cm^3, glucose is excreted by the kidneys. A blood glucose level below 40 mg/100 cm^3 affects the brain cells adversely, leading to convulsions and coma. By helping keep the glucose concentration between 80 and 150 mg, the liver prevents these undesirable effects and so contributes to the homeostasis (see below) of the body. (See Fig. 10 on p. 120 for the circulatory supply to liver.)

Production of bile

Cells in the liver make bile continuously and this is stored in the gall-bladder until it is discharged through the bile duct into the duodenum. The green colour of the bile results from a pigment, **bilirubin**, which comes from the breakdown of haemoglobin from worn-out red blood cells.

The bile also contains bile salts which assist the digestion of fats as described on p. 108.

A large proportion of the bile salts is reabsorbed in the ileum along with the fats they have helped to emulsify. Bile salts are also absorbed in the colon.

Deamination

The amino acids not needed for making proteins are converted to glycogen in the liver. During this process, the nitrogen-containing, amino part (NH_2) of the amino acid is removed and changed to **urea**, which is later excreted by the kidneys (see p. 139).

When the -NH$_2$ group is removed from certain amino acids it forms ammonia, NH$_3$ (or, more strictly, the ammonium ion -NH$_4^+$). Ammonia is very poisonous to the body cells, and the liver converts it at once to urea, (NH$_2$)$_2$CO, which is a comparatively harmless substance.

Storage of iron
Millions of red blood cells break down every day. The iron from their haemoglobin (p. 116) is stored in the liver.

Manufacture of plasma proteins
The liver makes most of the proteins found in blood plasma, including fibrinogen which plays an important part in the clotting action of the blood (p. 125).

Detoxication
Poisonous compounds, produced in the large intestine by the action of bacteria on amino acids, enter the blood, but on reaching the liver are converted to harmless substances, later excreted in the urine. Many other chemical substances normally present in the body or introduced as drugs are modified by the liver before being excreted by the kidneys. Hormones, for example, are converted to inactive compounds in the liver, so limiting their period of activity in the body.

Storage of vitamins
The fat-soluble vitamins A and D are stored in the liver. This is the reason why animal liver is a valuable source of these vitamins in the diet.

QUESTION
14 Explain how the liver exercises control over the substances coming from the intestine and entering the general blood circulation.

Homeostasis
A complete account of the functions of the liver would involve a very long list. It is most important, however, to realize that the one vital function of the liver, embodying all the details outlined above, is that it helps to maintain the concentration and composition of the body fluids, particularly the blood.

Within reason, a variation in the kind of food eaten will not produce changes in the composition of the blood.

If this **internal environment**, as it is called, were not so constant, the chemical changes that maintain life would become erratic and unpredictable so that with quite slight changes of diet or activity the whole organization might break down. The maintenance of the internal environment within narrow limits is called **homeostasis** and is discussed again on pp. 124, 143 and 187.

PRACTICAL WORK

1 The action of salivary amylase on starch
Rinse the mouth with water to remove traces of food. Collect saliva* in two test-tubes, labelled A and B, to a depth of about 15 mm (see Fig. 14). Heat the saliva in tube B over a small flame until it boils for about 30 seconds and then cool the tube under the tap. Add about 2 cm^3 of a 2 per cent starch solution to each tube; shake each tube and leave them for 5 minutes.

Share the contents of tube A between two clean test-tubes. To one of these add some iodine solution. To the other add some Benedict's solution and heat in a water bath as described on p. 38. Test the contents of tube B in exactly the same way.

*If there is some objection to using your own saliva, use a 5 per cent solution of commercially prepared amylase instead.

Figure 14 Salivary amylase acting on starch

Results The contents of tube A fail to give a blue colour with iodine, showing that the starch has gone. The other half of the contents, however, gives a red or orange precipitate with Benedict's solution, showing that sugar is present.

The contents of tube B still give a blue colour with iodine but do not form a red precipitate on heating with Benedict's solution.

Interpretation The results with tube A suggest that something in saliva has converted starch into sugar. The fact that the boiled saliva in tube B fails to do this, suggests that it was an enzyme in saliva which brought about the change (see p. 15), because enzymes are proteins and are destroyed by boiling. If the boiled saliva had changed starch to sugar, it would have ruled out the possibility of an enzyme being responsible.

> This interpretation assumes that it is something in saliva which changes starch into sugar. However, the results could equally well support the claim that starch can turn unboiled saliva into sugar. Our knowledge of (1) the chemical composition of starch and saliva and (2) the effect of heat on enzymes, makes the first interpretation more plausible.

2 The action of pepsin on egg-white protein

A cloudy suspension of egg-white is prepared by stirring the white of one egg into 500 cm³ tap-water, heating it to boiling point and filtering it through glass wool to remove the larger particles.

Label four test-tubes A, B, C and D and place 2 cm³ egg-white suspension in each of them. Then add pepsin solution and/or dilute hydrochloric acid to the tubes as follows (Fig. 15):

Figure 15 Pepsin acting on egg-white

A Egg-white suspension – 1 cm³ pepsin solution (1%)
B Egg-white suspension – 3 drops dilute hydrochloric acid (HCl)
C Egg-white suspension – 1 cm³ pepsin + 3 drops HCl
D Egg-white suspension – 1 cm³ boiled pepsin + 3 drops HCl

Place all four tubes in a beaker of warm water at 35 °C for 10–15 minutes.

Result The contents of tube C go clear. The rest remain cloudy.

Interpretation The change from a cloudy suspension to a clear solution shows that the solid particles of egg protein have been digested to soluble products. The failure of the other three tubes to give clear solutions shows that:

(a) Pepsin will only work in acid solutions.

(b) It is the pepsin and not the hydrochloric acid which does the digestion.

(c) Pepsin is an enzyme, because its activity is destroyed by boiling.

3 The action of lipase

Place 5 cm³ milk and 7 cm³ dilute (M/20) sodium carbonate solution in each of three test-tubes labelled 1 to 3 (Fig. 16). Add six drops of phenolphthalein to each to turn the contents pink. Add 1 cm³ of 3 per cent bile salts solution to tubes 2 and 3. Add 1 cm³ of 5 per cent lipase solution to tubes 1 and 3, and an equal volume of boiled lipase to tube 2.

Result In 10 minutes or less, the colour of the liquids in tubes 1 and 3 will change to white, tube 3 changing first. The liquid in tube 2 will remain pink.

Interpretation Lipase is an enzyme that hydrolyses fats to fatty acids and glycerol. When lipase acts on milk fats, the fatty acids so produced react with the alkaline sodium carbonate and make the solution more acid. In acid conditions the pH indicator, phenolphthalein, changes from pink to colourless. The presence of bile salts in tube 3 seems to accelerate the reaction, although bile salts with the denatured enzyme in tube 2 cannot bring about the change on their own.

For experiments investigating the effect of temperature and pH on enzyme action see pp. 17 and 18.

Figure 16 The action of lipase

These experiments and others are fully described in *Experimental Work in Biology*, Combined Edition. (See p. vi.)

QUESTIONS

15 In experiments with enzymes, the control often involves the boiled enzyme. Suggest why this type of control is used.

16 In Experiment 2, why does the change from cloudy to clear suggest that digestion has occurred?

17 How would you modify Experiment 2 if you wanted to find the optimum temperature for the action of pepsin on egg-white?

18 Experiment 2 is really two experiments combined because there are two independent variables. Identify
  a the two independent variables,
  b the dependent variable, and
  c the controlled variable.

19 It was suggested that an alternative interpretation of the result in Experiment 1 might be that starch has turned saliva into sugar. From what you know about starch, saliva and the design of the experiment, explain why this is a less acceptable interpretation.

20 Write down the menu for your breakfast and lunch (or supper). State the main food substances present in each item of the meal. State the final digestion product of each.

Checklist

- Digestion is the process which changes insoluble food into soluble substances.
- Digestion takes place in the alimentary canal.
- The changes are brought about by chemicals called digestive enzymes.
  - Maltose and sucrose are changed to glucose by enzymes in the epithelium of the villi.
- The ileum absorbs amino acids, glucose and fatty acids and glycerol.
- Amino acids and glucose are carried in the bloodstream first to the liver and then to all parts of the body.
- Internal folds, villi and microvilli greatly increase the absorbing surface of the small intestine.
- The digested food is used or stored in the following ways: Glucose is (1) oxidized for energy or (2) changed to glycogen or fat and stored. Amino acids are (1) built up into proteins or (2) deaminated to urea and glycogen and used for energy. Fats are (1) oxidized for energy or (2) stored.
- Glycogen in the liver and muscles acts as a short-term energy store; fat in the fat depots acts as a long-term energy store.
- The liver stores glycogen and changes it to glucose and releases it into the bloodstream to keep a steady level of blood sugar.
  - The liver exercises control over many other aspects of blood composition and so helps maintain chemical stability in the body.

Region of alimentary canal	Digestive gland	Digestive juice produced	Enzymes in the juice	Class of food acted upon	Substances produced
Mouth	salivary glands	saliva	salivary amylase	starch	maltose
Stomach	glands in stomach lining	gastric juice	pepsin	proteins	peptides
Duodenum	pancreas	pancreatic juice	proteases	proteins and peptides	peptides and amino acids
			amylase	starch	maltose
			lipase	fats	fatty acids and glycerol

12 THE BLOOD CIRCULATORY SYSTEM

The previous chapter explained how food is digested to amino acids, glucose, etc., which are absorbed in the small intestine. These substances are needed in all living cells in the body such as the brain cells, leg muscle cells and kidney cells. The substances are carried from the intestine to other parts of the body by the blood system. In a similar way, the oxygen taken in by the lungs is needed by all the cells and is carried round the body in the blood.

Composition of blood

Blood consists of red cells, white cells and platelets floating in a liquid called **plasma**. There are between 5 and 6 litres of blood in the body of an adult, and each cubic centimetre contains about 5 billion red cells.

Red cells

These are tiny, disc-like cells (Figs 1(a) and 2) which do not have nuclei. They are made of spongy cytoplasm enclosed in an elastic cell membrane. In their cytoplasm is the red pigment, **haemoglobin**, a protein combined with iron. Haemoglobin combines with oxygen in places where there is a high concentration of oxygen, to form **oxyhaemoglobin**. Oxyhaemoglobin is an unstable compound. It breaks down and releases its oxygen in places where the oxygen concentration is low (Fig. 3). This makes haemoglobin very useful in carrying oxygen from the lungs to the tissues.

Blood which contains mainly oxyhaemoglobin is said to be **oxygenated**. Blood with little oxyhaemoglobin is called **deoxygenated**.

Each red cell lives for about 4 months, after which it breaks down. The red haemoglobin changes to a yellow pigment, bilirubin, which is excreted in the bile. The iron from the haemoglobin is stored in the liver. About 200 000 million red cells

wear out and are replaced each day. This is about 1 per cent of the total. Red cells are made by the red bone marrow of certain bones in the skeleton; in the ribs, vertebrae and breastbone for example.

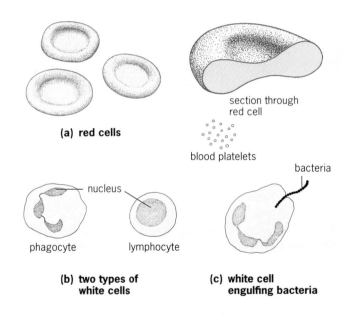

(a) red cells

section through red cell

blood platelets

bacteria

nucleus

phagocyte lymphocyte

(b) two types of white cells

(c) white cell engulfing bacteria

Figure 1 Blood cells

White cells

There are several different kinds of white cell (Figs 1(b) and 2). Most are larger than the red cells, and they all have a nucleus. There is 1 white cell to every 600 red cells and they are made in the same bone marrow that makes red cells. Many of them undergo a process of maturation and development in the thymus gland, lymph nodes or spleen (p. 123). The two most numerous types of white cells are **phagocytes** and **lymphocytes**.

The phagocytes can move about by a flowing action of their cytoplasm and can escape from the blood capillaries into the tissues by squeezing between the cells of the capillary walls. They collect at the site of an infection, engulfing (**ingesting**) and digesting harmful bacteria and cell debris (Fig. 1(c)). In this way they prevent the spread of infection through the body.

One of the functions of lymphocytes is to produce antibodies (p. 125).

Figure 2 Red and white cells from human blood (x2000). The large nucleus can be clearly seen in the white cells.

Platelets

These are pieces of special blood cells budded off in the red bone marrow. They help to clot the blood at wounds and so stop the bleeding (p. 125).

Plasma

The liquid part of the blood is called plasma. It is water with a large number of substances dissolved in it. The ions of sodium, potassium, calcium, chloride and hydrogen carbonate, for example, are present. Proteins such as fibrinogen, albumin and globulins constitute an important part of the plasma. Fibrinogen is needed for clotting (p. 125), and the globulin proteins include the antibodies which combat bacteria and other foreign matter (p. 125). The plasma will also contain varying amounts of food substances such as amino acids, glucose and lipids (fats). There may also be hormones (p. 185) present, depending on the activities taking place in the body. The excretory product, urea, is dissolved in the plasma.

The liver and kidneys keep the composition of the plasma more or less constant, but the amount of digested food, salts and water will vary within narrow limits according to food intake and body activities.

QUESTIONS

1 In what ways are white cells different from red cells in **a** their structure, **b** their function?
2 Where, in the body, would you expect haemoglobin to be combining with oxygen to form oxyhaemoglobin?
3 In what parts of the body would you expect oxyhaemoglobin to be breaking down to oxygen and haemoglobin?
4 Why is it important for oxyhaemoglobin to be an unstable compound, i.e. easily changed to oxygen and haemoglobin?
　　What might be the effect on a person whose diet contained too little iron?

oxygenated blood

HIGH OXYGEN CONCENTRATION

LOW OXYGEN CONCENTRATION

picks up oxygen

releases oxygen

key

oxygenated blood

deoxygenated blood

red blood cell

deoxygenated blood

Figure 3 The function of the red cells

The heart

The heart pumps blood through the circulatory system all round the body. The appearance of the heart from the outside is shown in Fig. 4, Fig. 5 shows the left side cut open, while Fig. 6 is a diagram of a vertical section to show its internal structure. Since the heart is seen as if in a dissection of a person facing you, the left side is drawn on the right.

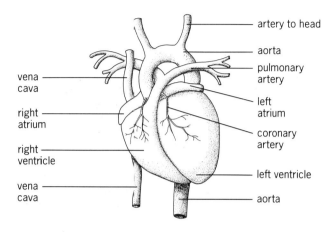

Figure 4 External view of the heart

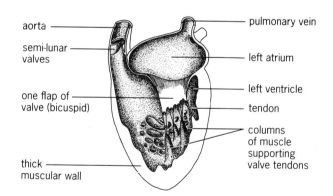

Figure 5 Diagram of the heart cut open (left side)

If you study Fig. 6 you will see that there are four chambers. The upper, thin-walled chambers are the **atria** (singular = atrium) and each of these opens into a thick-walled chamber, the **ventricle**, below.

Blood enters the atria from large veins. The **pulmonary vein** brings oxygenated blood from the lungs into the left atrium. The **vena cava** brings deoxygenated blood from the body tissues into the right atrium. The blood passes from each atrium to its corresponding ventricle, and the ventricle pumps it out into the arteries.

The artery carrying oxygenated blood to the body from the left ventricle is the **aorta**. The **pulmonary artery** carries deoxygenated blood from the right ventricle to the lungs.

In pumping the blood, the muscle in the walls of the atria and ventricles contracts and relaxes (Fig. 7). The walls of the atria

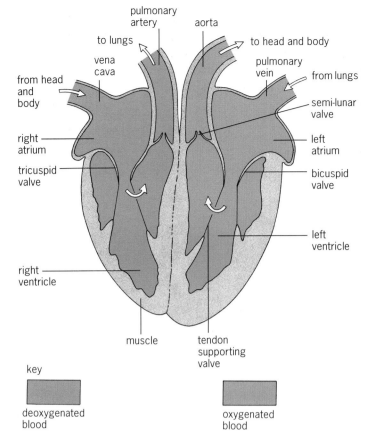

Figure 6 Diagram of the heart, longitudinal section

contract first and force blood into the two ventricles. Then the ventricles contract and send blood into the arteries. The blood is stopped from flowing backwards by four sets of valves. Between the right atrium and the right ventricle is the **tricuspid** (= three flaps) valve. Between the left atrium and left ventricle is the **bicuspid** (= two flaps) valve. The flaps of these valves are shaped rather like parachutes, with 'strings' called **tendons** or **cords** to prevent their being turned inside out.

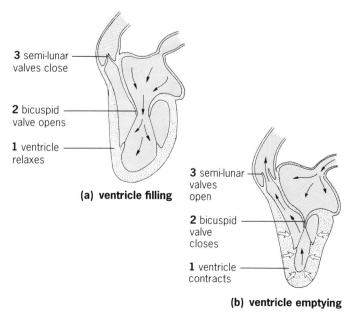

Figure 7 Diagram of heart beat (only the left side is shown)

In the pulmonary artery and aorta are the **semi-lunar** (= half-moon) valves. These each consist of three pockets which are pushed flat against the artery walls when blood flows one way. If blood tries to flow the other way, the 'pockets' fill up and meet in the middle to stop the flow of blood (Fig. 8).

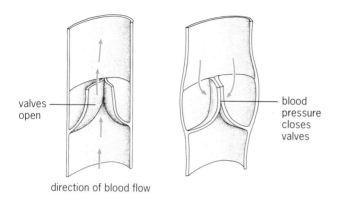

valves open

blood pressure closes valves

direction of blood flow

Figure 8 Action of the semi-lunar valves

When the ventricles contract, blood pressure closes the bicuspid and tricuspid valves and these prevent blood returning to the atria. When the ventricles relax, the blood pressure in the arteries closes the semi-lunar valves so preventing the return of blood to the ventricles.

The heart muscle is supplied with food and oxygen by the **coronary arteries** (Fig. 4).

From the description above, it may seem that the ventricles are filled with blood as a result of the contraction of the atria. In fact, when the ventricles relax, their internal volume increases and they draw in blood from the pulmonary vein or vena cava through the relaxed atria. Atrial contraction then forces the final amount of blood into the ventricles just before ventricular contraction.

Control of heart beat

Heart muscle has a natural rhythmic contraction of its own, about 40 contractions per minute. However, it is supplied by nerves which maintain a faster rate which can be adjusted to meet the body's needs for oxygen. At rest, the normal heart rate may lie between 50 and 100 beats per minute, according to age, sex and other factors. During exercise, the rate may increase to 200 per minute.

The heart beat is initiated by the '**pacemaker**', a small group of specialised muscle cells at the top of the right atrium. The pacemaker receives two sets of nerves from the brain. One group of nerves speeds up the heart rate and the other group slows it down. By this means, the heart rate is adjusted to meet the needs of the body at times of rest, exertion and excitement.

Artificial pacemaker

Some heart diseases lead to serious disturbances of the heart's rhythmic beat. This can be corrected in many cases by an artificial pacemaker. A small electronic device is implanted in the abdomen and wires are connected to the right ventricle. The apparatus produces regular pulses of electricity which stimulate the ventricles to contract at the right rate.

The batteries and the apparatus may need replacing every two or three years.

QUESTIONS

5 Which parts of the heart **a** pump blood into the arteries, **b** stop blood flowing the wrong way?
6 Put the following in the correct order: **a** blood enters arteries, **b** ventricles contract, **c** atria contract, **d** ventricles relax, **e** blood enters ventricles, **f** semi-lunar valves close, **g** tri- and bicuspid valves close.
7 Why do you think that **a** the walls of the ventricles are more muscular than the walls of the atria and **b** the muscle of the left ventricle is thicker than that of the right ventricle? (Consult Fig. 11.)
8 Which important veins are not shown in Fig. 4?
9 Why is a person whose heart valves are damaged by disease unable to take part in active sport?

The circulation

The blood, pumped by the heart, travels all round the body in blood vessels. It leaves the heart in arteries and returns in veins. The route of the circulation is shown in Fig. 9 as a diagram. The blood passes twice through the heart during one complete circuit; once on its way to the body and again on its way to the lungs. The circulation through the lungs is called the **pulmonary** circulation; the circulation round the rest of the body is called the **systemic** circulation. On average, a red cell would go round the whole circulation in 45 seconds. A more detailed diagram of the circulation is shown in Fig. 11.

Arteries

These are fairly wide vessels (Figs 10(a) and 15) which carry blood from the heart to the limbs and organs of the body (Fig. 17). The blood in the arteries, except for the pulmonary arteries, is oxygenated.

Arteries have elastic tissue and muscle fibres in their thick walls. The large arteries, near the heart, have a greater proportion of elastic tissue which allows these vessels to stand up to the surges of high pressure caused by the heart beat. The ripple of pressure which passes down an artery as a result of the heart beat can be felt as a 'pulse' when the artery is near the surface of the

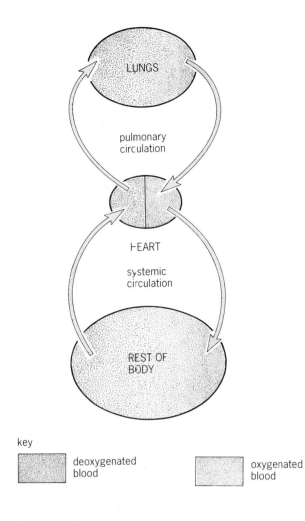

key

deoxygenated blood   oxygenated blood

Figure 9 Blood circulation

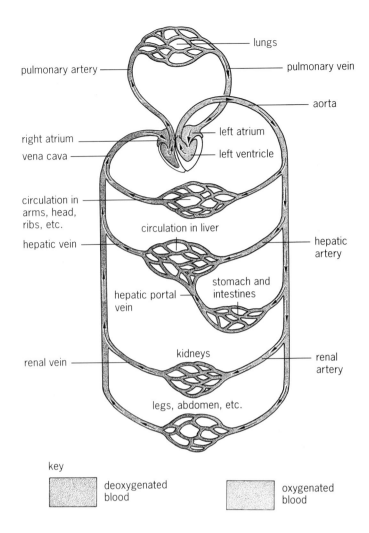

Figure 11 Diagram of human circulation

key

deoxygenated blood   oxygenated blood

body. You can feel the pulse in your radial artery by pressing the finger-tips of one hand on the wrist of the other (Fig. 12).

The arteries divide into smaller vessels called **arterioles**. The small arteries and the arterioles have proportionately less elastic tissue and more muscle fibres than the great arteries. When the muscle fibres of the arterioles contract, they make the vessels narrower and restrict the blood flow. In this way, the distribution of blood to different parts of the body can be regulated. (See p. 148 for an example of this.)

The arterioles divide repeatedly to form a branching network of microscopic vessels passing between the cells of every living tissue. These final branches are called **capillaries**.

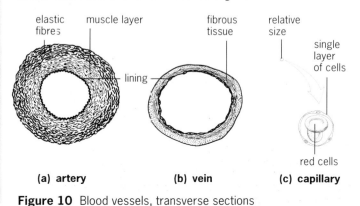

(a) artery   **(b) vein**   **(c) capillary**

Figure 10 Blood vessels, transverse sections

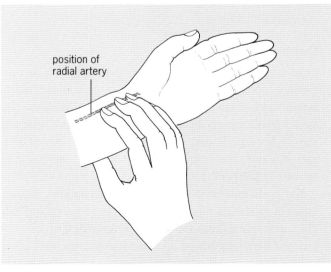

Figure 12 Taking the pulse

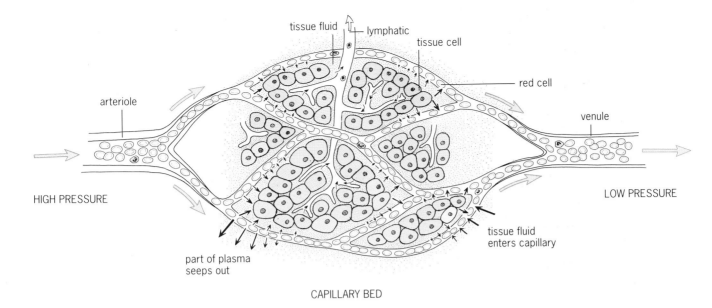

Figure 13 Relationship between capillaries, cells and lymphatics

Capillaries

These are tiny vessels, often as little as 0.001 mm in diameter and with walls only 1-cell thick (Figs 10(c) and 14). Although the blood as a whole cannot escape from the capillary, the thin capillary walls allow some liquid to pass through, i.e. they are permeable. Blood pressure in the capillaries forces part of the plasma out through the walls. The fluid which escapes is not blood, nor plasma, but **tissue fluid**. Tissue fluid is similar to plasma but contains less protein. This fluid bathes all the living cells of the body and since it contains dissolved food and oxygen from the blood, it supplies the cells with their needs (Figs 13 and 16). The tissue fluid eventually seeps back into the capillaries, having given up its oxygen and dissolved food to the cells, but it has now received the waste products of the cells, such as carbon dioxide, which are carried away by the bloodstream.

The capillary network is so dense that no living cell is far from a supply of oxygen and food. The capillaries join up into larger vessels, called **venules**, which then combine to form **veins**.

Veins

Veins return blood from the tissues to the heart (Fig. 17). The blood pressure in them is steady and is less than that in the arteries. They are wider and their walls are thinner, less elastic and less muscular than those of the arteries (Figs 10(b) and 15). They also have valves in them similar to the semi-lunar valves (Fig. 8).

Contraction of body muscles, particularly in the limbs, compresses the thin-walled veins. The valves in the veins prevent the blood flowing backwards when the vessels are compressed in this way.

The blood in most veins is deoxygenated and contains less food but more carbon dioxide than the blood in most arteries. This is because respiring cells have used the oxygen and food and produced carbon dioxide. The pulmonary veins, which return blood from the lungs to the heart, are an exception. They contain oxygenated blood and a reduced level of carbon dioxide.

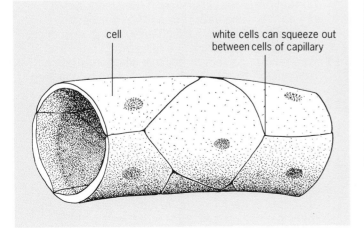

Figure 14 Diagram of blood capillary

Figure 15 Transverse section through a vein and artery. The vein is on the right, the artery on the left. Notice that the wall of the artery is much thicker than that of the vein. The material filling the artery is formed from coagulated red blood cells. These are also visible in two regions of the vein.

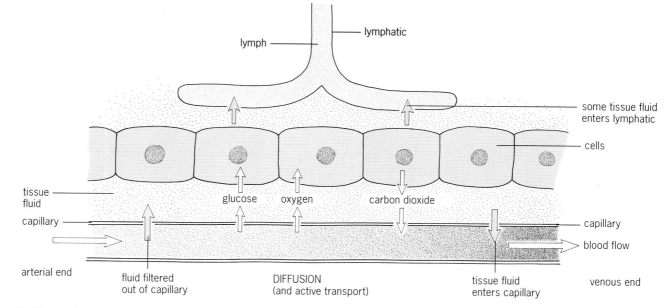

Figure 16 Blood, tissue fluid and lymph

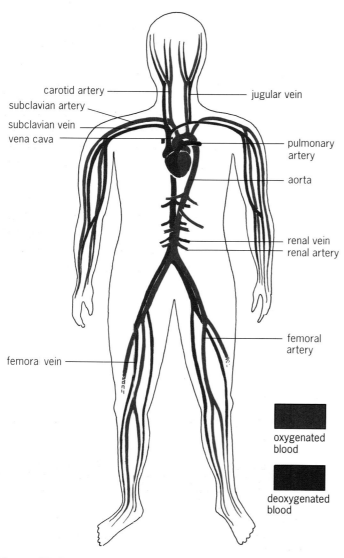

Figure 17 The main arteries and veins of the body

Blood pressure

The pumping action of the heart produces a pressure which drives blood round the circulatory system. In the arteries, the pressure fluctuates with the heart beat, and the pressure wave can be felt as a pulse. The millions of tiny capillaries offer resistance to the blood flow and, by the time the blood enters the veins the surges due to the heart beat are lost and the blood pressure is greatly reduced.

In medical terms, 'blood pressure' means the two pressures, as measured in the arteries, when the heart is contracting and relaxing. Blood pressure is usually measured by inflating a 'cuff' wrapped round the upper arm and observing the air pressure needed to just stop the pulse in the radial artery.

QUESTIONS

10 Starting from the left atrium, put the following in the correct order for circulation of the blood: **a** left atrium, **b** vena cava, **c** aorta, **d** lungs, **e** pulmonary artery, **f** right atrium, **g** pulmonary vein, **h** right ventricle, **i** left ventricle.

11 Why is it not correct to say that all arteries carry oxygenated blood and all veins carry deoxygenated blood?

12 How do veins differ from arteries in **a** their function, **b** their structure?

13 How do capillaries differ from other blood vessels in **a** their structure, **b** their function?

14 Describe the path taken by a molecule of glucose, from the time it is absorbed in the small intestine, and the path taken by a molecule of oxygen absorbed in the lungs, to the time when they both meet in a muscle cell of the leg. (Use Fig. 11.)

The lymphatic system

Not all the tissue fluid returns to the capillaries. Some of it enters blindly ending, thin-walled vessels called **lymphatics** (Fig. 13). The lymphatics from all parts of the body join up to make two large vessels which empty their contents into the blood system as shown in Fig. 18.

The lacteals from the villi in the small intestine (p. 109) join up with the lymphatic system, so most of the fats absorbed in the intestine reach the circulation by this route. The fluid in the lymphatic vessels is called **lymph** and is similar in composition to tissue fluid.

Some of the larger lymphatics can contract, but most of the lymph flow results from the vessels being compressed from time to time when the body muscles contract in movements such as walking or breathing. There are valves in the lymphatics (Fig. 19) like those in the veins, so that when the lymphatics are squashed, the fluid in them is forced in one direction only: towards the heart.

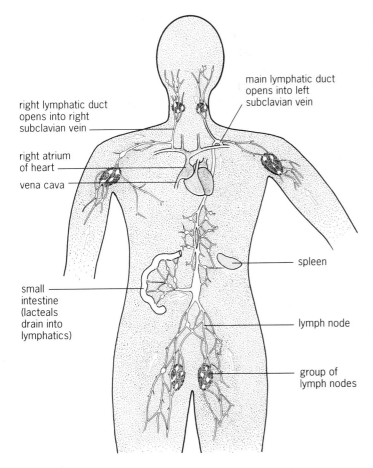

Figure 18 The main drainage routes of the lymphatic system

right lymphatic duct opens into right subclavian vein

main lymphatic duct opens into left subclavian vein

right atrium of heart

vena cava

small intestine (lacteals drain into lymphatics)

spleen

lymph node

group of lymph nodes

At certain points in the lymphatic vessels there are swellings called **lymph nodes** (Fig. 18). Lymphocytes are stored in the lymph nodes and released into the lymph to reach, eventually, the blood system. There are also phagocytes in the lymph nodes. If

bacteria enter a wound and are not ingested by the white cells of the blood or lymph, they will be carried in the lymph to a lymph node and white cells there will ingest them. The lymph nodes thus form part of the body's defence system against infection.

direction of lymph flow

valve

Figure 19 Lymphatic vessel cut open to show valves

Spleen

The spleen is the largest organ in the adult lymphatic system. It is a solid, deep red body about 12 cm long and lies in the left side of the upper abdomen, between the lower ribs and the stomach. It contains lymphatics and blood vessels. Its main functions are to (1) remove worn-out red cells, bacteria and cell fragments from the blood, (2) produce lymphocytes and antibodies (p. 125).

If bacteria or their antigens (p. 125) reach the spleen, the lymphocytes there start to make antibodies against them.

Thymus

The thymus gland lies at the top of the thorax, partly over the heart and lungs. It is an important lymphoid organ particularly in the newborn where it controls the development of the spleen and lymph nodes. The thymus produces lymphocytes and is the main centre for providing immunity against harmful micro-organisms.

After puberty, the thymus becomes smaller but is still an important immunological organ. White cells from the bone marrow are stored in the thymus. Here they undergo cell division to produce a large population of lymphocytes which can be 'programmed' to make antibodies against specific micro-organisms.

QUESTIONS

15 List the things you would expect to find if you analysed a sample of lymph.

16 Describe the course taken by the products of fat digestion from the time they are absorbed in the small intestine to the time they reach the liver to be oxidized for energy. (Use Fig. 10 on p. 110, Fig. 11 on p. 120 and Fig. 18 on this page.)

17 Read pp. 125–6 and then explain why the spleen and thymus are described as 'immunological organs'.

Functions of the blood

It is convenient, at this point, to distinguish between the functions of the blood (1) as the agent replenishing the tissue fluid surrounding the cells, i.e. its role in homeostasis (p. 113), (2) as a circulatory transport system, and (3) as a defence mechanism against harmful bacteria, viruses and foreign proteins.

Homeostatic functions

All the cells of the body are bathed by tissue fluid which is derived from plasma. Tissue fluid supplies the cells with the food and oxygen necessary for their living chemistry. It also removes unwanted substances produced by the cell's metabolism.

The composition of the blood plasma is regulated by the liver and kidneys so that, within narrow limits, the living cells are soaked in a liquid of almost unvarying composition. This provides them with the environment they need and enables them to live and grow in the most favourable conditions. By delivering oxygen and nutrients to the tissue fluid and removing the excretory products, the blood fulfils a homeostatic function (p. 113), maintaining the constancy of the internal environment. (See p. 143 for further details.)

Transport

Transport of oxygen from the lungs to the tissues

In the lungs, the concentration of oxygen is high and so the oxygen combines with the haemoglobin in the red cells, forming oxy-haemoglobin. The blood is now said to be **oxygenated**. When this oxygenated blood reaches tissues where oxygen is being used up, the oxyhaemoglobin breaks down and releases its oxygen to the tissues. Oxygenated blood is a bright red colour; **deoxygenated** blood is dark red.

Transport of carbon dioxide from the tissues to the lungs

The blood picks up carbon dioxide from actively respiring cells and carries it to the lungs. In the lungs, the carbon dioxide escapes from the blood and is breathed out (see p. 134).

The carbon dioxide is carried in the form of hydrogen carbonate ions (HCO_3^-). Some of the hydrogen carbonate is carried in the red cells, but most of it is dissolved in the plasma.

Transport of digested food from the intestine to the tissues

The soluble products of digestion pass into the capillaries of the villi lining the small intestine (p. 109). They are carried in solution by the plasma and, after passing through the liver, enter the main blood system. Glucose, salts, vitamins and some proteins pass out of the capillaries and into the tissue fluid. The cells bathed by this fluid take up the substances they need for their living processes.

Transport of nitrogenous waste from the liver to the kidneys

When the liver changes amino acids into glycogen (p. 113), the amino part of the molecules (NH_2) is changed into the nitrogenous waste product, urea. This substance is carried away in the blood circulation. When the blood passes through the kidneys, much of the urea is removed and excreted (p. 139).

Transport of hormones

Hormones are chemicals made by certain glands in the body (see p. 185). The blood carries these chemicals from the glands which make them, to the organs (target organs) where they affect the rate of activity. For example, a hormone called **insulin**, made in the pancreas, is carried by the blood to the liver and controls how much glucose is stored as glycogen (p. 111).

Table 1 Transport by the blood system

Substance	From	To
oxygen	lungs	whole body
carbon dioxide	whole body	lungs
urea	liver	kidneys
hormones	glands	target organs
digested food	intestine	whole body
heat	abdomen and muscles	whole body

Note that the blood is not directed to a particular organ. A molecule of urea may go round the circulation many times before it enters the renal artery, by chance, and is removed by the kidneys.

Transport of heat

The limbs and head lose heat to the surrounding air. Chemical changes elsewhere in the body produce heat. The blood carries the heat from the warm places to the cold places and so helps to keep an even temperature in all regions. The regions which produce most heat are the abdominal organs, the brain and active muscles.

In addition, the blood system helps to control the body temperature by opening or closing blood vessels in the skin (see p. 148).

QUESTION

18 What substance would the blood **a** gain, **b** lose, on passing through **i** the kidneys, **ii** the lungs, **iii** an active muscle? Remember that respiration (p. 20) is taking place in all these organs.

Defence against infection

Clotting

When tissues are damaged and blood vessels cut, platelets clump together and block the smaller capillaries. The platelets and damaged cells at the wound also produce a substance which acts, through a series of enzymes, on the plasma protein called **fibrinogen**. As a result of this action, the fibrinogen is changed into **fibrin**, which forms a network of fibres across the wound. Red cells become trapped in this network and so form a blood clot. The clot not only stops further loss of blood, but also prevents the entry of harmful bacteria into the wound (Figs 20 and 21).

Figure 20 A red cell trapped in a fibrin network (x8800)

White cells

The phagocytes at the site of the wound, in the blood capillaries or in lymph nodes (p. 123) may ingest harmful bacteria and so stop them entering the general circulation. White cells can squeeze through the walls of capillary vessels and so attack bacteria which get into the tissues, even though the capillaries themselves are not damaged.

The lymphocytes produce antibodies against foreign proteins or cells.

Antibodies

On the surface of all cells there are chemical substances called **antigens**. Lymphocytes produce proteins called **antibodies**, which attack the antigens of bacteria or any alien cells or proteins which invade the body. The antibodies may attach to the surface of the bacteria and make it easier for the phagocytes to ingest them, they may clump the bacteria together or they may neutralise the poisonous proteins (**toxins**) that the bacteria produce.

Each antibody is very **specific**. This means that an antibody which attacks a typhoid bacterium will not affect a pneumonia bacterium. This is illustrated in the form of a diagram in Fig. 22.

Some of the lymphocytes which produced the specific antibodies remain in the circulation for some time and divide rapidly and make antibodies if the same antigen gets into the body again. This means that the body has become **immune** to the disease caused by the antigen and explains why, once you have recovered from measles or chickenpox, for example, you are very unlikely to catch the same disease again. This is called **natural acquired immunity**. You may inherit some forms of immunity or acquire antibodies from your mother's milk (p. 157). This is **innate immunity**.

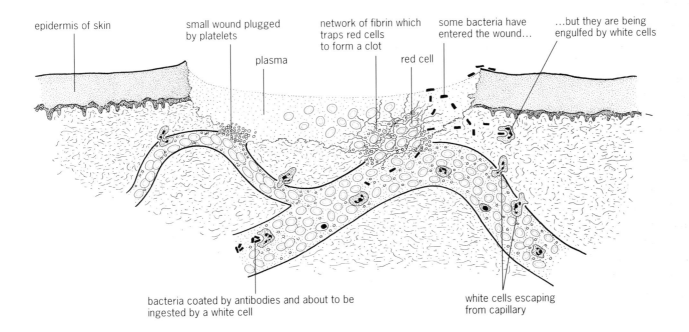

Figure 21 The defence against infection. An area of skin has been damaged and two capillaries broken open.

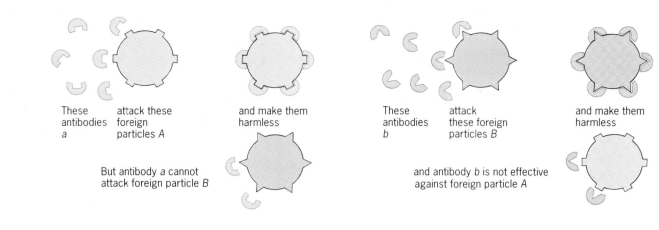

These antibodies *a* attack these foreign particles *A* and make them harmless

But antibody *a* cannot attack foreign particle *B*

These antibodies *b* attack these foreign particles *B* and make them harmless

and antibody *b* is not effective against foreign particle *A*

Figure 22 Antibodies are specific

When you are **inoculated** (vaccinated) against a disease, a harmless form of the bacteria or viruses is introduced into your body. The white cells make the correct antibodies, so that if the real micro-organisms get into the blood, the antibody is already present or very quickly made by the blood.

The material which is injected or swallowed is called a **vaccine** and is either:

1 a harmless form of the micro-organism, e.g. the BCG inoculation against tuberculosis and the Sabin oral vaccine against polio (oral, in this context, means 'taken by mouth');
2 the killed micro-organisms, e.g. the Salk anti-polio vaccine and the whooping cough vaccine;
3 a **toxoid**, i.e. the inactivated toxin from the bacteria, e.g. the diphtheria and tetanus vaccines. (A toxin is the poisonous substance produced by certain bacteria and which causes the disease symptoms.)

The immunity produced by a vaccine is called **artificial acquired immunity**. Immunity acquired as a result of catching a disease or receiving a vaccine is called **active immunity** because the person produces his or her own antibodies.

Some diseases can be prevented or cured by injecting the patient with a serum from a person who has recovered from the disease. Serum is plasma with the fibrinogen removed. Sera are prepared from the plasma given by blood donors. People who have recently received an anti-tetanus inoculation will have made anti-tetanus antibodies in their blood. Some of these people volunteer to donate their blood, but their plasma is separated at once and the red cells returned to their circulation. The anti-tetanus antibodies are then extracted from the plasma and used to treat patients who are at risk of contracting tetanus, as a result of an accident, for example. In a similar way, antibodies against chickenpox and rabies can be produced.

The temporary immunity conferred by these methods is called **passive immunity**, because the antibodies have not been produced by the patient.

B and T lymphocytes

These are the two main types of lymphocyte. Both types recognize antigens and undergo rapid cell division but their responses are different. B cells develop into **plasma cells** and **memory cells**. The plasma cells produce antibodies which attack the antigen in one of the ways already described. Memory cells remain in the circulation and reproduce rapidly if the same antigens invade the body again.

There are several types of T cell. Some activate the B cells and the phagocytes, others attack and destroy foreign cells, e.g. transplanted cells, cancer cells or body cells which have been invaded by viruses.

Transplants and transfusions

Transplants

White cells, lymph nodes, the spleen and the thymus make up the **immune system**, because they have the ability to produce antibodies. When the immune system comes into play the body is said to be making an **immune response**. This response protects the body against invaders such as bacteria, viruses and foreign proteins.

In transplant surgery the donated tissue, such as a kidney or a heart, would be treated as foreign material by the immune system and attacked and destroyed by white cells. Skin grafts present fewer problems because the donated skin is taken from another part of the patient's own body and is recognized as 'self' rather than 'foreign'.

Transplanted kidneys and hearts, however, have to be obtained from donors. To prevent these organs being attacked and 'rejected' by the immune system the patient is given **immunosuppressive drugs**. As the name implies, these

126

drugs suppress the patient's immune response but, in doing so, leave him or her vulnerable to attack by disease-causing bacteria or viruses.

The drug **Cyclosporin A** is more selective than most and suppresses only those antibodies being produced in response to the 'new' invaders. The rest of the immune system functions normally.

There is also a reduced risk of rejection if the donated tissue comes from a close relative.

Transfusions

If somebody loses a lot of blood as a result of an injury or surgical operation, he or she can be given a blood transfusion. Blood taken from a healthy person, the **donor**, is fed into one of the patient's veins. For a transfusion to be successful the blood type of the donor has to match the blood of the patient. If the two blood types do not match, the donor's red cells are clumped in the patient's blood vessels and cause serious harm. The red cells are clumped because they carry antigens on their cell membranes and if the blood types do not match, the antibodies in the patient's blood will act on the donor's red cells and clump them together (Fig. 23).

Figure 23 The red cells are being clumped (x800)

For the purposes of transfusion, people can be put into one of four groups called group A, group B, group AB and group O. The red cells of group A people have antigen A on their cell membranes and the plasma contains anti-B antibodies. Since antibodies are specific, as explained on p. 125, the anti-B antibodies do not attack the A antigens on the red cells.

The red cells of group B people carry the B antigen and their plasma has the anti-A antibody. So if group A cells are introduced into a group B person, they will be clumped by the anti-A antibody. Group AB people have both A and B antigens on their cells but no A or B antibodies in their plasma. The red cells of group O people have neither A nor B antigens but their plasma contains both anti-A and anti-B antibodies (Table 2).

Table 2 Antigens and antibodies

Group	Antigen on cells	Antibody in plasma
A	A	anti-B
B	B	anti-A
AB	A and B	neither
O	neither	anti-A and anti-B

The red cells from group O people can be given to any other group because they have neither the A nor B antigens and so cannot be clumped. Group O people, on the other hand, can receive blood only from their own group because their plasma contains both anti-A and anti-B antibodies.

Group AB people, having neither anti-A nor anti-B antibodies in their plasma, can receive blood from any group. Table 3 shows the acceptable pattern of giving and receiving for the four groups.

Table 3 Blood transfusion

Group	Can donate blood to	Can receive blood from
A	A and AB	A and O
B	B and AB	B and O
AB	AB	all groups
O	all groups	O

It is possible to find out a person's blood group by mixing a drop of his or her blood with anti-A serum and anti-B serum (Fig. 24). Group AB cells will clump in both anti-sera; group O cells will clump in neither; A cells will clump only in anti-A; and B cells only in anti-B.

Figure 24 Blood group test. The subject's blood group is A because it has clumped in B serum.

A donor gives 420 cm^3 of blood from a vein in the arm (Fig. 25). The blood is led into a sterilised bag containing sodium citrate which prevents clotting.

Figure 25 Blood donor. The veins in the upper arm are compressed by using an inflatable 'cuff'. The donor's blood is then tapped from a vein near the inside of the elbow. It takes 5–10 minutes to fill the bag.

The blood is then stored at 5 °C for 10 days, or longer if glucose is added. Before blood is transfused, even though both groups are known, it is carefully tested against the patient's blood to make sure of a good match. Then it is fed into one of the patient's arm veins at the correct rate and temperature.

In a few hours, the donor will have made up his or her blood to the normal volume and in a week or two the red cells will have been replaced.

QUESTIONS

19 What part do white cells play in the defence of the body against infection?
20 Why is it necessary to inoculate people against a disease before they catch it rather than wait until they catch it?
21 Give an example of **a** active immunity, **b** passive immunity.
22 One of the ABO blood groups is sometimes called the 'universal donor'. Which group do you think this is and why?
23 A drop of a person's blood shows clumping in anti-B serum but not in anti-A. What is this person's blood group?

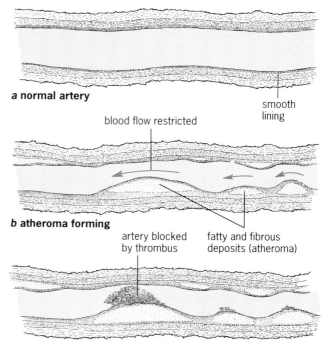

a normal artery

smooth lining

blood flow restricted

b atheroma forming

artery blocked by thrombus

fatty and fibrous deposits (atheroma)

c thrombus forming

Figure 26 Atheroma and thrombus formation

Coronary heart disease

In the lining of the large and medium arteries, deposits of a fatty substance, called **atheroma**, are laid down in patches. This happens to everyone and the patches get more numerous and extensive with age but until one of them actually blocks an important artery, the effects are not noticed. It is not known how or why the deposits form. Some doctors think that fatty substances in the blood pass into the lining. Others believe that small blood clots form on damaged areas of the lining and are covered over by the atheroma patches. The patches may join up to form a continuous layer which reduces the internal diameter of the vessel (Fig. 26).

The surface of a patch of atheroma sometimes becomes rough and causes fibrinogen in the plasma to deposit fibrin on it, so causing a blood clot (a **thrombus**) to form. If the blood clot blocks the **coronary artery** (Fig. 4) which supplies the muscles of the ventricles with blood, it starves the muscles of oxygenated blood and the heart may stop beating. This is a severe heart attack from **coronary thrombosis**. A thrombus might form anywhere in the arterial system, but its effects in the coronary artery and in parts of the brain are the most drastic.

In the early stages of coronary heart disease, the atheroma may partially block the coronary artery and reduce the blood supply to the heart. This can lead to **angina**, i.e. a pain in the chest which occurs during exercise or exertion. This is a warning to the person that he or she is at risk and should take precautions to avoid a coronary heart attack.

Causes of heart disease

Atheroma and thrombus formation are the immediate causes of a heart attack but the long-term causes, which give rise to these conditions, are not well understood.

There is an inherited tendency towards the disease but the disease has increased very significantly in affluent countries in recent years. This makes us think that some features of 'Western' diets or lifestyles might be causing it. Although there is very little direct evidence, the main factors are thought to be smoking, fatty diet, stress and lack of exercise.

Smoking

Statistical studies suggest that smokers are 2–3 times more likely to die from a heart attack than are non-smokers of a similar age (Fig. 27). The carbon monoxide and other chemicals in cigarette smoke may damage the lining of the arteries, allowing atheroma to form, but there is not much direct evidence for this.

Fatty diet

The atheroma deposits contain cholesterol, which is present, combined with proteins, in the blood. Cholesterol plays an essential part in our physiology, but it is known that people with high levels of blood cholesterol are more likely to suffer from heart attacks than people with low cholesterol levels.

Blood cholesterol can be influenced, to some extent, by the amount and type of fat in the diet. Many doctors and dieticians believe that animal fats (milk, cream, butter, cheese, egg yolk, fatty meat) are more likely to raise the blood cholesterol than are the vegetable oils which contain a high proportion of unsaturated fatty acids (p. 99). This is still a matter of some controversy.

Stress

Emotional stress often leads to a raised blood pressure. High blood pressure may increase the rate at which atheroma is formed in the arteries.

Lack of exercise

There is some evidence that regular, vigorous exercise reduces the chances of a heart attack. This could be the result of an improved coronary blood flow. A sluggish blood flow, resulting from lack of exercise, may allow atheroma to form in the arterial lining but, once again, the direct evidence for this is slim.

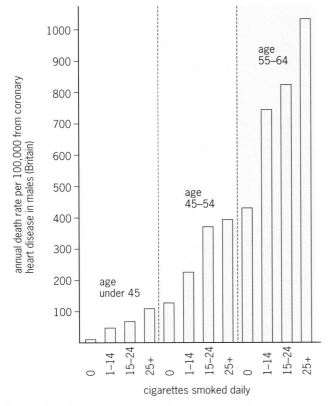

Figure 27 Smoking and heart disease. Obviously, as you get older you are more likely to die from a heart attack, but notice that in any age group, the more you smoke, the higher your chances of dying from heart disease.

Correlation and cause

It is not possible or desirable to conduct experiments on humans to find out, more precisely, the causes of heart attack. The evidence has to be collected from long-term studies on populations of individuals, e.g. smokers and non-smokers. Statistical analysis of these studies will often show a **correlation**, e.g. more smokers, within a given age band, suffer heart attacks than do non-smokers of the same age. This correlation does *not* prove that smoking **causes** heart attacks. It could be argued that people who are already prone to heart attacks for other reasons (e.g. high blood pressure) are more likely to take up smoking. This may strike you as implausible,

but until it can be shown that substances in tobacco smoke do cause an increase in atheroma, the correlation cannot be used on its own to claim a cause and effect.

Nevertheless, there are so many other correlations between smoking and ill-health (bronchitis, emphysema, lung cancer) that the circumstantial evidence against smoking is very strong.

Another example of a positive correlation is between the possession of a television set and heart disease. Nobody would seriously claim that television sets cause heart attacks. The correlation probably reflects an affluent way of life, associated with over-eating, fatty diets, lack of exercise and other factors which may contribute to coronary heart disease.

QUESTIONS

24 **a** What positive steps could you take and **b** what things should you avoid to reduce your risk of coronary heart disease in later life?

25 About 95 per cent of patients with disease of the leg arteries are cigarette smokers. Arterial disease of the leg is the most frequent cause of leg amputation. **a** Is there a correlation between smoking and leg amputation? **b** Does smoking cause leg amputation? **c** In what way could smoking be a possible cause of leg amputation?

26 Fig. 28 shows the relative increase in the rates of four body processes in response to vigorous exercise. **a** How are the changes related physiologically to one another? **b** What other physiological changes are likely to occur during exercise? **c** Why do you think that the increase in blood flow in muscle is less than the total increase in the blood flow?

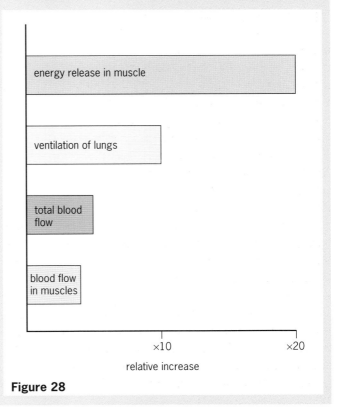

Figure 28

Checklist

- Blood consists of red cells, white cells and platelets suspended in plasma.
- Plasma contains water, proteins, salts, glucose and lipids.
- The red cells carry oxygen. The white cells attack bacteria.
- The heart is a muscular pump with valves, which sends blood round the circulatory system.
- The left side of the heart pumps oxygenated blood round the body.
- The right side of the heart pumps deoxygenated blood to the lungs.
- Blood pressure is essential in order to pump blood round the body.
- Arteries carry blood from the heart to the tissues.
- Veins return blood to the heart from the tissues.
- Capillaries form a network of tiny vessels in all tissues. Their thin walls allow dissolved food and oxygen to pass from the blood into the tissues, and carbon dioxide and other waste substances to pass back into the blood.

- All cells in the body are bathed in tissue fluid which is derived from plasma.
- Lymph vessels return tissue fluid to the lymphatic system and finally into the blood system.
- One function of the blood is to carry substances round the body, e.g. oxygen from lungs to body, food from intestine to body and urea from the liver to the kidneys.
- Lymph nodes, the spleen and the thymus are important immunological organs.
- Antibodies are chemicals made by white cells in the blood. They attack any micro-organisms or foreign proteins which get into the body.
- In blood transfusions it is essential to match the A, B, O blood group of donor and recipient.
- Blockage of the coronary arteries in the heart leads to a heart attack.
- Smoking, fatty diets, stress and lack of exercise may contribute to heart disease.

13 BREATHING

All the processes carried out by the body, such as movement, growth and reproduction, require energy. In animals, this energy can be obtained only from the food they eat. Before the energy can be used by the cells of the body, it must be set free from the chemicals of the food by a process called 'respiration' (see p. 20). Aerobic respiration needs a supply of oxygen and produces carbon dioxide as a waste product. All cells, therefore, must be supplied with oxygen and must be able to get rid of carbon dioxide.

In humans and other mammals, the oxygen is obtained from the air by means of the lungs. In the lungs, the oxygen dissolves in the blood and is carried to the tissues by the circulatory system (p. 117).

Lung structure

The lungs are enclosed in the thorax (chest region) (see Fig. 5, p. 107). They have a spongy texture and can be expanded and compressed by movements of the thorax in such a way that air is sucked in and blown out. The lungs are joined to the back of the mouth by the windpipe or **trachea** (Fig. 1). The trachea divides into two smaller tubes, **bronchi** (singular = bronchus), which enter the lungs and divide into even smaller branches. When these branches are only about 0.2 mm in diameter, they are called **bronchioles** (Fig. 2(a)). These fine branches form a mass of little, thin-walled, pouch-like air sacs called **alveoli** (Figs 2(b), 2(c) and 3).

Rings of gristle (cartilage) stop the trachea and bronchi collapsing when we breathe in. The **epiglottis** and other structures at the top of the trachea stop food and drink from entering the air passages when we swallow (see p. 107).

The epithelium which lines the inside of the trachea, bronchi and bronchioles consists of ciliated cells (p. 7). There are also cells which secrete mucus. The mucus forms a thin film over the internal lining. Dust particles and bacteria become trapped in the sticky mucus film and the mucus is carried upwards, away from the lungs, by the flicking movements of the cilia. In this way, harmful particles

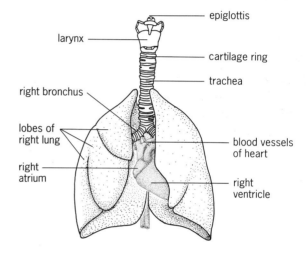

Figure 1 Diagram of lungs, showing position of heart

are prevented from reaching the alveoli. When the mucus reaches the top of the trachea, it passes down the gullet during normal swallowing.

The alveoli have thin elastic walls, formed from a single-cell layer or **epithelium**. Beneath the epithelium is a dense network of capillaries (Fig. 2(c)) supplied with deoxygenated blood (p. 122). This blood, from which the body has taken oxygen, is pumped from the right ventricle, through the pulmonary artery (see Fig. 11, p. 118). In humans, there are about 350 million alveoli, with a total absorbing surface of about 90 m^2. This large absorbing surface makes it possible to take in oxygen and give out carbon dioxide at a rate to meet the body's needs.

QUESTIONS

1 Place the following structures in the order in which air will reach them when breathing in: bronchus, trachea, nasal cavity, alveolus.
2 One function of the small intestine is to absorb food (p. 109). One function of the lungs is to absorb oxygen. Point out the basic similarities in these two structures which help to speed up the process of absorption.

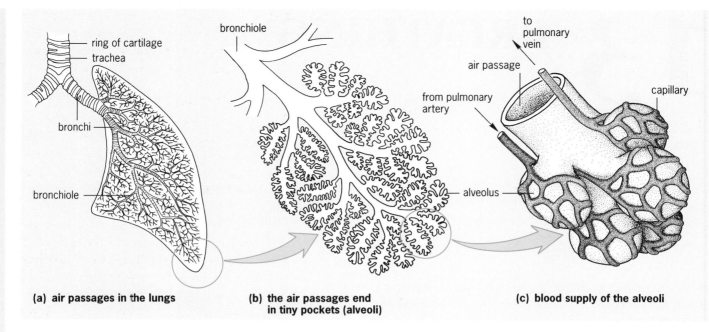

(a) air passages in the lungs

(b) the air passages end in tiny pockets (alveoli)

(c) blood supply of the alveoli

Figure 2 Lung structure

Figure 3 Small piece of lung tissue (x40). The capillaries have been injected with red and blue dye. The networks surrounding the alveoli can be seen.

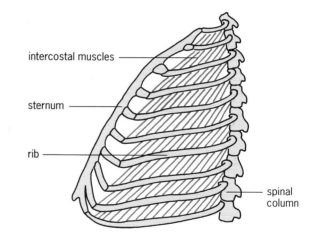

Figure 4 Rib cage seen from left side, showing intercostal muscles

Ventilation of the lungs

The movement of air into and out of the lungs, called **ventilation**, renews the oxygen supply in the lungs and removes the surplus carbon dioxide from them. The lungs contain no muscle fibres and are made to expand and contract by movements of the ribs and diaphragm.

The **diaphragm** is a sheet of tissue which separates the thorax from the abdomen (see Fig. 5, p. 107). When relaxed, it is domed slightly upwards. The ribs are moved by the **intercostal muscles** which run from one rib to the next (Fig. 4). In Fig. 5 the contraction of the intercostal muscles making the ribs move upwards is shown.

Figure 5 Model to show action of intercostal muscles

Inhaling

1 The diaphragm muscles contract and pull it down (Fig. 7(a)).
2 The intercostal muscles contract and pull the rib cage upwards and outwards (Fig. 6(a)).

These two movements make the space in the thorax bigger, so forcing the lungs to expand and draw air in through the nose and trachea.

Exhaling

1 The diaphragm muscles relax, allowing the diaphragm to return to its domed shape (Fig. 7(b)).
2 The intercostal muscles relax, allowing the ribs to move downwards under their own weight (Fig. 6(b)).

The lungs are elastic and shrink back to their relaxed size, forcing air out again.

The outside of the lungs and the inside of the thorax are lined with a smooth membrane called the **pleural membrane**. This produces a thin layer of liquid called **pleural fluid** which reduces the friction between the lungs and the inside of the thorax.

Lung capacity and breathing rate

The total volume of the lungs when fully inflated is about 5 litres in an adult. However, in quiet breathing, when asleep or at rest, you normally exchange only about 500 cm^3 (Fig. 8). During exercise you can take in and expel an extra 3 litres. There is a residual volume of 1.5 litres which cannot be expelled no matter how hard you breathe out.

At rest, you normally inhale and exhale about 16 times per minute. During exercise, the breathing rate may rise to 20 or 30 breaths per minute. The increased rate and depth of breathing during exercise, allows more oxygen to dissolve in the blood and supply the active muscles. The extra carbon dioxide which the muscles put into the blood will be removed by the faster, deeper breathing.

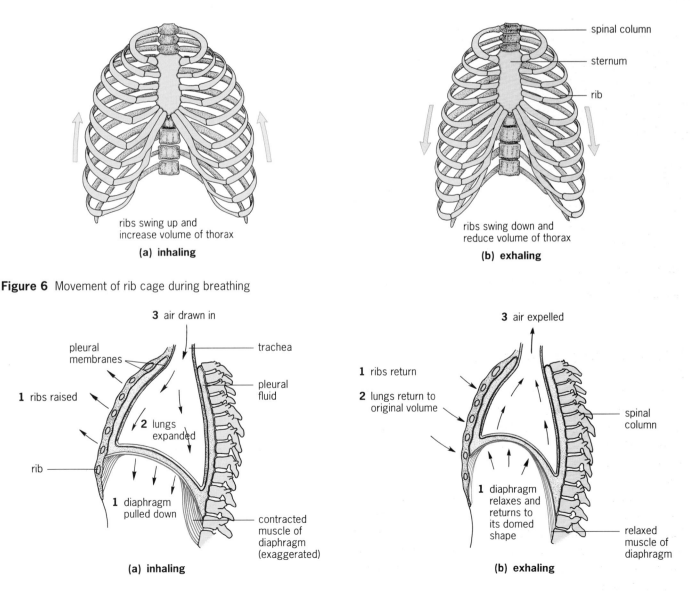

Figure 6 Movement of rib cage during breathing

Figure 7 Diagrams of thorax to show mechanism of breathing

Figure 8 A spirometer. This instrument measures the volume of air breathed in and out of the lungs and can be used to measure oxygen consumption.

QUESTIONS

3 What are the two principal muscular contractions which cause air to be inhaled?
4 Place the following in the correct order: lungs expand, ribs rise, air enters lungs, intercostal muscles contract, thorax expands.
5 During inhalation, which parts of the lung structure would you expect to expand most?

Gaseous exchange

Ventilation refers to the movement of air into and out of the lungs. Gaseous exchange refers to the exchange of oxygen and carbon dioxide which takes place between the air and the blood vessels in the lungs.

The 1.5 litres of residual air in the alveoli is not exchanged during ventilation and the oxygen has to reach the blood capillaries by the slower process of diffusion. The diagram in Fig. 9 shows how oxygen reaches the red blood cells and how carbon dioxide escapes from the blood.

The oxygen combines with the haemoglobin in the red blood cells, forming oxyhaemoglobin (p. 114). The carbon dioxide in the plasma is released when the hydrogencarbonate ions ($-HCO_3$) break down to CO_2 and H_2O.

The capillaries carrying oxygenated blood from the alveoli join up to form the pulmonary vein (see Fig. 11, p. 118), which returns blood to the left atrium of the heart. From here it enters the left ventricle and is pumped all round the body, so supplying the tissues with oxygen.

The process of gaseous exchange in the alveoli does not remove all the oxygen from the air. The air breathed in contains about 21 per cent of oxygen; the air breathed out still contains 16 per cent of oxygen (see Table 1).

Table 1 Changes in the composition of breathed air

	Inhaled %	Exhaled %
Oxygen	21	16
Carbon dioxide	0.04	4
Water vapour	variable	saturated

The remaining 79 per cent of the air consists mainly of nitrogen, whose percentage composition does not change significantly during breathing.

The lining of the alveoli is coated with a film of moisture in which the oxygen dissolves. Some of this moisture evaporates into the alveoli and saturates the air with water vapour. The air you breathe out, therefore, always contains a great deal more water vapour than the air you breathe in. The exhaled air is warmer as well, so in cold and temperate climates, you lose heat to the atmosphere by breathing.

Sometimes the word **respiration** or **respiratory** is used in connection with breathing. The lungs, trachea and bronchi are called the **respiratory system**; a person's rate of breathing may be called his or her **respiration rate**. This use of the word should not be confused with the biological meaning of respiration, namely the release of energy in cells (p. 20). This chemical process is sometimes called **tissue respiration** or **internal respiration** to distinguish it from breathing.

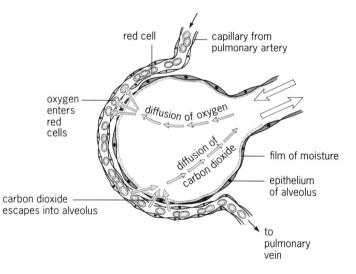

Figure 9 Gaseous exchange in the alveolus

Characteristics of respiratory surfaces

The exchange of oxygen and carbon dioxide across a respiratory surface, as in the lungs or over the gills of a fish, depends on the diffusion of these two gases. Diffusion occurs more rapidly if (1) there is a large surface exposed to the gas, (2) the distance across which diffusion has to take place is small, (3) there is a big difference in the concentrations of the gas at two points brought about by ventilation and (4) there is a rich supply of blood capillaries.

Large surface

The presence of millions of alveoli in the lungs provides a very large surface for gaseous exchange. The many branching filaments in a fish's gills have the same effect.

Thin epithelium

There is only a 2-cell layer, at the most, separating the air in the alveoli from the blood in the capillaries (Fig. 9). Thus, the distance for diffusion is very short.

Ventilation

Ventilation of the lungs helps to maintain a steep diffusion gradient (p. 29), between the air at the end of the air passages and the alveolar air. The concentration of the oxygen in the air at the end of the air passages is high, because the air is constantly replaced by the breathing actions.

Capillary network

The continual removal of oxygen by the blood in the capillaries lining the alveoli keeps its concentration low. In this way, a steep diffusion gradient is maintained which favours the rapid diffusion of oxygen from the air passages to the alveolar lining.

The continual delivery of carbon dioxide from the blood, into the alveoli and its removal from the air passages by ventilation, similarly maintains a diffusion gradient which promotes the diffusion of carbon dioxide from the alveolar lining into the bronchioles.

The respiratory surfaces of land-dwelling mammals are invariably moist. Oxygen has to dissolve in the thin film of moisture before passing across the epithelium.

QUESTIONS

6 Try to make a clear distinction between 'respiration' (p. 20), 'gaseous exchange' and 'ventilation'. Say how one depends on the other.

7 Describe the path taken by a molecule of oxygen from the time it is breathed in through the nose, to the time it enters the heart in some oxygenated blood.

8 Fig. 9 shows oxygen and carbon dioxide diffusing across an alveolus. What causes them to diffuse in opposite directions? (See p. 29.)

9 In 'mouth to mouth' resuscitation, air is breathed from the rescuer's lungs into the lungs of the person who has stopped breathing. How can this 'used' air help to revive the person?

10 What volume of air might you exchange in a minute **a** while resting **b** during exercise?

Smoking

The short-term effects of smoking cause the bronchioles to constrict and the cilia lining the air passages to stop beating. The smoke also makes the lining produce more mucus. Nicotine, the addictive component of tobacco smoke, produces an increase in the rate of the heart beat and a rise in blood pressure. It may, in some cases cause an erratic and irregular heart beat.

The long-term effects of smoking may take many years to develop but they are severe, disabling and often lethal.

Lung cancer

Although all forms of air pollution are likely to increase the chances of lung cancer, many scientific studies show, beyond all reasonable doubt, that the vast increase in lung cancer (4000 per cent in the last century) is almost entirely due to cigarette smoking (Fig. 10).

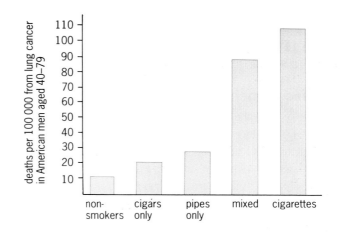

Figure 10 Smoking and lung cancer. Cigar and pipe smokers are probably at less risk because they often do not inhale. But notice that their death rate from lung cancer is still twice that of non-smokers.

There are at least 17 substances in tobacco smoke known to cause cancer in experimental animals, and it is now thought that 90 per cent of lung cancer is caused by smoking. Table 2 shows the relationship between smoking cigarettes and the risk of developing lung cancer.

Table 2 Cigarette smoking and lung cancer

Number of cigarettes per day	Increased risk of lung cancer
1–14	×8
15–24	×13
25 +	×25

Emphysema

Emphysema is a breakdown of the alveoli. The action of one or more of the substances in tobacco smoke weakens the walls of the alveoli. The irritant substances in the smoke cause a 'smokers' cough' and the coughing bursts some of the weakened alveoli. In time, the absorbing surface of the lungs is greatly reduced (Fig. 11). Then the smokers cannot oxygenate their blood properly and the least exertion makes them breathless and exhausted.

Figure 11 Emphysema. (a) Normal lung tissue showing a bronchiole and about 25 alveoli (x200).

(b) Lung tissue from a person with emphysema. This is the same magnification as (a). The alveoli have broken down leaving only about five air sacs which provide a much reduced absorbing surface.

Chronic bronchitis

The smoke stops the cilia in the air passages from beating and so the irritant substances in the smoke and the excess mucus collect in the bronchi. This leads to the inflammation known as **bronchitis**. Over 95 per cent of people suffering from bronchitis are smokers and they have a 20 times greater chance of dying from bronchitis than non-smokers.

Heart disease

Coronary heart disease is the leading cause of death in most developed countries. It results from a blockage of coronary arteries by fatty deposits. This reduces the supply of oxygenated blood to the heart muscle and sooner or later leads to heart failure (see p. 126). High blood pressure, diets with too much animal fat, and lack of exercise are also thought to be causes of heart attack, but about a quarter of all deaths due to coronary heart disease are thought to be caused by smoking (Fig. 27, p. 127).

The nicotine and carbon monoxide from cigarette smoke increase the tendency for the blood to clot and so block the coronary arteries, already partly blocked by fatty deposits. The carbon monoxide increases the rate at which the fatty material is deposited in the arteries.

Other risks

About 95 per cent of patients with disease of the leg arteries are cigarette smokers and this condition is the most frequent cause of leg amputations.

Strokes due to arterial disease in the brain are more frequent in smokers.

Cancer of the bladder, ulcers in the stomach and duodenum, tooth decay, gum disease and tuberculosis all occur more frequently in smokers.

Babies born to women who smoke during pregnancy are smaller than average, probably as a result of reduced oxygen supply caused by the carbon monoxide in the blood. In smokers, there is twice the frequency of miscarriages, a 50 per cent higher still-birth rate and a 26 per cent higher death rate of babies (Fig. 12).

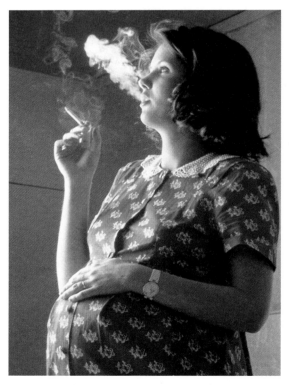

Figure 12 Cigarette smoke can harm the unborn baby. Pregnant women who smoke may have smaller babies and a higher chance of miscarriage or still-birth.

A recent estimate is that one in every three smokers will die as a result of their smoking habits. Those who do not die at an early age will probably be seriously disabled by one of the conditions described above.

Passive smoking

It is not only the smokers themselves who are harmed by tobacco smoke. Non-smokers in the same room are also affected. One study has shown that children whose parents both smoke, breathe in as much nicotine as if they were themselves smoking 80 cigarettes a year.

Statistical studies also suggest that the non-smoking wives of smokers have an increased chance of lung cancer.

Reducing the risks

By giving up smoking, a person who smokes up to 20 cigarettes a day will, after 10 years, be at no greater risk than a non-smoker of the same age. The pipe or cigar smoker, provided he or she does not inhale, is at less risk than a cigarette smoker but still at greater risk than a non-smoker. The risk of disease or death is also reduced by changing to low-tar cigarettes, leaving longer stubs, inhaling less and taking fewer puffs.

Correlations and causes

On p. 127 it was explained that a correlation between two variables does not prove that one of the variables causes the other. The fact that a higher risk of dying from lung cancer is correlated with heavy smoking does not actually prove that smoking is the cause of lung cancer. The alternative explanation is that people who become heavy smokers are, in some way, exposed to other potential causes of lung cancer, e.g. they live in areas of high air pollution or they have an inherited tendency to cancer of the lung. These alternatives are not very convincing, particularly when there is such an extensive list of ailments associated with smoking.

This is not to say that smoking is the only cause of lung cancer or that everyone who smokes will eventually develop lung cancer. There are likely to be complex interactions between life-styles, environments and genetic backgrounds which could lead, in some cases, to lung cancer. Smoking may be only a part, but a very important part, of these interactions.

QUESTIONS

11 What are **a** the immediate effects and **b** the long-term effects of tobacco smoke on the trachea, bronchi and lungs?
12 Why does a regular smoker get out of breath sooner than a non-smoker of similar age and build?
13 If you smoke 20 cigarettes a day, by how much are your chances of getting lung cancer increased?
14 Apart from lung cancer, what other diseases are probably caused by smoking?

PRACTICAL WORK

1 Oxygen in exhaled air

Place a large screw-top jar on its side in a bowl of water (Fig. 13(a)). Put a rubber tube in the mouth of the jar and then turn the jar upside-down, still full of water and with the rubber tube still in it. Start breathing out and when you feel your lungs must be about half empty, breathe the last part of the air down the rubber tubing so that the air collects in the upturned jar and fills it

(Fig. 13(b)). Put the screw top back on the jar under water, remove the jar from the bowl and place it upright on the bench.

Light the candle on the special wire holder (Fig. 13(c)), remove the lid of the jar, lower the burning candle into the jar and count the number of seconds the candle stays alight. Now take a fresh jar, with ordinary air, and see how long the candle stays alight in this.

(a) Lie the jar on its side under the water.

(b) Breathe out through the rubber tube and trap the air in the jar.

(c) Lower the burning candle into the jar until the lid is resting on the rim.

Figure 13 Testing exhaled air for oxygen

Results The candle will burn for about 15–20 seconds in a large jar of ordinary air. In exhaled air it will go out in about 5 seconds.

Interpretation Burning needs oxygen. When the oxygen is used up, the flame goes out. It looks as if exhaled air contains much less oxygen than atmospheric air.

2 Carbon dioxide in exhaled air

Prepare two large test-tubes, A and B, as shown in Fig. 14, each containing a little clear lime water. Put the ends of both rubber tubes at the same time in your mouth and breathe in and out gently through the tubes for about 15 seconds. Notice which tube is bubbling when you breathe out and which one bubbles when you breathe in.

If after 15 seconds there is no difference in the appearance of the lime water in the two tubes, continue breathing through them for another 15 seconds.

Results The lime water in tube B goes milky. The lime water in tube A stays clear.

Interpretation   Carbon dioxide turns lime water milky. Exhaled air passes through tube B. Inhaled air passes through tube A. Exhaled air must, therefore, contain more carbon dioxide than inhaled air.

Figure 14 Comparing the carbon dioxide content of inhaled and exhaled air

3 Volume of air in the lungs

Calibrate a large (5 litre) plastic bottle by filling it with water, half a litre at a time, and marking the water levels on the outside. Fill the bottle with water and put on the stopper. Put about 50 mm depth of water in a large plastic bowl. Hold the bottle upside-down with its neck under water and remove the screw top. Some of the water will run out but this does not matter. Push a rubber tube into the mouth of the bottle (Fig. 15) to position A, shown on the diagram. Take a deep breath and then exhale as much air as possible down the tubing into the bottle. The final water level inside the bottle will tell you how much air you can exchange in one deep breath.

Figure 15 Measuring the volume of air exhaled from the lungs. (A) shows the position of the tube when measuring the maximum usable lung volume. (B) is the position for measuring the volume exchanged in quiet breathing.

Now push the rubber tubing further into the bottle, to position B (Fig. 15), and blow out any water left in the tube. Support the bottle with your hand and breathe quietly in and out through the tube, keeping the water level inside and outside the bottle the same. This will give you an idea of how much air you exchange when breathing normally.

Checklist

- Ventilation is inhaling and exhaling air.
- The ribs, rib muscles and diaphragm make the lungs expand and contract. This causes inhaling and exhaling.
- Air is drawn into the lungs through the trachea, bronchi and bronchioles.
- The vast number of air pockets (alveoli) give the lungs an enormous internal surface area. This surface is moist and lined with capillaries.
- The blood in the capillaries picks up oxygen from the air in the alveoli and gives out carbon dioxide. This is called gaseous exchange.
- Ventilation exchanges the air in the air passages but not in the alveoli.
- Exchange of oxygen and carbon dioxide in the alveoli takes place by diffusion.
- The oxygen is carried round the body by the blood and used by the cells for their respiration.
- During exercise, the rate and depth of breathing increase. This supplies extra oxygen to the muscles and removes their excess carbon dioxide.
- Tobacco smoke causes the bronchioles to constrict, the cilia in their lining to stop beating and excessive mucus to be produced.
- Smoking is correlated with heart disease, bronchitis, emphysema and lung cancer.

14 EXCRETION AND THE KIDNEYS

A great number of chemical reactions take place inside the cells of an organism in order to keep it alive. The products of some of these reactions are poisonous and must be removed from the body. For example, the breakdown of glucose during respiration (p. 20) produces carbon dioxide. This is carried away by the blood and removed in the lungs. Excess amino acids are deaminated in the liver to form glycogen and **urea**, as explained on p. 112. The urea is removed from the tissues by the blood, and expelled by the kidneys.

Urea and similar waste products, like **uric acid**, from the breakdown of proteins, contain the element nitrogen. For this reason they are often called **nitrogenous waste products**.

During feeding, more water and salts are taken in with the food than are needed by the body. So these excess substances need to be removed as fast as they build up.

The hormones produced by the endocrine glands (p. 185) affect the rate at which various body systems work. Adrenaline, for example, speeds up the heart beat. When hormones have done their job, they are modified in the liver and excreted by the kidneys.

The nitrogenous waste products, excess salts and spent hormones are excreted by the kidneys as a watery solution called **urine**.

Excretion is the name given to the removal from the body of (1) the waste products of its chemical reactions, (2) the excess water and salts taken in with the diet, and (3) spent hormones.

Excretion also includes the removal of drugs or other foreign substances taken into the alimentary canal and absorbed by the blood. The term 'excretion' should not usually be applied to the passing out of faeces (p. 110), because most of the contents of the faeces, apart from the bile pigments, have not taken part in reactions in the cells of the body.

QUESTIONS

1 Write a list of the substances that are likely to be excreted from the body during the day.
2 Why do you think that urine analysis is an important part of medical diagnosis?

Excretory organs

Lungs
The lungs supply the body with oxygen, but they are also excretory organs because they get rid of carbon dioxide. They also lose a great deal of water vapour, but this loss is unavoidable and is not a method of controlling the water content of the body.

Kidneys
The kidneys remove urea and other nitrogenous waste from the blood. They also expel excess water, salts, hormones (p. 185) and drugs.

Liver
The yellow/green bile pigment, bilirubin, is a breakdown product of haemoglobin (see p. 112). Bilirubin is excreted with the bile into the small intestine and expelled with the faeces. The pigment undergoes changes in the intestine and is largely responsible for the brown colour of the faeces.

Skin
Sweat consists of water, with sodium chloride and traces of urea dissolved in it. When you sweat, you will expel these substances from your body and so, in one sense, they are being excreted. However, sweating is a response to a rise in temperature and not to a change in the blood composition. In this sense, therefore, skin is not an excretory organ like the lungs and kidneys.

	Excretory products	**Incidental losses**
Lungs	carbon dioxide	water
Kidneys	nitrogenous waste, water, salts, toxins, hormones, drugs	
Liver	bile pigments	
Skin		water, salt, urea

139

The kidneys

Structure

The two kidneys are fairly solid, oval structures. They are red-brown, enclosed in a transparent membrane and attached to the back of the abdominal cavity (Fig. 1). The **renal artery** branches off from the aorta and brings oxygenated blood to them. The **renal vein** takes deoxygenated blood away from the kidneys to the vena cava (see Fig. 11, p. 118). A tube, called the **ureter**, runs from each kidney to the bladder in the lower part of the abdomen.

The renal artery divides up into a great many arterioles and capillaries, mostly in the cortex (Fig. 3). Each arteriole leads to a **glomerulus**. This is a capillary repeatedly divided and coiled, making a knot of vessels (Fig. 4). Each glomerulus is almost entirely surrounded by a cup-shaped structure called a **renal capsule**, which leads to a coiled renal tubule. This tubule, after a series of coils and loops, joins a **collecting duct** which passes through the medulla to open into the pelvis (Fig. 5). There are thousands of glomeruli in the kidney cortex and the total surface area of their capillaries is very great.

A **nephron** is a single glomerulus with its renal capsule, renal tubule and blood capillaries (see Fig. 6).

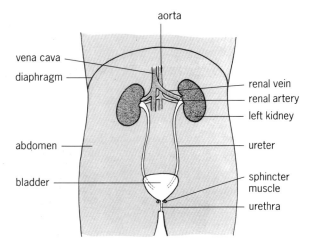

Figure 1 Position of the kidneys in the body

The kidney tissue consists of many capillaries and tiny tubes, called **renal tubules**, held together with connective tissue. If the kidney is cut down its length (sectioned), it is seen to have a dark, outer region called the **cortex** and a lighter, inner zone, the **medulla**. Where the ureter joins the kidney there is a space called the **pelvis** (Fig. 2).

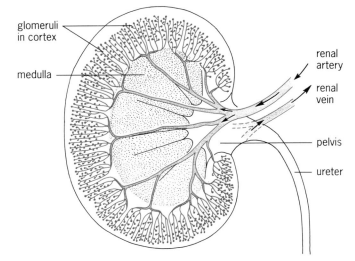

Figure 3 Section through kidney to show distribution of glomeruli

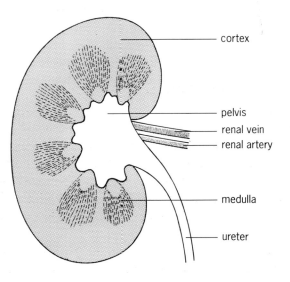

Figure 2 Section through the kidney to show regions

Figure 4 Glomeruli in the kidney cortex (x300). The three glomeruli are surrounded by kidney tubules sectioned at different angles. The light space round each glomerulus represents the renal capsule.

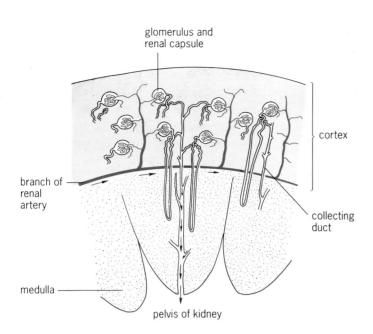

Figure 5 Section through cortex and medulla

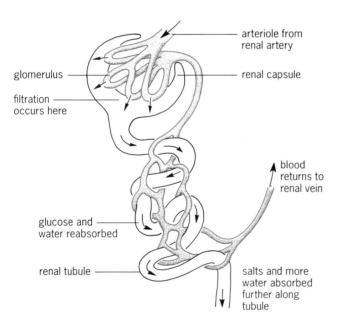

Figure 6 Part of a nephron (glomerulus, renal capsule and renal tubule)

Function of the kidneys

The blood pressure in a glomerulus causes part of the blood plasma to leak through the capillary walls. The red blood cells and the plasma proteins are too big to pass out of the capillary, so the fluid that does filter through is plasma without the protein, i.e. similar to tissue fluid (see p. 119). The fluid thus consists mainly of water with dissolved salts, glucose, urea and uric acid. The process by which the fluid is filtered out of the blood by the glomerulus is called **ultrafiltration**.

The filtrate from the glomerulus collects in the renal capsule and trickles down the renal tubule (Fig. 6). As it does so, the capillaries which surround the tubule absorb back into the blood those substances which the body needs. First, all the glucose is reabsorbed, with much of the water. Then some of the salts are taken back to keep the correct concentration in the blood. The process of absorbing back the substances needed by the body is called **selective reabsorption**.

Salts not needed by the body are left to pass on down the kidney tubule together with the urea and uric acid. So, these nitrogenous waste products, excess salts and water continue down the renal tube into the pelvis of the kidney. From here the fluid, now called **urine**, passes down the ureter to the bladder.

Table 1 shows some of the differences in composition between the blood plasma and the urine. The figures represent average values because the composition of the urine varies a great deal according to the diet, activity, temperature and intake of liquid.

Table 1 Composition of blood plasma and urine

	Plasma %	Urine %
Water	90–93	95.0
Urea	0.03	2.0
Uric acid	0.003	0.05
Ammonia	0.0001	0.05
Sodium	0.3	0.6
Potassium	0.02	0.15
Chloride	0.37	0.6
Phosphate	0.003	0.12

The **bladder** can expand to hold about 400 cm^3 urine. The urine cannot escape from the bladder because a band of circular muscle, called a sphincter, is contracted, so shutting off the exit. When this sphincter muscle relaxes, the muscular walls of the bladder expel the urine through the **urethra**. Adults can control this sphincter muscle and relax it only when they want to urinate. In babies, the sphincter relaxes by a reflex action (p. 181), set off by pressure in the bladder. By 3 years old, most children can control the sphincter voluntarily.

Water balance and osmoregulation

Your body gains water from food and drink. It loses water by evaporation, urination and defecation (p. 110). Evaporation from the skin takes place all the time but is particularly rapid when we sweat. Air from the lungs is saturated with water vapour which is lost to the atmosphere every time we exhale. Despite these gains and losses of water, the concentration of body fluids is kept within very narrow limits by the kidneys, which adjust the concentration of the blood flowing through them.

If the blood is too dilute (i.e. contains too much water), less water is reabsorbed from the renal tubules, leaving more to enter the bladder. Thus, after drinking a lot, a large volume of dilute urine is produced.

If the blood is too concentrated, more water is absorbed back into the blood from the kidney tubules. So, if the body is short of water, e.g. after sweating profusely, only a small quantity of concentrated urine is produced.

A rise in the blood concentration is thought to stimulate a 'thirst' centre in the brain. The drinking which follows this stimulation restores the blood to its correct concentration.

These regulatory processes keep the blood at a steady concentration and together are called **osmoregulation** because they regulate the osmotic strength (see p. 31) of the blood. Osmoregulation is one example of the process of **homeostasis** which is described on p. 143.

Changes in the concentration of the blood are detected by an area in the brain called the **hypothalamus**. If the blood passing through the brain is too concentrated, the hypothalamus stimulates the pituitary gland beneath it to secrete into the blood a hormone (p. 185) called antidiuretic hormone (ADH). When this hormone reaches the kidneys, it causes the kidney tubules to absorb more water from the glomerular filtrate back into the blood. Thus the urine becomes more concentrated and the further loss of water from the blood is reduced. If blood passing through the hypothalamus is too dilute, production of ADH from the pituitary is suppressed and less water is absorbed from the glomerular filtrate.

QUESTIONS

3 Why should a fall in blood pressure sometimes lead to kidney failure?

4 In what ways would you expect the composition of blood in the renal vein to differ from that in the renal artery? (Remember that the cells in the kidney will be respiring.)

5 Where, in the urinary system, do the following take place (answer as precisely as possible): filtration, reabsorption, storage of urine, transport of urine, osmoregulation?

6 In hot weather, when you sweat a great deal, you urinate less often and the urine is a dark colour. In cold weather, when you sweat little, urination occurs more often and the urine is pale in colour. Use your knowledge of kidney function to explain these observations.

7 Trace the path taken by a molecule of urea from the time it is produced in the liver, to the time it leaves the body in the urine (see also p. 118).

8 The heart pumps about 5 litres of blood per minute. At rest, about one-quarter of the heart's output passes through the kidneys. About 180 litres of blood-derived liquid (glomerular filtrate) passes out of the glomeruli, into the renal capsules each day. Use these figures to calculate the percentage volume of blood that is filtered by the kidneys.

The dialysis machine ('artificial kidney')

Kidney failure may result from an accident involving a drop in blood pressure, or from a disease of the kidneys. In the former case, recovery is usually spontaneous, but if it takes longer than 2 weeks, the patient may die as a result of a potassium imbalance in the blood, which causes heart failure. In the case of kidney disease, the patient can survive with only one kidney, but if both fail the patient's blood composition has to be regulated by a dialysis machine. Similarly, the accident victim can be kept alive on a dialysis machine until his or her blood pressure is restored.

In principle, a dialysis machine consists of a long cellulose tube coiled up in a water bath. The patient's blood is led from a vein in the arm and pumped through the cellulose (dialysis) tubing (Figs 7 and 8). The submicroscopic pores in the dialysis tubing allow small molecules, such as those of salts, glucose and urea to leak out into the water bath. Blood cells and protein molecules are too large to get through the pores (see Experiment 5, p. 35). This stage is similar to the filtration process in the glomerulus.

To prevent a loss of glucose and essential salts from the blood, the liquid in the water bath consists of a solution of salts and sugar of the correct composition, so that only the substances above this concentration can diffuse out of the blood into the bathing solution. Thus, urea, uric acid and excess salts are removed.

The bathing solution is also kept at body temperature and is constantly changed as the unwanted blood solutes accumulate in it. The blood is then returned to the patient's arm vein.

A patient with total kidney failure has to spend 2 or 3 nights each week connected to the machine (Fig. 8). With this treatment and a carefully controlled diet, the patient can lead a fairly normal life. A kidney transplant, however, is a better solution because the patient is not obliged to return to the dialysis machine every 3 days or so.

The problem with kidney transplants is to find enough suitable donors of healthy kidneys and to prevent the transplanted kidney from being rejected.

The donor may be a close relative who is prepared to donate one of his or her kidneys (you can survive adequately with one kidney). Alternatively, the donated kidney may be taken from a healthy person who dies, for example, as a result of a road accident. People willing for their kidneys to be used after their death can carry a kidney donor card but the relatives must give their permission for the kidneys to be used.

The problem with rejection is that the body reacts to any transplanted cells or tissues as it does to all foreign proteins and produces lymphocytes which attack and destroy them (see p. 124). This rejection can be overcome by (1) choosing a donor whose tissues are as similar as possible to those of the patient, e.g. a close relative, and (2) using immunosuppressive drugs which suppress the production of lymphocytes and their antibodies against the transplanted organ.

Figure 7 The principle of the kidney dialysis machine

Figure 8 Kidney dialysis machine. The patient's blood is sent by the pump (top right) to the dialyser (bottom right). The patient is adjusting the control box which regulates the temperature and concentration of the dialysing liquid.

Homeostasis

Homeostasis means 'staying similar'. It refers to the fact that the composition of the tissue fluid (p. 119) in the body is kept within narrow limits. The concentration, acidity and temperature of this fluid are being adjusted all the time to prevent any big changes.

On p. 13 it was explained that in living cells, all the chemical reactions are controlled by enzymes. The enzymes are very sensitive to the conditions in which they work. A slight fall in temperature or a rise in acidity (p. 15) may slow down or stop an enzyme from working and thus prevent an important reaction from taking place in the cell.

The cell membrane controls the substances which enter and leave the cell, but it is the tissue fluid which supplies or removes these substances, and it is therefore important to keep the composition of the tissue fluid as steady as possible. If the tissue fluid were too concentrated, it would withdraw water from the cells by osmosis (p. 31) and the body would be dehydrated. If the tissue fluid were too dilute, the cells would take up too much water from it by osmosis and the tissues would become waterlogged and swollen.

Many systems in the body contribute to homeostasis (Fig. 9). The obvious example is the kidneys, which remove substances that might poison the enzymes. The kidneys also control the level of salts, water and acids in the blood. The composition of the blood affects the tissue fluid which, in turn, affects the cells.

Another example of a homeostatic organ is the liver, which regulates the level of glucose in the blood (p. 112). The liver stores any excess glucose as glycogen, or turns glycogen back into glucose if the concentration in the blood gets too low. The brain cells are very sensitive to the glucose concentration in the blood and if the level drops too far, they stop working properly, and the person becomes unconscious and will die unless glucose is injected into the blood system. This shows how important homeostasis is to the body.

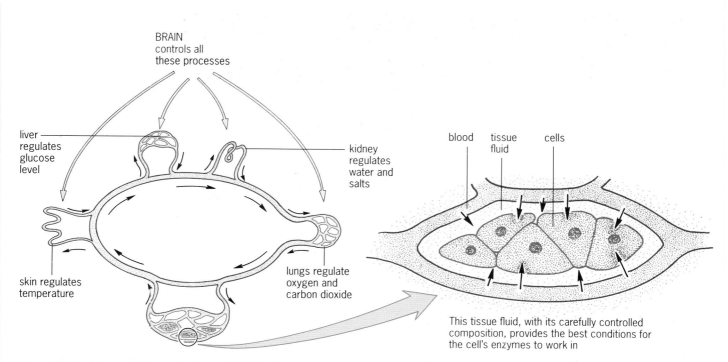

Figure 9 The homeostatic mechanisms of the body

The lungs (p. 131) play a part in homeostasis by keeping the concentrations of oxygen and carbon dioxide in the blood at the best level for the cells' chemical reactions, especially respiration.

The next chapter describes the way in which the skin regulates the temperature of the blood. If the cells were to get too cold, the chemical reactions would become too slow to maintain life. If they became too hot, the enzymes would be destroyed.

The brain has overall control of the homeostatic processes in the body. It checks the composition of the blood flowing through it and if it is too warm, too cold, too concentrated or has too little glucose, nerve impulses or hormones are sent to the organs concerned, causing them to make the necessary adjustments.

QUESTIONS

9 Where will the brain send nerve impulses or hormones if the blood flowing through it **a** has too much water, **b** contains too little glucose, **c** is too warm, **d** has too much carbon dioxide?

10 How does the dialysis machine **a** resemble and **b** differ from the nephron of a kidney in the way it functions?

Checklist

- Excretion is getting rid of unwanted substances produced by chemical reactions in the body or taken in with the diet.

- The lungs excrete carbon dioxide.

- The kidneys excrete urea, unwanted salts and excess water.

- Part of the blood plasma entering the kidneys is filtered out by the capillaries. Substances which the body needs, like glucose, are absorbed back into the blood. The unwanted substances are left to pass down the ureters into the bladder.

- The bladder stores urine, which is discharged at intervals.

- The kidneys help to keep the blood at a steady concentration by excreting excess salts and by adjusting the amounts of water (osmoregulation).

- The kidneys, lungs, liver and skin all help to keep the blood composition the same (homeostasis).

15 THE SKIN, AND TEMPERATURE CONTROL

Functions of the skin
Protection, sensitivity, temperature control.

Structure of the skin
Epidermis, dermis, hair, sweat glands.

Heat balance
Gain and loss of heat.

Temperature control
Vasodilation, vasoconstriction, sweating.

Homeostasis and negative feedback

Hypothermia

The skin forms a continuous layer over the entire body. It makes it difficult for harmful substances, bacteria and fungi to get into the body. It reduces the loss of water from the body and it helps to regulate the body temperature.

Functions of the skin

Protection

A brown or black pigment in the skin absorbs the harmful ultra-violet rays from sunlight. Skin containing only a little pigment may be damaged by these rays. This happens when white-skinned people become sunburned. Many white-skinned people, however, produce extra pigment and so acquire a 'sun-tan' which helps to protect the skin from the effects of ultraviolet light.

The outer layer of cells at the surface of the skin (1) stops harmful bacteria getting into the living tissues beneath and (2) greatly reduces the evaporation of water from the body, so helping to maintain the composition of the body fluids. The sensory endings also have a protective effect as described below.

Sensitivity

Scattered through the skin are a large number of tiny sense organs which give rise to sensations of touch, pressure, heat, cold and pain. These make us aware of changes in our surroundings and enables us to take action to avoid damage, to recognize objects by touch and to manipulate objects with our hands (see p. 171).

Temperature regulation

The skin helps to keep the body temperature within narrow limits. This is done by adjusting the flow of blood near the skin surface and by sweating. These processes are described more fully on p. 148.

QUESTION

1 To what dangers is the body exposed if **a** a small area of skin is damaged, **b** a large area of skin is damaged?

Structure of the skin

There are two main layers in the skin, an outer **epidermis** and an inner **dermis**. The thickness of these two layers depends on which part of the body they are covering. The skin on the palms of the hands and soles of the feet has a very thick epidermis and no hairs (Fig. 1). Over the rest of the body, the epidermis is thinner and has hairs (Fig. 2). The diagram of a section through the skin (Fig. 3) is a generalized one. It shows all the structures that are in skin as a whole, even though some may be absent in a particular area.

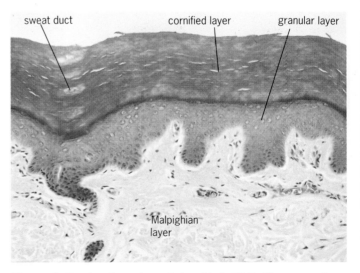

Figure 1 Section through non-hairy skin (x 300). The sweat ducts are contorted as they pass through the cornified layer.

145

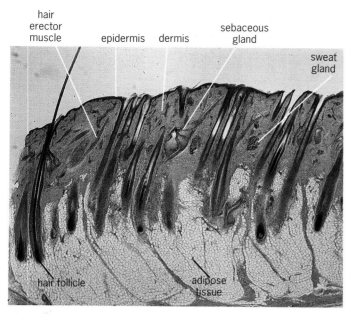

Figure 2 Section through hairy skin (x20)

Epidermis

Malpighian layer

This is the innermost layer of cells in the epidermis. Some of these cells contain the pigment, **melanin**, that gives the skin its colour and helps to absorb ultraviolet light from the sun. The cells of the Malpighian layer keep dividing and producing new cells which are pushed towards the outside of the skin, forming the granular layer.

Granular layer

The cells produced by the Malpighian layer move through the granular layer and then die, so forming the cornified layer.

Cornified layer

This consists of dead cells. It is the outermost layer of the epidermis which helps to cut down evaporation and keep bacteria out. The cells are constantly worn away and replaced from below by the Malpighian and granular layers (Fig. 4).

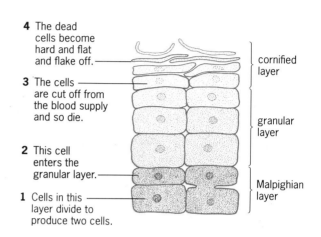

Figure 4 Growth of the epidermis

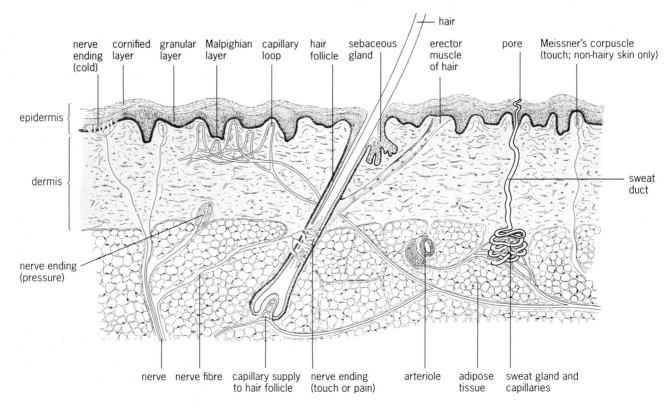

Figure 3 Generalized section through the skin

Dermis

The dermis is a layer of connective tissue containing capillaries, sensory nerve endings, lymphatics, sweat glands and hair follicles.

Capillaries

These bring oxygen and food to the skin and remove its carbon dioxide and nitrogenous waste. They supply the hair follicles and sweat glands. The capillary loops close to the surface, i.e. just beneath the epidermis, play an important part in regulating heat loss from the body (see p. 148).

Sweat glands

A sweat gland is a coiled tube deep in the dermis. When the body temperature is too high, the sweat gland takes up water from the capillaries around it. The water collects in the gland, travels up the **sweat duct**, comes out of a **pore** in the epidermis and on to the skin surface. When the sweat evaporates, it takes heat from the body and so cools it down. Unless the sweat evaporates, it has no cooling effect.

Sweat is mainly water but there are some dissolved salts and urea in it. Some people regard this as a form of excretion (p. 139), but sweating occurs in response to a rise in temperature and not because there is too much water or salt in the body.

Hair follicles

A hair follicle is a deep pit lined with granular and Malpighian cells. The Malpighian cells keep dividing and adding cells to the base of the hair, making it grow. A hair is a lot of cornified cells formed into a tube. The hair follicle has nerve endings which respond when the hair is touched, or give a sensation of pain if the hair is pulled.

In furry mammals, the hairs trap a layer of air close to the body. Air is a bad conductor of heat and so this layer of air insulates the body against heat loss. When the **hair erector muscle** contracts, it pulls the hair more upright. In mammals, this makes the fur stand up more and so provides a thicker layer of insulating air in cold weather. Most of our body is covered with short hairs, but they trap very little air and when the hair erector muscles contract, they produce only 'goose pimples'.

Sebaceous glands

The sebaceous glands open into the top of the hair follicles and produce an oily substance called **sebum**. The function of the sebum is not clear.

It has generally been thought to keep the skin supple and water-proof but there is not much evidence for this.

Sensory nerve endings

These are described more fully on p. 171. Stimulation of some of these sensory endings gives rise to a sensation of pain. This may be unwelcome but it does help us learn to avoid harmful situations.

Fat layer

The fat stored in the adipose tissue beneath the skin not only provides a store of food, but can also form an insulating layer and reduce the heat lost from the body.

QUESTIONS

2 Is a hair made by the epidermis or the dermis?
3 Describe the path taken by a water molecule from the time that it enters a capillary of the skin until the time it ends up in sweat on the surface of the skin.
4 Why do you think a robin looks fatter and fluffier in cold weather?

Heat balance

Body temperature

Different parts of the body are at different temperatures. The skin temperature is lower than the liver temperature and the feet and hands may be colder than the abdomen. However, 'body temperature' usually means the temperature deep inside the body. It is not possible to put a thermometer far inside the body so it is usually placed in the mouth, under the tongue, and held there for 2 minutes with the mouth closed.

The body temperature varies during the day and there is no single 'correct' body temperature. Variations in the range of 35.8–37.7 °C (96.4–99.8 °F) are normal. Body temperatures below 34 °C (93 °F) and above 40 °C (104 °F), if maintained for long, are considered to be dangerous (see p. 149).

The body may gain or lose heat as a result of internal changes or external influences as explained below.

Heat gain

Internal

Many of the chemical reactions in cells release heat. The chief heat producers are the abdominal organs, the brain and contracting muscles. An increase in the muscular activity of the body will result in more heat being produced.

External

Direct heat from the sun will be absorbed by the body. If the air temperature is above 37 °C, the body will absorb heat. Hot food and drink also add heat to the body.

Heat loss

Heat is lost to the air from the exposed surfaces of the body by conduction, convection and radiation. Evaporation from the skin and lungs takes place all the time and is a cause of heat loss. The cold air breathed into the lungs and cold food or drink taken into the stomach all absorb heat from the body.

To a large extent, the heat lost from the body is balanced by the heat absorbed or produced. However, changes in the temperature of the surroundings or in the rate of activity by the animal may upset this balance. In humans, any change in the temperature balance is regulated mainly by changes which take place in the skin.

We also control our heat loss and gain by taking conscious action. We remove clothing or move into the shade to cool down, or put on more clothes or take exercise to keep warm.

QUESTIONS

5 What conscious actions do we take to reduce the heat lost from the body?
6 What sort of chemical reaction in active muscle will produce heat? How does this heat get to other parts of the body? (See pp. 20 and 118.)
7 Draw up a balance sheet to show all the possible ways the human body can gain or lose heat. Make two columns, with 'Gains' on the left and 'Losses' on the right.

Temperature control

Skin structure

The structures in the skin which play a part in temperature control are the small blood vessels, the sweat glands and, in furry mammals, the hairs and hair muscles.

The arterioles and capillaries near the surface of the skin can increase or decrease in width and so increase or reduce the amount of blood flowing through them. An increase in width of the vessel is called **vasodilation**; a reduction in width is **vasoconstriction**.

The sweat glands are not active when the loss and gain of heat are balanced, but as soon as the body temperature starts to rise, the sweat glands of the trunk, limbs and face produce sweat as described above. Sweating from the hands, feet and armpits is a response to emotional stress rather than overheating.

Overheating

If the body gains or produces heat faster than it is losing it, the following processes occur:

Vasodilation   The widening of the blood vessels in the dermis allows more warm blood to flow near the surface of the skin and so lose more heat (Fig. 5(a)).
Sweating   The sweat glands pour sweat on to the skin surface. When this layer of liquid evaporates it takes heat (latent heat) from the body and so cools it down (Fig. 6).

Overcooling

If the body begins to lose heat faster than it can produce it, the following changes occur:

Sweat production stops   Thus the heat lost by evaporation is reduced.
Vasoconstriction   Vasoconstriction of the blood vessels in the skin reduces the amount of warm blood flowing near the surface (Fig. 5(b)).
Shivering   Uncontrollable bursts of rapid muscular contraction in the limbs release heat as a result of the chemical changes in the muscles.

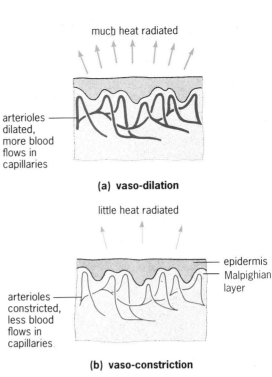

much heat radiated

arterioles dilated, more blood flows in capillaries

(a) vaso-dilation

little heat radiated

epidermis
Malpighian layer

arterioles constricted, less blood flows in capillaries

(b) vaso-constriction

Figure 5 Vasodilation and vasoconstriction

Figure 6 Sweating. During vigorous activity the sweat evaporates from the skin and helps to cool the body. When the activity stops, continued evaporation of sweat may over-cool the body unless it is towelled off.

As a result of these processes, the body temperature of an adult does not usually vary by more than 1 °C either side of 37 °C even when the external temperature changes. The average temperature recorded by a thermometer under the tongue is 36.7 °C and is called the 'core' temperature of the body. The temperature of the skin can change through very wide limits without affecting the core temperature. It is the sensory nerve endings in your dermis which make you feel hot or cold. You cannot consciously detect the small changes in your core temperature.

Homeostasis and negative feedback

Temperature regulation is an example of homeostasis (p. 143). Maintenance of a constant body temperature ensures that vital chemical reactions continue at a predictable rate and do not speed up or slow down when the surrounding temperature changes. Thus the constant-temperature or **homoiothermic** ('warm-blooded') animals, the birds and mammals, have an advantage over the variable-temperature or **poikilothermic** ('cold-blooded') animals. Poikilotherms such as reptiles and insects, can regulate their body temperature to some extent by, e.g. basking in the sun or seeking shade. Nevertheless, if their body temperature falls, their vital chemistry slows down and their reactions become more sluggish. They are then more vulnerable to predators.

The 'price' that homoiotherms have to pay is the intake of enough food to maintain their body temperature, usually above that of their surroundings.

In the hypothalamus (p. 184) of a homoiotherm's brain, there is a thermoregulatory centre. This centre monitors the temperature of the blood passing through it and also receives sensory nerve impulses from temperature receptors in the skin. A rise in body temperature is detected by the thermoregulatory centre and it sends nerve impulses to the skin which result in vasodilation and sweating. Similarly, a fall in body temperature will be detected and will promote impulses which produce vasoconstriction and shivering.

This system of control is called **negative feedback**. The outgoing impulses counteract the effects which produced the incoming impulses. For example, a rise in temperature triggers responses which counteract the rise.

Hypothermia

If the body loses heat faster than the tissues can produce it, the core temperature may fall. Below 35 °C there is an impairment of normal function (e.g. slurred speech, unreasonable behaviour, defective vision) and below 32 °C there is loss of consciousness. This lowering of core temperature is called **hypothermia** and unless the body is rewarmed in a controlled manner, it can result in death.

In young people, hypothermia may occur in conditions such as immersion in water or exposure in wet clothing. In elderly people, hypothermia may result simply from inactivity in inadequately heated surroundings, and from insufficient intake of food.

QUESTIONS

8 a Which structures in the skin of a furry mammal help to reduce heat loss?
b What changes take place in the skin of humans to reduce heat loss?
9 If your body temperature hardly changes at all, why do you sometimes feel hot and sometimes cold?
10 Sweating cools you down only if the sweat can evaporate. **a** In what conditions might the sweat be unable to evaporate from your skin? (See p. 65.) **b** What conditions might speed up the evaporation of sweat and so make you feel very cold?
11 Why, do you think, are most control systems examples of negative feedback rather than positive feedback?

Checklist

- Skin consists of an outer layer of epidermis and an inner dermis.
- The epidermis is growing all the time and has an outer layer of dead cells.
- The dermis contains the sweat glands, hair follicles, sense organs and capillaries.
- Skin (1) protects the body from bacteria and drying out, (2) contains sense organs which give us the sense of touch, warmth, cold and pain, and (3) controls the body temperature.
- Chemical activity in the body, and muscular contractions produce heat.
- Heat is lost to the surroundings by conduction, convection, radiation and evaporation.
- If the body temperature rises too much, the skin cools it down by sweating and vasodilation.
- If the body loses too much heat, vasoconstriction and shivering help to keep it warm.

16 REPRODUCTION

Reproductive systems
Male and female organs. Gamete production.

Fertilization and development
Mating and fertilization. Development of embryo. Placenta.
Antenatal care.
▌ Twins

Birth and parental care
Labour and delivery. Breast-feeding.

Puberty and the menstrual cycle
Hormones and secondary sexual characteristics. Menstruation and menopause.

Family planning and fertility
Birth control.

In vitro fertilization

Reproduction is the process of producing new individuals. Some single-celled creatures can reproduce by simply dividing into two. Some many-celled animals can produce offspring by a process of 'budding', in which part of their body breaks away and grows into a new individual. These are methods of asexual reproduction (see p. 321).

Most animals reproduce sexually. The two sexes, male and female, each produce special types of reproductive cells, called **gametes**. The male gametes are the **sperms** (or **spermatozoa**) and the female gametes are the **ova** (singular = ovum) or eggs (Fig. 1).

To produce a new individual, a sperm has to reach an ovum and join with it (**fuse** with it). The sperm nucleus then passes into the ovum and the two nuclei also fuse. This is called **fertilization**.

The cell formed after the fertilization of an ovum by a sperm is called a **zygote**. A zygote will grow by cell division to produce first an **embryo** and then a fully formed animal (Fig. 2).

The male animal always produces a large number (millions) of sperms, while the female produces a smaller number of eggs. In some animals, such as fish and frogs, many eggs are fertilized and there are a large number of offspring. In mammals, a small number of eggs are fertilized, from one to twenty. In humans, usually only one egg is fertilized at a time; two eggs being fertilized produces twins.

To bring the sperms close enough to the ova for fertilization to take place, there is an act of mating or **copulation**. In mammals this act results in sperms from the male animal being injected into the female. The sperms swim inside the female's reproductive system and fertilize any eggs which are present. The zygote then grows into an embryo inside the body of the female.

0.05 mm

cell membrane

cytoplasm

nucleus

(a) ovum

Figure 1 Human gametes

0.01 mm

tail

(b) sperm

nucleus

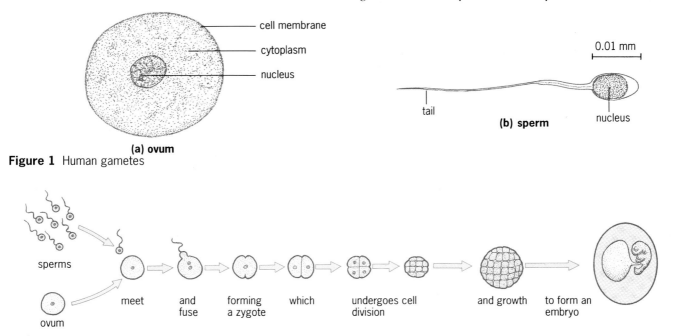

sperms

ovum

meet

and fuse

forming a zygote

which

undergoes cell division

and growth

to form an embryo

Figure 2 Fertilization and development

The human reproductive system

Female

The eggs are produced from the female reproductive organs called **ovaries**. These are two whitish oval bodies, 3–4 cm long. They lie in the lower half of the abdomen, one on each side of the **uterus** (Figs. 3(a) and (b)). Close to each ovary is the expanded, funnel-shaped opening of the **oviduct**, the tube down which the ova pass when released from the ovary. The oviduct is sometimes called the **Fallopian tube**.

The oviducts are narrow tubes that open into a wider tube, the uterus or womb, lower down in the abdomen. When there is no embryo developing in it, the uterus is only about 80 mm long. It leads to the outside through a muscular tube, the **vagina**. The **cervix** is a ring of muscle closing the lower end of the uterus where it joins the vagina. The urethra, from the bladder, opens into the **vulva** just in front of the vagina.

Male

Sperms are produced in the male reproductive organs (Figs 4 and 5), called the **testes** (singular = testis). These lie outside the abdominal cavity in a special sac called the **scrotum**. In this position they are kept at a temperature slightly below the rest of the body. This is the best temperature for sperm production.

The testes consist of a mass of sperm-producing tubes (Fig. 6). These tubes join to form ducts leading to the **epididymis**, a coiled tube about 6 metres long on the outside of each testis. The epididymis, in turn, leads into a muscular **sperm duct**. The two sperm ducts, one from each testis, open into the top of the urethra just after it leaves the bladder. A short, coiled tube called the **seminal vesicle** branches from each sperm duct just before it enters the **prostate gland**, which surrounds the urethra at this point.

The urethra passes through the **penis** and may conduct either urine or sperms at different times. The penis consists of connective tissue with many blood spaces in it. This is called **erectile tissue**.

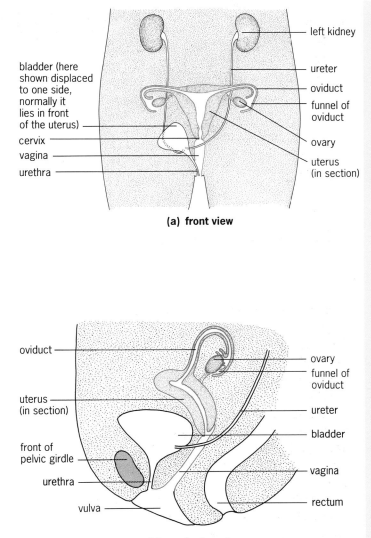

(a) front view

(b) vertical section

Figure 3 The female reproductive organs

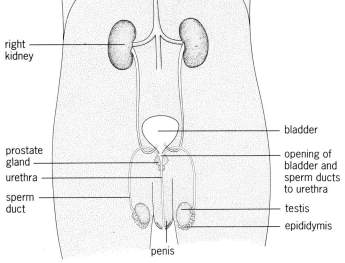

Figure 4 The male reproductive organs; front view

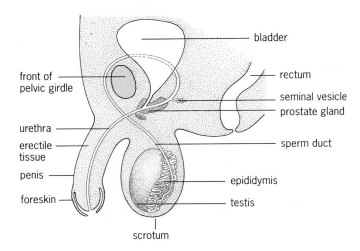

Figure 5 The male reproductive organs; side view

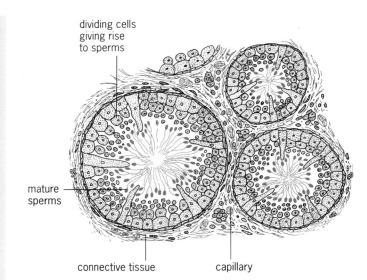

Figure 6 Section through tubules

dividing cells
giving rise
to sperms

mature
sperms

connective tissue   capillary

QUESTIONS

1 How do sperms differ from ova in their structure (see Fig. 1)?
2 List the structures, in the correct order, through which the sperms must pass from the time they are produced in the testis, to the time they leave the urethra.
3 What structures are shown in Fig. 5 which are not shown in Fig. 4?
4 In what ways does a zygote differ from any other cell in the body?

Production of gametes

Sperm production

The lining of the sperm-producing tubules in the testis consists of rapidly dividing cells (Fig. 6). After a series of cell divisions, the cells grow long tails and become sperms (Fig. 7) which pass into the epididymis.

During copulation, the epididymis and sperm ducts contract and force sperms out through the urethra. The prostate gland and seminal vesicle add fluid to the sperms. This fluid plus the sperms it contains is called **semen**, and the ejection of sperms through the penis is called **ejaculation**.

Ovulation

The egg cells (ova) are present in the ovary from the time of birth. No more are formed during the lifetime, but between the ages of 10 and 14 some of the egg cells start to mature and are released, one at a time about every 4 weeks from alternate ovaries. As each ovum matures, the cells round it divide rapidly and produce a fluid-filled sac. This sac is called a **follicle** (Fig. 8) and when mature, it projects from the surface of the ovary like a small blister (Fig. 9). Finally, the follicle bursts and releases the ovum with its coating of cells into the funnel of the oviduct. This is called **ovulation**. From here, the ovum is wafted down the oviduct by the action of cilia (p. 7) in the lining of the tube. If the ovum meets sperm cells in the oviduct, it may be fertilized by one of them.

The released ovum is enclosed in a jelly-like coat called the **zona pellucida** and is still surrounded by a layer of follicle cells. Before fertilization can occur, sperms have to get through this layer of cells and the successful sperm has to penetrate the zona pellucida with the aid of enzymes secreted by the head of the sperm.

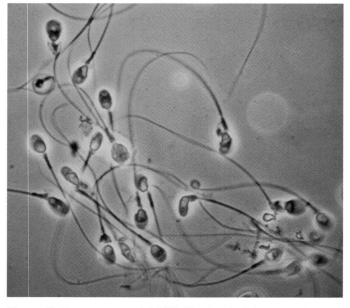

Figure 7 Human sperms (x1200). The head of the sperm has a slightly different appearance when seen in 'side' view or in 'top' view.

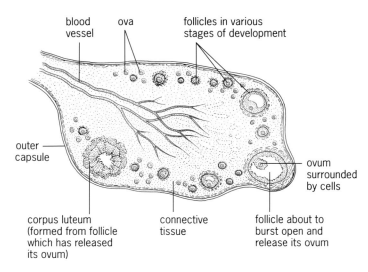

blood
vessel   ova   follicles in various
stages of development

outer
capsule

corpus luteum
(formed from follicle
which has released
its ovum)   connective
tissue   follicle about to
burst open and
release its ovum

ovum
surrounded
by cells

Figure 8 Section through an ovary

Figure 9 Mature follicle as seen in a section through part of an ovary (x30). The ovum is surrounded by follicle cells. These produce the fluid which occupies much of the space in the follicle.

Mating and fertilization

Mating

As a result of sexual stimulation, the male's penis becomes erect. This is due to blood flowing into the erectile tissue round the urethra. In the female, the lining of the vagina produces mucus which makes it possible for the penis to enter. The sensory stimulus (sensation) produced by copulation causes a reflex (p. 181) in the male which results in the ejaculation of semen into the top of the vagina.

The last paragraph is a very simple description of a biological event. In humans, however, the sex act has intense psychological and emotional importance. Most people feel a strong sexual drive, which has little to do with the need to reproduce. Sometimes the sex act is simply the meeting of an urgent physical need. Sometimes it is an experience that both man and woman enjoy together. At its 'highest' level it is both of these, and is also an expression of deeply felt affection within a lasting relationship.

Fertilization

The sperms swim through the cervix and into the uterus by wriggling movements of their tails. They pass through the uterus and enter the oviduct, but the method by which they do this is not known for certain. If there is an ovum in the oviduct, one of the sperms may bump into it and stick to its surface. The sperm then enters the cytoplasm of the ovum and the male nucleus of the sperm fuses with the female nucleus. This is the moment of fertilization and is shown in more detail in Fig. 10. Although a single ejaculation may contain about five hundred million sperms, only a few hundred will reach the oviduct and only one will fertilize the ovum. The function of the others is not understood.

The released ovum is thought to survive for about 24 hours; the sperms might be able to fertilize an ovum for about 2 or 3 days. So there is only a short period of about 4 days each month when fertilization might occur. If this fertile period can be estimated accurately, it can be used to either achieve or avoid fertilization (conception) (p.159).

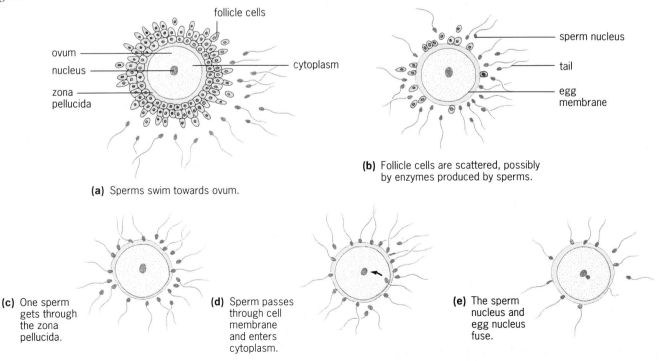

(a) Sperms swim towards ovum.

(b) Follicle cells are scattered, possibly by enzymes produced by sperms.

(c) One sperm gets through the zona pellucida.

(d) Sperm passes through cell membrane and enters cytoplasm.

(e) The sperm nucleus and egg nucleus fuse.

Figure 10 Fertilization of an ovum

5 If a woman starts ovulating at 13 and stops at 50, **a** how many ova are likely to be released from her ovaries, **b** about how many of these are likely to be fertilized?

6 List, in the correct order, the parts of the female reproductive system through which sperms must pass before reaching and fertilizing an ovum.

7 State exactly what happens at the moment of fertilization.

8 If mating takes place **a** 2 days before ovulation, **b** 2 days after ovulation, is fertilization likely to occur? Explain your answers.

Figure 11 Human embryo at the 5-cell stage (x230). The embryo is surrounded by the zona pellucida.

Pregnancy and development

The fertilized ovum first divides into two cells. Each of these divides again, so producing four cells. The cells continue to divide in this way to produce a solid ball of cells (Fig. 11), an early stage in the development of the **embryo**. This early embryo travels down the oviduct to the uterus. Here it sinks into the lining of the uterus, a process called **implantation** (Fig. 12(a)). The embryo continues to grow and produces new cells which form tissues and organs (Fig. 13). After eight weeks, when all the organs are formed, the embryo is called a **fetus**. One of the first organs to form is the heart, which pumps blood round the body of the embryo.

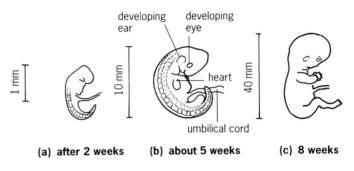

(a) after 2 weeks   **(b) about 5 weeks**   **(c) 8 weeks**

Figure 13 Human embryo: the first 8 weeks

As the embryo grows, the uterus enlarges to contain it. Inside the uterus the embryo becomes enclosed in a fluid-filled sac called the **amnion** or water sac, which protects it from damage and prevents unequal pressures from acting on it (Figs 12(b) and (c)). The oxygen and food needed to keep the embryo alive and growing are obtained from the mother's blood by means of a structure called the placenta.

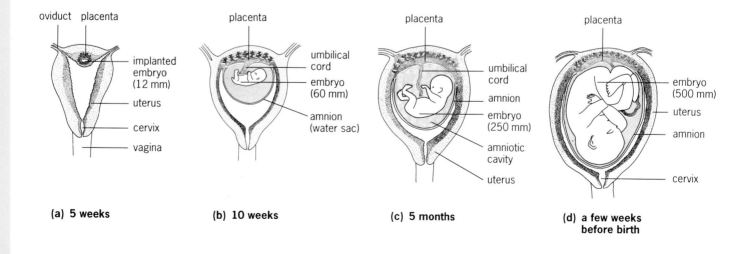

(a) 5 weeks   **(b) 10 weeks**   **(c) 5 months**   **(d) a few weeks before birth**

Figure 12 Growth and development in the uterus (not to scale)

Placenta

Soon after the ball of cells reaches the uterus, some of the cells, instead of forming the organs of the embryo, grow into a disc-like structure, the **placenta** (Fig. 12(c)). The placenta becomes closely attached to the lining of the uterus and is attached to the embryo by a tube called the **umbilical cord** (Fig. 12(c)). After a few weeks, the embryo's heart has developed and is circulating blood through the umbilical cord and placenta as well as through its own tissues. The blood vessels in the placenta are very close to the blood vessels in the uterus so that oxygen, glucose, amino acids and salts can pass from the mother's blood to the embryo's blood (Fig. 14(a)). So the blood flowing in the umbilical vein from the placenta carries food and oxygen to be used by the living, growing tissues of the embryo. In a similar way, the carbon dioxide and urea in the embryo's blood escape from the vessels in the placenta and are carried away by the mother's blood in the uterus (Fig. 14(b)). In this way the embryo gets rid of its excretory products.

There is no direct communication between the mother's blood system and the embryo's. The exchange of substances takes place across the thin walls of the blood vessels. In this way, the mother's blood pressure cannot damage the delicate vessels of the embryo and it is possible for the placenta to select the substances allowed to pass into the embryo's blood. The placenta can prevent some harmful substances in the mother's blood from reaching the embryo. It cannot prevent all of them, however, as is shown by the effects of cigarette smoke and alcohol described below.

Fig. 15 shows the human embryo at 7 weeks surrounded by the amnion and placenta.

> The placenta produces hormones, including oestrogens and progesterone. It is assumed that these hormones play an important part in maintaining the pregnancy and preparing for birth, but their precise function is not known. They may influence the development and activity of the muscle layers in the wall of the uterus and prepare the mammary glands in the breasts for milk production.

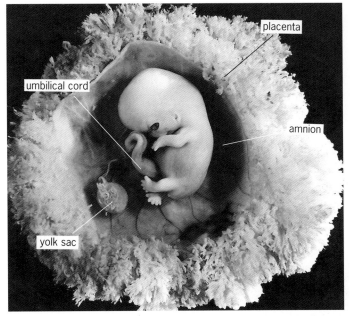

Figure 15 Human embryo, 7 weeks (x1.5). The embryo is enclosed in the amnion. Its limbs, eye and ear-hole are clearly visible. The amnion is surrounded by the placenta; the fluffy-looking structures are the placental villi which penetrate into the lining of the uterus. The umbilical cord connects the embryo to the placenta.

Antenatal care

'Antenatal' or 'prenatal' refers to the period before birth. Antenatal care is the way a woman should look after herself during pregnancy, so that the birth will be safe and her baby healthy.

The mother-to-be should make sure that she eats properly, and perhaps takes more iron and folic acid (p. 97) than she usually does to prevent anaemia. If her job is a light one, she may go on working for the first six months of pregnancy. She should not do heavy work, however, or repeated lifting or stooping.

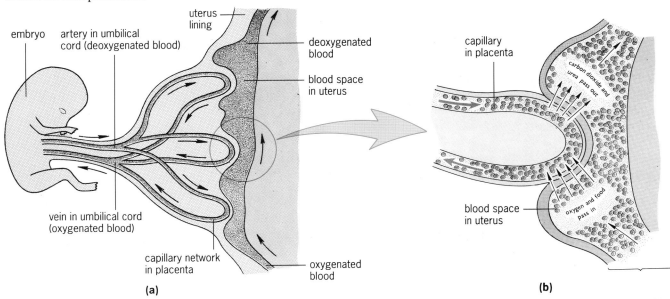

(a)

(b)

Figure 14 The exchange of substances between the blood of the embryo and the mother

Pregnant women who drink or smoke are more likely to have babies with low birth weights. These babies are more likely to be ill than babies of normal weight. Smoking may also make a miscarriage more likely. So a woman who smokes should give up smoking during her pregnancy. Heavy drinking is strongly suspected of damaging the developing brain of the fetus, so it is wise to avoid alcoholic drinks too.

During pregnancy, a woman should not take any drugs unless they are strictly necessary and prescribed by a doctor. In the 1950s, a drug called thalidomide was used to treat the bouts of early morning sickness which often occur in the first three months of pregnancy. Although tests had appeared to show the drug to be safe, it had not been tested on pregnant animals. About 20 per cent of pregnant women who took thalidomide had babies with deformed or missing limbs.

If a woman catches **rubella** (German measles) during the first 4 months of pregnancy, there is a danger that the virus may affect the fetus and cause abortion or still-birth. Even if the baby is born alive, the virus may have caused defects of the eyes (cataracts), ears (deafness) or nervous system. All girls should be vaccinated against rubella to make sure that their bodies contain antibodies to the disease (see p. 123).

Twins

Sometimes a woman releases two ova when she ovulates. If both ova are fertilized, they may form twin embryos, each with its own placenta and amnion. Because the twins come from two separate ova, each fertilized by a different sperm, it is possible to have a boy and a girl. Twins formed in this way are called **fraternal twins**. Although they are both born within a few minutes of each other, they are not more alike than other brothers or sisters.

Another cause of twinning is when a single fertilized egg, during an early stage of cell division, forms two separate embryos. Sometimes these may share a placenta and amnion. Twins formed from a single ovum and sperm must be the same sex, because only one sperm (X or Y, see p. 214) fertilized the ovum. These 'one-egg' twins are sometimes called **identical twins** because, unlike fraternal twins, they will closely resemble each other in every respect.

QUESTIONS

9 In what ways will the composition of the blood in the umbilical vein differ from that in the umbilical artery?

10 An embryo is surrounded with fluid, its lungs are filled with fluid and it cannot breathe. Why doesn't it suffocate?

11 If a mother gives birth to twin boys, does this mean that they are identical twins? Explain.

Birth

From fertilization to birth takes about 38 weeks in humans. This is called the **gestation** period. A few weeks before the birth, the fetus has come to lie head downwards in the uterus, with its head just above the cervix (Figs 12(d) and 16). When the birth starts, the uterus begins to contract rhythmically. This is the beginning of what is called 'labour'. These regular rhythmic contractions become stronger and more frequent. The opening of the cervix gradually widens enough to let the baby's head pass through and the contractions of the uterus are assisted by muscular contractions of the abdomen. The water sac breaks at some stage in labour and the fluid escapes through the vagina. Finally, the muscular contractions of the uterus and abdomen push the baby head-first through the widened cervix and vagina (Fig. 17). The umbilical cord which still connects the child to the placenta is tied and cut. Later, the placenta breaks away from the uterus and is pushed out separately as the 'afterbirth'.

The sudden fall in temperature felt by the newly born baby stimulates it to take its first breath and it usually cries. In a few days, the remains of the umbilical cord attached to the baby's abdomen shrivels and falls away, leaving a scar in the abdominal wall, called the navel.

Figure 16 Model of human fetus just before birth. The cervix and vagina seem to provide narrow channels for the baby to pass through but they widen quite naturally during labour and delivery.

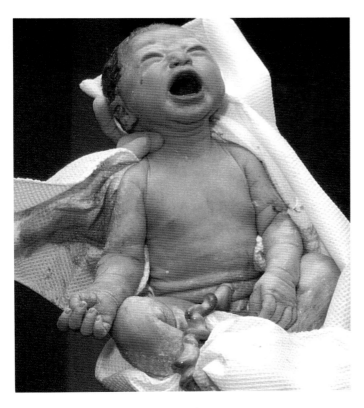

Figure 17 A new-born baby. The umbilical cord is still intact.

Induced birth

Sometimes, when a pregnancy has lasted for more than 38 weeks or when examination shows that the placenta is not coping with the demands of the fetus, birth may be induced. This means that it is started artificially.

This is often done by carefully breaking the membrane of the amniotic sac. Another method is to inject a hormone, **oxytocin**, into the mother's veins. Either of these methods brings on the start of labour. Sometimes both are used together.

Amniocentesis

As a woman gets older, she is more likely to have a baby with Down's syndrome (p. 205). This can be detected fairly early in the pregnancy (14–15 weeks), by taking a sample of amniotic fluid. Then the chromosomes of the fetal cells floating in it are counted under the microscope. The cells of a Down's syndrome fetus have 47 chromosomes in their nuclei, instead of the normal 46. The mother of a Down's syndrome fetus can then decide whether to end her pregnancy rather than give birth to a mentally handicapped child.

In amniocentesis, a tube is inserted through the wall of the abdomen, through the uterus and into the amnion. A local anaesthetic is used. In 1–2 per cent of cases, amniocentesis leads to a spontaneous abortion. The risk of this has to be balanced against the risk of having an unwanted child. For women over 40, amniocentesis is the lesser risk. Between 35 and 40, the risks are more evenly balanced.

The chromosomes of fetal cells can also be used to find out the sex of the fetus. This is useful where there is a family history of sex-linked genetic disease such as haemophilia (p. 215).

Analysis of the sample of amniotic fluid removed by amniocentesis can also detect certain chemicals which, if present, mean that the fetus has other defects.

A newer technique (chorionic villus sampling), takes a sample of the placental villi to look for abnormal chromosomes. One advantage is that the sampling tube can be passed through the cervix and does not enter the amnion. Placental sampling can also be carried out earlier in pregnancy than amniocentesis. So if an abnormality is found and the mother decides the pregnancy should be ended, the experience may be less distressing for her. The risks of this technique versus the possible benefit are still being studied and have recently been called into question.

Feeding and parental care

Within the first 24 hours after birth, the baby starts to suck at the breast. During pregnancy the mammary glands (breasts) enlarge as a result of an increase in the number of milk-secreting cells. No milk is secreted during pregnancy, but the hormones which start the birth process also act on the milk-secreting cells of the breasts. The breasts are stimulated to release milk by the first sucklings. The continued production of milk is under the control of hormones, but the amount of milk produced is related to the quantity taken by the child during suckling.

Milk contains the proteins, fats, sugar, vitamins and salts that babies need for their energy requirements and tissue-building, but there is too little iron present for the manufacture of haemoglobin. All the iron needed for the first weeks or months is stored in the liver of the fetus during gestation.

The liquid produced in the first few days is called **colostrum**. It is sticky and yellow, and contains more protein than the milk produced later. It also contains some of the mother's antibodies.

The mother's milk supply increases with the demands of the baby, up to 1 litre per day. It is gradually supplemented and eventually replaced entirely by solid food, a process known as **weaning**.

Cows' milk is not wholly suitable for human babies. It has more protein, sodium and phosphorus, and less sugar, vitamin A and vitamin C than human milk. Manufacturers modify the components of dried cows' milk to resemble human milk more closely and this makes it more acceptable if the mother cannot breast-feed her baby.

Cows' milk and proprietary dried milk both lack human antibodies, whereas the mother's milk contains antibodies to any diseases from which she has recovered. It also carries white cells which produce antibodies or ingest bacteria. These antibodies are important in defending the baby against infection at a time when its own immune responses are not fully developed. Breast-feeding provides milk free from bacteria, whereas bottle-feeding carries the risk of introducing bacteria which cause intestinal diseases. Breast-feeding also offers emotional and psychological benefits to both mother and baby (Fig. 18).

Figure 18 Breast-feeding helps to establish an emotional bond between mother and baby.

Most young mammals are independent of their parents after a few weeks or months even though they may stay together as a family group. In humans, however, the young are dependent on their parents for food, clothing and shelter for many years. During this long period of dependence, the young learn to talk, read and write and learn a great variety of other skills that help them to survive and be self-sufficient (Fig. 19).

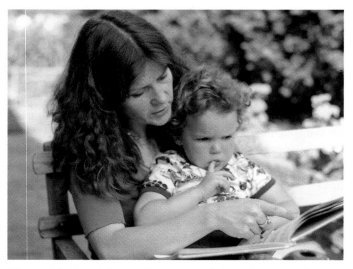

Figure 19 Parental care. Human parental care includes a long period of education.

Puberty and the menstrual cycle

Puberty

Although the ovaries of a young girl contain all the ova she will ever produce, they do not start to be released until she reaches an age of about 10–14 years. This stage in her life is known as **puberty**.

At about the same time as the first ovulation, the ovary also releases female sex hormones into the bloodstream. These hormones are called **oestrogens** and when they circulate round the body, they bring about the development of **secondary sexual characteristics**. In the girl these are the increased growth of the breasts, a widening of the hips and the growth of hair in the pubic region and in the armpits. There is also an increase in the size of the uterus and vagina. Once all these changes are complete, the girl is capable of having a baby.

Puberty in boys occurs at about the same age as in girls. The testes start to produce sperms for the first time and also release a hormone, called **testosterone**, into the bloodstream. The male secondary sexual characteristics which begin to appear at puberty are enlargement of the testes and penis, deepening of the voice, growth of hair in the pubic region, armpits, chest and, later on, the face. In both sexes there is a rapid increase in the rate of growth during puberty.

In addition to the physical changes at puberty, there are emotional and psychological changes associated with the transition from being a child to becoming an adult, i.e. the period of adolescence. Most people adjust to these changes smoothly and without problems. Sometimes, however, a conflict arises between having the status of a child but the sexuality and feelings of an adult.

The menstrual cycle

The ovaries release an ovum about every 4 weeks. As each follicle develops, the amount of oestrogens produced by the ovary increases. The oestrogens act on the uterus and cause its lining to become thicker and develop more blood vessels. These are changes which help an early embryo to implant as described on p. 154.

Two hormones, produced by the **pituitary gland** at the base of the brain (p. 187), promote ovulation. The hormones are **follicle-stimulating hormone** (FSH) and **luteinising hormone** (LH). They act on a ripe follicle and stimulate maturation and release of the ovum.

Once the ovum has been released, the follicle which produced it develops into a solid body called the **corpus luteum**. This produces a hormone called **progesterone**, which affects the uterus lining in the same way as the oestrogens, making it grow thicker and produce more blood vessels.

If the ovum is fertilized, the corpus luteum continues to release progesterone and so keeps the uterus in a state suitable for implantation. If the ovum is not fertilized, the corpus luteum stops producing progesterone. As a result, the thickened lining of the uterus breaks down and loses blood which escapes through the cervix and vagina. This is known as a **menstrual period**. The appearance of the first menstrual period is one of the signs of puberty in girls. The events in the menstrual cycle are shown in Fig. 20.

Menopause

Between the ages of 40 and 55, the ovaries cease to release ova or produce hormones. As a consequence, menstrual periods cease, the woman can no longer have children, and sexual desire is gradually reduced.

QUESTIONS

12 Apart from learning to talk, what other skills might young humans develop that would help them when they are no longer dependent on their parents in **a** an agricultural society, **b** an industrial society?

13 From the list of changes at puberty in girls, select those which are related to child-bearing and say what part you think they play.

14 One of the first signs of pregnancy is that the menstrual periods stop. Explain why you would expect this.

Family planning and fertility

As little as 4 weeks after giving birth, it is possible, though unlikely, that a woman may conceive again. Frequent breast-feeding may reduce the chances of conception. Nevertheless, it would be possible to have children at about 1-year intervals. Most people do not want, or cannot afford, to have as many children as this. All human communities, therefore, practise some form of birth control to space out births and limit the size of the family.

Natural methods of family planning

If it were possible to know exactly when ovulation occurred, intercourse could be avoided for 3–4 days before and 1 day after ovulation (see p. 152). At the moment, however, there is no simple, reliable way to recognize ovulation, though it is usually 12–16 days before the onset of the next menstrual period. By keeping careful records of the intervals between menstrual periods, it is possible to calculate a potentially fertile period of about 10 days in mid-cycle, when sexual intercourse should be avoided if children are not wanted.

On its own, this method is not very reliable but there are some physiological clues which help to make it more accurate. During or soon after ovulation, a woman's temperature rises about 0.5 °C. It

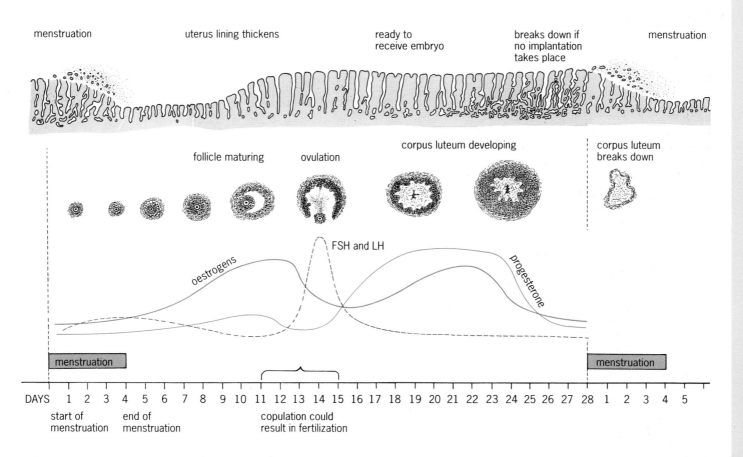

Figure 20 The menstrual cycle

is reasonable to assume that one day after the temperature returns to normal, a woman will be infertile. Another clue comes from the type of mucus secreted by the cervix and lining of the vagina. As the time for ovulation approaches, the mucus becomes more fluid. Women can learn to detect these changes and so calculate their fertile period.

By combining the 'calendar', 'temperature' and 'mucus' methods, it is possible to achieve about 80 per cent 'success', i.e. only 20 per cent unplanned pregnancies. Highly motivated couples may achieve better rates of success and, of course, it is a very helpful way of finding the fertile period for couples who do want to conceive.

Artificial methods of family planning

Contraception

The sheath or condom
A thin rubber sheath is placed on the erect penis before sexual intercourse. The sheath traps the sperms and prevents them from reaching the uterus.

The diaphragm
A thin rubber disc, placed in the vagina before intercourse, covers the cervix and stops sperms entering the uterus. Condoms and diaphragms, used in conjunction with chemicals that immobilize sperms, are about 95 per cent effective.

Spermicides
Spermicides are chemicals which, though harmless to the tissues, do kill or immobilize sperms. The spermicide, in the form of a cream, gel or foam is placed in the vagina. On their own, spermicides are not very reliable but, in conjunction with condoms or diaphragms, they are effective.

Intra-uterine device (IUD)
A small metal or plastic strip bent into a loop or coil is inserted and retained in the uterus, where it probably prevents implantation of a fertilized ovum. It is about 98 per cent effective but there is a small risk of developing uterine infections, particularly if sexual relationships are promiscuous.

The contraceptive pill
The pill contains chemicals which have the same effect on the body as the hormones oestrogen and progesterone. When mixed in suitable proportions these hormones suppress ovulation (see 'Feedback', p. 188) and so prevent conception. The pills need to be taken each day for the 21 days between menstrual periods.

There are many varieties of contraceptive pill in which the relative proportions of oestrogen- and progesterone-like chemicals vary. They are 99 per cent effective but long-term use of some types is thought to increase the risk of cancer of the breast and cervix.

Sterilization

Vasectomy
This is a simple and safe surgical operation in which the man's sperm ducts are cut and the ends sealed. This means that his semen contains the secretions of the prostate gland and seminal vesicle but no sperms and so cannot fertilize an ovum. Sexual desire, erection, copulation and ejaculation are quite unaffected.

The testis continues to produce sperms and testosterone. The sperms are removed by white cells as fast as they form. The testosterone ensures that there is no loss of masculinity.

The sperm ducts can be rejoined by surgery but this is not always successful.

Laparotomy
A woman may be sterilized by an operation in which her oviducts are tied, blocked or cut. The ovaries are unaffected. Sexual desire and menstruation continue as before, but sperms can no longer reach the ova. Ova are released, but break down in the upper part of the oviduct.

The operation cannot usually be reversed.

In vitro fertilization ('test-tube babies')
'*In vitro*' means literally 'in glass' or, in other words, the fertilization is allowed to take place in laboratory glassware rather than '*in vivo*', i.e. in the living body.

Some cases of female infertility can be treated by giving the fertility drugs FSH and LH (p. 158). If the oviducts are blocked, however, surgery may be used to eliminate the blockage.

When surgery is ineffective or inappropriate, the woman may be offered the chance of *in vitro* fertilization, a development which has received considerable publicity since the first 'test-tube baby' was born in 1978.

The woman may be given the fertility drugs FSH and LH, which cause her ovaries to release several mature ova simultaneously. These ova are then collected by laparoscopy, i.e. they are sucked up in a fine tube inserted through the abdominal wall. The ova are then mixed with the husband's seminal fluid and watched under the microscope to see if cell division takes place. (Fig. 11 on p. 154 is a photograph of such an '*in vitro*' fertilized ovum.)

One or more of the dividing zygotes are then introduced to the woman's uterus by means of a tube inserted through the cervix. Usually, only one (or none) of the zygotes develops though, occasionally, there are multiple births. The success rate for *in vitro* fertilization is between 12 and 40 per cent depending on how many embryos are transplanted. There is some controversy about the fate of the 'spare' embryos which are not returned to the uterus. Some people believe that since these embryos are potential human beings, they should not be destroyed or used for research. In some cases the 'spare' embryos have been frozen and used later if the first transplants did not work.

Checklist

- The male reproductive cells (gametes) are sperms. They are produced in the testes and expelled through the urethra and penis during mating.

- The female reproductive cells (gametes) are ova (eggs). They are produced in the ovaries. One is released each month. If sperms are present, the ovum may be fertilized as it passes down the oviduct to the uterus.

- Fertilization happens when a sperm enters an ovum and the sperm and egg nuclei join up (fuse).

- The fertilized ovum (zygote) divides into many cells and becomes embedded in the lining of the uterus. Here it grows into an embryo.

- The embryo gets its food and oxygen from its mother.

- The embryo's blood is pumped through blood vessels in the umbilical cord to the placenta, which is attached to the uterus lining. The embryo's blood comes very close to the mother's blood so that food and oxygen can be picked up and carbon dioxide and nitrogenous waste can be got rid of.

- When the embryo is fully grown, it is pushed out of the uterus through the vagina by contractions of the uterus and abdomen.

- Each month, the uterus lining thickens up in readiness to receive the fertilized ovum. If an ovum is not fertilized, the lining and some blood is lost through the vagina. This is menstruation.

- The release of ova and the development of an embryo are under the control of hormones like oestrogen, progesterone, follicle-stimulating hormone and luteinising hormone.

  - Twins may result from two ova being fertilized at the same time or from a zygote forming two embryos.

- At puberty, (1) the testes and ovaries start to produce mature gametes, (2) the secondary sexual characteristics develop.

- Human milk and breast-feeding are best for babies.

- Young humans are dependent on their parents for a long time. During this period much essential learning takes place.

- There are effective natural and artificial methods for spacing births and limiting the size of a family.

17 THE SKELETON, MUSCLES AND MOVEMENT

Structure of the skeleton

A human skeleton is shown in Fig. 1. It consists of a vertebral column (sometimes called the 'backbone', 'spine' or 'spinal column') which supports the skull. Twelve pairs of ribs are attached to the upper part of the vertebral column and the limbs are attached to it by means of **girdles**.

The hip girdle (pelvic girdle) is joined rigidly to the lower end of the vertebral column. The shoulder girdle (pectoral girdle) consists of a pair of collarbones and shoulder blades which are not rigidly fixed to the vertebral column but held in place by muscles.

The upper arm bone (humerus) fits into a socket in the shoulder blade; the thigh bone (femur) fits into a socket in the hip girdle.

Bone contains living cells and non-living fibres. The fibrous tissue between the cells becomes hardened by a deposit of calcium salts such as calcium phosphate.

Although bone contains a high proportion of non-living material it is penetrated by blood vessels which keep the cells alive and allow growth and repair to take place.

Vertebral column

The vertebral column forms the central supporting structure of the skeleton. It consists of 33 individual bones called **vertebrae** (singular = vertebra), separated by discs of fibrous cartilage. These discs allow the vertebrae to move slightly and so enable the vertebral column to bend backwards and forwards or from side to side.

The spinal cord (p. 184) runs through an arch of bone (**neural arch**) formed by the vertebrae. In this position the cord is protected from damage.

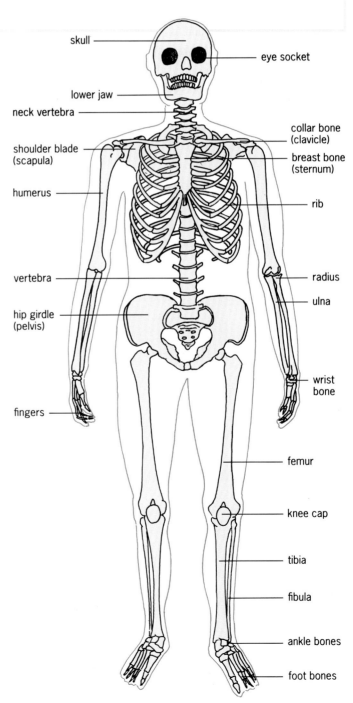

Figure 1 The skeleton

The skull

This is made up of many bony plates joined together (Fig. 2). It encloses and protects the brain and also carries and protects the main sense organs, the eyes, ears and nose. The upper jaw is fixed to the skull but the lower jaw is hinged to it in a way which allows chewing.

The base of the skull makes a joint with the top vertebra of the vertebral column. This joint allows the head to make nodding and rotational movements.

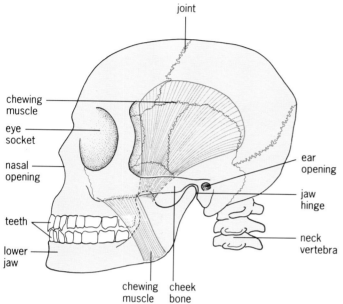

Figure 2 The skull and chewing muscles

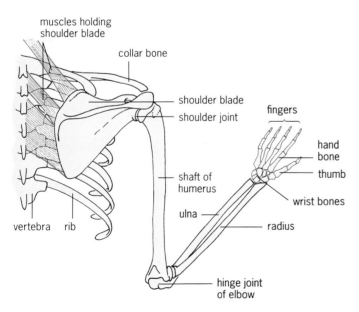

Figure 3 Skeleton of arm and shoulder

The limbs

Arm

The upper arm bone is the **humerus**. It is attached by a hinge joint to the lower arm bones, the **radius** and **ulna** (Fig. 3). These two bones make a joint with a group of small wrist bones which in turn join to a series of five hand and finger bones. The ulna and radius can partly rotate round each other so that the hand can be held palm up or palm down.

Leg

The thigh bone or **femur** is attached at the hip to the pelvic girdle by a ball joint and at the knee it makes a hinge joint with the **tibia**. The **fibula** runs parallel to the tibia but does not form part of the knee joint. The ankle, foot and toe bones are similar to those of the wrist, hand and fingers.

Functions of the skeleton

Support

The skeleton holds the body off the ground and keeps its shape even when muscles are contracting to produce movement.

Protection

The brain is protected from injury by being enclosed in the skull. The heart, lungs and liver are protected by the rib cage, and the spinal cord is enclosed inside the neural arches of the vertebrae.

Movement

Many bones of the skeleton act as levers. When muscles pull on these bones, they produce movements such as the raising of the ribs during breathing (see p. 133), or the chewing action of the jaws. For a skeletal muscle to produce movement, both its ends need to have a firm attachment. The skeleton provides suitable points of attachment for the ends of muscles.

Production of blood cells

The red marrow of some bones, e.g. the vertebrae, ribs, breastbone and the heads of the limb bones, produce both red and white blood cells (pp. 114 and 115).

QUESTION

1 Study Fig. 1 and then write the biological names of the following bones: upper arm bone, upper leg bone, hip bone, breastbone, 'backbone', lower arm bones.

Joints

Where two bones meet they form a joint. It may be a **fixed joint** as in the junction of the hip girdle and the vertebral column (Fig. 4(a)) or a **movable joint** as in the knee. Two important types of movable joint already mentioned are the **ball and socket joints** of the hip and the shoulder and the **hinge joints** of the elbow (Fig. 4(c)) and knee. The ball and socket joint allows movement forwards, backwards and sideways, while the hinge joint allows movement in only one direction.

(a) pelvic girdle
The hip bones form a fixed joint with the lower vertebrae of the vertebral column. The femur forms a ball and socket joint with the hip. Notice also the discs of cartilage between the vertebrae.

(b) ligaments of the hip joint

Figure 4 The joints of the hip and elbow

Where the surfaces of the bones in a joint rub over each other, they are covered with smooth cartilage which reduces the friction between them. Friction is also reduced by a thin layer of lubricating fluid called **synovial fluid** (Fig. 4(c)). (Movable joints are sometimes called **synovial joints**.) The bones forming the joint are held in place by tough bands of fibrous tissues called **ligaments** (Fig. 4(b)). Ligaments keep the bones together but do not stop their various movements.

(c) section through elbow joint (ligaments cannot be shown in this sectional drawing)

QUESTIONS

2 Apart from the elbow and knee, what other joints in the body function like hinge joints?
3 What types of joints are visible in Fig. 4(a)? Name the bones which form each type of joint that you mention.
4 Which parts of the skeleton are concerned with both protection and movement?

Muscle

There are three main types of muscle. One kind is called **skeletal muscle** (or striated, or voluntary muscle). Another kind is called **smooth muscle** (or unstriated, or involuntary muscle). A third kind occurs only in the heart.

Smooth muscle cells make layers of muscle tissue rather than distinct muscles, e.g. in the walls of the alimentary canal (p. 105), the uterus (p. 151) and the arterioles (p. 118).

Skeletal muscle
Skeletal muscle is made up of long fibres. Each fibre is formed from many cells, but the cells have fused together. The cell boundaries cannot be seen but the individual nuclei are still present (Fig. 5(c)).

The muscle fibres are arranged in bundles which form distinct muscles (Fig. 5(b)). Most of these are attached to bones, and produce movement as described below. Each muscle has a nerve supply. When a nerve impulse is sent to a muscle, it makes the muscle contract (i.e. get shorter and fatter). We can usually control most of our skeletal muscles. For this reason they are called **voluntary muscles**.

Muscle contraction
The fibres of skeletal muscle and the cells of smooth muscle have the special property of being able to contract, i.e. shorten, when stimulated by nerve impulses. However, the fibres and cells cannot elongate; they can only contract and relax. So, they have to be pulled back into their elongated shape by other muscles which work in the opposite direction (see 'Muscles and movement' below).

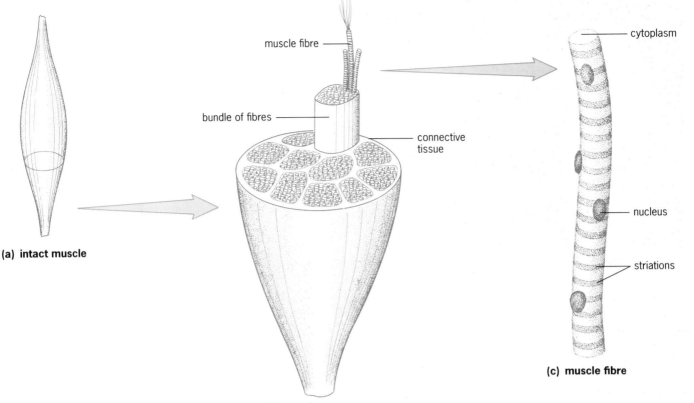

muscle fibre

bundle of fibres

connective tissue

cytoplasm

nucleus

striations

(a) intact muscle

(b) muscle cut across

(c) muscle fibre

Figure 5 Skeletal (striated) muscle

Movement and locomotion

Muscles and movement

The ends of the limb muscles are drawn out into **tendons** which attach each end of the muscle to the skeleton (Fig. 4(c)).

The way a muscle is attached to a limb to make it bend at the joint is shown in Fig. 6. The tendon at one end is attached to a non-moving part of the skeleton while the tendon at the other end is attached to the movable bone close to the joint.

When the muscle contracts it pulls on the bones and makes one of them move. The position of the attachment means that a small contraction of the muscle will produce a large movement at the end of the limb. A model showing how the shortening of muscle can move a limb is given in Fig. 7, and Fig. 6 shows how a contraction of the **biceps muscle** bends (or **flexes**) the arm at the elbow, while the **triceps** straightens (or **extends**) the arm.

The non-moving end of the biceps is attached to the shoulder blade while the moving end is attached to the ulna, near the elbow joint.

Limb muscles are usually arranged in pairs having opposite effects. This is because muscles can only shorten or relax, they cannot elongate, so the triceps is needed to pull the relaxed biceps back to its elongated shape after it has contracted. Pairs of muscles like this are called **antagonistic** muscles. Antagonistic muscles are also important in holding the limbs steady, both muscles keeping the same state of tension or 'tone'.

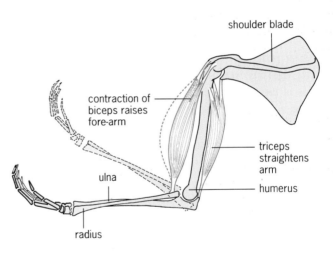

shoulder blade

contraction of biceps raises fore-arm

triceps straightens arm

humerus

ulna

radius

Figure 6 Antagonistic muscles of the fore-arm

elastic band shortens (= contracting muscle)

stretched elastic band (= relaxed muscle)

paper fastener

cardboard   slot

Figure 7 Model to show how muscles pull on bones to produce movement

165

The contraction of muscles is controlled by nerve impulses. The brain sends out impulses in the nerves so that the muscles are made to contract or relax in the right order to make a movement. For example, when a muscle contracts to bend the limb, its antagonistic muscle must be kept in a relaxed state.

There are many muscular activities which bring about movements but do not result in locomotion. Chewing, breathing, throwing, swallowing and blinking are examples of such movements.

Energy for muscle contraction

The energy for muscle contraction comes from ATP when it breaks down to ADP (p. 21). The ATP is reformed from ADP using the energy from the respiration of lipids and carbohydrates, principally glycogen.

Resting muscle has a store of glycogen but during exercise this has to be replenished from the glucose delivered to the muscle by its blood supply.

Aerobic respiration needs a supply of oxygen, also carried by the blood and delivered to the muscle. In addition, the blood carries away the waste products of respiration, carbon dioxide, water and, in the case of anaerobic respiration, lactic acid (p. 21).

During exercise, there is an increase in the volume of blood (**stroke volume**) expelled at each beat of the heart, and in its rate of beating. The breathing rate and volume of air exchanged at each breath also increase. Both these changes increase the supply of glucose and oxygen to the active muscles and speed up the removal of the waste products.

During vigorous exercise, there is insufficient oxygen to meet the demands of the contracting muscle, which increasingly comes to depend on anaerobic respiration.

If the lactic acid produced by this form of respiration is not removed quickly enough, it may lead to muscle fatigue (though this has not been established for certain).

The heat generated by the chemical reactions in muscle contraction is carried away by the blood and may trigger the thermoregulatory centre to initiate vasodilation and sweating (p. 148).

Locomotion

Locomotion is brought about by the limb muscles contracting and relaxing in an orderly (i.e. co-ordinated) manner. The photograph in Fig. 8 shows a sprinter at the start of a race, and Fig. 9 shows how some of his leg muscles are acting on the bones to thrust him forward. When muscle A contracts, it pulls the femur backwards. Contraction of muscle B straightens the leg at the knee. Muscle C contracts and pulls the foot down at the ankle. When these three muscles contract at the same time, the leg is pulled back and straightened and the foot is extended, pushing the foot downwards and backwards against the ground. If the ground is firm, the straightening of the leg pushes upwards against the pelvic girdle which in turn pushes the vertebral column and so lifts the whole body upwards and forwards.

Figure 8 Leg muscles used in running (compare with Fig. 9b)

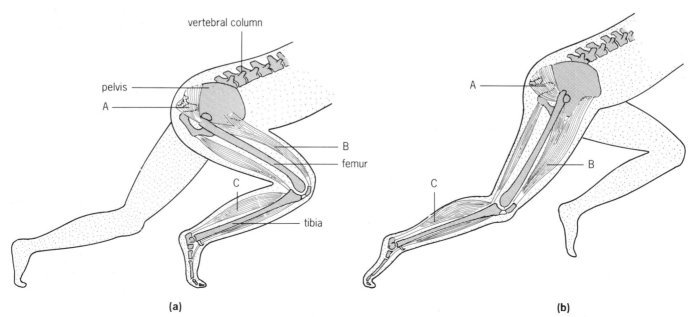

(a)

(b)

Figure 9 Action of leg muscles. There are many other muscles which are not shown here. A, B and C are all contracting to straighten the leg and foot.

While muscles A, B and C are contracting to extend the leg, their antagonistic muscles are kept in a state of relaxation. At the end of the extension movement, muscles A, B and C relax, and their antagonistic partners contract to flex the leg.

Long-term benefits of exercise

Some of the evidence for the benefits of exercise comes from experiments on laboratory animals and some from observations and measurements on the general population.

Whatever benefits exercise may bring, they do not persist unless you exercise regularly. Taking part in sport at school will help to make you fit, but you will not stay fit unless you keep up a regular pattern of exercise.

To be useful, exercise has to be strenuous, last about 20 minutes or more and be repeated at least three times per week. Examples are brisk walking, jogging, cycling, disco dancing, playing football, rugby, hockey, netball, tennis, squash and so on.

Regular strenuous exercise can produce the following physiological changes, which are beneficial:

1 *The resting heart rate goes down.* The ventricles enlarge and the heart muscle grows stronger, so that the stroke volume is increased. This means that when you take exercise, your heart can deliver more blood to the muscles without its rate of beating rising too far. Coronary blood flow may possibly increase, either because the vessels widen or because more branches develop. But the evidence for this is not very strong. (There is no evidence that exercise has any beneficial effect on the lungs or breathing mechanism.)

2 *The muscles used in exercise grow larger.* At first the muscle fibres grow thicker and then their number increases. The capillaries in the muscle develop more branches. Thus the muscles become stronger. But this only happens if the exercise makes them work to about 80 per cent of their maximum capacity.

3 *More enzymes are made in the muscle tissue.* These are the enzymes needed for breaking down glucose, glycogen or fatty acids. Thus the muscle is able to take up oxygen and food more rapidly from the blood and increase the rate of energy production. The muscles can also store more glycogen.

All these changes, (1) to (3), increase your strength, and also your stamina and endurance. That means that you can take more strenuous exercise, for longer periods, without getting tired or out of breath. Weight training improves strength but not stamina. Walking improves stamina but not strength.

Only the muscles taking part in the exercise will grow stronger. So training for one sport probably won't make you better at another. Improved stamina is useful in all active sports, however.

4 *Your ligaments and tendons become stronger.* This reduces the chance of injury during sudden vigorous activity. Back injury is less likely if the muscles and ligaments of the vertebral column are strong.

5 *Your joints become more flexible*, giving a greater range of movement with a lower risk of 'pulling' a muscle or spraining a joint.

The changes (1) to (5) help to postpone the effects of old age, but only if the pattern of exercise is maintained. Once regular exercise ceases, the muscles get thinner. All the improved physiological functions go back to their original level.

6 *Protection from heart attacks.* Studies have been carried out on large groups of adults, comparing those who take regular exercise with those who do not. Some of these suggest that people who take regular exercise are less likely to suffer prematurely from heart disease. Other studies show no benefit, or even the reverse. On balance, however, it does seem likely that regular exercise throughout life reduces the chance of an early heart attack.

The mechanism for this might be increased coronary circulation, a change in the ratio of lipids in the blood, a slowing down in the formation of atheroma (p. 126) in the arteries, a reduced tendency for the blood to form clots in the blood vessels, and long-term reduction in blood pressure. Increased efficiency of the muscles might reduce the strain on the heart.

The evidence for these changes comes mainly from animal studies rather than from humans, and is not all clear-cut. However, even if exercise does not make you live longer, you will certainly stay fitter.

7 *Well-being.* Most people agree that exercise, undertaken willingly, makes you 'feel good'. The 'scientific' evidence for this is lacking but it could be explained in physiological terms.

QUESTIONS

5 Say which of the following you would consider to be beneficial forms of exercise: running for a bus, climbing a mountain, playing badminton, cutting the grass, playing cricket, swimming. Give reasons for your answers.
6 Why do you have to exercise regularly in order to benefit?
7 Explain what part an oxygen debt (p. 21) might play in the time taken to recover from exercise.

Injuries to joints and muscles

Sprains

If a joint is forced beyond its normal degree of movement, some of the ligaments may be partially torn. This results in swelling, pain and a leakage of blood into the tissues (bruising).

A short period of immobilization followed by a return to normal movement as soon as possible is the recommended treatment. Turning your ankle is a common example of a sprain.

If a ligament is completely torn, a longer period of immobilization and possibly surgical repair may be necessary.

167

Strains

Violent contraction of a muscle may cause some of the fibres to tear ('pulling' a muscle). This is the case in the 'ham string' injury of the posterior thigh muscle. Often, a long period of immobilization is needed for recovery.

Dislocations

This happens when the bones at a joint become displaced as a result of a violent movement. Usually the ligaments are damaged as well. The shallow ball and socket joint of the shoulder is vulnerable to dislocation. The bones have to be manipulated (by a qualified practitioner) back into place. Then the treatment is as for a severe sprain.

Fractures

When a limb bone is broken, the limb has to be restored to its correct shape and immobilized in a plaster cast (pod) until new bone grows and joins up the fractured ends.

PRACTICAL WORK

1 The structure of bone

Obtain two small bones, e.g. limb bones from a chicken. Place one of them in a test-tube or beaker, cover it with dilute hydrochloric acid and leave it for 24 hours, during which time the acid will dissolve most of the calcium salts in the bone. After 24 hours pour away the acid, wash the bone thoroughly with water and then try to bend it.

Take the second bone and hold one end of it in a hot Bunsen flame, heating it strongly for 2 minutes. At first the bone will char and then glow red as the fibrous organic material burns away, but it will retain its shape. Allow the bone to cool and then try crushing the heated end against the bench with the end of a pencil.

Result   Both bones retain their shape after treatment, but the bone whose calcium salts have been dissolved in acid is rubbery and flexible because only the organic, fibrous connective tissue is left. The bone that had this fibrous tissue burned away is still hard but very brittle and easily shattered.

Interpretation   The combination in bone of mineral salts and organic fibres produces a hard, strong and resilient structure.

QUESTIONS

8 What is the difference between the functions of a ligament and a tendon?

9 What is the main action of
   a your calf muscle,
   b the muscles in the front of your thigh and
   c the muscles in your forearm?
   If you don't already know the answer, try making the muscles contract and feel where the tendons are pulling.

10 In Fig. 9: **a** To what bone is the non-moving end of muscle B attached?   **b** Which muscle is the antagonistic partner to C?

Teeth

Function of teeth

Before food can be swallowed, pieces have to be bitten off which are small enough to pass down the gullet into the stomach. Digestion is made easier if these pieces of food are crushed into even smaller particles by the action of chewing. Biting and chewing are actions carried out by our teeth, jaws and muscles (see Fig. 2, p. 163).

The teeth are given different names according to their position in the jaw (Figs 10 and 11). In the front of each of the upper and lower jaws there are four **incisors**. Our top incisors pass in front of our bottom incisors and cut pieces off the food, as when biting into an apple or taking a bite out of a sandwich.

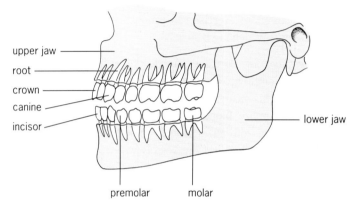

Figure 10 Human teeth and jaws

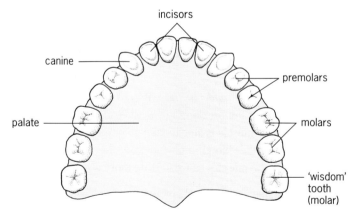

Figure 11 Teeth in the upper jaw

On each side of the incisors there is a **canine** tooth. In carnivorous mammals, like dogs, the canines are long and pointed (Fig.6, p.312) but in humans they are similar to the incisors but a little more pointed. They function rather like extra incisors.

At the side of each jaw are the **premolars**, two on each side. They are larger than the canines and have one or two blunt points or **cusps**. At the back of each jaw are two or three **molars**. The molars are larger than the premolars and have four or more cusps.

The premolars and molars are similar in function. Their knobbly surfaces meet when the jaws are closed, and crush the food into small pieces.

As well as breaking the food into a suitable size for swallowing, the crushing action of the molars and premolars reduces food to quite small particles. Small particles of food are easier to digest than large pieces because they offer a greater surface area to the digestive enzymes.

Structure of teeth

The structure of an incisor or canine tooth as it would appear in vertical section (i.e. cut from top to bottom) is shown in Fig. 12(a), and Fig. 12(b) shows a section through a molar tooth.

Enamel

Enamel covers the exposed part, or **crown**, of the tooth and makes a hard biting surface. It is a non-living substance containing 97 per cent calcium salts and only 3 per cent organic matter. Although the enamel is laid down before the tooth emerges from the gum, it can be thickened and strengthened by salts deposited from the saliva or from food and drink. Fluoride ions taken up by the enamel increase its resistance to decay.

Dentine

This is rather like bone and softer than enamel. It is a living tissue with threads of cytoplasm running through it. The hardness of both enamel and dentine depends on there being enough calcium in the diet and sufficient vitamin D to help the absorption of calcium in the intestine.

Pulp

In the centre of the tooth is soft connective tissue. It contains cells which make the dentine and keep the tooth alive. In the pulp are blood vessels which bring food and oxygen, so that the tooth can grow at first and then remain alive when growth has stopped. There are also sensory nerve endings in the pulp, which are sensitive to heat and cold but give only the sensation of pain. If you plunge your teeth into an ice-cream, they do not feel cold but they do hurt.

Cement

This is a bone-like substance which covers the root of the tooth. In the cement are embedded tough fibres which pass into the bone of the jaw and hold the tooth in place.

Milk teeth and permanent teeth

Mammals have two sets of teeth in their lifetimes. In humans, the first set, or **milk teeth**, grow through the gum during the first year of life and consist of four incisors, two canines and four molars in each jaw. Between the ages of 6 and 12 years, these milk teeth gradually fall out (Fig. 13) and are replaced by the **permanent teeth**, including four premolars and six molars in each jaw. The last of these molars, the 'wisdom teeth', may not grow until the age of 17 or later. In some cases they do not appear at all. If the permanent teeth are lost for any reason, they do not grow again.

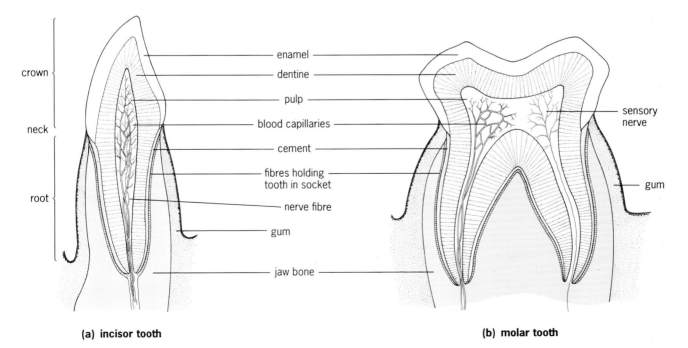

(a) incisor tooth

(b) molar tooth

Figure 12 Longitudinal sections through incisor and molar teeth

Figure 13 Permanent teeth coming through. The first two incisors in the bottom jaw have emerged. The next two incisors are just breaking through the gum.

QUESTIONS

14 Which of the permanent teeth are not represented in the set of milk teeth?

15 What are the differences between dentine and enamel?

Checklist

Skeleton

- The vertebral column ('backbone') is made up of 33 vertebrae.
- The vertebral column forms the main support for the body and also protects the spinal cord.
- The legs are attached to the vertebral column by the hip girdle.
- The shoulder blades and collar bones form the shoulder girdle.
- The arms are attached to the shoulder blades.
- The skull protects the brain, eyes and ears.
- The ribs protect the lungs, heart and liver, and also play a part in breathing.
- The limb joints are either ball and socket (e.g. hip and shoulder) or hinge (e.g. knee and elbow).
- The surfaces of the joints are covered with cartilage and lubricated with synovial fluid.

Muscles and movement

- Skeletal muscles are formed from fibres and can be consciously controlled.
- Smooth muscle is formed from layers of cells and cannot be consciously controlled.
- Limb muscles are attached to the bones by tendons.
- When the limb muscles contract, they pull on the bones and so bend and straighten the limb or move it forwards and backwards.
- Most limb muscles are arranged in antagonistic pairs, e.g. one bends and one straightens the limb.
- By straightening the leg and thrusting it against the ground, an animal can propel itself forwards.
- Exercise can have long-term benefits but only if maintained throughout life.

Teeth

- The roots of our teeth are held in the jawbone.
- The crowns of the teeth are covered with enamel and project into the mouth.
- The pulp and the dentine are kept alive by blood vessels bringing food and oxygen.
- We have incisor, canine, premolar and molar teeth.
- The incisors and canines bite off pieces of food.
- The premolars and molars crush the food ready for swallowing.
- We have two sets of teeth in our lifetime, milk teeth (20) and permanent teeth (32).

18 THE SENSES

Our senses make us aware of changes in our surroundings and in our own bodies. We have sense cells which respond to stimuli (singular = stimulus). A **stimulus** is a change in light, temperature, pressure, etc., which produces a reaction in a living organism. Structures which detect stimuli are called **receptors**. Some of these receptors are scattered through the skin while others are concentrated into special organs such as the eye and the ear.

The special property of sensory cells and sense organs is that they are able to convert one form of energy to another. Structures which can do this are called energy **transducers**. The eyes can convert light energy into the electrical energy of a nerve impulse. The ears convert the energy in sound vibrations into nerve impulses. The forms of energy which make up the stimuli may be very different, e.g. mechanical, chemical, light, but they are all transduced into pulses of electrical energy in the nerves.

Skin senses

There are a great many sensory nerve endings in the skin which respond to the stimuli of touch, pressure, heat and cold, and some which cause a feeling of pain. These sensory nerve endings are very small; they can be seen only in sections of the skin when studied under the microscope (Fig. 1), and the functions of some of them have not yet been identified.

Some of the sensory nerve endings are **encapsulated**, i.e. they are enclosed in a capsule, e.g. the Meissner's corpuscles (touch) and the Pacinian corpuscles (pressure) (Fig. 2). Other sensory endings seem to consist only of fine branches from the nerve fibre, e.g. the temperature sensors or the hair plexuses. These are called **'free' nerve endings**.

Certain regions of the skin have a greater concentration of sense organs than others. The finger-tips, for example, have a large number of touch organs, making them particularly sensitive to touch. The front of the upper arm is sensitive to heat and cold. Some areas of the skin have relatively few sense organs and can be pricked or burned in certain places without any sensation being felt.

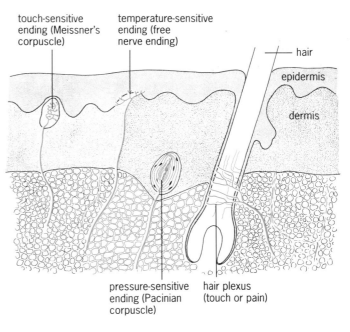

Figure 1 The sense organs of the skin (generalized diagram)

Figure 2 Pacinian corpuscle in human skin (x60)

When the nerve ending receives a stimulus, it sends a nerve impulse to the brain which makes us aware of the sensation. Generally, each type of nerve ending responds to only one kind of stimulus. For example, a heat receptor would send off a nerve impulse if its temperature were raised but not if it were touched.

QUESTIONS

1 What sensation would you expect to feel if a warm pin-head was pressed on to a touch receptor in your skin? Explain.
2 If a piece of ice is pressed on to the skin, which receptors are likely to send impulses to the brain?

Taste and smell

Taste
In the lining of the nasal cavity and on the tongue are groups of sensory cells, called **chemoreceptors**, which respond to chemicals. On the tongue, these groups are called **tastebuds** and they lie mostly in the grooves of the tongue. The receptor cells in the tastebuds can recognize only four classes of chemicals. These are the chemicals which give the taste sensations of sweet, sour, salt or bitter. Nearly all acids, for example, give the taste sensation we call 'sour', but a wide variety of quite different chemicals give the sensation of 'sweet'. Generally, the taste cells are sensitive to only one or two of these classes of chemical. For a substance to produce a sensation of taste, it must be able to dissolve in the film of water covering the tongue.

Smell
The epithelium lining the top of the nasal cavity contains chemoreceptors. Fine processes from these chemoreceptors extend into the film of mucus which lines the nasal epithelium. A wide range of airborne chemicals stimulates the nasal receptors and they send nerve impulses to the brain.

Our sense of smell is able to distinguish a great many more chemicals than our sense of taste. However, there is not yet a satisfactory classification of types of smell we can recognize, nor an explanation of how they are distinguished.

The sensation we call 'flavour' (as distinct from 'taste') is the result of the vapours, from the food in the mouth, reaching the chemoreceptors in the nose. Flavour is, therefore, largely 'smell' and we lose some of the sensation of flavour when the nose is blocked, as during a heavy cold.

The sense of smell is easily fatigued, that is, a smell experienced for a long period ceases to give any sensation and we become unaware of it, though a newcomer may detect it at once.

QUESTIONS

3 Which types of tastebuds are likely to be stimulated by lemonade?
4 Apart from the cells which detect chemicals, what other types of receptor must be present in the tongue?
5 What is the difference between taste, smell and flavour?

Sight

The eye
The structure of the eye is shown in Figs 3 and 4. The **sclera** is the tough, white outer coating. The front part of the sclera is clear and allows light to enter the eye. This part is called the **cornea**. The **conjunctiva** is a thin epithelium which lines the inside of the eyelids and the front of the sclera, and is continuous with the epithelium of the cornea.

The eye contains a clear liquid whose outward pressure on the sclera keeps the spherical shape of the eyeball. The liquid behind the lens is jelly-like and called **vitreous humour**. The **aqueous humour** in front of the lens is watery.

The **lens** is a transparent structure, held in place by a ring of fibres called the **suspensory ligament**. Unlike the lens of a camera or a telescope, the eye lens is flexible and can change its shape. In front of the lens is a disc of tissue called the **iris**. It is the iris we refer to when we describe the colour of the eye as brown or blue. There is a hole in the centre of the iris called the **pupil**. This lets in light to the rest of the eye. The pupil looks black because all the light entering the eye is absorbed by the black pigment in the **choroid**. The choroid layer, which contains many blood vessels, lies between the retina and the sclera. In the front of the eyeball, it forms the iris and the **ciliary body** (see p. 174). The ciliary body produces aqueous humour.

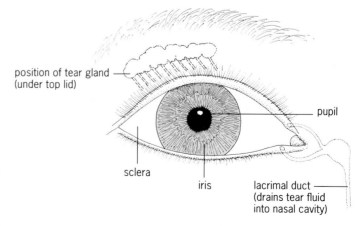

position of tear gland (under top lid)

pupil

sclera

iris

lacrimal duct (drains tear fluid into nasal cavity)

Figure 3 Appearance of right eye from the front

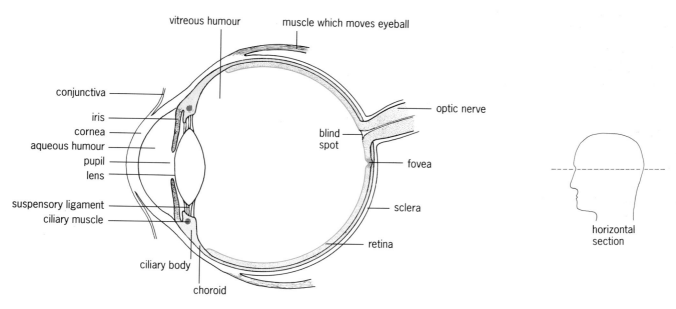

Figure 4 Horizontal section through left eye

The internal lining at the back of the eye is the **retina** and it consists of many thousands of cells which respond to light. When light falls on these cells, they send off nervous impulses which travel in nerve fibres, through the **optic nerve**, to the brain and so give rise to the sensation of sight.

Tear glands under the top eyelid produce tear fluid. This is a dilute solution of sodium chloride and sodium hydrogen carbonate. The fluid is spread over the eye surface by the blinking of the eyelids, keeping the surface moist and washing away any dust particles or foreign bodies. Tear fluid also contains an enzyme, **lysozyme**, which attacks bacteria.

Vision

Light from an object produces a focused **image** on the retina (like a 'picture' on a cinema screen) (Figs 5 and 7). The curved surfaces of the cornea and lens both 'bend' the light rays which enter the eye, in such a way that each 'point of light' from the object forms a 'point of light' on the retina. These points of light will form an image, upside-down and smaller than the object.

The cornea and the aqueous and vitreous humours are mainly responsible for the 'bending' (refraction) of light. The lens makes the final adjustments to the focus (Fig. 5(b)).

The pattern of sensory cells stimulated by the image will produce a pattern of nerve impulses sent to the brain. The brain interprets this pattern, using its past experience and learning, and so forms an impression of the real size, distance and upright nature of the object.

Blind spot

At the point where the optic nerve leaves the retina, there are no sensory cells and so no information reaches the brain about that part of the image which falls on this **blind spot** (see Fig. 6).

Figure 6 The blind spot. Hold the book about 50 cm away. Close the left eye and concentrate on the cross with the right eye. Slowly bring the book closer to the face. When the image of the dot falls on the blind spot it will seem to disappear.

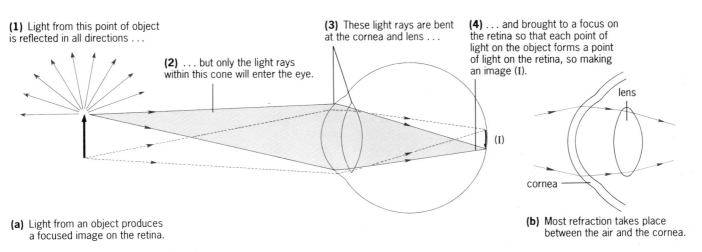

(1) Light from this point of object is reflected in all directions . . .

(2) . . . but only the light rays within this cone will enter the eye.

(3) These light rays are bent at the cornea and lens . . .

(4) . . . and brought to a focus on the retina so that each point of light on the object forms a point of light on the retina, so making an image (I).

(a) Light from an object produces a focused image on the retina.

(b) Most refraction takes place between the air and the cornea.

Figure 5 Image formation on the retina

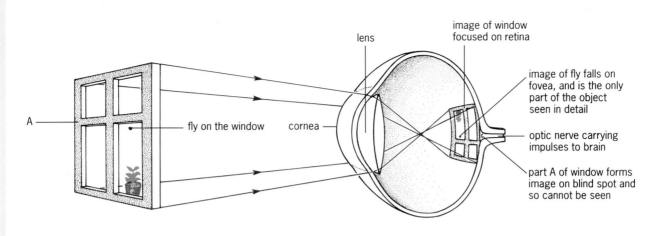

Figure 7 Image formation in the eye

Retina

The millions of light-sensitive cells in the retina are of two kinds, the **rods** and the **cones** (according to shape). The cones enable us to distinguish colours. There are thought to be three types of cone cell. One type responds best to red light, one to green and one to blue. If all three types are equally stimulated we get the sensation of white. The cone cells are concentrated in a central part of the retina, called the **fovea** (Fig. 4), and when you study an object closely, you are making its image fall on the fovea.

Fovea

Only in the fovea is it possible for the eye and brain to make a really detailed analysis of the colour and shape of an object. Although we can see objects included in a zone of about 100° from each eye, only those objects within a 2° zone can be seen in detail. This is much less than people imagine and means, for example, that only about two letters in any word on this page can be studied in detail. It is the constant movement of the eyes which enables you to build up an accurate picture of a scene.

Accommodation (focusing)

The eye can produce a focused image of either a near object or a distant object. To do this the lens changes its shape, becoming thinner for distant objects and fatter for near objects. This change in shape is caused by contracting or relaxing the **ciliary muscle** which forms a circular band of muscle in the ciliary body (Figs 8 and 9). When the ciliary muscle is relaxed, the outward pressure of the humours on the sclera pulls on the suspensory ligament and stretches the lens to its thin shape. The eye is now accommodated (i.e. focused) for distant objects (Figs 9(a) and 10(a)). To focus a near object, the ciliary muscle contracts to a smaller circle and this takes the tension out of the suspensory ligament (Figs 9(b) and 10(b)). The lens is elastic and flexible and so is able to change to its fatter shape. This shape is better at bending the light rays from a close object.

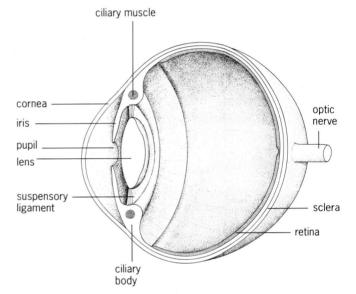

Figure 8 Vertical section through the left eye

Control of light intensity

The amount of light entering the eye is controlled by altering the size of the pupil. If the light intensity is high, it causes a contraction in a ring of muscle fibres (a sphincter) in the iris. This reduces the size of the pupil and cuts down the intensity of light entering the eye. High intensity light can damage the retina, so this reaction has a protective function.

In low light intensities, the sphincter muscle of the iris relaxes and muscle fibres running radially (i.e. like wheel spokes) contract. This makes the pupil enlarge and admits more light (see Experiment 4, p. 178).

The change in size of the pupil is caused by an automatic reflex action (p. 181); you cannot control it consciously.

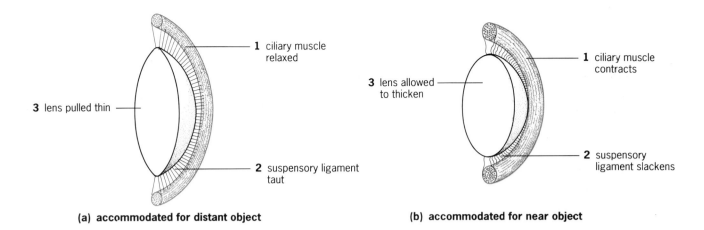

Figure 9 How accommodation is brought about

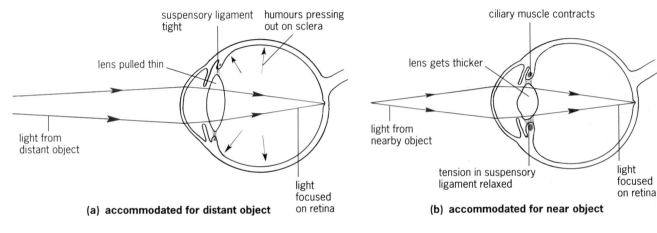

Figure 10 Accommodation

3-D vision and distance judgement

When we look at an object, each eye forms its own image of the object. So, two sets of impulses are sent to the brain. The brain somehow combines these impulses so that we see only one object and not two. However, because the eyes are spaced apart, they do not have the same view of the object. The left eye sees more of the left side and the right eye sees more of the right side. When the brain combines the information from both eyes, it gives the impression that the object is three-dimensional (3-D) rather than flat.

In order to look at a nearby object, the eyes have to turn inwards slightly. (Trying to focus on a very close object will make you look 'cross-eyed'.) Stretch receptors in the eye muscles will send impulses to the brain and make it aware of how much the eyes are turning inwards. This is one way by which the brain may be able to judge how close an object is, but it is not likely to be much use for objects beyond 15 metres. We probably judge greater distances by comparing the sizes of familiar objects. The further away an object is, the smaller will be its image on the retina. We use many other clues to help us to judge distances.

We are able to experience 3-D vision and to judge distances effectively because our eyes are set in the front of our head. This gives us what is called **binocular vision**. Animals with eyes set at the side of the head, e.g. rabbits and many birds, do not have binocular vision.

QUESTIONS

6 In Fig. 7, what structures of the eye are not shown in the diagram?

7 In Fig. 5, explain what the broken lines are meant to represent.

8 **a** If your ciliary muscles are relaxed, are your eyes focused on a near or a distant object? Explain. **b** If you dissected a cow's eye, would you expect the lens to be at its thinnest or its fattest shape? Explain.

9 How do you think animals without binocular vision are able to judge distances?

Hearing

The hearing apparatus is enclosed in bone on each side of the skull, just behind the jaw hinge. Sound vibrations travel through a short, wide tube (the outer ear) and are converted to nerve impulses by the apparatus in the middle ear and inner ear. The structure of the ear is shown in Fig. 11.

Outer ear

Sound is the name we give to the sensation we get as a result of vibrations in the air. These vibrations are pulses of compressed air. They enter the tube of the outer ear and hit the **ear drum**, a thin membrane like a drum-skin across the inner end of the tube. The air vibrations cause the ear drum to vibrate backwards and forwards. If there are 200 pulses of compressed air every second, the ear drum will move backwards and forwards at the same rate.

The ear **pinnae** are the flaps of skin and elastic cartilage which project from the sides of the head and which we usually call our 'ears'. The ear pinnae of mammals, such as dogs and cats, help to direct sound vibrations into the outer ear and enable the animal to locate the source of a sound. It is possible that our ear pinnae have a similar function.

Middle ear

This is a cavity with air in it. It contains a chain of tiny bones or **ossicles**. The first of these ossicles, the **malleus**, is attached to the ear drum, and the inner ossicle, the **stapes**, fits into a small hole in the skull called the **oval window**. The malleus is connected to the stapes by the **incus** (Fig. 12). When the ear drum vibrates back and forth, it forces the ossicles to vibrate in the same way, so that the stapes moves rapidly backwards and forwards like a tiny piston in the oval window. The way the ossicles are attached to each other increases the force of the vibrations and so amplifies the sound.

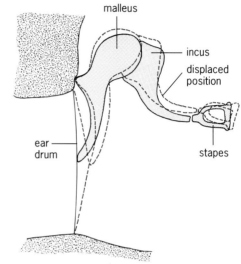

Figure 12 Movement of the ear ossicles in transmitting sound

Figure 11 Diagram of the ear

cochlea is coiled something like this

Eustachian tube (connects to nasal cavity and keeps air pressure on inside of ear drum the same as that outside)

(coloured arrows represent movements of ear drum, ossicles and ear liquids)

A narrow tube, called the **Eustachian tube**, connects the nasal cavity (see Fig. 6, p. 107) to the middle ear. Although the Eustachian tube is normally closed, it does open to admit or release air if the air pressure on the outside of the ear drum changes. Air pressure falls when going up a mountain or in an aircraft. The action of swallowing often causes the Eustachian tube to open and gives a 'popping' sound if there has been a change of pressure.

Inner ear

This is where the vibrations are changed into nerve impulses. The inner ear contains liquid, and the vibrations of the ossicles are passed to this liquid. The sensitive part of the inner ear is the **cochlea**, a coiled tube with sensory nerve endings in it. When the liquid in the cochlea is made to vibrate, the nerve endings send off impulses to the brain. The nerve endings at the inner (top) end of the cochlea are sensitive to low-frequency vibration (low notes) and those at the outer end are sensitive to high-frequency vibrations (high notes). So, if the brain receives nerve impulses coming from the first part of the cochlea, it interprets this as a high-pitched noise or a high musical note. If the impulses come from the top end of the cochlea, the brain recognizes them as being caused by a low note.

In fact, the way in which vibrations of different frequencies are converted to nerve impulses by the cochlea has not been fully worked out. It is certainly more complicated than the simplified account given here.

Balance

The **semicircular canals** and **utriculus** are organs of balance. They contain sensory receptors which detect movements of the head. The semicircular canals are stimulated mainly by rotational movements, whereas the utriculus responds to positional changes.

QUESTIONS

10 What is the function of **a** the ear drum, **b** the ossicles, **c** the liquid of the inner ear, **d** the cochlea?

11 a How does the function of a sensory cell in the retina differ from the function of a sensory cell in the cochlea? **b** If nerve impulses from the cochlea were fed into the optic nerve, what sensations would you expect if somebody clapped their hands?

12 Sometimes, as a result of catching a cold, the middle ear becomes filled with a clear sticky fluid. Why do you think this causes deafness? (The fluid usually drains away through the Eustachian tube when the cold goes.)

13 a How do you think we are able to tell the direction from which a sound is coming?
b Suggest an experiment which could test your hypothesis.
c In owls, the external ear on one side of the head is higher than on the other. How, do you think, this helps them to locate their prey?

PRACTICAL WORK

All the experiments suggested here are best done by three people working together: an *experimenter* who applies or offers the stimuli, the *subject* who is given the stimuli, and a *recorder* who watches and makes a note of the responses made by the subject.

1 Sensitivity to touch

The experimenter marks a regular pattern of dots on the back of the subject's hand. To do this she uses an ink-pad and a rubber stamp like the one in Fig. 13. She also stamps the same pattern onto a piece of paper for the recorder to use. The experimenter now tests the sensitivity of the subject's skin by pressing a fine bristle onto each dot in turn. The bristle, e.g. a horsehair, is glued to a wooden handle or held in forceps and pressed onto each mark until the bristle just starts to bend. The subject must not look, and simply says 'yes' if he feels the stimulus. The recorder marks on his pattern how many spots in each row were sensitive to touch.

The sensitivity of the back of the hand can now be compared with the sensitivity of the finger-tips or the back of the neck by repeating the experiment in these regions.

Rubber stamp for marking area in Experiment 1. This can be made by sticking a piece of 'finger cone' on a wooden block.

Figure 13 Testing sensitivity to touch

2 Ability to distinguish between two touch stimuli

A piece of wire is bent to a hairpin shape like the one in Fig. 14(a). The experimenter adjusts the hairpin so that the points are exactly 5 mm apart and presses *one* of the points or *both at once* (Fig. 14(b)) onto the skin of the subject's hand, just enough to dent the skin. The subject must not watch the experiment and simply says 'one' or 'two' if he thinks he is being touched by one or two points. The recorder notes how many times the subject is correct. The experimenter must use one point or two points in a random way so that the subject does not recognize a pattern. The recorder should make up a programme of ten stimuli—for example, 1.1.2.1.1.2.2.2.1.2, with five single and five double stimuli. The experimenter carries out this programme and the recorder ticks the correctly recognized stimuli on his plan.

If the subject gets them all right, the experimenter should move the points closer together and try again. If the subject makes a lot of mistakes, the points should be opened out to 10 mm and the experiment repeated. The idea is to find the least distance the points have to be apart for the subject to recognize every time whether one or two points are touching him. Once this distance has been discovered for the back of the hand, the experiment is repeated for the finger-tips and the back of the neck.

Figure 14 Double and single stimulus

3 To find which eye is used more

A pencil is held at arm's length in line with a distant object. First one eye is closed and opened and then the other. With one the pencil will seem to jump sideways. This shows which eye was used to line up the pencil in the first place.

4 The iris diaphragm

For this experiment the room needs to be darkened, but not blacked out.

If you are working in pairs, sit facing each other and take it in turns to hold a bench lamp or torch about 10 cm from the *side* of your partner's face. Switch on the torch or bench lamp and watch the pupil of the eye very carefully. Try switching the light on and off at about 3- or 4-second intervals.

If you are working on your own, you must look closely in a mirror while switching the light on and off.

There are more experiments on human senses in *Experimental Work in Biology*, Combined Edition. (See p. vi.)

Checklist

- Our senses detect changes in ourselves and in our surroundings.
- The skin is a sense organ which detects heat, cold, touch and pressure.
- Sensory endings in the nose respond to chemical substances in the air and give us the sense of smell.
- The tongue responds to chemicals in food and drink and gives the sense of taste.

Sight

- The lens focuses light from the outside world to form a tiny image on the retina.
- The sensory cells of the retina are stimulated by the light and send nerve impulses to the brain.
- The brain interprets these nerve impulses and so gives us the sense of vision.
- The eye can focus on near or distant objects by changing the thickness of the lens.

Sound

- The ear drum is made to vibrate backwards and forwards by sound waves in the air.
- The vibrations are passed on to the inner ear by the three tiny ear bones.
- Sensory nerve endings in the inner ear respond to the vibrations and send impulses to the brain.
- The brain interprets these impulses as sound.

Balance

- Movements of the head are detected by the semicircular canals and utriculus.

19 CO-ORDINATION

Co-ordination is the way all the organs and systems of the body are made to work efficiently together (Fig. 1). If, for example, the leg muscles are being used for running, they will need extra supplies of glucose and oxygen. To meet this demand, the lungs breathe faster and deeper to obtain the extra oxygen and the heart pumps more rapidly to get the oxygen and glucose to the muscles more quickly.

The brain detects changes in the oxygen and carbon dioxide content of the blood and sends nervous impulses to the diaphragm, intercostal muscles and heart. In this example, the co-ordination of the systems is brought about by the **nervous system**.

The extra supplies of glucose needed for running come from the liver. Glycogen in the liver is changed to glucose which is released into the bloodstream (p. 112). The conversion of glycogen to glucose is stimulated by, among other things, a chemical called adrenaline (p. 186). Co-ordination by chemicals is brought about by the **endocrine system**.

The nervous system works by sending electrical impulses along nerves. The endocrine system depends on the release of chemicals, called **hormones**, from **endocrine glands**. Hormones are carried by the bloodstream. For example, insulin (p. 187), is carried from the pancreas to the liver by the circulatory system.

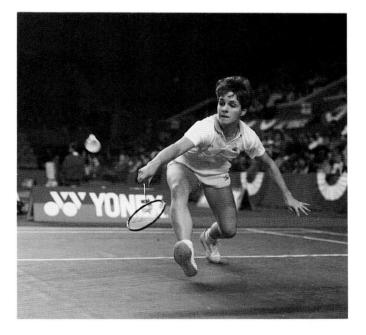

Figure 1 Co-ordination. The badminton player's brain is receiving sensory impulses from her eyes, semicircular canals, utriculus and muscle stretch receptors. Using this information, the brain co-ordinates the muscles of her limbs so that even while running or leaping she can control her stroke.

The nervous system

The human nervous system is shown in Fig. 2. The brain and spinal cord together form the **central nervous system**. Nerves carry electrical impulses from the central nervous system to all parts of the body, making muscles contract or glands produce enzymes or hormones.

Glands and muscles are called **effectors** because they go into action when they receive nerve impulses or hormones. The biceps muscle (p. 165) is an effector which flexes the arm; the salivary gland is an effector which produces saliva when it receives a nerve impulse from the brain.

The nerves also carry impulses back to the central nervous system from receptors in the sense organs of the body. These impulses from the eyes, ears, skin, etc. make us aware of changes in our surroundings or in ourselves. Nerve impulses from the sense organs to the central nervous system are called **sensory impulses**; those from the central nervous system to the effectors, resulting in action of some kind, are called **motor impulses**.

The nerves which connect the body to the central nervous system make up the **peripheral nervous system**.

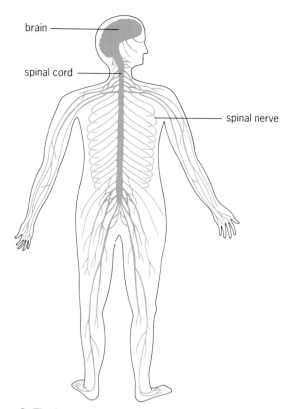

Figure 2 The human nervous system

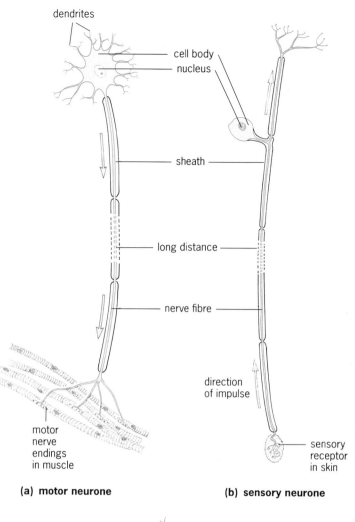

(a) motor neurone　　　**(b) sensory neurone**

(c) multi-polar neurone (in brain or spinal cord)

Figure 3 Nerve cells (neurones)

Nerve cells (neurones)

The central nervous system and the peripheral nerves are made up of nerve cells, called **neurones**. Three types of neurone are shown in Fig. 3. The **motor neurones** carry impulses from the central nervous system to muscles and glands. The **sensory neurones** carry impulses from the sense organs to the central nervous system. The **multi-polar neurones** are neither sensory nor motor but make connections to other neurones inside the central nervous system.

Each neurone has a **cell body** consisting of a nucleus surrounded by a little cytoplasm. Branching fibres, called **dendrites**, from the cell body make contact with other neurones. A long filament of cytoplasm, surrounded by an insulating sheath, runs from the cell body of the neurone. This filament is called a **nerve fibre** (Figs 3(a) and (b)). The cell bodies of the neurones are mostly located in the brain or in the spinal cord and it is the nerve fibres which run in the nerves. A **nerve** is easily visible, white, tough and stringy and consists of hundreds of microscopic nerve fibres bundled together (Fig. 4). Most nerves will contain a mixture of sensory and motor fibres. So a nerve can carry many different impulses. These impulses will travel in one direction in sensory fibres and in the opposite direction in motor fibres.

Some of the nerve fibres are very long. The nerve fibres to the foot have their cell bodies in the spinal cord and the fibres run inside the nerves, without a break, down to the skin of the toes or the muscles of the foot. Thus a single nerve cell may have a fibre about 1 metre long.

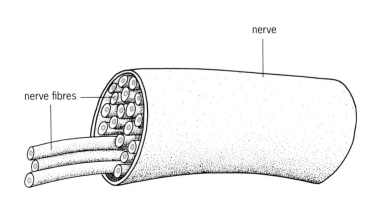

Figure 4 Nerve fibres grouped into a nerve

QUESTIONS

1 What is the difference between a nerve and a nerve fibre?
2 In what ways are sensory neurones and motor neurones similar **a** in structure, **b** in function? How do they differ?
3 Can **a** a nerve fibre, **b** a nerve, carry both sensory and motor impulses? Explain your answers.

Synapse

Although nerve fibres are insulated, it is necessary for impulses to pass from one neurone to another. An impulse from the finger-tips has to pass through at least three neurones before reaching the brain and so produce a conscious sensation. The regions where impulses are able to cross from one neurone to the next are called **synapses**.

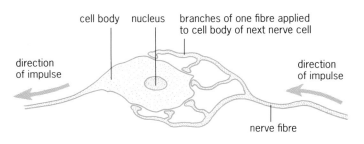

Figure 5 Diagram of synapses

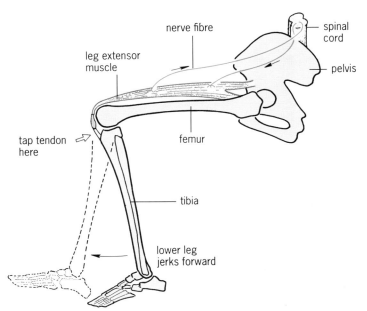

Figure 6 The reflex knee jerk

At a synapse, a branch at the end of one fibre is in close contact with the cell body or dendrite of another neurone (Fig. 5). When an impulse arrives at the synapse, it releases a tiny amount of a chemical substance (a **neurotransmitter**) which sets off an impulse in the next neurone. Sometimes several impulses have to arrive at the synapse before enough chemical is released to cause an impulse to be fired off in the next neurone.

The nerve impulse

The nerve fibres do not carry sensations like pain or cold. These sensations are felt only when a nerve impulse reaches the brain. The impulse itself is a series of electrical pulses which travel down the fibre. Each pulse lasts about 0.001 second and travels at speeds of up to 100 metres per second. All nerve impulses are similar; there is no difference between nerve impulses from the eyes, ears or hands.

We are able to tell where the sensory impulses have come from and what caused them only because the impulses are sent to different parts of the brain. The nerves from the eye go to the part of the brain concerned with sight. So when impulses are received in this area, the brain recognizes that they have come from the eyes and we 'see' something.

QUESTIONS

4 Look at Fig. 7. **a** How many cell bodies are drawn? **b** How many synapses are shown? Look at Fig. 9 and answer the same questions.
5 If you could intercept and 'listen to' the nerve impulses travelling in the spinal cord, could you tell which ones came from pain receptors and which from temperature receptors? Explain your answer.

The reflex arc

One of the simplest situations where impulses cross synapses to produce action is in the reflex arc. A **reflex action** is an automatic response to a stimulus. When a particle of dust touches the cornea of the eye, you will blink; you cannot prevent yourself from blinking. A particle of food touching the lining of the windpipe will set off a coughing reflex which cannot be suppressed. When a bright light shines in the eye, the pupil contracts (see p. 174). You cannot stop this reflex and you are not even aware that it is happening.

The nervous pathway for such reflexes is called a **reflex arc**. In Fig. 6 the nervous pathway for a well-known reflex called the 'knee-jerk' reflex is shown.

One leg is crossed over the other and the muscles are totally relaxed. If the tendon just below the kneecap of the upper leg is tapped sharply, a reflex arc makes the thigh muscle contract and the lower part of the leg swings forward.

The pathway of this reflex arc is traced in Fig. 7. Hitting the tendon stretches the muscle and stimulates a stretch receptor. The receptor sends off impulses in a sensory fibre. These sensory impulses travel in the nerve to the spinal cord.

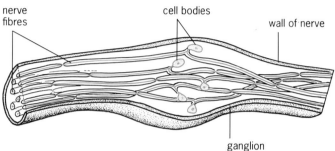

(which means that sensory impulses must be reaching the brain), but there is nothing you can do to stop it.

Spinal cord

In Fig. 7 the spinal cord is drawn in transverse section. The spinal nerve divides into two 'roots' at the point where it joins the spinal cord. All the sensory fibres enter through the **dorsal root** and the motor fibres all leave through the **ventral root**, but both kinds of fibre are contained in the same spinal nerve. This is like a group of insulated wires in the same electric cable. The cell bodies of all the sensory fibres are situated in the dorsal root and they make a bulge called a **ganglion** (Fig. 8).

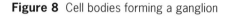

Figure 8 Cell bodies forming a ganglion

Figure 7 The reflex arc. This reflex arc needs only one synapse for making the response. Most reflex actions need many more synapses (a) to adjust other muscles in the body and (b) to send impulses to the brain.

In the central region of the spinal cord, the sensory fibre passes the impulse across a synapse to a motor neurone which conducts the impulse down the fibre, back to the thigh muscle. The arrival of the impulses at the muscle makes it contract, and jerk the lower part of the limb forward. You are aware that this is happening

In even the simplest reflex action, many more nerve fibres, synapses and muscles are involved than are described here. In Fig. 9 the reflex arc which would result in the hand being removed from a painful stimulus is illustrated. On the left side of the spinal cord, an incoming sensory fibre makes its first synapse with a **connector neurone** (or association neurone). This can pass the

Figure 9 Reflex arc (withdrawal reflex)

impulse on to many other motor neurones, although only one is shown in the diagram. On the right side of the spinal cord, some of the incoming sensory fibres are shown making synapses with neurones which send nerve fibres to the brain, thus keeping the brain informed about events in the body. Also, nerve fibres from the brain make synapses with motor neurones in the spinal cord so that 'commands' from the brain can be sent to muscles of the body.

Reflexes

The reflex just described is a **spinal reflex**. The brain, theoretically, is not needed for it to happen. Responses which take place in the head, such as blinking, coughing and iris contraction, have their reflex arcs in the brain, but may still not be consciously controlled.

Bright light stimulates the light-sensitive cells of the retina. The nerve impulses in the sensory fibres from these receptors travel through the optic nerve to the brain. In the mid-brain (p. 184) the fibres synapse with connector and motor fibres which carry impulses back through the optic nerve to the sphincter muscle of the iris and stimulate it to contract.

The reflex closure of the iris (p. 174) protects the retina from bright light; the withdrawal reflex removes the hand from a dangerously hot object; the coughing reflex dislodges a foreign particle from the windpipe. Thus, these reflexes have a protective function.

There are many other reflexes going on inside our bodies. We are usually unaware of these, but they maintain our blood pressure, breathing rate, heart beat, etc. and so maintain the body processes.

Innate behaviour and learning

A day-old chick will peck at objects on the ground when it sees its mother pecking. It does not have to learn this behaviour; it is born with the nerve 'circuits' that produce this behaviour already connected. This is **innate** behaviour (sometimes called 'instinct') and it may involve a simple reflex such as pecking, or more complex reactions such as crouching and remaining motionless in response to the parent's warning cries.

At first, the chick pecks indiscriminately at small objects on the ground. Some of these will taste unpleasant and the pecking reflex is modified to avoid these and select only palatable material. This is **learned** behaviour. The innate reflex has been modified, by trial and error and the chick has been **conditioned** by its experience.

Learned behaviour may become very complex, e.g. titmice and squirrels learning to overcome complicated obstacles in order to obtain food (Fig. 10). In humans, learning may involve conditioned reflexes but it also involves laying down a memory store in the brain. Human learned behaviour depends largely on being able to draw on the memory store either consciously or unconsciously.

Survival value

The ability to learn by conditioned reflex, or by any other process, has survival value (see p. 222). An animal which has learned to avoid danger or to take advantage of a favourable situation, is likely to live longer and to leave more offspring.

Figure 10 The blue tit has learned to pull out the matchstick to release the peanuts

Voluntary actions

A voluntary action starts in the brain. It may be the result of external events, such as seeing a book on the floor, but any resulting action, such as picking up the book, is entirely voluntary. Unlike a reflex action it does not happen automatically; you can decide whether or not you carry out the action.

The brain sends motor impulses down the spinal cord in the nerve fibres. These make synapses with motor fibres which enter spinal nerves and make connections to the sets of muscles needed to produce effective action. Many sets of muscles in the arms, legs and trunk would be brought into play in order to stoop and pick up the book, and impulses passing between the eyes, brain and arm would direct the hand to the right place and 'tell' the fingers when to close on the book.

One of the main functions of the brain is to co-ordinate these actions so that they happen in the right sequence and at the right time and place.

QUESTIONS

6 Put the following in the correct order for a simple reflex arc: **a** impulse travels in motor fibre, **b** impulse travels in sensory fibre, **c** effector organ stimulated, **d** receptor organ stimulated, **e** impulse crosses synapse.

7 Which receptors and effectors are involved in the reflex actions of **a** sneezing, **b** blinking, **c** contraction of the iris?

8 Explain why the tongue may be considered to be both a receptor and an effector organ.

9 Discuss whether coughing is a voluntary or reflex action.

10 Give some examples of behaviour patterns in the domesticated dog which you would consider to be **a** innate, **b** learned.

11 What is the possible survival value of an animal learning to remain motionless ('freeze') when a predator is nearby?

Central nervous system

Spinal cord

Like all other parts of the nervous system, the spinal cord consists of thousands of nerve cells. The structure of the spinal cord is shown in Figs 7, 9 and 11.

Figure 11 Section through spinal cord (x7). The light area is the white matter, consisting largely of nerve fibres running to and from the brain. The darker central area is the grey matter, consisting largely of nerve cell bodies.

All the cell bodies, apart from those in the dorsal root ganglia, are concentrated in the central region called the **grey matter**. The **white matter** consists of nerve fibres. Some of these will be passing from the grey matter to the spinal nerves and others will be running along the spinal cord connecting the spinal nerve fibres to the brain. The spinal cord is thus concerned with (a) reflex actions involving body structures below the neck, (b) conducting sensory impulses from the skin and muscles to the brain and (c) carrying motor impulses from the brain to the muscles of the trunk and limbs.

The brain

The brain may be thought of as the expanded front end of the spinal cord. Certain areas are greatly enlarged to deal with all the information arriving from the ears, eyes, tongue, nose and semicircular canals. A simplified diagram of the main regions of the brain as seen in vertical section is given in Fig. 12(d). The **medulla** is concerned with regulation of the heart beat, body temperature and breathing rate. The **cerebellum** controls posture, balance and co-ordinated movement. The **mid-brain** deals with reflexes involving the eye. The largest part of the brain, however, consists of the **cerebrum**, made up of two **cerebral hemispheres**. These are very large and highly developed in mammals, especially

(a) The front end of the spinal cord develops three bulges: the fore-, mid- and hind-brain. Each region receives impulses mainly from sense organs in the head.

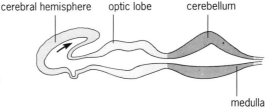

(b) The roofs of the fore-, mid- and hind-brain become thicker and form the cerebral hemispheres, optic lobes and cerebellum. The floor of the hind-brain thickens to form the medulla.

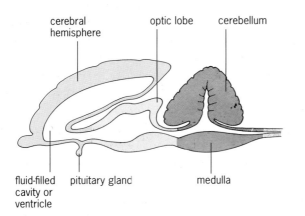

(c) A rabbit's brain would look something like this in vertical section.

(d) The same regions are present in a human brain, but because of our upright position, the brain is bent through 90°.

Figure 12 Development of the brain of a mammal (vertical sections) humans, and are thought to be the regions concerned with intelligence, memory, reasoning ability and acquired skills.

In the cerebral hemispheres and the cerebellum, there is an outer layer of grey matter, the **cortex**, with hundreds of thousands of multi-polar neurones (Fig. 3(c)) forming the outer layers and making possible an enormous number of synapse connections between the dendrites (Fig. 13).

Figure 13 Multi-polar neurones in the cerebral cortex (x350). The cell bodies and their branching fibres have been darkly stained to make them show up.

Association centres

The cerebral cortex, and many other regions of the brain, function as association centres. In these centres, sensory information from a variety of sources (eyes, ears, muscles, skin, etc.) is fed in. This information is correlated with data from the memory centres and a course of action (or inaction) is computed. Motor impulses are then sent to the various brain centres concerned with action.

Functions of the brain

To sum up:

1 The brain receives impulses from all the sensory organs of the body.
2 As a result of these sensory impulses, it sends off motor impulses to the glands and muscles, causing them to function accordingly.
3 In its association centres it correlates the various stimuli from the different sense organs and the memory.
4 The association centres and motor areas co-ordinate bodily activities so that the mechanisms and chemical reactions of the body work efficiently together.
5 It 'stores' information so that behaviour can be modified according to past experience.

QUESTIONS

12 Would you expect synapses to occur in grey matter or in white matter? Explain your answer.
13 Look at Fig. 2. If the spinal cord were damaged at a point about one-third of the way up the vertebral column, what effect would you expect this to have on the bodily functions?
14 a With which senses are the fore-, mid- and hind-brain mainly concerned? **b** Which part of the brain seems to be mainly concerned with keeping the basic body functions going?
15 Describe the biological events involved when you hear a sound and turn your head towards it.

The endocrine system

Co-ordination by the nervous system is usually rapid and precise. Nerve impulses, travelling at up to 100 metres per second, are delivered to specific parts of the body and produce an almost immediate response. A different kind of co-ordination is brought about by the endocrine system. This system depends on chemicals, called **hormones**, which are released from special glands, called **endocrine glands**, into the bloodstream. The hormones circulate round the body in the blood and eventually reach certain organs, called **target organs**. Hormones speed up or slow down or alter the activity of those organs. After being secreted, hormones do not remain permanently in the blood but are changed by the liver into inactive compounds and excreted by the kidneys.

Unlike the digestive glands, the endocrine glands do not deliver their secretions through ducts. For this reason, the endocrine glands are sometimes called the 'ductless glands'. The hormones are picked up directly from the glands by the blood circulation.

Responses of the body to hormones are much slower than responses to nerve impulses. They depend, in the first instance, on the speed of the circulatory system and then on the time it takes for the cells to change their chemical activities. Many hormones affect long-term changes such as growth rate, puberty and pregnancy. Nerve impulses often cause a response in a very limited area of the body, such as an eye-blink or a finger movement. Hormones often affect many organ systems at once.

Serious deficiencies or excesses of hormone production give rise to illnesses. Small differences in hormone activity between individuals probably contribute to differences of personality and temperament.

The position of the endocrine glands in the body is shown in Fig. 14. Notice that the pancreas and the reproductive organs have a dual function.

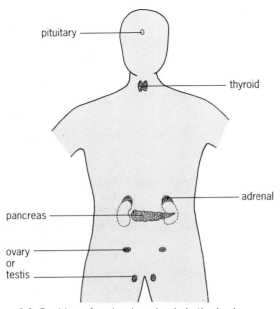

pituitary

thyroid

adrenal

pancreas

ovary
or
testis

Figure 14 Position of endocrine glands in the body

Thyroid gland

The thyroid gland is situated in the front part of the neck and lies in front of the windpipe. It produces a hormone called **thyroxine**, which is formed from an amino acid and iodine (p. 96). Thyroxine has a stimulatory effect on the metabolic rate of nearly all the body cells. It controls our level of activity, promotes normal skeletal growth and is essential for the normal development of the brain.

Adrenal glands

These glands are attached to the back of the abdominal cavity, one above each kidney. Each adrenal gland is made up of two distinct regions with different functions. There is an outer layer called the

adrenal cortex and an inner zone called the **adrenal medulla**. The medulla receives nerves from the brain and produces the hormone **adrenaline**. The cortex has no nerve supply and produces a number of hormones called **corticosteroids**. The corticosteroids help to control the metabolism of carbohydrates, fats, proteins, salts and water.

Adrenaline, from the medulla, has less important but more obvious effects on the body. In response to a stressful situation, nerve impulses are sent from the brain to the adrenal medulla which releases adrenaline into the blood. As adrenaline circulates round the body it affects a great many organs, as shown in Table 1 below.

All these effects make us more able to react quickly and vigorously in dangerous situations that might require us to run away or put up a struggle. However, in many stressful situations, such as taking examinations or giving a public performance, vigorous activity is not called for. So the extra adrenaline in our bodies just makes us feel tense and anxious. You will recognize the sensations described in column 4 of Table 1 as characteristic of fear and anxiety.

Most of the systems affected by adrenaline are also controlled by a section of the nervous system called the **sympathetic nervous system** which affects internal organs other than voluntary muscles. It is difficult to tell whether it is adrenaline or the sympathetic nervous system which is mainly responsible for the stress reactions. It is known, however, that loss of the adrenal medulla seems to cause no ill-effects.

Adrenaline is quickly converted by the liver to a less active compound which is excreted by the kidneys. All hormones are similarly altered and excreted, some within minutes, others within days. Thus their effects are not long-lasting. The long-term hormones, such as thyroxine, are secreted continuously to maintain a steady level.

Table 1 Responses to adrenaline

Target organ	Effects of adrenaline	Biological advantage	Effect or sensation
Heart	beats faster	sends more glucose and oxygen to the muscles	thumping heart
Breathing centre of the brain	faster and deeper breathing	increased oxygenation of the blood; rapid removal of carbon dioxide	panting
Arterioles of the skin	constricts them (see p. 148)	less blood going to the skin means more is available to the muscles	person goes paler
Arterioles of the digestive system	constricts them	less blood for the digestive system allows more to reach the muscles	dry mouth
Muscles of alimentary canal	relax	peristalsis and digestion slow down; more energy available for action	'hollow' feeling in stomach
Muscles of body	tenses them	ready for immediate action	tense feeling; shivering
Liver	conversion of glycogen to glucose	glucose available in blood for energy production	no sensation
Fat depots	conversion of fats to fatty acids	fatty acids available in blood, for muscle contraction	

The pancreas

The pancreas is a digestive gland which secretes enzymes into the duodenum through the pancreatic duct (p. 108). It is also an endocrine (ductless) gland. Most of the pancreas cells produce digestive enzymes but some of them produce hormones. The hormone-producing cells are arranged in small isolated groups called **islets** (Fig. 15) and secrete their hormones directly into the bloodstream. One of the hormones is called **glucagon** and the other is **insulin**.

Figure 15 Section of pancreas tissue showing islets (x250)

If the level of sugar in the blood falls, the islets release glucagon into the bloodstream. Glucagon acts on the cells in the liver and causes them to convert some of their stored glycogen into glucose and so restore the blood sugar level (p. 112).

Insulin has the opposite effect to glucagon. If the concentration of blood sugar increases (e.g. after a meal rich in carbohydrate), insulin is released from the islet cells. When the insulin reaches the liver it stimulates the liver cells to take up glucose from the blood and store it as glycogen.

Insulin has many other effects; it increases the uptake of glucose in all cells; it promotes the conversion of carbohydrates to fats and slows down the conversion of protein to carbohydrate.

All these changes have the effect of regulating the level of glucose in the blood to within narrow limits—a very important example of homeostasis (p. 143).

If anything goes wrong with the production or function of insulin, the person will show the symptoms of **diabetes**.

Diabetes

This may result from a failure of the islet cells to produce sufficient insulin or from the reduced ability of the body cells to use it. Two forms of diabetes are recognized, juvenile-onset and adult-onset diabetes.

Juvenile-onset diabetes is the less common form. It mainly affects young people and results from the islets producing too little insulin. There is a slight inherited tendency towards the disease, but it may be triggered by some event, possibly a virus infection, which causes the body's immune system to attack the islet cells which produce insulin. It is therefore classed as an **autoimmune** disease. The outcome is that the patient's blood is deficient in insulin and he or she needs regular injections of the hormone in order to control blood sugar level and so lead a normal life. This form of the disease is, therefore, sometimes called 'insulin-dependent' diabetes.

Adult-onset diabetes usually affects people after the age of 40. The level of insulin in their blood is often not particularly low but it seems that their bodies are unable to use the insulin properly. This condition can be controlled by careful regulation of the diet and does not usually require insulin injections.

In both forms of diabetes, the patient is unable to regulate the level of glucose in the blood. It may rise to such a high level that it is excreted in the urine or fall so low that the brain cells cannot work properly and the person goes into a coma.

All diabetics need a carefully regulated diet to reduce the intake of carbohydrates and keep the blood sugar within reasonable limits.

Reproductive organs

These produce hormones as well as gametes (sperms and ova) and their effects have been described on p. 158.

The hormones from the ovary, **oestrogen** and **progesterone**, both prepare the uterus for the implantation of the embryo, by making its lining thicker and increasing its blood supply.

The hormones **testosterone** (from the testes) and oestrogen (from the ovaries) play a part in the development of the secondary sexual characters as described on p. 158.

During pregnancy, the placenta produces a hormone which has effects similar to those of progesterone.

Pituitary gland

This gland is attached to the base of the brain (Fig. 12(d)). It produces many hormones. One of these (**antidiuretic hormone**, ADH) acts on the kidneys and regulates the amount of water reabsorbed in the kidney tubules (p. 141). Another pituitary hormone (growth hormone) affects the growth rate of the body as a whole and the skeleton in particular. Several of the pituitary hormones act on the other endocrine glands and stimulate them to produce their own hormones. For example, the pituitary releases into the blood a **follicle-stimulating hormone** (FSH) which, when it reaches the ovaries, makes one of the follicles start to mature and to produce oestrogen. **Luteinising hormone** (LH) is also produced from the pituitary and, together with FSH, induces ovulation.

A **thyroid-stimulating hormone** (TSH) acts on the thyroid gland and makes it produce thyroxine.

Homeostasis and feedback

Homeostasis

The endocrine system plays an important part in maintaining the composition of the body fluids (homeostasis, see p. 143).

A rise in blood sugar after a meal stimulates the pancreas to produce insulin. The insulin causes the liver to remove the extra

187

glucose from the blood and store it as glycogen (p. 112). This helps to keep the concentration of blood sugar within narrow limits.

The brain monitors the concentration of the blood passing through it. If the concentration is too high, the pituitary gland releases ADH (antidiuretic hormone). When this reaches the kidneys (the target organs) it causes them to reabsorb more water from the blood passing through them (p. 141). If the blood is too dilute, production of ADH is suppressed and less water is absorbed in the kidneys. Thus ADH helps to maintain the amount of water in the blood at a fairly constant level.

Feedback

Some of the endocrine glands are themselves controlled by hormones. For example, pituitary hormones such as LH (luteinising hormone) affect the endocrine functions of the ovaries. In some cases, the output of hormones is regulated by a process of **negative feedback**.

The negative feedback mechanism of temperature control is described on p. 149.

The feedback between the pituitary and the ovaries produces fluctuations which cause the menstrual cycle (p. 158).

When the level of oestrogen in the blood rises, it affects the pituitary gland, suppressing its production of FSH (follicle-stimulating hormone). A low level of FSH in the blood reaching the ovary will cause the ovary to slow down its production of oestrogen. With less oestrogen in the blood, the pituitary is able to resume its production of FSH which, in turn, makes the ovary start to produce oestrogen again (Fig. 16). This cycle of events takes about 1 month and is the basis of the monthly menstrual cycle.

The oestrogen and progesterone in the female contraceptive pill act on the pituitary and suppress the production of FSH. If there is not enough FSH, none of the follicles in the ovary will grow to maturity and so no ovum will be released.

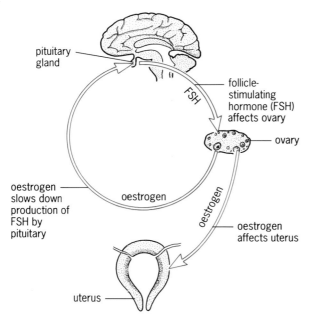

oestrogen slows down production of FSH by pituitary

pituitary gland

FSH

follicle-stimulating hormone (FSH) affects ovary

ovary

oestrogen

oestrogen

oestrogen affects uterus

uterus

Figure 16 Feedback

Table 2 Endocrine and nervous control compared

Endocrine	Nervous
transmission of chemicals	transmission of electrical impulses
transmission via blood	transmission in nerves
slow transmission	rapid transmission
hormones dispersed throughout body	impulse sent directly to target organ
long-term effects	short-lived effects

Therapeutic and industrial uses of hormones

In humans, hormones can be used to treat various conditions. A pituitary growth hormone has been used successfully to promote the growth of abnormally small children; oestrogen is used to prevent the thinning of bones (osteoporosis) in women after menopause (hormone replacement therapy or HRT).

LH and FSH help to cure certain cases of infertility; progesterone and oestrogen form the basis of the contraceptive pill. Injection of insulin is effective in treating juvenile-onset diabetes.

All these forms of hormone therapy have their drawbacks (side effects). HRT may increase the risk of cancer; LH may lead to multiple births. Some treatments, such as insulin injection, have been tried, tested and refined over many years; others are still in the early stages of development and may eventually be superseded by safer methods of treatment.

In agriculture, hormones may be used to increase productivity. Anabolic steroids (p. 22) are given to beef cattle to increase their meat production; bovine somatotrophin (BST) given to dairy cows can increase their milk yields by up to 25 per cent. There is still a good deal of controversy over the use of these hormones and BST is not licensed for use in Britain. It is claimed that the hormone makes cows more susceptible to disease and there are worries about the levels of the hormone or its residues in the milk.

The American Food and Drug Administration stated in February 1994 that milk produced with the aid of BST was indistinguishable from milk produced without it. However, the West does not really need more milk so the use of the hormone is related to profitability (for large-scale dairy farms and drug companies) rather than to food production.

Checklist

- The body systems are made to work efficiently together by the nervous system and the endocrine system.

Nervous system

- The nervous system consists of the brain, the spinal cord and the nerves.
- The nerves consist of bundles of nerve fibres.
- Each nerve fibre is a thin filament which grows out of a nerve cell body.
- The nerve cell bodies are mostly in the brain and spinal cord.
- Nerve fibres carry electrical impulses from sense organs to the brain or from the brain to muscles and glands.
- A reflex is an automatic nervous reaction that cannot be consciously controlled.
- A reflex arc is the nervous pathway which carries the impulses causing a reflex action.

- The simplest reflex involves a sensory nerve cell and a motor nerve cell, connected by synapses in the spinal cord.
- The brain and spinal cord contain millions of nerve cells.
- The millions of possible connections between the nerve cells in the brain allow complicated actions, learning, memory and intelligence.

Endocrine system

- The thyroid, adrenal and pituitary are all endocrine glands.
- The testes, ovaries and pancreas are also endocrine glands in addition to their other functions.
- The endocrine glands release hormones into the blood system.
- When the hormones reach certain organs they change the rate or kind of activity of the organ.
- Too much or too little of a hormone can cause a metabolic disorder.
- Under-production of insulin causes diabetes.

20 PERSONAL HEALTH

Diet; dental health

Smoking; exercise

Mood-influencing drugs
Tolerance and dependence.
  Stimulants; depressants
Alcohol. Hallucinogens. Narcotics.

Solvent misuse

Health is not merely the absence of disease but a state of physical and mental well-being. However, you are not usually aware of your state of health until you are unwell. Therefore you may need to make a conscious effort to maintain good health.

The general rules for health are:

Do   eat a balanced diet
       take regular vigorous exercise
       develop a positive attitude to life

Don't smoke
       drink too much
       misuse drugs.

If you feel this recipe for health is 'boring', ask yourself whether being fat, having bronchitis, craving drugs and alcohol, and looking forward to premature heart failure are likely to make you happy and content.

You may know middle-aged or elderly people who claim to have smoked, got drunk, over-eaten and taken no exercise all their lives, and still remain healthy. You will also have heard of fit people who exercise regularly and do not smoke but die suddenly from a heart attack.

These stories prove nothing. We inherit part of our 'constitution' from our parents. So some people are more able than others to resist disease and to ill-use their bodies. You can do nothing about your genes, but you can do a lot to make the best of the constitution you have inherited.

You can find physiological reasons for some of the 'dos and don'ts' listed in the preceding chapters. This chapter will refer to them only briefly. In some cases, such as smoking and drug abuse, there is overwhelming evidence to support the advice. In others, such as diet and exercise, the evidence is less strong, but the advice given represents most expert opinion.

Diet

To remain healthy, you must eat enough food to meet all your energy requirements. It must contain protein and fat to make new cells. You also need vitamins and salts (see pp. 95–97).

A healthy diet should contain only a little salt and refined sugar (if any) and a low proportion of fat. The diet also needs to include plenty of vegetable fibre (Fig. 1). A low fat intake may help reduce the chances of arterial and heart disease (p. 128). A high fibre intake probably helps to prevent diseases of the large intestine.

In some people, over-eating can lead to obesity with its attendant problems (such as high blood pressure and diabetes).

A healthy digestive system also depends on regular meals eaten in a relaxed atmosphere. Hasty snacks and irregular heavy meals can lead to digestive disorders.

The health aspects of diet are discussed more fully on pp. 99–100.

Figure 1 Some components of a healthy diet

Dental health

Sugary food and the neglect of oral hygiene can lead to toothache, gum disease and, ultimately, the loss of the teeth and the need to wear dentures.

Dental decay (dental caries)

Decay begins when small holes (cavities) appear in the enamel. The cavities are caused by bacteria on the tooth surface. The bacteria produce acids which dissolve the calcium salts in the tooth enamel. The enamel and dentine are dissolved away in patches, forming cavities. The cavities reduce the distance between the outside of the tooth and the nerve endings. The acids produced by the bacteria irritate the nerve endings and cause toothache. If the cavity is not cleaned and filled by a dentist, the bacteria will get into the pulp cavity and cause a painful abscess at the root. Often, the only way to treat this is to have the tooth pulled out.

Although some people's teeth are more resistant to decay than others, it seems that it is the presence of refined sugar (sucrose) that contributes to decay.

Western diets contain a good deal of refined sugar and children suck sweets between one meal and the next. The high level of dental decay in Western society is thought to be caused mainly by keeping sugar in the mouth for long periods of time.

The graph in Fig. 2(a) shows how the pH in the mouth falls (i.e. becomes more acid) when a single sweet is sucked. The pH below which the enamel is attacked is called the **critical pH** (between 5.5 and 6). In this case, the enamel is under acid attack for about 10 minutes.

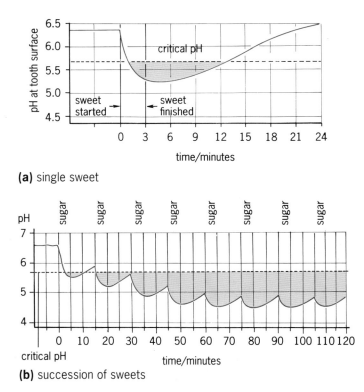

(a) single sweet

(b) succession of sweets

Figure 2 pH in the mouth when sweets are sucked

The graph in Fig. 2(b) shows the effect of sucking sweets at the rate of four an hour. In this case the teeth are exposed to acid attack almost continually.

The best way to prevent tooth decay, therefore, is to avoid eating sugar at frequent intervals either in the form of sweets or in sweet drinks such as orange squash or cola drinks.

Brushing the teeth or rinsing the mouth does little to prevent dental decay. However, if a fluoride toothpaste is used it does help to increase the resistance of enamel to bacterial acids. Adding fluoride to drinking water also reduces the incidence of dental decay but it is a controversial measure. Brushing the teeth is very important in the prevention of gum disease, which causes more tooth loss than caries does.

Gum disease (periodontal disease)

There is usually a layer of saliva and mucus over the teeth. This layer contains bacteria which live on the food residues in the mouth, building up a coating on the teeth, called **plaque**. If the plaque is not removed, mineral salts of calcium and magnesium are deposited on it, forming a hard layer of 'tartar' or **calculus**. If the bacterial plaque which forms on teeth is not removed regularly, it spreads down the tooth into the narrow gap between the gum and enamel. Here it causes inflammation, called **gingivitis**, which leads to redness and bleeding of the gums and to bad breath. It also causes the gums to recede and expose the cement. If gingivitis is not treated, it progresses to **periodontitis**; the fibres holding the tooth in the jaw are destroyed, so the tooth becomes loose and falls out or has to be pulled out.

There is evidence that cleaning the teeth does help to prevent gum disease. It is best to clean the teeth about twice a day using a toothbrush. No one method of cleaning has proved to be any better than any other but the cleaning should attempt to remove all the plaque from the narrow crevice between the gums and the teeth.

Drawing a waxed thread ('dental floss') between the teeth helps to remove plaque in these regions.

Smoking

There is a long list of diseases associated with smoking, including lung cancer, bronchitis, emphysema, arterial disease, stomach ulcers and bladder cancer. This does not mean that all smokers will develop these diseases. But the chances of their doing so are far higher than they are for non-smokers.

See pp. 126–7 and 135–6 for more detail.

QUESTIONS

1 List the features of your diet over the last two days which might be considered to make it **a** healthy, **b** unhealthy.
2 What are the most important things to do to avoid dental decay and gum disease?

Exercise

Exercise increases stamina, improves flexibility, makes muscles stronger and more efficient, and helps to keep your weight down (Fig. 3). Exercise may also reduce your chance of a premature coronary heart attack, though the evidence is not clear-cut.

Most people agree that exercise makes you 'feel good'. But to enjoy the benefits, you need to take exercise throughout your life.

Exercise is discussed in more detail on p. 167.

Figure 3 Exercise. Regular exercise contributes to good health.

Mood-influencing drugs

Any substance used in medicine to help our bodies fight illness or disease is called a **drug**. One group of drugs helps to control pain and ease feelings of distress. These are the mood-influencing drugs.

Sensations of excitement, well-being or depression are the products of our nervous and endocrine systems. The mood-influencing drugs mimic the effects of our hormones or neuro-transmitters (p. 181) and produce corresponding states of mind.

If these drugs are used wisely and under medical supervision, they can be very helpful. A person who feels depressed to the point of wanting to commit suicide may be able to lead a normal life with the aid of an antidepressant drug which removes the feeling of depression. However, if drugs are used for trivial reasons, to produce feelings of excitement or calm, they may be extremely dangerous. This is because they can cause **tolerance** and **dependence**.

Tolerance

This means that if the substance is taken over a long period, the dosage has to keep increasing in order to have the same effect. People who take sleeping pills containing barbiturates may need to increase the dose from one to two or three tablets in order to get to sleep. People who drink alcohol in order to relieve anxiety may find that they have to keep drinking more and more before they feel relaxed. If the dosage continues to increase it will become so large that it causes death.

Dependence

This is the term used to describe the condition in which the user cannot do without the substance.

Sometimes a distinction is made between emotional and physical dependence, though the distinction is not always clear. A person with emotional dependence may feel a craving for the substance, may be bad-tempered, anxious or depressed without it, and may commit crimes in order to obtain it. Cigarette smoking is one example of emotional dependence.

Physical dependence involves the same experiences. But there are also physical symptoms, called **withdrawal symptoms**, when the substance is withheld. These may be nausea, vomiting, diarrhoea, muscular pain, uncontrollable shaking and hallucinations. Physical dependence is sometimes called **addiction**.

Not everyone who takes a mood-influencing drug develops tolerance or becomes dependent on it. Millions of people can take alcoholic drinks in moderation with no obvious physical or mental damage. Those who become dependent cannot drink in moderation. Their bodies seem to develop a need for permanently high levels of alcohol and dependent people (**alcoholics**) get withdrawal symptoms if they do not drink.

Physical and emotional dependence are very distressing states. Getting hold of the substance becomes the centre of the addicts' lives, and they lose interest in their personal appearance, their jobs and their families. Because the substances they need cannot be obtained legally or because they need the money to buy them, they turn to crime. Cures are slow, difficult and usually unpleasant.

There is no way of telling in advance which person will become dependent and which will not. Dependence is much more likely with some drugs than with others, however. These are, therefore, prescribed with great caution. Experimenting with drugs for the sake of emotional excitement is extremely unwise. Some people may become dependent on almost any substance which gives them a conscious sensation.

Some of the mood-influencing drugs will now be considered.

Stimulants

The **caffeine** in coffee, tea and cocoa is a mild stimulant and makes you more wakeful. In normal use there is no build up of tolerance or dependence.

Amphetamines reduce fatigue and increase alertness but they also reduce accuracy and give a false sense of confidence.

If taken to improve athletic performance they can cause dangerously high blood pressure. Also they are quite useless for helping examination candidates because although they increase confidence, they also reduce accuracy. Use of amphetamines can lead to tolerance and addiction.

Cocaine gives a temporary feeling of excitement but this is soon followed by depression and listlessness. It is very addictive and prolonged use constricts the arteries and causes mental disorders.

Depressants

Depressants, e.g. sedatives, act on the central nervous system to decrease emotional tension and anxiety. Different types of sedative probably affect different areas of the brain but they all lead to relaxation and, in sufficient doses, to sleep or anaesthesia. In excessive doses, they suppress the breathing centre of the brain and cause death.

Tranquillizers

Some tranquillizers have been extremely valuable in treating severe mental illnesses such as schizophrenia and mania. Many thousands of mentally ill patients have been able to leave hospital and live normal lives as a result of using the tranquillizing drug **chlorpromazine**.

Nowadays, tranquillizers are being prescribed in their millions for the relief of anxiety and tension. Some people think that these drugs are being used merely to escape the stresses of everyday life that could be overcome by a little more will-power and determination. Others think that there is no reason why people should suffer the distress of acute anxiety when drugs are available for its relief. On the other hand, some degree of anxiety is probably needed for mental and physical activity. These activities are unlikely to be very effective in people who tranquillize themselves every time a problem crops up. In some cases, use of certain tranquillizers has led to addiction.

Alcohol

The alcohol in wines, beer and spirits is a depressant of the central nervous system. Small amounts give a sense of well-being, with a release from anxiety. However, this is accompanied by a fall-off in performance in any activity requiring skill. It also gives a misleading sense of confidence in spite of the fact that one's judgement is clouded. The drunken driver usually thinks he or she is driving extremely well.

Even a small amount of alcohol in the blood increases our reaction time (the interval between receiving a stimulus and making a response). In some people, the reaction time is doubled even when the alcohol in the blood is well below the legal limit laid down for car drivers (Fig. 4). This can make a big difference in the time needed for a driver to apply the brakes after seeing a hazard such as a child running into the road.

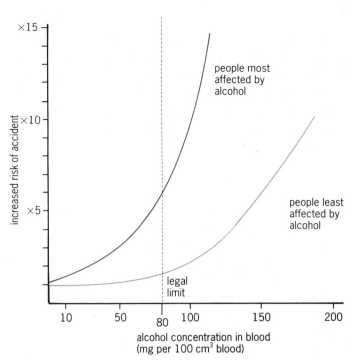

Figure 4 Increased risk of accidents after drinking alcohol. People vary in their reactions to alcohol. Body weight, for example, makes a difference.

Alcohol causes vasodilation in the skin, giving a sensation of warmth but in fact leading to a greater loss of body heat (see p. 148). A concentration of 500 mg of alcohol in 100 cm³ of blood results in unconsciousness. More than this will cause death because it stops the breathing centre in the brain. Ninety per cent of alcohol taken in is **detoxified** in the liver. That is, it is oxidized to carbon dioxide and water. Only 10 per cent is excreted by the kidneys. On average, the liver can oxidize about 75 mg alcohol per 1 kg body weight per hour. This rate varies considerably from one individual to the next but it indicates that it would take about 3 hours to oxidize the alcohol in a pint of beer or a glass of wine. If the alcohol intake exceeds this rate of oxidation, the level of alcohol in the blood builds up to toxic proportions; that is, it leads to **intoxication**.

Some people build up a tolerance to alcohol and this may lead to both emotional and physical dependence (alcoholism). High doses of alcohol can cause the liver cells to form too many fat droplets, leading to the disease called **cirrhosis**. A cirrhotic liver is less able to stop poisonous substances in the intestinal blood from reaching the general circulation (p. 113).

Pregnancy

Alcohol can cross the placenta and damage the fetus. Pregnant women who take as little as one alcoholic drink a day are at risk of having babies with lower than average birth weights. These underweight babies are more likely to become ill.

Heavy drinking during pregnancy can lead to deformed babies. All levels of drinking are thought to increase the risk of miscarriage.

Behaviour

Alcohol reduces inhibitions, because it depresses that part of the brain which causes shyness. This may be considered an advantage in 'breaking the ice' at parties. But it can also lead to irresponsible behaviour such as vandalism and aggression.

Moderate drinking

A moderate intake of alcoholic drink seems to do little physiological harm (except in pregnant women). But what is a 'moderate' intake?

A variety of drinks which all contain the same amount of alcohol is shown in Fig. 5. Beer is a fairly dilute form of alcohol. Whisky, however, is about 40 per cent alcohol. Even so, half a pint of beer contains the same amount of alcohol as a single whisky. This amount of alcohol can be called a 'unit'. It is the number of units of alcohol, not the type of drink, which has a physiological effect on the body. The Health Education Council recommends upper limits of 21 units for men and 14 units for women over a one-week period.

½ pint of beer or cider   1 glass of wine   1 glass of sherry   a single whisky

Figure 5 Alcohol content of drinks. All these drinks contain the same amount of alcohol (1 unit). Although the alcohol is more dilute in the beer than in the whisky, it has the same effect on the body.

Hallucinogens

Cannabis

Cannabis and other extracts of Indian hemp are chewed or smoked to produce a sense of well-being, detachment and sometimes hallucinations. There does not seem to be much evidence of tolerance or emotional and physical dependence. But unstable individuals looking for more 'exciting' experiences are thought to be likely to move on from cannabis to the 'hard drugs' such as morphine and heroin.

It is difficult to conduct research into the effects of an illegal drug. Nevertheless, evidence from animal studies and human surveys is accumulating and shows that cannabis-smoking can have harmful effects on the lungs, central nervous system, immune sys-

tem and reproductive function. A particular worry is the long time that the products persist in the body (days rather than hours) so that the user may be unaware of the impairment of performance, days after the last smoke.

Animal and human studies show that sperm count and sexual drive are diminished, the immune system is depressed, and learning ability and memory are impaired. These effects are in addition to the lung damage which is known to occur.

As in the case of tobacco, it may take many years to establish the full extent of the harmful side-effects of cannabis use.

Narcotics

Morphine, codeine and heroin are narcotics made from opium. Morphine and heroin relieve severe pain and produce short-lived feelings of well-being and freedom from anxiety. They can both lead to tolerance and physical dependence within weeks, and so they are prescribed with caution.

The illegal use of heroin has terrible effects on the unfortunate addict. The overwhelming dependence on the drug leads many addicts into prostitution and crime in order to obtain the money to buy it. There are severe withdrawal symptoms and a 'cure' is a long and often unsuccessful process.

An additional hazard is the blood poisoning, hepatitis and AIDS that may result from the use of unsterilized needles when injecting the drug.

Codeine is a less effective analgesic than morphine, but does not lead to dependence so easily. It is still addictive if used in large enough doses.

Solvent misuse

Solvent misuse or 'glue-sniffing' is the inhaling of vapours from various organic solvents (not only glue) in order to become intoxicated. The vapours produce effects similar to drunkenness, but they are more intense and last for only a few minutes. They include dizziness and stupor, followed by a loss of co-ordination and control and eventually by unconsciousness.

The short-term after-effects are headache, nausea, vomiting and, in some cases, convulsions. After-effects which last for several weeks include runny nose, bloodshot eyes, an acne-like rash round the mouth, irritability, lethargy and depression. In the long term, the liver and kidneys may be damaged.

The number of deaths resulting from solvent misuse has been increasing steadily in the last few years. It is estimated that 39 people died in the United Kingdom in 1981, 62 in 1982, 117 in 1985 and 122 in 1991[*]. Since then figures show a decrease. Death may result from (1) accidents while the person is intoxicated and out of control, (2) suffocation by the plastic bag used for inhaling, (3) choking on vomit while unconscious and (4) toxic effects of the solvent.

* Data from St George's Hospital Medical School

Solvent misusers are mostly adolescents of 14 or 15 years old or younger. They may experiment with 'glue-sniffing' as a form of rebellion against authority, to 'keep in' with their friends, to relieve boredom or for 'kicks' (risk-taking for the sake of a thrill). These reasons can, of course, be motives for other forms of drug misuse and destructive behaviour.

It is not clear whether solvent misuse leads to tolerance and dependence. Most 'glue-sniffers' do not continue once they are old enough to buy alcoholic drinks. But they are likely to drink too much, and to be tempted to move on to other drugs.

QUESTIONS

3 What is the difference between **a** becoming tolerant of a drug and **b** becoming dependent on a drug? Which of these do you think is meant by being 'hooked' on a drug?

4 Why are amphetamine stimulants unsuitable for improving performance in **a** athletics, **b** examinations?

5 Why should drinking alcohol make you 'feel' warm, but cause you to lose heat?

6 Why is it dangerous to take alcoholic drinks before driving?

7 If morphine and heroin make addicts 'feel good', why can't they keep taking steady low doses to stay in this 'happy' state?

Checklist

- A good diet and regular exercise contribute to good health.

- Smoking and excessive drinking contribute to ill-health.

- Mood-influencing drugs may be useful for treating certain illnesses but are dangerous if used for other purposes.

- Tolerance means that the body needs more and more of a particular drug to produce the same effect.

- Dependence means that a person cannot do without a particular drug.

- Withdrawal symptoms are unpleasant physical effects experienced by an addict when the drug is not taken.

- Alcohol is a depressant drug which slows down reaction time and reduces inhibitions.

- Alcohol in a pregnant woman's blood can damage her fetus.

- Heroin and morphine are strongly addictive drugs.

- Solvent sniffing produces intoxication and has caused an increasing number of deaths.

EXAMINATION QUESTIONS

Section 3 Human Physiology

Do not write on this page. Where necessary copy drawings, tables or sentences.

1 All arteries
A have valves   **C** carry oxygenated blood
B carry blood away   **D** have walls one cell thick
  from the heart   *[N]*

2 Explain how each of the following helps to preserve food.
(a) Refrigeration.   (2)
(b) Pickling.   (3)
                                                [S]

3 Give an account of the structures and processes involved in the digestion, absorption and assimilation of a piece of starchy food.   (15)
                                                [L]

4 Explain how attention to diet, exercise and hygiene can help keep your body healthy. Marks will be given both for showing knowledge and understanding and for the way your account is organised and expressed.   (8)
                                                [N]

5 The diagram shows a kidney and the tubes joined to it.

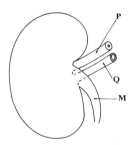

(a) (i) Name the tube labelled M.   (ii) Name the liquid carried by tube M.   (iii) To which other structure is tube M joined?
(b) (i) Which one of the blood vessels, P or Q, is an artery?   (ii) Give a reason for your answer in (b) (i).   (5)
                                                [L]

6 Compare nervous and hormonal communication. In your answer refer to FOUR of the following: the nature of the message; the speed of the message; the duration of the effect; the accuracy of the message and the speed of response.   (8)
                                                [M]

7 The diagram shows the teeth in the lower jaw of a human adult.
(a) Name the types of teeth labelled A and B.   (2)
(b) Describe how a molar tooth would look different in a carnivore (e.g. a cat) and in a herbivore (e.g. a sheep).   (2)
                                                [S]

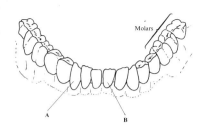

8 The diagram below shows a human uterus during pregnancy.

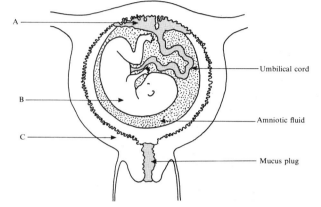

(i) Name the parts labelled A, B and C.   (3)
(ii) Give ONE function for each of the following: amniotic fluid; umbilical cord.   (2)
(iii) Name TWO substances that pass from the mother's blood to the blood of the fetus.   (2)
(iv) Use the diagram to suggest THREE changes that will occur before the birth of the fetus.   (3)
                                                [NI]

9 The diagram shows a section through a human hip joint.
(a) On the diagram, name the parts labelled A,B and C.   (3)
(b) Explain the function of the part labelled D.   (2)
                                                [N]

10 Excess proteins are broken down in the body. The waste nitrogen compounds produced from these proteins are excreted. Describe the processes involved. Your answer should include a description of how and where the waste nitrogen compounds are produced, how they are transported and how they are excreted. Marks will be given BOTH for showing knowledge and understanding AND for the way in which your account is organised and expressed.   (10)
                                                [N]

196

Genetics and Heredity

21 CELL DIVISION, CHROMOSOMES AND GENES

Heredity and genetics
Explanation of terms.

Chromosomes and mitosis
Mitosis: the movement of chromosomes at cell division. Function of chromosomes: genes on the chromosomes control the cell's physiology and structure. Number of chromosomes: a fixed number for each species.

Gamete production and chromosomes
Meiosis: the chromosomes are shared between the gametes. Meiosis and mitosis compared.

Genes
The structure of a gene. The role of DNA. The genetic code. Replication of DNA. Mutations.

Genetic engineering
Transfer of genes between species. Applications.

Practical work
Observing chromosomes in plant cells.

Heredity and genetics

We often talk about people inheriting certain characteristics: 'John has inherited his father's curly hair', or 'Mary has inherited her mother's blue eyes'. We expect tall parents to have tall children. The inheritance of such characteristics is called **heredity** and the branch of biology which studies how heredity works is called **genetics**.

Genetics also tries to forecast what sort of offspring are likely to be produced when plants or animals reproduce sexually. What will be the eye colour of children whose mother has blue eyes and whose father has brown eyes? Will a mating between a black mouse and a white mouse produce grey mice, black-and-white mice or some black and some white mice?

To understand the method of inheritance, we need to look once again at the process of sexual reproduction and fertilization. In sexual reproduction, a new organism starts life as a single cell called a **zygote** (p. 150). This means that you started from a single cell. Although you were supplied with oxygen and food in the uterus, all your tissues and organs were produced by cell division from this one cell. So, the 'instructions' that dictated which cells were to become liver, or muscle, or bone must all have been present in this first cell. The 'instructions' which decided that you should be tall or short, dark or fair, male or female must also have been present in the zygote.

To understand how these 'instructions' are passed from cell to cell, we need to look in more detail at what happens when the zygote divides and produces an organism consisting of thousands of cells. This type of cell division is called **mitosis**. It does not take place only in a zygote but occurs in all growing tissues.

QUESTION

1 a What are gametes? What are the male and female gametes of **i** plants and **ii** animals called, and where are they produced?
b What happens at fertilization?
c What is a zygote and what does it develop into?
See pp.75 and 150.

Chromosomes and mitosis

Mitosis

When a cell is not dividing, there is not much detailed structure to be seen in the nucleus even if it is treated with special dyes called stains. Just before cell division, a number of long, thread-like structures appear in the nucleus and show up very clearly when the nucleus is stained (Figs 1(a) and 2). These thread-like structures are called **chromosomes**. Although they are present in the nucleus all the time, they show up clearly only at cell division because at this time they get shorter and thicker.

Each chromosome is seen to be made up of two parallel strands, called **chromatids**. When the nucleus divides into two, one chromatid from each chromosome goes into each daughter nucleus. The chromatids in each nucleus now become chromosomes and later they will make copies of themselves ready for the next cell division. The process of copying is called **replication** because each chromosome makes a replica (an exact copy) of itself. As Fig. 1 is a diagram of mitosis, only two chromosomes are shown, but there are always more than this. Human cells contain 46 chromosomes.

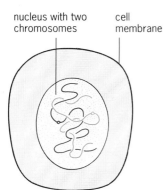

nucleus with two chromosomes

cell membrane

(a) Just before the cell divides, chromosomes appear in the nucleus.

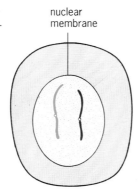

nuclear membrane

(b) The chromosomes get shorter and thicker.

two chromatids

two chromatids

(c) Each chromosome is now seen to consist of two chromatids.

fibres pull chromatids apart

(d) The nuclear membrane disappears and the chromatids are pulled apart to opposite ends of the cell

nuclear membrane forms

(e) a nuclear membrane forms round each set of chromatids, and the cell starts to divide.

nucleus with two chromosomes

'daughter' cell

(f) Cell division completed, giving two 'daughter' cells, each containing the same number of chromosomes as the parent cell.

Figure 1 Mitosis. Only one pair of chromosomes is shown. Three of the stages described here are shown in Fig. 2.

Figure 2 Mitosis in a root tip (x 500). The letters refer to the stages described in Fig. 1. (The tissue has been squashed to separate the cells.)

Mitosis will be taking place in any part of a plant or animal which is producing new cells for growth or replacement. Bone marrow produces new blood cells by mitosis; the epidermal cells of the skin are replaced by mitotic divisions in the Malpighian layer; new epithelial cells lining the alimentary canal are produced by mitosis; growth of muscle or bone in animals, and root, leaf, stem or fruit in plants, results from mitotic cell divisions.

An exception to this occurs in the final stages of gamete production in the reproductive organs of plants and animals. The cell divisions which give rise to gametes are not mitotic but meiotic, as explained on p. 201.

Cells which are not involved in the production of gametes are called **somatic cells**. Mitosis takes place only in somatic cells.

QUESTIONS

2 In the nucleus of a human cell just before cell division, how many chromatids will there be?

3 Why can chromosomes not be seen when a cell is not dividing?

4 Look at Fig. 4 on p. 146. Where would you expect mitosis to be occurring most often?

5 In which human tissues would you expect mitosis to be going on in **a** a five-year-old, **b** an adult?

The function of chromosomes

When a cell is not dividing, its chromosomes become very long and thin. Along the length of the chromosome is a series of chemical structures called **genes** (Fig. 3). The chemical which forms the genes is called DNA (which is short for deoxyribonucleic acid, p. 202). Each gene controls some part of the chemistry of the cell. It is these genes which provide the 'instructions' mentioned at the beginning of the chapter. For example, one gene may 'instruct' the cell to make the pigment which is formed in the iris of brown eyes. On one chromosome there will be a gene which causes the cells of the stomach to make the enzyme pepsin. When the chromosome replicates, it builds up an exact replica of itself, gene by gene (Fig. 4). When the chromatids separate at mitosis, each cell will receive a full set of genes. In this way, the chemical instructions in the zygote are passed on to all cells of the body. All the chromosomes, all the genes and, therefore, all the 'instructions' are faithfully reproduced by mitosis and passed on complete to all the cells.

Which of the 'instructions' are used depends on where a cell finally ends up. The gene which causes brown eyes will have no effect in a stomach cell and the gene for making pepsin will not function in the cells of the eye. So a gene's chemical instructions are carried out only in the correct situation.

3 The chromosomes have different shapes and sizes and can be recognized by a trained observer.

4 The chromosomes are always in pairs (Fig. 5), e.g. two long ones, two short ones, two medium ones. This is because when the zygote is formed, one of each pair comes from the male gamete and one from the female gamete. Your 46 chromosomes consist of 23 from your mother and 23 from your father.

5 The number of chromosomes in each body cell of a plant or animal is called the **diploid number**. Because the chromosomes are in pairs, it is always an even number.

chromosome

gene for
eye colour

gene

genes for
hair colour

genes for
tallness

Figure 3 Relationship between chromosomes and genes. The drawing does not represent real genes or a real chromosome. There are probably thousands of genes on a chromosome.

kangaroo (12)

human (46)

domestic fowl (36)

fruit fly (8)

Figure 5 Chromosomes of different species. Note that the chromosomes are always in pairs.

(a) a chromosome builds up a replica of itself.

(b) When the cell divides, the original and the replica are called chromatids.

(c) Mitosis separates the chromatids. Each new cell gets a full set of genes.

Figure 4 Replication. (A, B, C, etc. represent genes.)

The chromosomes of each pair are called **homologous** chromosomes. In Fig. 7(b), the two long chromosomes form one homologous pair and the two short chromosomes form another.

Number of chromosomes

1 There is a fixed number of chromosomes in each species. Human body cells each contain 46 chromosomes, mouse cells contain 40, and garden pea cells 14 (see also Fig. 5).

2 The number of chromosomes in a species is the same in all of its body cells. There are 46 chromosomes in each of your liver cells, in every nerve cell, skin cell and so on.

QUESTIONS

6 How many chromosomes would there be in the nucleus of **a** a human muscle cell, **b** a mouse kidney cell, **c** a human skin cell that has just been produced by mitosis?

7 What is the diploid number in humans?

Gamete production and chromosomes

The genes on the chromosomes carry the 'instructions' which turn a single-cell zygote into a bird, or a rabbit or an oak tree. The zygote is formed at fertilization, when a male gamete fuses with a female gamete. Each gamete brings a set of chromosomes to the zygote. The gametes, therefore, must each contain only half the diploid number of chromosomes, otherwise the chromosome number would double each time an organism reproduced sexually. Each human sperm cell contains 23 chromosomes and each human ovum has 23 chromosomes. When the sperm and ovum fuse at fertilization (p. 150), the diploid number of 46 (23 + 23) chromosomes is produced (Fig. 6).

Meiosis

In a cell which is going to divide and produce gametes, the diploid number of chromosomes shorten and thicken as in mitosis. The pairs of homologous chromosomes, e.g. the two long ones and the two short ones in Fig. 7(b), lie alongside each other and, when the nucleus divides for the first time, it is the chromosomes and not the chromatids which are separated. This results in only half the total number of chromosomes going to each daughter cell. In Fig. 7(c) the diploid number of four chromosomes is being reduced to two chromosomes prior to the first cell division.

By now (Fig. 7(d)), each chromosome is seen to consist of two chromatids and there is a second division of the nucleus (Fig. 7(e)) which separates the chromatids into four distinct nuclei (Fig. 7(f)).

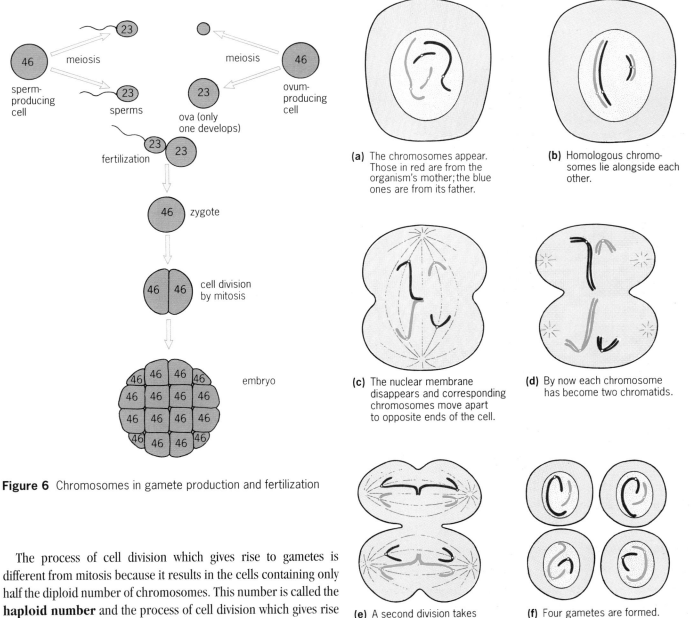

Figure 6 Chromosomes in gamete production and fertilization

(a) The chromosomes appear. Those in red are from the organism's mother; the blue ones are from its father.

(b) Homologous chromosomes lie alongside each other.

(c) The nuclear membrane disappears and corresponding chromosomes move apart to opposite ends of the cell.

(d) By now each chromosome has become two chromatids.

(e) A second division takes place to separate the chromatids.

(f) Four gametes are formed. Each contains only half the original number of chromosomes.

The process of cell division which gives rise to gametes is different from mitosis because it results in the cells containing only half the diploid number of chromosomes. This number is called the **haploid number** and the process of cell division which gives rise to gametes is called **meiosis**.

Meiosis takes place only in reproductive organs.

Figure 7 Meiosis

This gives rise to four gametes, each with the haploid number of chromosomes. In the anther of a plant, four haploid pollen grains are produced when a pollen mother cell divides by meiosis (Fig. 8). In the testis of an animal, meiosis of each sperm-producing cell forms four sperms. In the cells of the ovule of a flowering plant or in the ovary of a mammal, meiosis gives rise to only one mature female gamete. Although four gametes may be produced initially, only one of them turns into an egg cell which can be fertilized.

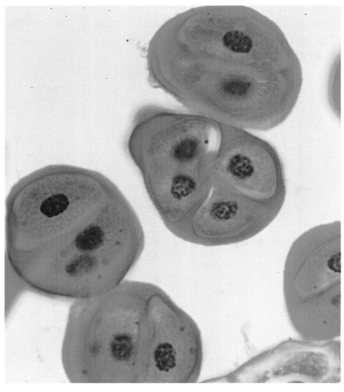

Figure 8 Meiosis in an anther (x1000). The last division of meiosis in the anther of a flower produces four pollen grains.

QUESTIONS

8   What is the haploid number for **a** a human, **b** a fruit fly?
9   Which of the following cells would be haploid and which diploid: white blood cell, male cell in pollen grain, guard cell, root hair, ovum, sperm, skin cell, egg cell in ovule?
10 Where in the body of **a** a human male, **b** a human female and **c** a flowering plant, would you expect meiosis to be taking place?
11 How many chromosomes would be present in **a** a mouse sperm cell, **b** a mouse ovum?
12 Why are organisms, which are produced by asexual reproduction, identical to each other?

Genes

The structure of the gene

The diagram of genes and a chromosome given in Fig. 3 is greatly over-simplified. Chromosomes consist of a protein framework, with a long DNA molecule coiled round the framework in a complicated way (Fig. 9). It is the DNA part of the chromosome which controls the inherited characters, and it is sections of the DNA molecule which constitute the genes.

A DNA molecule is a long chain of nucleotides. A **nucleotide** is a 5-carbon sugar molecule joined to a phosphate group (-PO$_3$) and an organic base (Fig. 10). In DNA the sugar is deoxyribose, and the organic base is either **adenine** (A), **thymine** (T), **cytosine** (C) or **guanine** (G). The nucleotides are joined by their phosphate groups to form a long chain, often thousands of nucleotides long. The phosphate and sugar molecules are the same all the way down the chain but the bases may be any one of the four listed above (Fig. 11).

Mitosis and meiosis compared

Mitosis

Occurs during cell division of somatic cells.

A full set of chromosomes is passed on to each daughter cell. This is the diploid number of chromosomes.

The chromosomes and genes in each daughter cell are identical.

If new organisms are produced by mitosis in asexual reproduction (e.g. bulbs, p. 79) they will all resemble each other and their parents. They are said to form a 'clone'.

Meiosis

Occurs in the final stages of cell division leading to production of gametes.

Only half the chromosomes are passed on to the daughter cells, i.e. the haploid number of chromosomes.

The homologous chromosomes and their genes are randomly assorted between the gametes. (See p. 220 for a fuller explanation of this.)

New organisms produced by meiosis in sexual reproduction will show variations from each other and from their parents.

Figure 9 Simplified model of chromosome structure. This is a '1974' model, which has been superseded by something much more complicated.

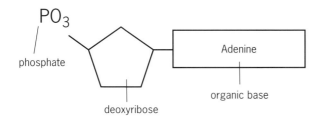

Figure 10 A nucleotide (adenosine monophosphate)

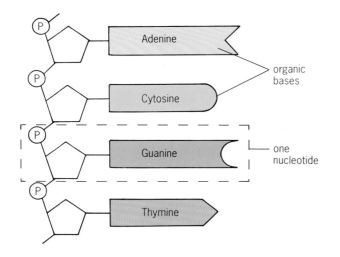

Figure 11 Part of a DNA molecule with four nucleotides

The sequence of bases down the length of the DNA molecule forms a code which instructs the cell to make particular proteins. Proteins are made from amino acids linked together (p. 11). The type and sequence of the amino acids joined together will determine the kind of protein formed. For example, one protein molecule may start with the sequence *alanine–glycine–glycine*. . . . A different protein may start *glycine–serine–alanine*. . . .

It is the sequence of bases in the DNA molecule which decides which amino acids are used and in which order they are joined. Each group of three bases stands for one amino acid, e.g. the triplet of bases *cytosine–guanine–adenine* (CGA) specifies the amino acid *alanine*; the base triplet *cytosine–adenine–thymine* (CAT) specifies the amino acid *valine*, and the triplet (CCA) stands for *glycine*. The tripeptide, *valine–glycine–alanine* would be specified by the DNA code CAT–CCA–CGA (Fig. 12).

The DNA base sequence . . . determines . . . the sequence of amino acids in a peptide

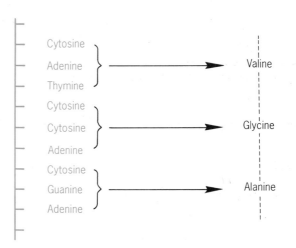

Figure 12 The genetic code (triplet code)

A gene, then, is a sequence of triplets of the four bases, which specifies an entire protein. Insulin is a small protein with only 51 amino acids. A sequence of 153 (i.e. 3×51) bases in the DNA molecule would constitute the gene which makes an islet cell in the pancreas produce insulin. Most proteins are much larger than this and most genes contain a thousand or more bases.

203

The chemical reactions which take place in a cell determine what sort of a cell it is and what its functions are. These chemical reactions are, in turn, controlled by enzymes. Enzymes are proteins. It follows, therefore, that the genetic code of DNA, in determining which proteins, particularly enzymes, are produced in a cell, also determines the cell's structure and function. In this way, the genes also determine the structure and function of the whole organism.

Replication of DNA

The DNA in a chromosome consists of two chains of nucleotides held together by chemical bonds between the bases. The size of the molecules ensures that adenine always pairs with thymine and cytosine pairs with guanine. The double strand is twisted to form a helix (like a twisted rope ladder with the base pairs representing the rungs) (Figs 13 and 14).

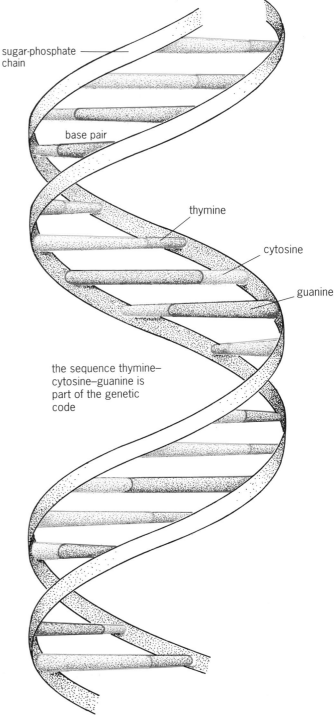

Figure 14 The drawing shows schematically part of a DNA molecule

Figure 13 Model of the structure of DNA

Before cell division can occur, the DNA of the chromosome has to replicate, i.e. make replicas of itself. To do this, enzymes make the double strands of DNA unwind and separate into two single strands, rather like undoing a zip (Fig. 15(a)).

Nucleotides are brought to the 'unzipped' DNA and joined to the exposed bases with the aid of enzymes (Fig. 15(b)). The adenine of an arriving nucleotide always joins to the thymine of the DNA; the cytosine of a nucleotide always joins to the guanine of the

DNA. Similarly, the thymine or guanine on a nucleotide will join to the adenine or cytosine, respectively, of the DNA. The pairing is always A with T, and C with G. Thus an exposed sequence of C–C–A–T–T–G–C–A on the single strand of DNA would build up a corresponding sequence of G, G, T, A, A, C, G, T nucleotides.

The new nucleotides join up to form a chain attached to the exposed strand. This happens all the way along each DNA strand. Since this is happening in both strands of DNA, the double helix is replicated and the full set of genetic instructions is passed to both daughter cells at cell division (Fig. 15(c)).

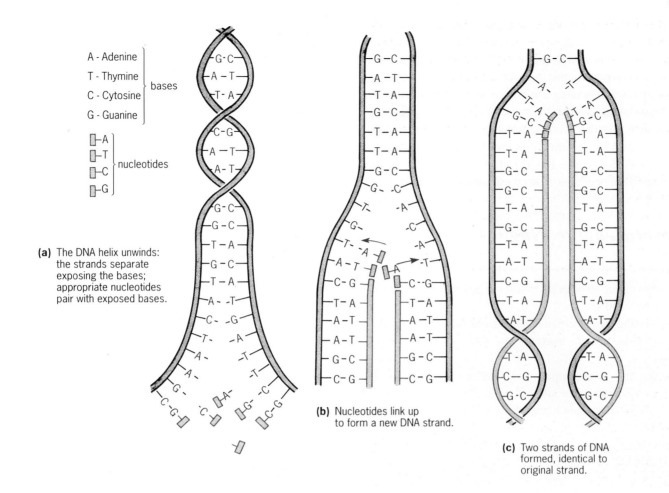

A - Adenine
T - Thymine
C - Cytosine
G - Guanine
} bases

▯–A
▯–T
▯–C
▯–G
} nucleotides

(a) The DNA helix unwinds: the strands separate exposing the bases; appropriate nucleotides pair with exposed bases.

(b) Nucleotides link up to form a new DNA strand.

(c) Two strands of DNA formed, identical to original strand.

Figure 15 Replication of DNA

Mutations

A mutation is a spontaneous change in a gene or a chromosome. Any change in a gene or chromosome usually has a harmful effect on the cell in which it occurs. If the mutation occurs in a gamete, it will affect all the cells of the individual which develops from the gamete. Thus, the whole organism may be affected. If the mutation occurs in a somatic cell (body cell), it will affect only those cells produced, by mitosis, from the affected cell.

Thus, a mutation in a gamete may result in a genetic disorder, e.g. haemophilia (p. 215) or cystic fibrosis. Mutations in somatic cells may give rise to cancers by promoting uncontrolled cell division in the affected tissue. For example, skin cancer results from uncontrolled cell division in the Malpighian layer.

A mutation may be as small as the substitution of one organic base for another in the DNA molecule, or as large as the breakage, loss or gain of a chromosome.

A disease called **sickle-cell anaemia** (p. 223) results from a defective haemoglobin molecule which causes the red blood cells to distort when subjected to a low oxygen concentration. The defective haemoglobin molecule differs from normal haemoglobin by only one amino acid, i.e. *valine* replaces *glutamic acid*. This could be the result of faulty replication at meiosis. When the relevant parental chromosome replicated at gamete formation, the DNA could have produced the triplet –CAT– (which specifies *valine*) instead of –CTT– (which specifies *glutamic acid*).

An inherited form of mental and physical retardation, known as Down's syndrome, results from a chromosome mutation in which the ovum carries an extra chromosome. The affected child, therefore, has 47 chromosomes in his or her cells instead of the normal 46.

Mutations in bacteria often produce resistance to drugs. Bacterial cells reproduce very rapidly, perhaps as often as once every 20 minutes. Thus a mutation, even if it occurs only rarely, is likely to appear in a large population of bacteria. If a population of bacteria, containing one or two drug-resistant mutants, is subjected to that particular drug, the non-resistant bacteria will be killed, but the drug-resistant mutants survive. Mutant genes are inherited in the same way as normal genes, so when the surviving mutant bacteria reproduce, all their offspring will be resistant to the drug.

Mutations are comparatively rare events; perhaps only one in every 100 000 replications results in a mutation. Nevertheless they do occur naturally all the time. (There are bound to be some mutants in the 500 million sperms produced in a human ejaculate.) It is known, however, that exposure to ultraviolet light, X-rays, atomic radiation and certain chemicals (carcinogens) does raise the rate of mutation. Any increase in radiation can increase the risk of mutation. What is less certain is whether there is a minimum dose below which there is negligible risk. It is possible that repeated exposure to low doses of radiation is as harmful as one exposure to a high dose. It has become clear in recent years that in light-skinned people, unprotected exposure to the ultra-violet radiation from the sun can cause a form of skin cancer.

Generally speaking, however, exposure to natural and medical sources of radiation carries less risk than smoking cigarettes or driving a car but it is sensible to keep exposure to a minimum.

Genetic engineering

Genetic engineering involves the transfer of genes from one organism to an unrelated species.

To understand the principles of genetic engineering you need to know something about **bacteria** (p. 299) and **restriction enzymes**.

Bacteria are microscopic single-celled organisms with cytoplasm, cell membranes and cell walls, but without a proper nucleus. Genetic control in a bacterium is exercised by a double strand of DNA in the form of a circle, but not enclosed in a nuclear membrane. This circular DNA strand carries the genes that control bacterial metabolism.

In addition, there are present in the cytoplasm a number of small, circular pieces of DNA called **plasmids**. The plasmids often carry genes which give the bacterium resistance to particular antibiotics such as tetracycline and ampicillin.

Restriction enzymes are produced by bacteria. They 'cut' DNA molecules at specific sites, e.g. between the A and the T in the sequence GAA–TTC. Restriction enzymes can be extracted from bacteria and purified. By using a selected restriction enzyme, DNA molecules extracted from different organisms can be cut at predictable sites and made to produce lengths of DNA which contain specific genes.

DNA from human cells can be extracted and restriction enzymes used to 'cut' out a sequence of DNA which includes a gene, e.g. the gene for production of insulin. Plasmids are extracted from bacteria and 'cut open' with the same restriction enzyme. If the human DNA is then added to a suspension of the plasmids, some of the human DNA will attach to some of the plasmids which will then close up again, given suitable enzymes. The DNA in these plasmids is called **recombinant DNA**.

The bacteria can be induced to take up the plasmids and, by ingenious culture methods using antibiotics, it is possible to select the bacteria which contain the recombinant DNA. The human DNA in the plasmids continues to produce the same protein as it did in the human cells. In the example mentioned, this would be the protein, insulin (p. 187). The plasmids are said to be the **vectors** which carry the human DNA into the bacteria and the technique is sometimes called **gene-splicing**.

Given suitable nutrient solutions, bacteria multiply rapidly and produce vast numbers of offspring. The bacteria reproduce by mitosis and so each daughter bacterium will contain the same DNA and the same plasmids as the parent. The offspring, therefore, form a **clone** and the insulin gene is said to be **cloned** by this method.

The bacteria are cultured in special vessels called **fermenters** (p. 344) and the insulin which they produce can be extracted from the culture medium and purified for use in treating diabetes (p. 187).

This is only one type of genetic engineering. The vector may be a virus rather than a plasmid; the DNA may be inserted directly, without a vector; the donor DNA may be synthesized from nucleotides rather than extracted from cells; yeast may be used instead of bacteria. The outcome, however, is the same. DNA from one species is inserted into a different species and made to produce its normal proteins (Fig. 16).

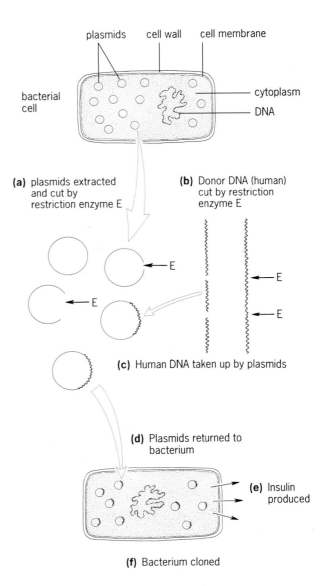

Figure 16 The principles of genetic engineering

Some applications of genetic engineering

The following are a few examples of achievements in genetic engineering. Many of these have been accomplished in the laboratory but have not yet been commercially developed. In the past 15 years, about 1300 biotechnology companies were set up to exploit the possibilities of genetic engineering. Some were successful but many failed. In some cases the products were too expensive or showed little advantage over conventionally produced substances. All the companies have to convince the regulatory bodies that their products are safe enough for clinical trials in the case of drugs, or harmless to the environment in the case of plants.

Drugs, hormones and vaccines

Insulin   This can be produced by genetically engineered bacteria and has been in use since 1982. Theoretically, the human insulin produced in this way should be better than the insulin prepared from pigs or cattle as it avoids allergic or immunological reactions to these 'foreign' proteins.

Alpha-anti-trypsin   This enzyme inhibitor is used to treat hereditary emphysema (p. 135). The gene for its production has been transferred to the milk-producing cells of sheep so that they secrete the anti-enzyme in their milk (Fig. 17). One sheep has produced 35 g per litre. Clinical trials are being planned.

Figure 17 Transgenic sheep. The three sheep sharing the milking platform are all daughters of a transgenic ram and all three produce alpha-anti-trypsin in their milk.

Cancer   Researchers are working on, for example, lung cancer, breast cancer and skin cancer. One technique involves taking white blood cells from the patient and introducing a gene for **tumour necrosis factor** (TNF). The lymphocytes are returned to the patient and when they reach the cancer cells, they release the TNF which kills the cancer cells but does not harm normal cells. Clinical trials are taking place.

Rabies vaccine   Genes from the rabies vaccine are transferred to a harmless virus (the one that was used to vaccinate people against smallpox). When this altered virus is injected, it stimulates the immune system to produce antibodies. This is still in the experimental stage.

Hepatitis B   The gene for the protein coat (p. 301) of the hepatitis virus is inserted into yeast cells. The yeast cells are cultured and produce the protein which is used as an antigen (vaccine) to promote the production of antibodies. The method is used in Singapore and Malaysia. It is a safer method of production than handling human blood plasma which contains the active virus.

Cholera   Trials of a genetically engineered vaccine started in Chile in 1991.

Gene therapy

Human hereditary diseases are caused by one or more defective genes (p. 215). Gene therapy aims to replace the defective genes with normal genes. These treatments are either in the experimental stage or undergoing clinical trials.

Adenosine deaminase (ADA)   If the gene for producing this enzyme is faulty, the patient cannot make normal white cells and so the immune system doesn't function. Samples of bone marrow are taken from the patient and the normal gene is introduced to the blood-producing cells using a virus vector. Clinical trials are under way in Britain and USA.

Cystic fibrosis   Absence of a gene for a cell membrane protein leads to the production of a thick, sticky mucus in the airways and makes the patient very susceptible to infections (Fig. 18). Trials are now taking place in which the genetically engineered gene is sprayed into the patient's nose or lungs. The vector is either a recombinant virus or microscopic oil drops (liposomes). The virus used is the one which causes the common cold but it is genetically 'disabled' so that it cannot reproduce. Nevertheless, there are some worries about the safety of this method.

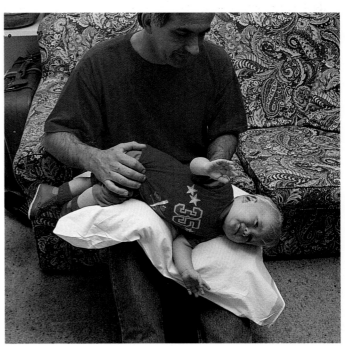

Figure 18 Physiotherapy for a young cystic fibrosis patient: regular treatment is required to move copious viscous mucus from the lung, to help breathing

Genetically engineered plants

Pest resistance The bacterium, *Bacillus thuringiensis*, produces a toxin which kills caterpillars and other insect larvae. The toxin has been in use for some years as an insecticide. The gene for the toxin has been successfully introduced into some plant species using a bacterial vector. The plants produce the toxin and show increased resistance to attack by insect larvae. The gene is also passed on to the plant's offspring. Unfortunately there are signs that insects are developing immunity to the toxin. The effect of a gene for mould resistance is shown in Fig. 19.

Figure 20 Genetically engineered tomatoes. In the engineered tomatoes (labelled antisense), biologists have deleted the gene which produces the enzyme which makes fruit go soft. (a) and (c) control tomatoes before and after storage. (b) and (d) genetically engineered tomatoes before and after storage.

Figure 19 Oil seed rape. The leaf on the right is from a plant with introduced genes which confer resistance to a mould.

Herbicide resistance Some of the safest and most effective herbicides are those, such as glyphosate, which kill any green plant but become harmless as soon as they reach the soil. These herbicides cannot be used on crops because they kill the crop plants as well as the weeds. A gene for an enzyme which breaks down glyphosate can be introduced into a plant cell culture (p. 81) which can then be used to produce resistant plants. When a crop of these plants is sprayed, only the weeds are killed. This is still in the experimental stage.

Modifying plant products A gene introduced to oilseed rape and other oil-producing plants can change the nature of the oils they produce to make them more suitable for commercial processes, e.g. detergent production. This might be very important when stocks of petroleum run out. It could be a renewable source of oil which would not contribute to global warming (p. 253).

The tomatoes in Fig. 20 have been modified to improve their keeping qualities.

Possible hazards of genetic engineering

In the past, when an alien species has been introduced into a foreign environment, the results have often been disastrous. Examples are the prickly pear and rabbit in Australia and rats in Malaysia. Some people fear that genetically engineered organisms might upset the existing balance of organisms in a similar way.

It is also possible that the altered genes which benefit one organism might reach other organisms where they would prove harmful. If, for example, the genes for herbicide resistance got into the weed plants, it would render the herbicides useless. Genetically

engineered viruses which kill pests might transfer their altered genes to other viruses which had hitherto lived harmlessly in a beneficial organism.

For these reasons there are strict rules governing the release of genetically engineered organisms.

In practice, biologists ensure that the organisms which carry modified genes are disabled so that they cannot survive outside their target organism. Some are altered so that they have very short life-spans or a very low rate of reproduction. Some are made sensitive to ultraviolet light so that if they escape from their host, they are promptly killed by daylight.

The ethics of genetic engineering

It is difficult to see any ethical objection to the use of genetic engineering to treat disease or improve food production. However, it is possible to delete or alter genes in laboratory animals so that they suffer from diseases similar to those in humans, such as cancer or cystic fibrosis. This makes them excellent subjects for trying out therapeutic techniques which could benefit humans but is this an ethical procedure?

Another ethical 'minefield' is the possibility of introducing into human zygotes genes for intelligence, musical skill or disease resistance, but this is very remote. These characteristics are under the control of many genes (not yet identified) and are also the product of the interaction between the environment and the genetic constitution of the individual. Nevertheless, at present it is forbidden to conduct genetic engineering on human gametes or embryos so recombinant DNA cannot be passed from one generation to another.

PRACTICAL WORK

Squash preparation of chromosomes using acetic orcein

Material *Allium cepa* (onion) root tips. Support onions over beakers or jars of water. Keep the onions in darkness for several days until the roots growing into the water are 2–3 cm long. Cut off about 5 mm of the root tips, place them in a watch glass and:

1 Cover them with 9 drops acetic orcein and 1 drop molar hydrochloric acid.
2 Heat the watch glass gently over a very small Bunsen flame till the steam rises from the stain, but do not boil.
3 Leave the watch glass covered for at least 5 minutes.
4 Place one of the root tips on a clean slide, cover with 45 per cent ethanoic (acetic) acid and cut away all but the terminal 1 mm.
5 Cover this root tip with a clean cover-slip and make a squash preparation as described below.

Making the squash preparation Squash the softened, stained root tips by lightly tapping on the cover-slip with a pencil: hold the pencil vertically and let it slip through the fingers to strike the cover-slip (Fig. 21). The root tip will spread out as a pink mass on the slide; the cells will separate and the nuclei, many of them with chromosomes in various stages of mitosis (because the root tip is a region of rapid cell division), can be seen under the high power of the microscope (×400).

Figure 21 Tap the cover-slip gently to squash the tissue

QUESTIONS

13 What peptide is specified by the DNA sequence CGA–CGA–CAT–CCA–CAT?
14 A mutation in the DNA sequence in question 13 produces a valine in place of the glycine. What change in the genetic code could have produced this result?
15 State briefly the connection between genes, enzymes and cell structure.
16 Why is it particularly important to prevent radiation from reaching the reproductive organs?
17 What is the potential disadvantage of using viruses as vectors for recombinant DNA? How can this disadvantage be minimized?
18 Can you think of any ethical objection to sheep being used to produce a drug for treating humans?

Checklist

- In the nuclei of all cells there are thread-like structures called 'chromosomes'.

- The chromosomes are in pairs; one of each pair comes from the male and one from the female parent.

- On these chromosomes are carried the genes.

- The genes control the chemical reactions in the cells and, as a result, determine what kind of organism is produced.

- Each species of plant or animal has a fixed number of chromosomes in its cells.

- When cells divide by mitosis, the chromosomes and genes are copied exactly and each new cell gets a full set.

- At meiosis, only one chromosome of each pair goes into the gamete.

- The DNA molecule is coiled along the length of the chromosome.

- A DNA molecule is made up of a double chain of nucleotides in the form of a helix.

- The nucleotide bases in the helix pair up adenine–thymine (A–T) and cytosine–guanine (C–G).

- Triplets of bases control production of the specific amino acids which make up a protein.

- Genes consist of particular lengths of DNA.

- Most genes control the type of enzyme that a cell will make.

- At replication, DNA strands separate and build up new chains on the exposed bases.

- A mutation is a spontaneous change in a gene or chromosome. Most mutations produce harmful effects.

- Genetic engineering involves transferring lengths of DNA from one species to another.

22 HEREDITY

Patterns of inheritance

A knowledge of mitosis and meiosis allows us to explain, at least to some extent, how heredity works. The gene in a mother's body cells which causes her to have brown eyes may be present on one of the chromosomes in each ovum she produces. If the father's sperm cell contains a gene for brown eyes on the corresponding chromosome, the zygote will receive a gene for brown eyes from each parent. These genes will be reproduced by mitosis in all the embryo's body cells and when the embryo's eyes develop, the genes will make the cells of the iris produce brown pigment and the child will have brown eyes.

In a similar way, the child may receive genes for curly hair and Fig. 1 shows this happening. But it does not, of course, show all the other chromosomes with thousands of genes for producing the enzymes, making different types of cell and all the other processes which control the development of the organism.

Single-factor inheritance

Because it is impossible to follow the inheritance of the thousands of characteristics controlled by genes, it is usual to start with the study of a single gene which controls one characteristic. We have used eye colour as an example so far. Probably more than one gene pair is involved, but the simplified example will serve our purpose. It has already been explained how a gene for brown eyes from each parent results in the child having brown eyes. Suppose, however, that the mother has blue eyes and the father brown eyes. The child might receive a gene for blue eyes from its mother and a gene for brown eyes from its father (Fig. 2). If this happens, the child will, in fact, have brown eyes. The gene for brown eyes is said to be **dominant** to the gene for blue eyes. Although the gene for blue eyes is present in all the child's cells, it does not contribute to the eye colour. It is said to be **recessive** to brown.

Eye colour is a useful 'model' for explaining inheritance but it is not wholly reliable because 'blue' eyes vary in colour and sometimes contain small amounts of brown pigment.

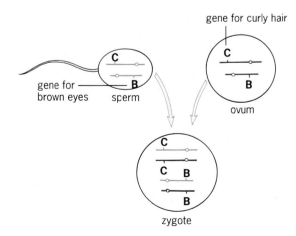

Figure 1 Fertilization. Fertilization restores the diploid number of chromosomes and combines the genes from the mother and father.

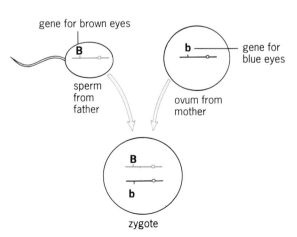

Figure 2 Combination of genes in the zygote (only one chromosome is shown). The zygote has both genes for eye colour; the child will have brown eyes.

This example illustrates the following important points:

1 There is a pair of genes for each characteristic, one gene from each parent.
2 Although the gene pairs control the same character, e.g. eye colour, they may have different effects. One tries to produce blue eyes, the other tries to produce brown eyes.
3 Often one gene is dominant over the other.
4 The genes of each pair are on corresponding chromosomes and occupy corresponding positions. For example, in Fig. 1 the genes for eye colour are shown in the corresponding position on the two short chromosomes and the genes for hair curliness are in corresponding positions on the two long chromosomes. In diagrams and explanations of heredity:

  a genes are represented by letters;
  b genes controlling the same characteristic are given the same letter; and
  c the dominant gene is given the capital letter.

For example, in rabbits, the dominant gene for black fur is labelled **B**. The recessive gene for white fur is labelled **b** to show that it corresponds to **B** for black fur. If it were labelled **w**, we would not see any connection between **B** and **w**. **B** and **b** are obvious partners. In the same way **L** could represent the gene for long fur and **l** the gene for short fur.

Breeding true

A white rabbit must have both the recessive genes **b** and **b**. If it had **B** and **b,** the dominant gene for black (**B**) would override the gene for white (**b**) and produce a black rabbit. A black rabbit, on the other hand, could be either **BB** or **Bb** and, by just looking at the rabbit, you could not tell the difference. When the male black rabbit **BB** produces sperms by meiosis, each one of the pair of chromosomes carrying the **B** genes will end up in different sperm cells. Since the genes are the same, all the sperms will have the **B** gene for black fur (Fig. 3(a)).

The black rabbit **BB** is called a true-breeding black and is said to be **homozygous** for black coat colour ('homo-' means 'the same'). If this rabbit mates with another black (**BB**) rabbit, all the babies will be black because all will receive a dominant gene for

black fur. When all the offspring have the same characteristic as the parents, this is called 'breeding true' for this characteristic.

When the **Bb** black rabbit produces gametes by meiosis, the chromosomes with the **B** genes and the chromosomes with the **b** genes will end up in different gametes. So 50 per cent of the sperm cells will carry **B** genes and 50 per cent will carry **b** genes (Fig. 3(b)). Similarly, in the female, 50 per cent of the ova will have a **B** gene and 50 per cent will have a **b** gene. If a **b** sperm fertilizes a **b** ovum, the offspring, with two **b** genes (**bb**), will be white. The black **Bb** rabbits are not true-breeding because they may produce some white babies as well as black ones. The **Bb** rabbits are called **heterozygous** ('hetero-' means 'different').

The black **BB** rabbits are homozygous dominant.

The white **bb** rabbits are homozygous recessive.

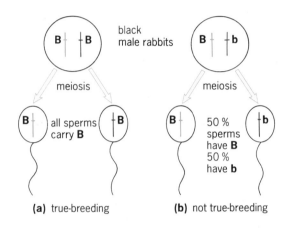

Figure 3 Breeding true

Genotype and phenotype

The two kinds of black rabbit **BB** and **Bb** are said to have the same **phenotype**. This is because their coat colours look exactly the same. However, because they have different gene pairs for coat colour they are said to have different **genotypes**, i.e. different combinations of genes. One genotype is **BB** and the other is **Bb**.

You and your brother might both be brown-eyed phenotypes but your genotype could be **BB** and his could be **Bb**. You would be homozygous dominant for brown eyes; he would be heterozygous for eye colour.

Alleles

The genes which occupy corresponding positions on homologous chromosomes and control the same characteristic are called **allelomorphic genes** or **alleles**. The word 'allelomorph ' means 'alternative form'. The genes **B** and **b** are alternative forms of a gene for eye colour. **B** and **b** are alleles.

There are often more than two alleles of a gene. The human ABO blood groups (p. 125) are controlled by three alleles, I^A, I^B and **i**, though only two of these can be present in one genotype.

The three to one ratio

The result of a mating between a true-breeding (homozygous) black mouse (**BB**), and a true-breeding (homozygous) brown mouse (**bb**) is shown in Fig. 4(a). The illustration is greatly simplified because it shows only one pair of the 20 pairs of mouse chromosomes and only one pair of alleles on the chromosomes.

Because black is dominant to brown, all the offspring from this mating will be black phenotypes, because they all receive the dominant allele for black fur from the father. Their genotypes, however, will be **Bb** because they all receive the recessive **b** allele from the mother. They are heterozygous for coat colour. The offspring resulting from this first mating are called the F_1 **generation**.

Fig. 4 (b) shows what happens when these heterozygous, F_1 black mice are mated together to produce what is called the F_2 **generation**. Each sperm or ovum produced by meiosis can contain only one of the alleles for coat colour, either **B** or **b**. So there are two kinds of sperm cell, one kind with the **B** allele and one kind with the **b** allele. There are also two kinds of ovum, with either **B** or **b** alleles. When fertilization occurs, there is no way of telling whether a **b** or a **B** sperm will fertilize a **B** or a **b** ovum, so we have to look at all the possible combinations as follows:

1 a **b** sperm fertilizes a **B** ovum. Result: **bB** zygote.
2 a **b** sperm fertilizes a **b** ovum. Result: **bb** zygote.
3 a **B** sperm fertilizes a **B** ovum. Result: **BB** zygote.
4 a **B** sperm fertilizes a **b** ovum. Result: **Bb** zygote.

There is no difference between **bB** and **Bb**, so there are three possible genotypes in the offspring—**BB**, **Bb** and **bb**. There are only two phenotypes—black (**BB** or **Bb**) and brown (**bb**). So, according to the laws of chance, we would expect three black baby mice and one brown. Mice usually have more than four offspring and what we really expect is that the **ratio** (proportion) of black to brown will be close to 3:1.

If the mouse had 13 babies, you might expect nine black and four brown, or eight black and five brown. Even if she had 16 babies you would not expect to find exactly 12 black and four brown because whether a **B** or **b** sperm fertilizes a **B** or **b** ovum is

a matter of chance. If you spun ten coins, you would not expect to get exactly five heads and five tails. You would not be surprised at six heads and four tails or even seven heads and three tails. In the same way, we would not be surprised at 14 black and two brown mice in a litter of 16.

To decide whether there really is a 3:1 ratio, we need a lot of results. These may come either from breeding the same pair of mice together for a year or so to produce many litters, or from mating 20 black and 20 brown mice, crossing the offspring and adding up the number of black and brown babies in the F_2 families (see also Fig. 5).

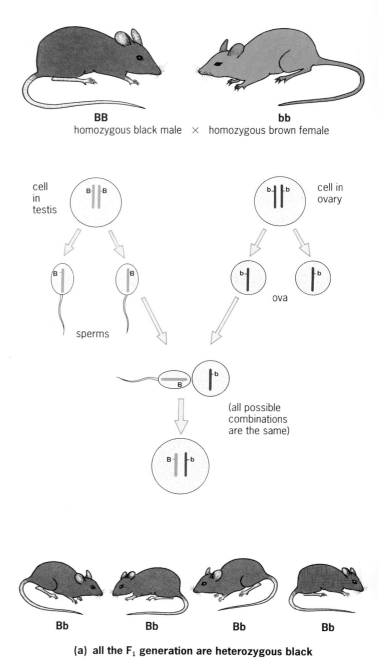

(a) all the F_1 generation are heterozygous black

Figure 4 Inheritance of coat colour in mice

PARENTS

Bb × **Bb**

heterozygous black male × heterozygous black female

MEIOSIS

cell in testis

cell in ovary

GAMETES

sperms (two possibilities)

ova (two possibilities)

FERTILIZATION

(four possible combinations)

POSSIBLE ZYGOTES

OFFSPRING

BB **bB** **Bb** **bb**

(b) the probable ratio of coat colours in the F₂ generation is 3 black:1 brown

Figure 5 F$_2$ maize cobs. This is the offspring from a breeding experiment using maize instead of mice. One of the alleles for colour gives yellow grains, the other gives dark grains. What was the colour of the seeds which produced the plant with this cob?

QUESTIONS

7 Look at Fig. 4(a). Why is there no possibility of getting a **BB** or a **bb** combination in the offspring?

8 In Fig. 4(b) what proportion of the F$_2$ black mice are true-breeding?

9 Two black guinea-pigs are mated together on several occasions and their offspring are invariably black. However, when their black offspring are mated with white guinea-pigs, half of the matings result in all black litters and the other half produce litters containing equal numbers of black and white babies.

From these results, deduce the genotypes of the parents and explain the results of the various matings, assuming that colour in this case is determined by a single pair of alleles.

213

The recessive test-cross (back-cross)

A black mouse could have either the **BB** or the **Bb** genotype. One way to find out which, is to cross the black mouse with a known homozygous recessive mouse, **bb**. The **bb** mouse will produce gametes with only the recessive **b** allele. A black homozygote, **BB**, will produce only **B** gametes. Thus, if the black mouse is **BB**, all the offspring from the cross will be black heterozygotes, **Bb**.

Half the gametes from a black, **Bb**, mouse would carry the **B** allele and half would have the **b** allele. So, if the black mouse is **Bb**, half of the offspring from the cross will, on average, be brown homozygotes, **bb**, and half will be black heterozygotes, **Bb**.

The term 'back-cross' refers to the fact that, in effect, the black, mystery mouse is being crossed with the same genotype as its brown grandparent, the **bb** mouse in Fig. 4(a). Mouse ethics and speed of reproduction makes the use of the actual grandparent quite feasible!

QUESTION

10 Two black rabbits thought to be homozygous for coat colour were mated and produced a litter which contained all black babies. The F_2, however, resulted in some white babies which meant that one of the grandparents was heterozygous for coat colour. How would you find out which parent was heterozygous?

Codominance and incomplete dominance

Codominance

If both genes of an allelomorphic pair produce their effects in an individual (i.e. neither allele is dominant to the other) the alleles are said to be codominant.

The inheritance of the human ABO blood groups includes an example of codominance. On p. 125 it was explained that, in the ABO system, there are four phenotypic blood groups, A, B, AB and O. The alleles for groups A and B are codominant. If a person inherits alleles for group A and group B, his or her red cells will carry both antigen A and antigen B.

However, the alleles for groups A and B are both completely dominant to the allele for group O. (Group O people have neither A nor B antigens on their red cells.)

Table 1 shows the genotypes and phenotypes for the ABO blood groups. (Note that the allele for group O is sometimes represented as **i** and sometimes as **I°**.)

Table 1 The ABO blood groups

Genotype	Blood group (phenotype)
I^AI^A or I^Ai	A
I^BI^B or I^Bi	B
I^AI^B	AB
ii	O

Since the genes for groups A and B are dominant to that for group O, a group A person could have the genotype I^AI^A or I^Ai. Similarly a group B person could be I^BI^B or I^Bi. There are no alternative genotypes for groups AB and O.

Incomplete dominance

This term is sometimes taken to mean the same as 'codominance' but, strictly, it applies to a case where the effect of the recessive allele is not completely masked by the dominant allele.

An example occurs with sickle-cell anaemia (p. 223). If a person inherits both recessive genes (Hb^SHb^S) for sickle-cell haemoglobin, then he or she will show manifest signs of the disease, i.e. distortion of the red cells leading to severe bouts of anaemia.

A heterozygote (Hb^AHb^S), however, will have a condition called '**sickle-cell trait**'. Although there may be mild symptoms of anaemia the condition is not serious or life-threatening. In this case, the normal haemoglobin allele (Hb^A) is not completely dominant over the recessive (Hb^S) allele.

QUESTIONS

11 What are the possible blood groups likely to be inherited by children born to a group A mother and a group B father? Explain your reasoning.

12 A woman of blood group A claims that a man of blood group AB is the father of her child. A blood test reveals that the child's blood group is O. Is it possible that the woman's claim is correct? Could the father have been a group B man? Explain your reasoning.

13 A red cow has a pair of alleles for red hairs. A white bull has a pair of alleles for white hairs. If a red cow and a white bull are mated, the offspring are all 'roan', i.e. they have red and white hairs equally distributed over their body.
a Is this an example of codominance or incomplete dominance?
b What coat colours would you expect among the offspring of a mating between two roan cattle?

Determination of sex

Whether you are a male or female depends on one particular pair of chromosomes called the 'sex chromosomes'. In females, the two sex chromosomes, called the X chromosomes, are the same size as each other. In males, the two sex chromosomes are of different sizes. One corresponds to the female sex chromosomes and is called the X chromosome. The other is smaller and is called the Y chromosome. So the female genotype is XX and the male genotype is XY.

When meiosis takes place in the female's ovary, each ovum receives one of the X chromosomes, so all the ova are the same for this. Meiosis in the male's testes results in 50 per cent of the sperms getting an X chromosome and 50 per cent getting a Y chromosome (Fig. 6). If an X sperm fertilizes the ovum, the zygote will be XX and will grow into a girl. If a Y sperm fertilizes the ovum, the zygote will

FEMALE

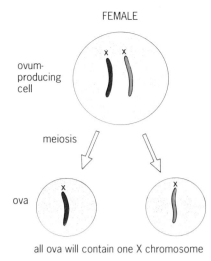

ovum-producing cell

meiosis

ova

all ova will contain one X chromosome

MALE

sperm-producing cell

meiosis

sperms

half the sperms will contain an X chromosome and half will carry a Y chromosome

Figure 6 Determination of sex. Note that:
  i only the X and Y chromosomes are shown
  ii the Y chromosome is not smaller than the X in all organisms
  iii details of meiosis have been omitted
  iv in fact, four gametes are produced in each case, but two are sufficient to show the distribution of X and Y chromosomes

be XY and will develop into a boy. There is an equal chance of an X or Y chromosome fertilizing an ovum, so the numbers of girl and boy babies are more or less the same.

QUESTION

14 A married couple has four girl children but no boys. This does not mean that the husband produces only X sperms. Explain why not.

Hereditary diseases

Most of the characteristics we think of as important are controlled by more than one pair of alleles, and can be influenced by our environment. Your height, for example, is probably determined by several allelomorphic pairs. It will also depend on whether you get enough food while you are growing.

The most striking single-factor characteristics are those associated with inherited defects such as sickle-cell anaemia, haemophilia, cystic fibrosis and colour blindness.

These four diseases are caused by recessive genes. It follows that people who develop the disease, must have received both recessive alleles, one from each of their parents. This, in turn, means that both parents must be heterozygous for the gene. If the father, for example, had been homozygous normal, he would have passed the dominant gene to all his children and none would be affected.

A few genetic diseases are caused by dominant alleles, which means that even heterozygotes will develop the disease.

Albinism

An albino lacks a gene for producing the pigment melanin (p. 146). As a result the iris of the eye is pink, the hair is white or very pale yellow, and the skin is unpigmented and easily damaged by sunlight. The albinism allele is recessive to the pigment-producing allele.

Haemophilia

Haemophilia is a genetic disease in which blood clots very slowly. The blood of people with the disease lacks one of the plasma proteins, called factor VIII, which plays a part in clotting. The production of factor VIII is controlled by a single gene. A person who lacks this gene will be a haemophiliac. Quite minor cuts tend to bleed for a long time and internal bleeding may occur.

Although haemophilia is a genetically controlled disease, symptoms vary from mild to severe in different individuals. The condition can be largely controlled by daily or weekly injections of factor VIII. (See also 'Sex linkage', p. 216.)

Cystic fibrosis

Cystic fibrosis is the commonest inherited disease in white people, affecting about one in every 2500 children. The disease occurs when a person inherits two copies of a recessive allele which controls the production of a protein in cell membranes. The protein normally controls the passage of chloride ions across the membrane but is defective in homozygous recessive individuals. As a consequence, the epithelial cells lining the respiratory passages and digestive glands produce a thick, sticky mucus which makes the person very susceptible to infections and interferes with digestion.

Trials are taking place, with genetically engineered genes, which may alleviate the condition (p. 207).

Sickle-cell anaemia

(See 'Incomplete dominance', p. 214.)

215

Sex linkage

Some genetic disorders affect many more males than females. Haemophilia and colour blindness are examples. Such diseases are said to be sex-linked. Sex linkage results from the fact that the X chromosome is longer than the Y chromosome. Thus there are genes on the X chromosome which have no corresponding alleles on the Y chromosome (Fig. 7).

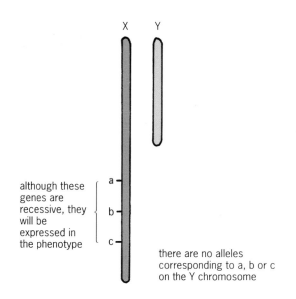

although these genes are recessive, they will be expressed in the phenotype

there are no alleles corresponding to a, b or c on the Y chromosome

Figure 7 Sex linkage. Genes a, b and c are sex-linked to the X chromosome

The alleles for red–green colour blindness are recessive. But since they lie on the section of the X chromosome which is not matched by the Y, there is no chance of their effects being suppressed by a dominant allele.

The outcome of a cross between a normal man with a normal woman who carries a recessive allele for colour blindness is shown in Fig. 8. Although the woman has normal colour vision she is a 'carrier' for the disorder.

There is a 1 in 2 chance that any son will be colour blind. None of the daughters will be affected, but half of them can be expected to carry the allele.

The same pattern of inheritance applies to haemophilia; women are the carriers but men are affected.

Parents:	XN Y		XN Xn	
	normal man		carrier woman	
Gametes:	XN Y		XN Xn	
Possible	XN XN	XN Xn	XN Y	Xn Y
combinations	normal	carrier	normal	colour-blind
of gametes	girl	girl	boy	boy

Figure 8 Inheritance of colour blindness

QUESTIONS

15 A haemophiliac man marries a normal woman. What are the chances of their having an affected son
   a if the woman is a carrier of the recessive allele,
   b if she is homozygous for the normal allele?
16 A man with inherited red–green colour-blindness marries a woman with normal colour vision. What are the chances of affected sons and daughters
   a if the woman is homozygous for colour vision,
   b if she is heterozygous for this condition? (See Fig. 9.)
17 A woman has the sickle-cell *trait* (p. 214). What are the chances of her children inheriting   **i** the sickle-cell trait   **ii** sickle-cell disease if she marries
   a a normal man
   b a man with sickle-cell trait
   c a man with sickle-cell disease?

Figure 9 Ishihara colour vision test card. People with normal colour vision see a figure 8. People with red-green colour blindness see a figure 3. Totally colour blind people cannot see a numeral. Note that for accurate testing the **original card must be used.**

Genetic counselling

If normal parents have a child affected by an inherited disorder caused by a single gene, they can be sure that they are both heterozygous (**Nn**) for the gene.

Thus there is a 1 in 4 chance that any one of their later children will be affected too (see p. 212). If the condition is serious, the parents may decide that the risk of having another affected child is too high and therefore they will have no more children.

If one member of a couple has a genetic disorder, the chances that they will have an affected child can be worked out. For an albino man (**nn**) marrying a normal woman, this is done as follows:

1 About 1 person in 70 carries the recessive allele for albinism.

2 Thus there is a 1 in 70 chance that the woman is a carrier (heterozygous for albinism, **Nn**).

3 The chance of an affected child being born to an affected man (**nn**) and a heterozygous woman (**Nn**) is 1 in 2.

4 Combining probabilities 2 and 3 gives a chance of $\frac{1}{2} \times \frac{1}{70} = \frac{1}{140}$ that any one child will be an albino.

The parents may decide that this risk is low, and that it is worth starting a family.

Genetic counselling would be more effective if scientists could detect heterozygotes. Unfortunately, this can be done for only a few conditions. For others, they have to wait until an affected child is born.

An inherited disease called **phenylketonuria** (PKU) is caused by a recessive gene. Children who are homozygous for PKU have such a high level of the amino acid phenylalanine in their blood that their brains are damaged and they become mentally handicapped. However, by using a diet which is free from phenylalanine in the first six years, the damage can be avoided. This is one of the few diseases where heterozygotes can be detected. However, all babies are given a blood test at birth to see if their phenylalanine level is normal.

Anyone with a genetic disorder, or having a family history of one, should seek genetic counselling before starting a family. This can usually be arranged through the family doctor.

Applied genetics

It is possible for biologists to use their knowledge of genetics to produce new varieties of plants and animals. For example, suppose one variety of wheat produces a lot of grain but is not resistant to a fungus disease. Another variety is resistant to the disease but has only a poor yield of grain. If these two varieties are cross-pollinated (Fig. 10), the F_1 offspring should be disease-resistant and give a good yield of grain (assuming that the useful characteristics are controlled by dominant genes).

R represents a dominant allele for resistance to disease, and **r** is the recessive allele for poor resistance. **H** is a dominant allele for high yield and **h** is the recessive allele for low yield. The high-yield/low-resistance variety (**HHrr**) is crossed with the low-yield/high-resistance variety (**hhRR**). Each pollen grain from the **HHrr** plant will contain one **H** and one **r** allele (**Hr**). Each ovule from the **hhRR** plant will contain an **h** and an **R** allele (**hR**). The seeds will, therefore, all be **HhRr**. The plants which grow from these seeds will have dominant alleles for both high yield and good disease resistance.

The offspring from crossing two varieties are called **hybrids**. If the F_1 hybrids from this cross bred true, they could give a new variety of disease-resisting, high-yielding wheat. As you learned on p. 212, the F_1 generation from a cross does not necessarily breed true. The F_2 generation of wheat may contain:

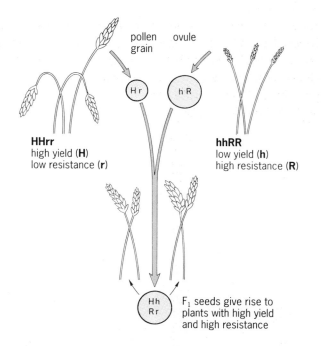

Figure 10 Combining useful characteristics

1 high-yield, disease-resistant

2 low-yield, disease-prone

3 low-yield, disease-resistant } parental

4 high-yield, disease-prone } types

This would not give such a successful crop as the F_1 plants.

With some commercial crops, the increased yield from the F_1 seed makes it worthwhile for the seedsman to make the cross and sell the seed to the growers. The hybrid corn (maize) grown in America is one example. The F_1 hybrid gives nearly twice the yield of the standard varieties.

In other cases it is possible to work out a cross-breeding programme to produce a hybrid which breeds true (Fig. 11). If, instead of the **HhRr** in Fig. 10 an **HHRR** could be produced, it would breed true.

Selective breeding

An important part of any breeding programme is the selection of the desired varieties. The largest fruit on a tomato plant might be picked and its seeds planted next year. In the next generation, once again only seeds from the largest tomatoes are planted. Eventually it is possible to produce a true-breeding variety of tomato plant which forms large fruits. Fig. 10 (p. 224) shows the result of such selective breeding.

The same technique can be used for selecting other desirable qualities, such as flavour.

Similar principles can be applied to farm animals. Desirable characteristics, such as high milk yield and resistance to disease, may be combined. Stock-breeders will select calves from cows which give large quantities of milk. These calves will be used as breeding stock to build a herd of high yielders. A characteristic

Figure 11 The genetics of bread wheat. A primitive wheat (a) was crossed with a wild grass (b) to produce a better-yielding hybrid wheat (c). The hybrid wheat (c) was crossed with another wild grass (d) to produce one of the varieties of wheat (e) which is used for making flour and bread.

such as milk yield is probably under the control of many genes. At each stage of selective breeding the farmer, in effect, is keeping the beneficial genes and discarding the less useful genes from his or her animals.

Selective breeding in farm stock can be slow and expensive because the animals often have small numbers of offspring and breed only once a year.

By producing new combinations of genes, selective breeding achieves the same objectives as genetic engineering but it takes much longer and is less precise.

In selective breeding, the transfer of genes takes place between individuals of the same, or closely related species. Genetic engineering involves transfer between unrelated species.

QUESTIONS

18 Suggest some good characteristics that an animal-breeder might try to combine in sheep by mating different varieties together.

19 A variety of barley has a good ear of seed but has a long stalk and is easily blown over. Another variety has a short, sturdy stalk but a poor ear of seed.

Suggest a breeding programme to obtain and select a new variety which combined both of the useful characteristics.

Choose letters to represent the genes and show the genotypes of the parent plants and their offspring.

Checklist

- In breeding experiments, the effect of only one or two genes (out of thousands) is studied, e.g. colour of fur in rabbits or mice.

- The genes are in pairs (allelomorphic pairs), because the chromosomes are in pairs.

- Although each pair of alleles controls the same characteristic, the alleles do not necessarily have the same effect. For example, of a pair of alleles controlling fur colour, one may try to produce black fur and the other may try to produce white fur.

- Usually, one allele is dominant over the other, e.g. the allele (**B**) for black fur is dominant over the allele (**b**) for white fur.

- This means that a rabbit with the alleles **Bb** will be black even though it has a gene for white fur.

- Although **BB** rabbits and **Bb** rabbits are both black, only the **BB** rabbits will breed true.

- **Bb** black rabbits mated together are likely to have some white babies.

- The expectation is that, on average, there will be one white baby rabbit to every three blacks.

- Meiosis is the kind of cell division that leads to production of gametes.

- Only one of each chromosome pair goes into a gamete.

- A **Bb** rabbit would produce two kinds of gametes for coat colour; 50 per cent of the gametes would have the **B** gene and 50 per cent would have the **b** gene.

- In some cases, neither one of a pair of alleles is fully dominant over the other. This may be called incomplete dominance or co-dominance.

- Sex, in mammals, is determined by the X and Y chromosomes. Males are XY; females are XX.

- Hereditary diseases result from a person inheriting two recessive alleles for the disease.

- Some hereditary diseases are sex-linked.

- A knowledge of genetics enables breeders to produce new varieties of plants or animals.

- Cross-breeding two varieties enables beneficial genes from both to be brought together in a new variety.

23 VARIATION, SELECTION AND EVOLUTION

Variation
New gene combinations and mutations. Meiosis and new gene combinations. Discontinuous variation: distinct differences. Continuous variation: graduated differences. Interaction of genes and environment.

Natural selection
The selection of more efficient varieties. Peppered moth: selection of the different forms. Sickle-cell anaemia: selection in different environments. Artificial selection for improved breeds.

Evolution
First life; increasing complexity; vertebrate evolution. Evidence for evolution; fossil record. Extinction.

Variation

The term 'variation' refers to observable differences within a species. All domestic cats belong to the same species, i.e. they can all interbreed, but there are many variations of size, coat colour, eye colour, fur length, etc.

Those variations which can be inherited are determined by genes. They are genetic or heritable variations.

There are also variations which are not heritable, but determined by factors in the environment. A kitten which gets insufficient food will not grow to the same size as its litter mates. A cat with a skin disease may have bald patches in its coat. These conditions are not heritable. They are caused by environmental effects. Similarly, a fair-skinned person may be able to change the colour of his or her skin by exposing it to the sun, so getting a tan. The tan is an **acquired characteristic**. You cannot inherit a sun-tan. The dark skin of a Negro, on the other hand, is an **inherited characteristic**.

Many features in plants and animals are a mixture of acquired and inherited characteristics (Fig. 1). For example, some fair-skinned people never go brown in the sun, they only become sunburned. They have not inherited the genes for producing the extra brown pigment in their skin. A fair-skinned person with the genes for producing pigment will only go brown if he exposes himself to sunlight. So his tan is a result of both inherited and acquired characteristics.

Heritable variation may be the result of mutations (p. 205), or new combinations of genes in the zygote.

New combinations of genes

If a grey cat with long fur is mated with a black cat with short fur, the kittens will all be black with short fur. If these offspring are mated together, in due course the litters may include four varieties: black–short, black–long, grey–short and grey–long. Two of these

north side, upper branches   south side, upper branches

north side, lower branches   south side, lower branches

Figure 1 Aquired characteristics. These apples have all been picked from different parts of the same tree. All the apples have similar genotypes, so the differences in size must have been caused by environmental effects.

are different from either of the parents. (See 'Meiosis and new combinations of characteristics' below.)

Mutations

Many of the coat variations mentioned above may have arisen, in the first place, as mutations in a wild stock of cats. A recent variant produced by a mutation is the 'rex' variety, in which the coat has curly hairs.

Many of our high-yielding crop plants have arisen as a result of mutations in which the whole chromosome set has been doubled.

Meiosis and new combinations of characteristics

On p. 201 it was explained that, during meiosis, homologous chromosomes pair up and then at the first nuclear division, separate again.

One of the homologous chromosomes comes from the male parent and the other from the female parent. The alleles for a particular characteristic occupy identical positions on the homologous chromosomes but they do not necessarily control the characteristic in the same way.

The allele for eye colour will be in the same position on the maternal and paternal chromosome but, in one case, it may be the allele for brown eyes and in the other case, for blue eyes. Separation of homologous chromosomes at meiosis means that the alleles for blue and brown eyes will end up in different gametes (Fig. 2).

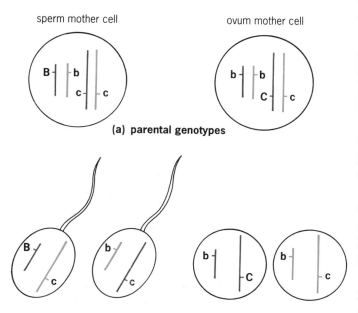

(a) parental genotypes

(b) possible gene combinations in gametes

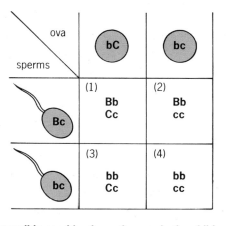

(c) possible combinations of genes in the children

Figure 2 New combinations of characteristics

On a second pair of homologous chromosomes there may be allelomorphic genes for hair curliness (**C** = curly; **c** = straight). These chromosomes and their alleles will also be separated at the first division of meiosis.

Suppose a father has brown eyes and straight hair, and a mother has blue eyes and curly hair. Also suppose that the father is heterozygous for eye colour (**Bb**) and the mother is heterozygous for hair curliness (**Cc**).

In the mother's ovary, the **b** and **b** alleles will be separated at the first division of meiosis and so will the **C** and **c** alleles. The **b** allele could finish up in the same gamete as either the **C** or **c** allele. So, the genotype of the ovum could be either **bC** or **bc**.

Similarly, meiosis in the father's testes will produce equal numbers of **Bc** and **bc** gametes.

At fertilization, it is a matter of chance which of the two types of sperm fertilizes which of the two types of ovum. (Although usually only one ovum is released, there is a 50:50 chance of its being **bC** or **bc**.)

The grid (**Punnett square**), in Fig 2(c) shows the possible genotypes of the children in the family. Offspring (2) and (3) would have the same combination of characteristics as their parents: (2) has brown eyes and straight hair (father's phenotype), and (3) has blue eyes and curly hair (mother's phenotype). Offspring (1) and (4), however, would have different combinations of these two characteristics; combinations which are not present in either parent, namely, (1) brown eyes and curly hair, (4) blue eyes and straight hair.

The separation of parental chromosomes at meiosis and their recombination at fertilization has thus introduced the possibility of new combinations of characteristics. It occurs because the homologous chromosomes derived from one parent do not all go into the same gamete, but move independently of each other.

This recombination of characteristics as a result of meiosis is important for plant and animal breeding programmes as described on p. 217. It is also important as a source of variation for natural selection to act on, as described on p. 222.

Meiosis takes place only at gamete formation and this is an essential feature of sexual reproduction. It can be claimed, therefore, that one of the important biological advantages of sexual reproduction is the production of new varieties which might be more successful than existing varieties.

Crossing over

At an early stage in meiosis, when homologous chromosomes are lying alongside each other, the chromatids break at various points. They may then rejoin with their original chromatid or with a chromatid from their homologous partner. As a result, portions of the maternal chromatids become attached to paternal chromatids and vice versa (Fig. 3). Thus there will be new combinations of genes on these chromatids when they separate and produce gametes. This is a further source of genetic variation which results from meiosis.

QUESTIONS

1 Which of the following do you think are **a** mainly inherited characteristics, **b** mainly acquired characteristics or **c** a more or less equal mixture: manual skills, facial features, body build, language, athleticism, ability to talk?

2 **a** Bearing in mind the role of the sex chromosomes, suggest two other new variations which might occur among the children in Fig. 2.
b In Fig. 2, if the mother had been homozygous for hair curliness and the father had been homozygous for eye colour, is there a possibility of new combinations of those characters in the offspring?

3 What new combinations of characters are possible as a result of crossing a tall plant with yellow seeds (**TtYy**) with a dwarf plant with green seeds (**ttyy**)?

4 What are the environmental effects which might have caused the variation in apple size in Fig. 1?

5 What new combination of genes has arisen from the crossing over in Fig. 3?

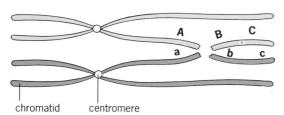

(a) the chromatids break at corresponding points

(b) they rejoin but to their opposite partner

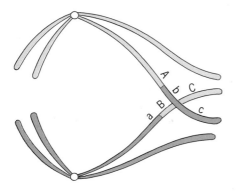

(c) when they separate they carry new gene combinations into the gametes

Figure 3 Crossing over

Discontinuous variations

These are variations under the control of a single pair of alleles or a small number of genes. The variations take the form of distinct, alternative phenotypes with no intermediates. The mice in Fig. 4 on p. 212 are either black or brown; there are no intermediates. You are either male or female. Apart from a small number of abnormalities, sex is inherited in a discontinuous way.

Discontinuous variations cannot usually be altered by the environment. You cannot change your blood group or eye colour by altering your diet. A genetic dwarf cannot grow taller by eating more food.

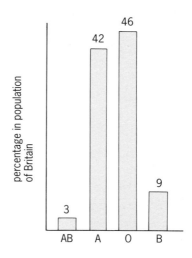

Figure 4 Discontinuous variation. Frequencies of ABO blood groups in Britain. The figures could not be adjusted to fit a smooth curve because there are no intermediates.

Continuous variation

There are no distinct categories of height; people are not either tall or short. There are all possible intermediates between very short and very tall. This is a case of continuous variation (Fig. 5).

Continuously variable characteristics are usually controlled by several pairs of alleles. There might be five pairs of alleles for height—(**Hh**), (**Tt**), (**Ll**), (**Ee**) and (**Gg**)—each dominant allele adding 4 cm to your height. If you inherited all ten dominant genes (**HH, TT**, etc.) you could be 40 cm taller than a person who inherited all ten recessive genes (**hh, tt**, etc.).

The actual number of genes which control height, intelligence and even the colour of hair and skin, is not known.

Continuously variable characteristics are greatly influenced by the environment. A person may inherit genes for tallness and yet not get enough food to grow tall. A plant may have the genes for large fruits but not get enough water, minerals or sunlight to produce large fruits. Continuous variations in human populations, such as height, physique and intelligence are always the result of interaction between the genotype and the environment.

There are many characteristics which are difficult to classify as either wholly continuous or discontinuous variations. Human eye colour has already been mentioned. People can be classified roughly as having blue eyes or brown eyes, but there are also categories described as grey, hazel or green eyes. Probably there is a small number of genes for eye colour and a dominant gene for brown eyes which overrides all the others when it is present. Similarly, red hair is a discontinuous variation but it is masked by genes for other colours and there is a continuous range of hair colour from blond to black.

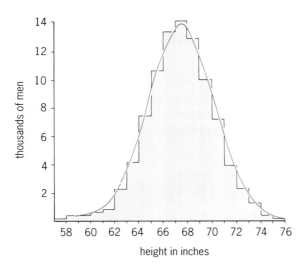

Figure 5 Continuous variation. Heights of 90 000 recruits in 1939. The apparent 'steps' in the distribution are the result of arbitrarily chosen categories, differing in height by one inch. But heights do not differ by exactly one inch. If measurements could be made accurately to the nearest millimetre, there would be a smooth curve like the one shown in colour.

QUESTION

6 Which of the following would you expect to be inherited in **a** a continuous way and **b** a discontinuous way: presence of horns in sheep, black colour in cats, number of seeds in an ear of wheat, size of apples, colour of roses, ability to sing, physical strength?

Natural selection

Theories of evolution have been put forward in various forms for hundreds of years. In 1858, Charles Darwin and Alfred Russel Wallace published a theory of evolution by natural selection which is still an acceptable theory today.

The theory of natural selection suggests that

1 Individuals within a species are all slightly different from each other (Fig. 6). These differences are called **variations**.

2 If the climate or food supply changes, some of these variations

may be better able to survive than others. A variety of animal that could eat the leaves of shrubs as well as grass would be more likely to survive a drought than one which fed only on grass.

3 If one variety lives longer than others, it is also likely to leave behind more offspring. A mouse that lives for 12 months may have ten litters of five babies (50 in all). A mouse that lives for 6 months may have only five litters of five babies (25 in all).

4 If some of the offspring inherit the variation that helped the parent survive better, they too will live longer and have more offspring.

5 In time, this particular variety will outnumber and finally replace the original variety.

Thomas Malthus, in 1798, suggested that the increase in the size of the human population would outstrip the rate of food production. He predicted that the number of people would eventually be regulated by famine, disease and war. When Darwin read the Malthus essay, he applied its principles to other populations of living organisms.

He observed that animals and plants produce vastly more offspring than can possibly survive to maturity and he reasoned that, therefore, there must be a 'struggle for survival'.

For example, if a pair of rabbits had eight offspring which grew up and formed four pairs, eventually having eight offspring per pair, in four generations the number of rabbits stemming from the original pair would be 512 (i.e. $2 \rightarrow 8 \rightarrow 32 \rightarrow 128 \rightarrow 512$). The population of rabbits, however, remains more or less constant. Many of the offspring in each generation must, therefore, have failed to survive to reproductive age.

Competition and selection

There will be **competition** between members of the rabbit population for food, burrows and mates. If food is scarce, space is short and the number of potential mates limited, then only the healthiest, most vigorous, most fertile and otherwise well-adapted rabbits will survive and breed.

The competition does not necessarily involve direct conflict. The best adapted rabbits may be able to run faster from predators, digest their food more efficiently, have larger litters or grow coats which camouflage them better or more effectively reduce heat losses. These rabbits will survive longer and leave more offspring. If the offspring inherit the advantageous characteristics of their parents, they may give rise to a new race of faster, different coloured, thicker furred and more fertile rabbits which gradually replace the original, less well-adapted varieties. The new variations are said to have **survival value**.

This is natural selection; the better adapted varieties are 'selected' by the pressures of the environment (**selection pressures**).

For natural selection to be effective, the variations have to be heritable. Variations which are not heritable are of no value in natural selection. Training may give athletes more efficient muscles, but this characteristic will not be passed on to their children.

Evolution

Most biologists believe that natural selection, among other processes, contributes to the evolution of new species and that the great variety of living organisms on the Earth is the product of millions of years of evolution, involving natural selection (p. 224).

The peppered moth

An example of natural selection is provided by a species of moth called the peppered moth. The common form is speckled but there is also a variety which is black. The black variety was rare in 1850, but by 1895 in the Manchester area its numbers had risen to 98 per cent of the population of peppered moths. Observation showed that the light variety was concealed better than the dark variety when they rested on tree-trunks covered with lichens (Fig. 7). In the Manchester area, pollution had caused the death of the lichens and the darkening of the tree-trunks with soot. In this industrial area the dark variety was the better camouflaged (hidden) of the two and was not picked off so often by birds. So the dark variety survived better, left more offspring and nearly replaced the light form.

The selection pressure, in this case, was presumed to be mainly predation by birds. The adaptive variation which produced the selective advantage was the dark colour.

(In fact, the story is more complex than this and some of the evidence has been challenged.)

Sickle-cell anaemia

This condition has already been mentioned on p. 214. A person with sickle-cell disease has inherited both recessive alleles ($\mathbf{Hb^S Hb^S}$) for defective haemoglobin. The distortion and destruction of the red cells which occurs in low oxygen concentrations leads to bouts of severe anaemia (Fig. 8).

Figure 8 Sickle-cell anaemia. At low oxygen concentration the red cells become distorted.

Figure 6 Variation. The garden tiger moths in this picture are all from the same family. There is a lot of variation in the pattern on the wings.

Figure 7 Selection for varieties of the peppered moth

(a) and (b) show the two forms against the background on which they are best concealed.

(c) and (d) show the forms on alien backgrounds.

In many African countries, sufferers have a reduced chance of reaching reproductive age and having a family. There is thus a selection pressure which tends to remove the homozygous recessives from the population. In such a case, you might expect the harmful Hb^S allele to be selected out of the population altogether. However, the heterozygotes (Hb^AHb^S) have virtually no symptoms of anaemia but do have the advantage that they are more resistant to malaria than the homozygotes Hb^AHb^A.

The selection pressure of malaria, therefore, favours the heterozygotes over the homozygotes and the potentially harmful Hb^S allele is kept in the population (Fig. 9).

When Africans migrate to countries where malaria does not occur, the selective advantage of the Hb^S allele is lost and the frequency of this allele in the population diminishes.

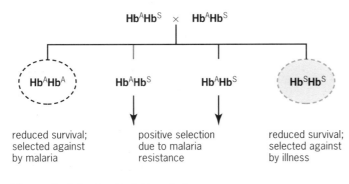

$$Hb^AHb^S \times Hb^AHb^S$$

Hb^AHb^A	Hb^AHb^S	Hb^AHb^S	Hb^SHb^S
reduced survival; selected against by malaria	positive selection due to malaria resistance		reduced survival; selected against by illness

Figure 9 Selection in sickle-cell disease

Artificial selection

Human communities practise a form of selection when they breed plants and animals for specific characteristics. The many varieties of dog that you see today have been produced by selecting individuals with short legs, curly hair, long ears, etc. One of the puppies in a litter might vary from the others by having longer ears. This individual, when mature, is allowed to breed. From the offspring, another long-eared variant is selected for the next breeding stock, and so on, until the desired or 'fashionable' ear length is established in a true-breeding population.

More important are the breeding programmes to improve agricultural livestock or crop plants. Animal-breeders will select cows for their high milk yield, sheep for their wool quality, pigs for their long backs (more bacon rashers). Plant-breeders will select varieties for their high yield and resistance to fungus diseases (Fig. 10). (See p. 217.)

QUESTIONS

7 What features of a bird's appearance and behaviour do you think might help it compete for a mate?

8 What selection pressures do you think might be operating on the plants in a lawn?

Figure 10 Selective breeding in tomatoes. Different breeding programmes have selected genes for fruit size, colour and shape. Similar processes have given rise to most of our cultivated plants and domesticated animals.

The theory of evolution

On p. 223 it was pointed out that Darwin's theory of natural selection in 1858 provided a good explanation for the process of evolution, but theories of evolution were in existence long before that.

Aristotle (384–322 BC) suggested a classification system which implied a progression from simple to complex creatures. However, the spread of Judaeo–Christian religion effectively deterred people from putting forward explanations that differed from the biblical story of Creation until the 18th and 19th centuries.

Jean Baptiste Lamarck (1744–1829) was a professor in zoology in Paris. He was perhaps the first person to openly suggest that species could change. His explanation, however, depended on the selection of characteristics acquired during the lifetime of the organism rather than selection of characteristics that arose genetically by chance (p. 222).

The first life

The theory, as it stands at present, suggests that at one time there were no living organisms of any kind on the Earth, possibly because it was too hot. As the Earth cooled, the theory supposes that conditions were just right for certain kinds of chemical reaction to take place in the water. These chemical reactions might have produced compounds like amino acids and enzymes, which could make other reactions take place. If these chemicals somehow came together inside a membrane, they would form the first single-celled creatures, such as bacteria, which could feed, grow and reproduce. There is a great deal of argument among biologists about how this could have happened or whether it could have happened at all. It is possible to make amino acids (p. 11) in laboratory experiments in the conditions suggested, but it is impossible to be confident that these really were the conditions on the Earth millions of years ago.

Organisms became more numerous and more complicated

The first single-celled creatures might have given rise to many-celled creatures if the cells stuck together after cell division. The many-celled creatures could have started as a simple ball of cells. Some cells might have become specialized to carry food, conduct nerve impulses, or contract to produce movement. As a result, the creatures would become more complicated and different forms would arise (Fig. 16, p. 228). Some could stick to rocks and produce tentacles like sea anemones. Some might swim in the surface waters like shrimps or burrow in the sand like worms.

Organisms came onto the land

If evolution has occurred, it most likely started in the water. Gradually some plants and animals became able to survive on land for longer and longer periods. Until green plants developed, there was probably no oxygen in the air. The plants' photosynthesis produced oxygen and this would have made it possible for animals to evolve.

Evolution of vertebrates

Some biologists think that most evolutionary change takes place as a result of a series of very small changes ('gradualism'). Others claim that there were long periods with no change and then short periods of rapid change ('punctuated equilibrium'). In either case it would take millions of years for a fish-like creature to turn into a frog-like creature. Over a period of about 400 million years it is thought that some fish-like creatures developed legs and lungs and so became amphibia, like newts. Some amphibian ancestors could have developed scales and the ability to lay eggs on land and so become reptiles. One group of ancient reptiles might have given rise to the birds by developing feathers and wings and another group, by developing fur and producing milk, could have become the mammals.

The theory of evolution does not suggest that frogs turned into lizards or that lizards turned into birds. Frogs and lizards are themselves thought to be the products of millions of years of evolution from ancestral amphibia and reptiles. It is supposed that some of the ancestral amphibia gave rise to two main groups of descendants. One of these groups developed into primitive reptiles and the other continued evolving to become frogs and toads (Fig. 11).

Evidence for evolution

Fossils

Fossils are the remains of animals and plants preserved in various ways and Fig. 12 shows how a skeleton of the fish might become embedded in mud or sand which was settling down on the bottom of a lake. After a few million years, with the pressure of more layers building up on top of it, the mud or sand becomes rock. Massive earth movements may raise the rocks above the water and

(a) 'Correct' representation of vertebrate evolution.

(b) 'Incorrect' representation of vertebrate evolution. It seems to suggest that present-day fish gave rise to present-day amphibia, and so on.

Figure 11 The theory of evolution. This says that fish and amphibia, for example, share a common ancestor. It does *not* say that fish give rise to amphibia. However, we do not know for sure what the common ancestors were like.

when they form cliffs or are quarried, the preserved skeletons can be found and chipped out of the rock.

This is not the only method of fossilization. Any environment that prevents rapid decay may produce fossils. Insects have been found trapped in amber which is formed from resin exuding from trees; pollen grains are preserved in the anaerobic conditions of bogs and their fossil remains are found in peat; whole animals have been found frozen in ancient ice.

It is usually the hard tissues which are preserved, e.g. bones, shells, wood, seeds, etc. These decay more slowly than the softer tissues and, in many cases, the organic matter is entirely replaced by minerals which do not decay.

The skeletons, shells, tree-trunks and other fossil remains give us some idea of the animals and plants that were living millions of years ago. The deepest layers of rock are likely to contain the oldest fossils (Fig. 12). So by studying the fossil record, the theory of evolution can be put to the test.

The extent of the fossil remains of the five vertebrate classes in rock layers of increasing depth is shown in Fig. 13. The width of the coloured bands represents the amount of fossil remains found. For example, 100 million years ago, it looks as if there were many more reptiles about than there are today. Also you will notice that fossils of mammals do not appear at all in rocks which are older than about 300 million years.

The fossil record appears to support the idea that fish-like creatures could have given rise to mammals because, 300 million years ago, there were plenty of fish but no mammals. Most of the organisms preserved in the fossil record appear to be different from the organisms we know today. They are obviously related but different in important ways. For example, Fig. 14 shows a fish that lived 350 million years ago. It is clearly a fish, but quite different from any fish living today.

So the fossil record shows that plants and animals have changed from one form to another over a long period—or does it?

Could it be that 350 million years ago there were all the fish, reptiles, amphibia, birds and mammals that we know today plus all those we know as fossils? Perhaps there were so few mammals at that time that we have not yet found any in the rocks of that period.

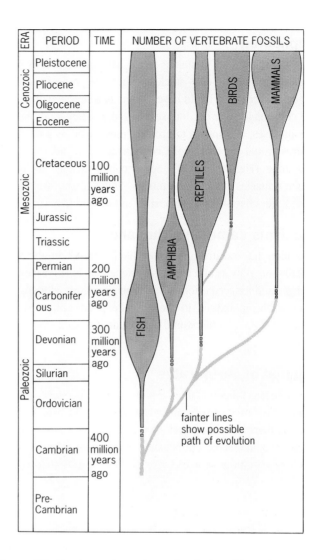

Figure 13 Numbers of vertebrate fossils found in rocks

Figure 14 An extinct 'armour-plated' fish of 350 million years ago (reconstructed from fossil remains)

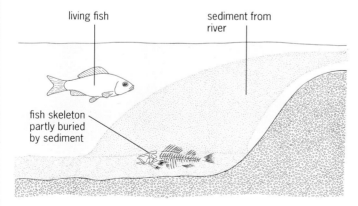

Figure 12 Formation of fossils

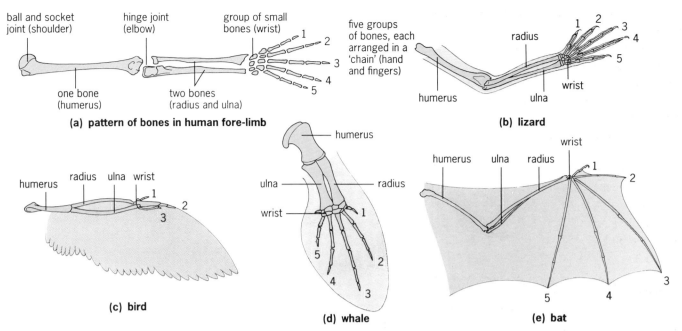

Figure 15 Skeletons of five vertebrate limbs

In such a case, all that the fossil record would show is that some creatures have become more abundant, some have become scarce and some have died out altogether.

This seems a very improbable interpretation of the fossil evidence and fails to offer any explanation of the origin of life or the diversity of organisms.

Evidence from anatomy

The skeleton of the front limb of five types of vertebrate are shown in Fig. 15. Although the limbs have different functions such as grasping, flying, running and swimming, the arrangement and number of the bones is almost the same in all five. There is a single top bone (humerus), with a ball and socket joint at one end and a hinge joint at the other. It makes a joint with two other bones (radius and ulna) which join to a group of small wrist bones. The limb skeleton ends with five groups of bones (hand and fingers), although some of these groups are missing in the bird.

The argument for evolution says that, if these animals are not related, it seems very odd that such a similar limb skeleton should be used to do such different things as flying, running and swimming. If, on the other hand, all the animals came from the same ancestor, the ancestral skeleton could have changed by small stages in different ways in each group. So we would expect to find that the basic pattern of bones was the same in all these animals. There are many other examples of this kind of evidence among the vertebrate animals.

QUESTIONS

9 According to Fig. 13 when were the amphibia most abundant?

10 What features of Fig. 14 suggest that the fossil animal is related to present-day fishes?

11 In Fig. 16 why do you think that mammals and birds are drawn at the top of the diagram and worms and insects lower down?

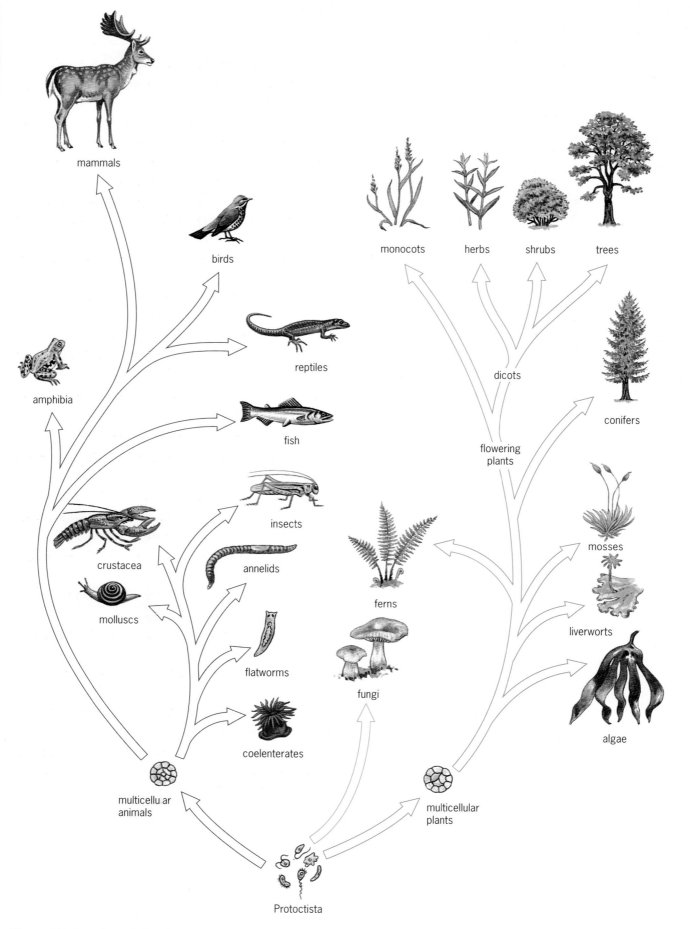

mammals

birds

reptiles

fish

amphibia

insects

crustacea

annelids

molluscs

flatworms

coelenterates

multicellular animals

ferns

fungi

Protoctista

monocots

herbs

shrubs

trees

dicots

flowering plants

conifers

mosses

liverworts

algae

multicellular plants

Figure 16 Organic evolution. Only a few of the main types of organism are shown.

Extinction

From the fossil record, it is obvious that a great many organisms have become extinct. It is not usually possible to be sure of the causes of extinction in the distant past. Perhaps there were drastic changes in climate, outbreaks of disease or an invasion of more successful competitors. There have been many periods of mass extinction, one of which included the decline of the dinosaurs 65 million years ago. One theory suggests that the impact of a giant asteroid produced such a vast dust cloud that the Earth became dark and cold. As a result, plants and the animals which fed on them, died out.

More recent extinctions usually result from human activities. The dodo, a flightless bird on the island of Mauritius was driven to extinction in 1680 by hunting and the introduction of pigs. The Polynesians are thought to have exterminated half the population of birds on the islands which they colonized in the 4th and 5th centuries.

The expanding human population is hunting some animals to the verge of extinction and destroying their habitats by cutting down forests for timber or agriculture. The tiger and the giant panda are obvious examples.

The World Conservation Monitoring Centre estimates that over 300 species have become extinct since 1600 and Friends of the Earth claim that one species per day is being lost for ever. The exact figures may be disputed but the message is clear.

Checklist

- Variations within a species may be inherited or acquired.
- Inherited variations arise from different combinations of genes or from mutations.
- At meiosis the maternal and paternal chromosomes are randomly distributed between the gametes.
- Because the gametes do not carry identical sets of genes, new combinations of genes may arise at fertilization.
- Discontinuous variation results, usually, from the effects of a single pair of alleles, and produces distinct and consistent differences between individuals.
- Discontinuous variations cannot be changed by the environment.
- Continuous variations are usually controlled by a number of genes affecting the same characteristic.
- Continuous variation can be influenced by the environment.
- Members of a species compete with each other for food and mates.
- Some members of a species may have variations which enable them to compete more effectively.
- These variants will live longer and leave more offspring.
- If the beneficial variations are inherited, the offspring will also survive longer.
- The new varieties may gradually replace the older varieties.
- Natural selection involves the elimination of less well-adapted varieties by environmental pressures.
- Artificial selection is used to improve commercially useful plants and animals.
- The theory of evolution tries to explain how present-day plants and animals came into existence.
- It supposes that simple living organisms, such as bacteria, were formed from non-living matter.
- These simple organisms changed over hundreds of millions of years to become many-celled and more complicated.
- They also became more varied, giving rise to many different groups of plants and animals.
- The evidence from the fossil record supports the theory of evolution because it shows animals and plants becoming more varied and more complicated as time goes on.

EXAMINATION QUESTIONS

Section 4 Genetics and Heredity

Do not write on this page. Where necessary copy drawings, tables or sentences.

1 The nucleus of a body cell of an animal has 16 pairs of chromosomes. When it divides the nucleus of each cell will contain:
A 8 different chromosomes **C** 8 pairs of chromosomes
B 16 different chromosomes **D** 16 pairs of chromosomes *[N]*

2 (a) Scientists wished to breed a certain fruit for improved flavour and a more attractive colour. Explain, with full details, how they could achieve this. (9)
(b) Suggest, with an example, how artificial selection might be of importance to a farmer. (3)
[L]

3 (a) Artificial selection can be used to improve crop plants and animals.
(i) Give THREE different examples of improvements brought about by artificial selection. (3)
(ii) How does a plant or animal breeder produce a new variety by artificial selection? (2)
(b) Describe the theory of evolution by natural selection. The following words may help you: variation; over population; competition; adaptation; selection. (5)
[M]

4 The diagram shows a bacterial chromosome into which the human gene for the production of growth hormone has been transplanted. The genetic instructions in the transplanted gene cause the bacterium to produce the growth hormone.

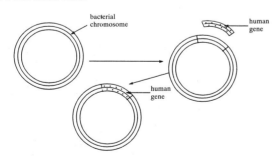

bacterial chromosome
human gene
human gene

(a) What happens to the transplanted gene when the bacterial cell divides? (1)
(b) Give two advantages of this method of hormone production. (2)
[NI]

5 (a) In the nucleus of a human liver cell there are 46 chromosomes. How many chromosomes would you expect to find in the following human cells?
(i) A cell from the lining of the cheek. (ii) A cell from the lining of the womb. (iii) An egg. (iv) A zygote. (4)
(b) Name the type of cell division which causes the zygote to become an embryo. (1)
[W]

6 Cystic fibrosis is a serious disease of humans in which the lungs clog with mucus. The disease is inherited and involves a single gene. The gene exists as two alleles. Two healthy parents had three children. The third was found to have cystic fibrosis although the other two children were normal.
(a) Is the allele that causes cystic fibrosis dominant or recessive? Explain the reason for your answer. (1)
(b) What is the genotype of the parents? (1)
(c) (i) What is the chance of a fourth child having the disease? (1)
(ii) Use a genetic diagram to explain your answer. (3)
[N]

7 A black mouse was mated to a white mouse. All the offspring were black. Three pairs of these offspring, A, B and C were mated together. These were the results.

Pair	Number of black young	Number of white young
A	6	2
B	7	3
C	5	1

(a) What is the total number of young black mice? (1)
(b) What is the ratio of black mice to white mice? (1)
(c) State a conclusion that you can make from these results. (1)
(d) Which colour must be homozygous? (1)
(e) If the young mice from the first mating were all grey, what would you suspect about the nature of the genes for black and white colour? (1)
(f) Explain why it is important to repeat the test before you can form a conclusion. (1)
[W]

8 A potato grower wanted to produce a new variety of potato which grows quickly and makes good chips. She has one variety which grows quickly but makes poor chips, and another which grows more slowly but makes good chips. When plants of these two varieties produce flowers she crosses the two varieties. Later she collects the seeds and plants them.
(a) To cross the two varieties, the grower pollinates a flower of one variety with pollen from the other. Describe how she should do this. (2)
(b) From the seeds she collects, she finds that one of the new varieties grows very quickly and produces potatoes which make good chips. How would she produce a crop of potatoes which are exactly the same? (2)
[N]

9 Genetic engineering sometimes involves the introduction into a bacterium of a foreign:
A Chromosome **C** Gene
B Nucleus **D** Nucleotide

Organisms and
their Environment

24 THE INTER-DEPENDENCE OF LIVING ORGANISMS

> *Food chains and food webs*
> Pyramids of numbers, recycling.
>
> *The carbon cycle and nitrogen cycle*
>
> *Agriculture*
>
> *Decay*
>
> *Energy flow in an ecosystem*

'Interdependence' means the way in which living organisms depend on each other in order to remain alive, grow and reproduce. For example, bees depend for their food on pollen and nectar from flowers. Flowers depend on bees for pollination (p. 73). Bees and flowers are, therefore, interdependent.

Food chains and food webs

One important way in which organisms depend on each other is for their food. Many animals, such as rabbits, feed on plants. Such animals are called **herbivores**. Animals called **carnivores** eat other animals. A **predator** is a carnivore which kills and eats other animals. A fox is a predator which preys on rabbits. **Scavengers** are carnivores which eat the dead remains of animals killed by predators. These are not hard and fast definitions. Predators will sometimes scavenge for their food and scavengers may occasionally kill living animals.

Food chains

Basically, all animals depend on plants for their food. Foxes may eat rabbits, but rabbits feed on grass. A hawk eats a lizard, the lizard has just eaten a grasshopper but the grasshopper was feeding on a grass blade. This relationship is called a **food chain** (Fig. 1).

The organisms at the beginning of a food chain are usually very numerous while the animals at the end of the chain are often large

Figure 1 A food chain. The caterpillar eats the leaf; the blue tit eats the caterpillar but may fall prey to the kestrel.

(a) land	(b) water
owl	perch
blue tits	minnows
caterpillars	water fleas — consumers
plant leaves	microscopic algae — producers

Figure 2 Examples of food pyramids

232

(a) phytoplankton (x100)
These microscopic algae form the basis of a food pyramid in the water.

(b) zooplankton (x20)
These crustacea will eat microscopic algae.

Figure 3 Plankton. The microscopic organisms which live in the surface waters of the sea or fresh water are called collectively, plankton. The single-celled algae (see p. 295) are the phytoplankton. They are surrounded by water, salts and dissolved carbon dioxide. Their chloroplasts absorb sunlight and use its energy for making food by photosynthesis. The phytoplankton is eaten by small animals in the zooplankton, mainly crustacea (see p. 290). Small fish will eat the crustacea.

and few in number. The **food pyramids** in Fig. 2 show this relationship. There will be millions of microscopic, single-celled algae in a pond (Fig. 3(a)). These will be eaten by the larger but less numerous water fleas and other crustacea (Fig. 3(b)), which in turn will become the food of small fish, like minnow and stickleback.

The hundreds of small fish may be able to provide enough food for only four or five large carnivores, like pike or perch.

The organisms at the base of the food pyramids in Fig. 2 are plants. Plants produce food from carbon dioxide, water and salts (see 'Photosynthesis', p. 37), and are, therefore, called **producers**. The animals which eat the plants are called **primary consumers**, e.g. grasshoppers. Animals which prey on the plant-eaters are called **secondary consumers**, e.g. shrews, and these may be eaten by **tertiary consumers**, e.g. weasels or kestrels (Fig. 4).

The position of an organism in a food pyramid is sometimes called its **trophic level**.

Pyramids of numbers and biomass

The width of the bands in Fig. 2 is meant to represent the relative number of organisms at each trophic level. So, the diagrams are sometimes called **pyramids of numbers**.

However, you can probably think of situations where a pyramid of numbers would not show the same effect. For example, a single sycamore tree may provide food for thousands of greenfly. One oak tree may feed hundreds of caterpillars. In these cases the pyramid of numbers is upside-down.

The way round this problem is to consider not the single tree, but the mass of the leaves that it produces in the growing season, and the mass of the insects which can live on them. **Biomass** is the term used when the mass of living organisms is being considered, and pyramids of biomass can be constructed as in Fig. 16, p. 240.

An alternative is to calculate the energy available in a year's supply of leaves and compare this with the energy needed to maintain the population of insects which feed on the leaves. This would produce a **pyramid of energy**, with the producers at the bottom having the greatest amount of energy. Each successive trophic level would show a reduced amount of energy.

Later in the chapter, the recycling of matter is discussed. The elements which make up living organisms are recycled, i.e. they are used over and over again. This is not the case with energy which flows from producers to consumers and is eventually lost to the atmosphere as heat. (See 'Energy flow', p. 239.)

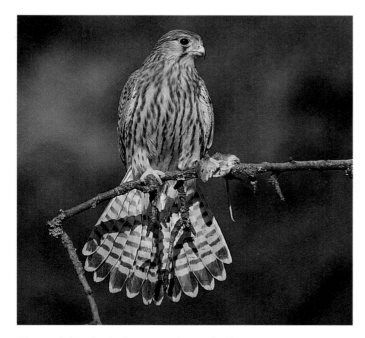

Figure 4 The kestrel; a secondary or tertiary consumer

Food webs

Food chains are not really as straightforward as described above, because most animals eat more than one type of food. A fox, for example, does not feed entirely on rabbits but takes beetles, rats and voles in its diet. To show these relationships more accurately, a **food web** can be drawn up (Fig. 5 and Fig. 1, p. 243).

The food webs for land, sea and fresh water, or for ponds, rivers and streams will all be different. Food webs will also change with the seasons when the food supply changes.

If some event interferes with a food web, all the organisms in it are affected in some way. For example, if the rabbits in Fig. 5 were to die out, the foxes, owls and stoats would eat more beetles and rats. Something like this happened in 1954 when the disease myxomatosis wiped out nearly all the rabbits in England. Foxes ate more voles, beetles and blackberries, and attacks on lambs and chickens increased. Even the vegetation was affected because the tree seedlings which the rabbits used to nibble off were able to grow. As a result, woody scrubland started to develop on what had been grassy downs. A similar effect is shown in Fig. 6. (See also pp. 242–3.)

Dependence on sunlight

If you take the idea of food chains one step further you will see that all living organisms depend on sunlight and photosynthesis (p. 37). Green plants make their food by photosynthesis, which needs sunlight. Since all animals depend, in the end, on plants for their food, they therefore depend indirectly on sunlight. A few examples of our own dependence on photosynthesis are given below.

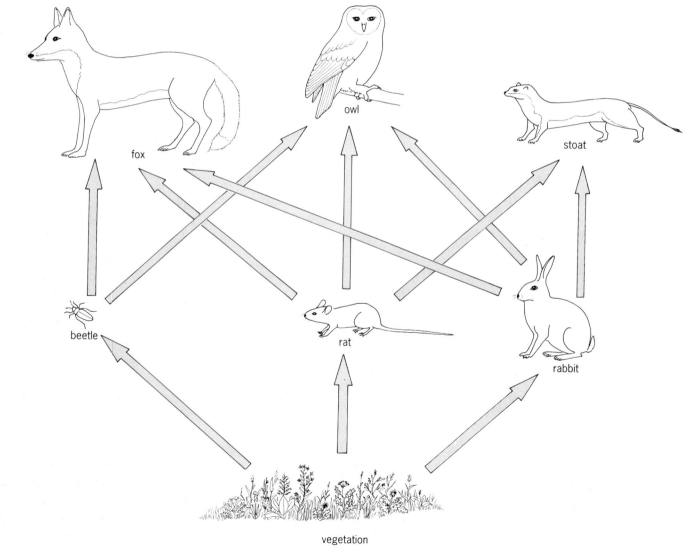

Figure 5 A food web

(a)   **(b)**

Figure 6 Effect of grazing
a Sheep have eaten any seedlings which grew under the trees
b Ten years later, the fence has kept the sheep off and the tree seedlings have grown

Nearly all the energy released on the Earth can be traced back to sunlight. Coal comes from tree-like plants, buried millions of years ago. These plants absorbed sunlight for their photosynthesis when they were alive. Petroleum was formed, also millions of years ago, probably from the partly decayed bodies of microscopic algae which lived in the sea. These, too, had absorbed sunlight for photosynthesis.

Today it is possible to use mirrors and solar panels to collect energy from the sun directly, but the best way, so far, of trapping and storing energy from sunlight is to grow plants and make use of their products, such as starch, sugar, oil, alcohol and wood, for food or as energy sources. For example, sugar from sugar-cane can be fermented (p. 21) to alcohol, and used as a motor fuel instead of petrol.

Recycling

There are a number of organisms which have not been fitted into the food webs or food chains described so far. Among these are the **saprotrophs**. Saprotrophs do not obtain their food by photosynthesis, nor do they kill and eat living animals or plants. Instead they feed on dead and decaying matter such as dead leaves in the soil or rotting tree-trunks (see Fig. 7). The most numerous examples are the fungi, such as mushrooms, toadstools or moulds, and the bacteria, particularly those which live in the soil. They produce extra-cellular enzymes (p. 15) which digest the decaying matter and then they absorb the soluble products back into their cells. In so doing, they remove the dead remains of plants and animals which would otherwise collect on the Earth's surface. They also break these remains down into substances which can be used by other organisms. Some bacteria, for example, break down the protein of dead plants and animals and release nitrates which are taken up by plant roots and there built into new amino acids and proteins (p. 11). This use and re-use of materials in the living world is called **recycling**.

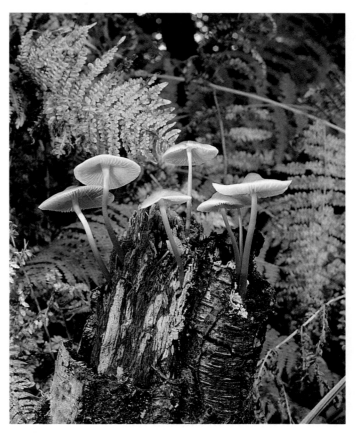

Figure 7 Saprotrophs. These toadstools are getting their food from the rotting tree stump.

The general idea of recycling is illustrated in Fig. 8. The green plants are the producers, and the animals which eat the plants and each other are the consumers. The bacteria and fungi, especially those in the soil, are called the **decomposers** because they break down the dead remains and release the chemicals for the plants to use again. Two examples of recycling, one for carbon and one for nitrogen, are described below.

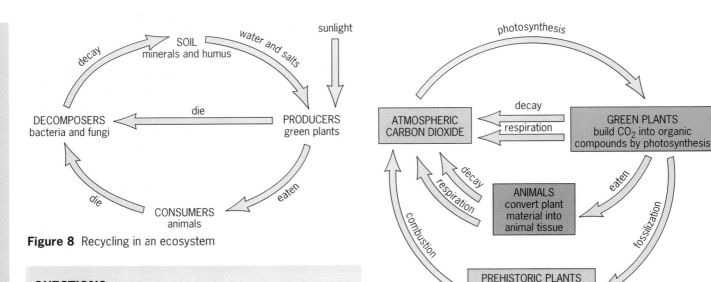

Figure 8 Recycling in an ecosystem

1 Try to construct a simple food web using the following: sparrow, fox, wheat seeds, cat, kestrel, mouse.
2 Describe briefly all the possible ways in which the following might depend on each other: grass, earthworm, blackbird, oak tree, soil.
3 Explain how the following foodstuffs are produced as a result of photosynthesis: wine, butter, eggs, beans.
4 An electric motor, a car engine and a race horse can all produce energy. Show how this energy could come, originally, from sunlight. What forms of energy on the Earth are *not* derived from sunlight?
5 How do you think evidence is obtained in order to place animals such as a fox and a pigeon in a food web?
6 When humans colonized islands they often introduced their domestic animals such as goats or cats. This usually had a devastating effect on the natural food webs. Suggest reasons for this.

The carbon cycle

Carbon is an element which occurs in all the compounds which make up living organisms. Plants get their carbon from carbon dioxide in the atmosphere and animals get their carbon from plants. The carbon cycle, therefore, is mainly concerned with what happens to carbon dioxide (Fig. 9).

Removal of carbon dioxide from the atmosphere

Photosynthesis

Green plants remove carbon dioxide from the atmosphere as a result of their photosynthesis. The carbon of the carbon dioxide is built first into a carbohydrate such as sugar. Some of this is changed into starch or the cellulose of cell walls, and the proteins, pigments and other compounds of a plant. When the plants are eaten by animals, the organic plant material is digested, absorbed and built into the compounds making up the animals' tissues. Thus the carbon atoms from the plant become part of the animal.

Figure 9 The carbon cycle

Addition of carbon dioxide to the atmosphere

Respiration

Plants and animals obtain energy by oxidizing carbohydrates in their cells to carbon dioxide and water (p. 20). The carbon dioxide and water are excreted and so the carbon dioxide returns once again to the atmosphere.

Decay

The organic matter of dead animals and plants is used by saprotrophs, especially bacteria and fungi, as a source of energy. These micro-organisms decompose the plant and animal remains and turn the carbon compounds into carbon dioxide.

Combustion (burning)

When carbon-containing fuels such as wood, coal, petroleum and natural gas are burned, the carbon is oxidized to carbon dioxide $C + O_2 \rightarrow CO_2$). The hydrocarbon fuels, such as coal and petroleum, come from ancient plants which have only partly decomposed over the millions of years since they were buried.

So, an atom of carbon which today is in a molecule of carbon dioxide in the air may tomorrow be in a molecule of cellulose in the cell wall of a blade of grass. When the grass is eaten by a cow, the carbon atom may become part of a glucose molecule in the cow's bloodstream. When the glucose molecule is used for respiration, the carbon atom will be breathed out into the air once again as carbon dioxide.

The same kind of cycling applies to nearly all the elements of the Earth. No new matter is created, but it is repeatedly rearranged. A great proportion of the atoms of which you are composed will, at one time, have been part of other organisms.

QUESTIONS

7 a Why do living organisms need a supply of carbon?
  b Give three examples of carbon-containing compounds which occur in living organisms (see pp. 11–12).
  c Where do **i** animals, **ii** plants get their carbon from?

8 Write three chemical equations **a** to illustrate that respiration produces carbon dioxide (see p. 20), **b** to show that burning produces carbon dioxide, **c** to show that photosynthesis uses up carbon dioxide (see p. 37).

9 Outline the events that might happen to a carbon atom in a molecule of carbon dioxide which entered the stoma in the leaf of a potato plant, and became part of a starch molecule in a potato tuber which was then eaten by a man. Finally the carbon atom is breathed out again in a molecule of carbon dioxide.

10 Construct a diagram, on the lines of the carbon cycle (Fig. 9) to show the cycling process for hydrogen (starting from the water used in photosynthesis).

The nitrogen cycle

When a plant or animal dies, its tissues decompose (p. 239), partly as a result of the action of saprotrophic bacteria. One of the important products of this decay, is **ammonia** (NH_3, a compound of nitrogen), which is washed into the soil.

The excretory products of animals contain nitrogenous waste products such as ammonia, urea and uric acid (p. 139). The organic matter in their droppings is also decomposed by soil bacteria.

Processes which add nitrates to soil

Nitrifying bacteria

These are bacteria living in the soil which use the ammonia from excretory products and decaying organisms as a source of energy (as we use glucose in respiration). In the process of getting energy from ammonia, the bacteria produce **nitrates**.

The 'nitrite' bacteria oxidize ammonium compounds to nitrites ($NH_4^- \rightarrow NO_2^-$).
'Nitrate' bacteria oxidize nitrites to nitrates ($NO_2^- \rightarrow NO_3^-$).

Although plant roots can take up ammonia in the form of its compounds, they take up nitrates more readily, so the nitrifying bacteria increase the fertility of the soil by making nitrates available to the plants.

Nitrogen-fixing bacteria

This is a special group of nitrifying bacteria which can absorb nitrogen as a gas from the air spaces in the soil, and build it into compounds of ammonia. Nitrogen gas cannot itself be used by plants. When it has been made into a compound of ammonia, however, it can easily be changed to nitrates by other nitrifying bacteria. The process of building the gas, nitrogen, into compounds of ammonia is called **nitrogen fixation**. Some of the nitrogen-fixing bacteria live freely in the soil. Others live in the roots of **leguminous plants** (peas, beans, clover), where they cause swellings called **root nodules** (Fig. 8, p. 313). These leguminous plants are able to thrive in soils where nitrates are scarce, because the nitrogen-fixing bacteria in their nodules make compounds of nitrogen available for them. Leguminous plants are also included in crop rotations to increase the nitrate content of the soil.

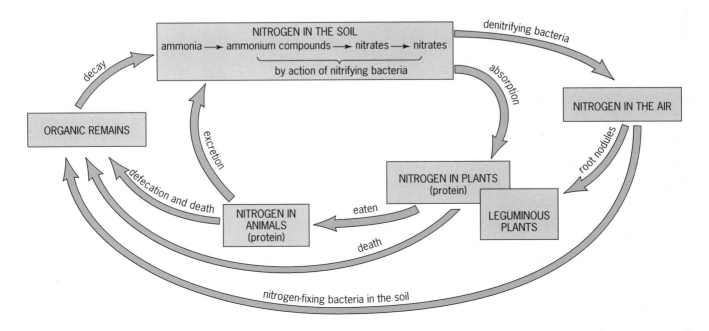

Figure 10 The nitrogen cycle

Lightning

The high temperature of lightning discharge causes some of the nitrogen and oxygen in the air to combine and form oxides of nitrogen. These dissolve in the rain and are washed into the soil as weak acids, where they form nitrates. Although several million tonnes of nitrate may reach the Earth's surface in this way each year, this forms only a small fraction of the total nitrogen being recycled.

Processes which remove nitrates from the soil

Uptake by plants

Plant roots absorb nitrates from the soil and combine them with carbohydrates to make proteins (p. 45).

Leaching

Nitrates are very soluble (i.e. dissolve easily in water), and as rain water passes through the soil it dissolves the nitrates and carries them away in the run-off or to deeper layers of the soil. This is called **leaching**.

Denitrifying bacteria

These are bacteria which obtain their energy by breaking down nitrates to nitrogen gas which then escapes from the soil into the atmosphere.

These processes are summed up in Fig. 10.

All the elements which make up living organisms are recycled, not just carbon and nitrogen. It would be possible to trace out cycles for hydrogen, oxygen, phosphorus, sulphur, iron, etc.

QUESTIONS

11 On a lawn growing on nitrate-deficient soil, the patches of clover often stand out as dark green and healthy against a background of pale green grass. Suggest a reason for this contrast. (See also p. 46.)

12 Very briefly explain the difference between nitrifying, nitrogen-fixing and denitrifying bacteria.

Agriculture

In a natural community of plants and animals, the processes which remove and replace mineral elements in the soil are in balance. In agriculture, most of the crop is usually removed so that there is little or no organic matter for nitrifying bacteria to act on. In a farm with animals, the animal manure, mixed with straw, is ploughed back into the soil or spread on the pasture. The manure thus replaces the nitrates and other minerals removed by the crop. It also gives the soil a good structure and improves its water-holding properties.

When animal manure is not available in large enough quantities, artificial fertilizers are used. These are mineral salts made on an industrial scale. Examples are ammonium sulphate (for nitrogen and sulphur), ammonium nitrate (for nitrogen) and compound NPK fertilizer for nitrogen, phosphorus and potassium (see p. 46). These are spread on the soil in carefully calculated amounts to provide the minerals, particularly nitrogen, phosphorus and potassium that the plants need. These artificial fertilizers increase the yield of crops from agricultural land, but they do little to maintain a good soil structure because they contain no organic matter (Figs 11 and 12).

Figure 11 Experimental plots of wheat. The rectangular plots have been treated with different fertilizers.

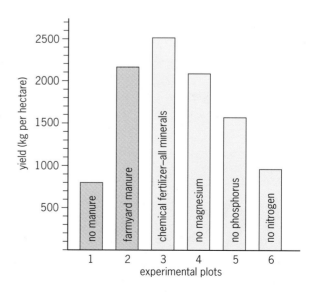

Figure 12 Average yearly wheat yields from 1852 to 1925, Broadbalk field, Rothamsted Experimental Station. Plot 1 received no manure or chemical fertilizer for 73 years. Plot 2 received an annual application of farmyard manure. Plot 3 received chemical fertilizer with all necessary minerals. Plots 4 to 6 received chemical fertilizer lacking one element.

QUESTIONS

13 To judge from Fig. 12, which mineral element seems to have the most pronounced effect on the yield of wheat? Explain your answer.
14 Draw up two columns headed A and B. In A list the processes which add nitrates to the soil and in B list those which remove nitrates from the soil. How might the activities of humans alter the normal balance between these two processes?
15 Suggest **a** some advantages, **b** some disadvantages of **i** organic manures (e.g. compost or farmyard manure) and **ii** chemical fertilizers.

Decay

A crucial factor in recycling is the process of decay. If it were not for decay, essential materials would not be released from dead organisms. When an organism dies, the enzymes in its cells, freed from normal controls, start to digest its own tissues (auto-digestion). Soon, scavengers appear on the scene and eat much of the remains; blowfly larvae devour carcases, earthworms consume dead leaves.

Finally the decomposers, fungi and bacteria (collectively called **micro-organisms**), arrive and invade the remaining tissues (Fig. 13). These saprotrophs secrete extracellular enzymes (p. 15) into the tissues and reabsorb the liquid products of digestion (p. 302). When the micro-organisms themselves die, auto-digestion takes place, releasing the products such as nitrates, sulphates, phosphates, etc. into the soil or the surrounding water to be taken up again by the producers in the ecosystem.

The speed of decay depends on the abundance of micro-organisms, temperature, presence of water and, in many cases, oxygen. High temperatures speed up decay because they speed up respiration of the micro-organisms. Water is necessary for all living processes and oxygen is needed for aerobic respiration of the bacteria and fungi. Decay can take place in anaerobic conditions but it is slow and incomplete as in the waterlogged conditions of peat bogs.

Energy flow in an ecosystem

An **ecosystem** is a community of living organisms and the habitat in which they live (p. 265). Most ecosystems are sustained by a constant input of energy from sunlight. This energy drives the chemistry of photosynthesis in the producers and passes to the other organisms through the food webs in the ecosystem. A pond is an ecosystem consisting of plants, animals, water, dissolved air, minerals and mud. The input of energy from the sun and a supply of water from rain is all that the pond community needs to maintain its existence. A forest is an ecosystem, so is an ocean. The whole of the Earth's surface may be considered as one vast ecosystem.

With the exception of atomic energy and tidal power, all the energy released on Earth is derived from sunlight. The energy released by animals comes, ultimately, from plants that they or their prey eat and the plants depend on sunlight for making their food (p. 234). The energy in organic fuels also comes ultimately from sunlight trapped by plants. Coal is formed from fossilized forests, and petroleum probably comes from the cells of ancient marine algae.

Use of sunlight

To try and estimate just how much life the Earth can support it is necessary to examine how efficiently the sun's energy is used. The amount of energy from the sun reaching the Earth's surface in 1 year ranges from 2 million to 8 million kilojoules per 1 m^2 ($2-8 \times 10^9$ Jm^{-2}y^{-1}) depending on the latitude. When this energy falls onto grassland, about 20 per cent is reflected by the vegetation, 39 per cent is used in evaporating water from the leaves (transpiration), 40 per cent warms up the plants, the soil and the air, leaving only about 1 per cent to be used in photosynthesis for making new organic matter in the leaves of the plants (Fig. 14).

This figure of 1 per cent will vary with the type of vegetation being considered and with climatic factors such as availability of water and the soil temperature. Sugar-cane grown in ideal conditions can

Figure 13 Mould fungus growing on an over-ripe orange

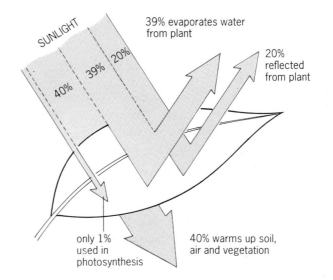

Figure 14 Absorption of sun's energy by plants

convert 3 per cent of the sun's energy into photosynthetic products; sugar-beet at the height of its growth has nearly a 9 per cent efficiency. Tropical forests and swamps are far more productive than grassland but it is difficult, and probably undesirable, to harvest and utilize their products (p. 247).

In order to allow crop plants to approach their maximum efficiency they must be provided with sufficient water and mineral salts. This can be achieved by irrigation and the application of fertilizer.

Energy transfer between organisms

Having considered the energy conversion from sunlight to plant products the next step is to study the efficiency of transmission of energy from plant products to primary consumers. On land, primary consumers eat only a small proportion of the available vegetation. In a deciduous forest only about 2 per cent is eaten; in grazing land, 40 per cent of the grass may be eaten by cows. In open water, however, where the producers are microscopic plants (phytoplankton, see Fig. 3(a)) and are swallowed whole by the primary consumers in the zooplankton (see Fig. 3(b)), 90 per cent or more may be eaten. In the land communities, the parts of the vegetation not eaten by the primary consumers will eventually die and be used as a source of energy by the decomposers.

A cow is a primary consumer; over 60 per cent of the grass it eats passes through its alimentary canal (p. 105) without being digested. Another 30 per cent is used in the cow's respiration to provide energy for its movement and other life processes. Less than 10 per cent of the plant material is converted into new animal tissue to contribute to growth (Fig. 15). This figure will vary with the diet and the age of the animal. In a fully grown animal all the digested food will be used for energy and replacement and none will contribute to growth. Economically it is desirable to harvest the primary consumers before their rate of growth starts to fall off.

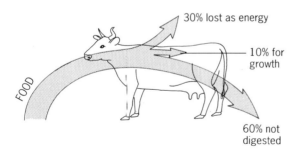

30% lost as energy

10% for growth

FOOD

60% not digested

Figure 15 Energy transfer from plants to animals

The transfer of energy from primary to secondary consumers is probably more efficient since a greater proportion of the animal food is digested and absorbed than is the case with plant material. The transfer of energy at each stage in a food chain may be represented by classifying the organisms in a community as producers, or primary, secondary or tertiary consumers, and showing their relative masses in a pyramid such as the one shown in Fig. 2 but on a more accurate scale. In Fig. 16 the width of the horizontal bands is proportional to the masses (dry weight) of the organisms in a shallow pond.

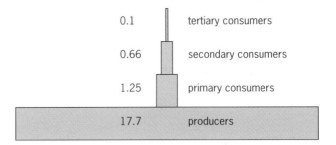

0.1	tertiary consumers
0.66	secondary consumers
1.25	primary consumers
17.7	producers

Figure 16 Biomass (dry weight) of living organisms in a shallow pond (grams per square metre)

Energy transfer in agriculture

In human communities, the use of plant products to feed animals which provide meat, eggs and dairy products is wasteful, because only 10 per cent of the plant material is converted to animal products. It is more economical to eat bread made from the wheat than to feed the wheat to hens and then eat the eggs and chicken meat. This is because eating the wheat as bread avoids using any part of its energy to keep the chickens alive and active. Energy losses can be reduced by keeping hens indoors in small cages where they lose little heat to the atmosphere and cannot use much energy in movement. The same principles can be applied in 'intensive' methods of rearing calves. However, many people feel that these methods are less than humane, and the saving of energy is far less than if the plant products were eaten directly by humans (Fig. 17).

Figure 17 Battery chickens. The hens are well fed but kept in crowded and cramped conditions with no opportunity to move about or scratch in the soil as they would normally do. Often the upper mandible of the beak is cut off to stop the birds pecking each other.

Consideration of the energy flow of a modern agricultural system reveals other sources in inefficiency. To produce 1 tonne of nitrogenous fertilizer takes energy equivalent to burning 5 tonnes of coal. Calculations show that if the energy needed to produce the fertilizer is added to the energy used to produce a tractor and to power it, the energy derived from the food so produced is less than that expended in producing it.

QUESTIONS

17 It can be claimed that the sun's energy is used indirectly to produce a muscle contraction in your arm. Trace the steps in the transfer of energy which would justify this claim.

18 Discuss the advantages and disadvantages of human attempts to exploit a food chain nearer to its source, e.g. the plankton in Fig. 3.

19 A group of explorers is stranded on a barren island where there is no soil and no vegetation. From their stores they have salvaged some living hens and some wheat. To make these resources last as long as possible should they
a eat the wheat and when it is finished, kill and eat the hens, or
b feed the wheat to the hens, collect and eat the eggs laid and when the wheat is gone, kill and eat the hens, or
c kill and eat the hens first and when they are finished, eat the wheat? Justify your answer.

Checklist

- All animals depend, ultimately, on plants for their source of food.

- Since plants need sunlight to make their food, all organisms depend, ultimately, on sunlight for their energy.

- Plants are the producers in a food web; animals may be primary, secondary or tertiary consumers.

- The materials which make up living organisms are constantly recycled.

- Plants take up carbon dioxide during photosynthesis; all living organisms give out carbon dioxide during respiration; the burning of carbon-containing fuels produces carbon dioxide.

- The uptake of carbon dioxide by plants balances the production of carbon dioxide from respiration and combustion.

- Soil nitrates are derived naturally from the excretory products of animals and the dead remains of living organisms.

- Nitrifying bacteria turn these products into nitrates which are taken up by plants.

- Nitrogen-fixing bacteria can make nitrogenous compounds from gaseous nitrogen.

- An ecosystem is a self-contained community of organisms.

- Only about 1 per cent of the sun's energy which reaches the Earth is trapped by plants during photosynthesis.

- At each step in a food chain, only a small proportion of the food is used for growth. The rest is used for energy to keep the organism alive.

25 THE HUMAN IMPACT ON THE ENVIRONMENT

Food webs
Hunting, agriculture, pesticides, eutrophication.

Forests
Erosion, flooding, climate, biodiversity.

Soil
Erosion, pesticides.

Water
Sewage, chemical pollution.

Air
Sulphur dioxide, oxides of nitrogen, lead, smog, carbon monoxide, chlorofluorocarbons, greenhouse effect.

A few thousand years ago, most of the humans on the Earth probably obtained their food by gathering leaves, fruits or roots and by hunting animals. The population was probably limited by the amount of food that could be collected in this way.

Human faeces, urine and dead bodies were left on or in the soil and so played a part in the nitrogen cycle (p. 237). Life may have been short, and many babies may have died from starvation or illness, but humans fitted into the food web and nitrogen cycle like any other animal.

Once agriculture had been developed, it was possible to support much larger populations and the balance between humans and their environment was upset.

An increasing population has three main effects on the environment.

1 Intensification of agriculture
Forests and woodland are cut down and the soil is ploughed up in order to grow more food. This destroys important wildlife habitats and may affect the climate.

Tropical rainforest is being cut down at the rate of 111 400 square kilometres per year. Since 1950, between 30 and 50 per cent of British deciduous woodlands have been felled to make way for farmland or conifer plantations.

The application of chemical fertilizers can cause deterioration of the soil structure and, in some cases, results in pollution of rivers and streams. Application of pesticides often kills beneficial creatures as well as pests.

2 Urbanization
The development of towns and cities makes less and less land available for wildlife. In addition, the crowding of growing populations into towns leads to problems of waste disposal. The sewage and domestic waste from a town of several thousand people can cause disease and pollution in the absence of effective means of disposal.

When fuels are burned for heating and transport, they produce gases which pollute the atmosphere.

3 Industrialization
In some cases, an increasing population is accompanied by an increase in manufacturing industries which produce gases and other waste products which damage the environment.

The effects of the human population on the environment are complicated and difficult to study. They are even more difficult to forecast. In their ignorance, humans have destroyed many plants and animals and great areas of natural vegetation. Unless we control our consumption of the Earth's resources, limit our own numbers and treat our environment with more care and understanding, we could make the Earth's surface impossible to live on and so cause our own extinction.

The account which follows mentions just some of the ways in which our activities damage the environment.

The human impact on food webs

The hunting of animals
One obvious way to upset a food web is to remove some of the animals or plants which form part of it. If the tawny owls were removed from the food web in Fig. 1, we would expect the numbers of shrews to increase because fewer were being eaten by the owls. The numbers of woodlice and earthworms might then go down because there were more shrews to eat them. The effect of the rabbit disease, myxomatosis, on a food web, has been described on p. 234.

In 1910, in the Grand Canyon National Game Reserve (USA), an attempt was made to protect the deer population by shooting the animals which ate them. These were cougars, wolves, bobcats and coyotes. After fourteen years, the deer population had increased from about 4000 to 100 000, and the environment could not support them. The grass was overgrazed, the trees and young shrubs were destroyed by browsing and the deer were dying in

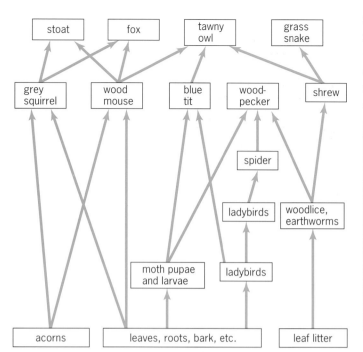

Figure 1 The food web of an oak tree (only a small sample of animals is shown)

large numbers from starvation. Ignorant human interference with the food web had not only destroyed hundreds of cougars, wolves and coyotes but threatened to lead to the destruction of the environment and the deer which lived there (Fig. 2).

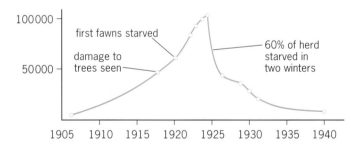

Figure 2 The result of human interference with a food web: changes in deer population after predators are killed

It is, of course, easy to be 'wise after the event', but it is not always obvious what should be done to conserve a population or its habitat. Arguments are currently taking place about the elephant population in some of Africa's game parks. Many conservationists believe that banning all trade in ivory will help reduce the level of poaching. Others think a ban will result in an increase in the price of ivory and an intensification of poaching.

Many animal populations are threatened because humans kill them for food, profit or 'sport'. Over-fishing has reduced some fish stocks to the point where they cannot reproduce fast enough to keep up their numbers.

Animals like the leopard and tiger have been reduced to dangerously low levels by hunting, in order to sell their skins or their bones for 'traditional medicine' (Fig. 3). The blue whale's

numbers have been reduced from about 2 000 000 to 6000 as a result of intensive hunting.

The World Wide Fund for Nature (WWF) believes that 15–20 per cent of all species on Earth will disappear by the year 2000 if we do not change our patterns of destruction and consumption. This is one thousand times faster than the natural extinction rate.

Figure 3 The rhinoceros is endangered because some people believe, mistakenly, that powdered rhino horn has medicinal properties, and others greatly prize rhino horn handles for their daggers.

Agriculture

Monoculture

The whole point of crop farming is to remove a mixed population of trees, shrubs, wild flowers and grasses and replace it with a dense population of only one species such as wheat or beans (Figs 4 and 5). This is called a **monoculture**.

Figure 4 Natural vegetation. Uncultivated land carries a wide variety of species.

In Fig. 1 a simplified diagram of a food web which can be supported by a single oak tree is shown. Similar food webs could be constructed for grasses, wild flowers and shrubs. Clearly, a field of wheat could not support such a mixed population of creatures.

243

Indeed, every attempt is made to destroy any organisms such as rabbits, insects or pigeons, which try to feed on the crop plant.

So, the balanced life of a natural plant and animal community is displaced from farmland and left to survive only in small areas of woodland, heath or hedgerow. We have to decide on a balance between the amount of land to be used for agriculture, roads or building and the amount of land left alone in order to keep a rich variety of wildlife on the Earth's surface.

Figure 5 A monoculture. Only wheat is allowed to grow. All competing plants are destroyed.

Pesticides

Monocultures, with their dense populations of single species, are very susceptible to attack by insects or the spread of fungus diseases. To combat these threats, pesticides are used. A pesticide is a chemical which destroys agricultural pests or competitors.

For a monoculture to be maintained, plants which compete with the crop plant for root space, soil minerals and sunlight are killed

Figure 6 Weed control by herbicide spraying. A young wheat crop is sprayed with herbicide to suppress weeds.

Figure 7 Effect of a herbicide spray. The crop has been sprayed except for a strip which the tractor driver missed.

by chemicals called **herbicides** (Figs 6 and 7). The crop plants are protected against fungus diseases by spraying them with chemicals called **fungicides** (Fig. 8). To destroy insects which eat and damage the plants, the crops are sprayed with **insecticides**.

Pesticide	kills
insecticide	insects
fungicide	parasitic fungi
herbicide	'weed' plants

The trouble with nearly all these pesticides is that they kill the harmless or beneficial organisms as well as the harmful ones.

Figure 8 Control of fungus disease. The tree bearing the apples on the right has been sprayed with a fungicide. The apples on the unsprayed tree have developed apple scab.

In about 1960, a group of chemicals, including **aldrin** and **dieldrin**, were used as insecticides to kill wireworms and other insect pests in the soil. Dieldrin was also used as a seed dressing. If seeds were dipped in the chemical before planting, it prevented certain insects from attacking the seedlings. This was thought to be better than spraying the soil with dieldrin which would have killed all the insects in the soil. Unfortunately pigeons, rooks, pheasants and partridges dug up and ate so much of the seed that the dieldrin poisoned them. Thousands of these birds were poisoned and, because they were part of a food web, birds of prey and foxes, which fed on them, were also killed.

The use of dieldrin and aldrin was restricted in 1981 and banned in 1992.

Pesticides in the food chain

The concentration of insecticide often increases as it passes along a food chain (Fig. 9). Clear Lake in California was sprayed with DDT to kill gnat larvae. The insecticide made only a weak solution of 0.015 parts per million (ppm) in the lake water. The microscopic plants and animals which fed in the lake water built up concentrations of about 5 ppm in their bodies. The small fish which fed on the microscopic animals had 10 ppm. The small fish were eaten by larger fish, which in turn were eaten by birds called grebes. The grebes were found to have 1600 ppm of DDT in their body fat and this high concentration killed large numbers of them.

A similar build-up of pesticides can occur in food chains on land. In the 1950s in the USA, DDT was sprayed onto elm trees to try and control the beetle which spread Dutch elm disease. The fallen leaves, contaminated with DDT, were eaten by earthworms. Because each worm ate many leaves, the DDT concentration in their bodies was increased ten times. When birds ate a large number of worms, the concentration of DDT in the birds' bodies reached lethal proportions and there was a 30–90 per cent mortality among robins and other song birds in the cities.

Even if DDT did not kill the birds, it caused them to lay eggs with thin shells. The eggs broke easily and fewer chicks were raised. In Britain, the numbers of peregrine falcons and sparrow hawks declined drastically between 1955 and 1965. These birds are at the top of a food web and so accumulate very high doses of the pesticides which are present in their prey, such as pigeons. After the use of DDT was restricted, the population of peregrines and sparrow hawks started to recover.

These new insecticides had been thoroughly tested in the laboratory to show that they were harmless to humans and other animals when used in low concentrations. It had not been foreseen that the insecticides would become more and more concentrated as they passed along the food chain.

Insecticides like this are called persistent because they last a long time without breaking down. This makes them good insecticides but they also persist for a long time in the soil, in rivers, lakes and the bodies of animals, including humans. This is a serious disadvantage.

Pesticides in food

Pesticides have to be poisonous in order to kill the target pests. In high doses they are also poisonous to humans. Many items of our food contain small amounts of residual pesticides. Some of these are suspected of causing cancer and other disorders but whether they do so in the very low doses we ingest is not certain. Some scientists think the levels are so low as to be negligible.

All pesticides have to be approved by the Ministry of Agriculture, Fisheries and Food (MAFF) and there are legal maximum residue limits for 62 of them. New European Union (EU) rules are being drawn up which will probably extend the list and lower the maximum residue limits, but hundreds of pesticides are not covered by these regulations. In 1990–91 MAFF tested food samples and found residues in 29 per cent of fruit and vegetables, 32 per cent of cereals, 48 per cent of potatoes and 55 per cent of milk, though only 1 per cent of all the samples had pesticides above the maximum residue limit.

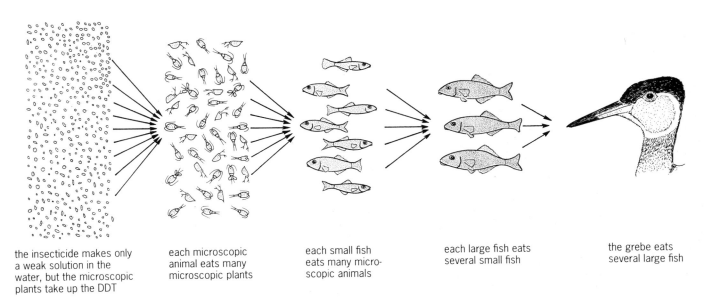

the insecticide makes only a weak solution in the water, but the microscopic plants take up the DDT

each microscopic animal eats many microscopic plants

each small fish eats many microscopic animals

each large fish eats several small fish

the grebe eats several large fish

Figure 9 Pesticides may become more concentrated as they move along a food chain. The intensity of colour represents the concentration of DDT.

Peeling apples and potatoes removes most of the surface pesticides but there is not much you can do to reduce any residues on the inside. Cooking seems to have variable effects, depending on the particular residue.

Eutrophication

On p. 45 it was explained that plants need a supply of nitrates for making their proteins, and a source of phosphates for many chemical reactions in their cells. The rate at which plants grow is often limited by how much nitrate and phosphate they can obtain. In recent years, the amount of nitrate and phosphate in our rivers and lakes has been greatly increased. This leads to an accelerated process of **eutrophication**.

Eutrophication is the enrichment of natural waters with nutrients which allow the water to support an increasing amount of plant life. This process takes place naturally in many inland waters but usually very slowly. The excessive enrichment which results from human activities leads to an overgrowth of microscopic algae (Fig. 10).

These aquatic algae are at the bottom of the food chain. The extra nitrates and phosphates from the processes listed below enable them to increase so rapidly that they cannot be kept in check by the microscopic animals which normally eat them. So they die and fall to the bottom of the river or lake. Here, their bodies are broken down by bacteria. The bacteria need oxygen to carry out this breakdown and the oxygen is taken from the water (Fig. 11). So much oxygen is taken that the water becomes deoxygenated and can no longer support animal life. Fish and other organisms die from suffocation (Fig. 12).

Figure 10 Growth of algae in a canal. Abundant nitrate and phosphate from treated sewage and from farmland make this growth possible.

Figure 12 Fish killed by pollution. The water may look clear but is so short of oxygen that the fish have died from suffocation.

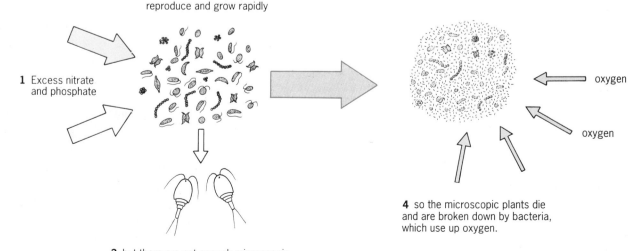

2 allow microscopic plants to reproduce and grow rapidly

1 Excess nitrate and phosphate

3 but there are not enough microscopic animals to eat the surplus plants

4 so the microscopic plants die and are broken down by bacteria, which use up oxygen.

oxygen

oxygen

Figure 11 Effects of eutrophication

The following processes are the main causes of eutrophication.

Discharge of treated sewage

In a sewage treatment plant, human waste is broken down by bacteria (p. 256) and made harmless, but the breakdown products include phosphates and nitrates. When the water from the sewage treatment is discharged into rivers it contains large quantities of phosphate and nitrate which allow the microscopic plant life to grow very rapidly (Fig. 10).

Use of detergents

Some detergents contain a lot of phosphate. This is not removed by sewage treatment and is discharged into rivers. The large amount of phosphate encourages growth of microscopic plants (algae).

Arable farming

Since the Second World War, more and more grassland has been ploughed up in order to grow arable crops such as wheat and barley. When soil is exposed in this way, the bacteria, aided by the extra oxygen and water, produce soluble nitrates which are washed into streams and rivers where they promote the growth of algae. If the nitrates reach underground water stores they may increase the nitrate in drinking water to levels considered 'unsafe' for babies.

Some people think that it is excessive use of artificial fertilizers which causes this pollution but there is not much evidence for this.

'Factory farming'

Chickens, calves and pigs are often reared in large sheds instead of in open fields. Their urine and faeces are washed out of the sheds with water forming 'slurry'. If this slurry gets into streams and rivers it supplies an excess of nitrates and phosphates for the microscopic algae.

The degree of pollution of river water is often measured by its **biochemical oxygen demand (BOD)**. This is the amount of oxygen used up by a sample of water in a fixed period of time. The higher the BOD, the more polluted the water is likely to be.

It is possible to reduce eutrophication by using
1 detergents with less phosphates;
2 agricultural fertilizers that do not dissolve so easily;
3 animal wastes on the land instead of letting them reach rivers.

QUESTIONS

1 What might be the effect of the removal of earthworms from the food web in Fig. 1 on p. 243?
2 Give five examples of monocultures.
3 What might be the effect on the food web of Fig. 1 of spraying the tree with an insecticide?
4 DDT is a fat-soluble compound, so it was often stored in the fat depots of the birds which ingested it. The harmful effects were often unnoticed until a spell of bad weather occurred. Suggest reasons for this.
5 Explain briefly why too much nitrate could lead to too little oxygen in river water.

Humans and forests

Forests have a profound effect on climate, water supply and soil maintenance. They have been described as environmental buffers. For example, they intercept heavy rainfall and release the water steadily and slowly to the soil beneath and to the streams and rivers that start in or flow through them. The tree roots hold the soil in place.

At present, we are destroying forests, particularly tropical forests, at a prodigious rate (*a*) for their timber, (*b*) to make way for agriculture, roads (Fig. 13) and settlements, and (*c*) for firewood. At the current rate of destruction, it is estimated that all tropical rainforests will have disappeared in the next 85 years.

Figure 13 Cutting a road through a tropical rainforest. The road not only destroys the natural vegetation, it also opens up the forest to further exploitation.

Removal of forests allows soil erosion, silting up of lakes and rivers, floods and the loss for ever of thousands of species of animals and plants.

Trees can grow on hillsides even when the soil layer is quite thin. When the trees are cut down and the soil is ploughed, there is less protection from the wind and rain. Heavy rainfall washes the soil off the hillsides into the rivers. The hillsides are left bare and useless and the rivers become choked up with mud and silt which can cause floods (Figs 14 and 17(a)). For example, Argentina spends 10 million dollars a year on dredging silt from the River Plate estuary to keep the port of Buenos Aires open to shipping. It has been found that 80 per cent of this sediment comes from a deforested and overgrazed region 1800 km upstream which represents only 4 per cent of the river's total catchment area. Similar sedimentation has halved the lives of reservoirs, hydroelectric schemes and irrigation programmes. The disastrous floods in India and Bangladesh in recent years may be attributed largely to deforestation.

The soil of tropical forests is usually very poor in nutrients. Most of the organic matter is in the leafy canopy of the tree tops. For a year or two after felling and burning, the forest soil yields good crops but the nutrients are soon depleted and the soil eroded. The agricultural benefit from cutting down forests is very short-lived, and the forest does not recover even if the impoverished land is abandoned.

Figure 14 Soil erosion. Removal of forest trees from steeply sloping ground has allowed the rain to wash away the topsoil.

Forests and climate

About half the rain which falls in tropical forests comes from the transpiration of the trees themselves. The clouds which form from this transpired water help to reflect sunlight and so keep the region relatively cool and humid. When areas of forest are cleared, this source of rain is removed, cloud cover is reduced and the local climate changes quite dramatically. The temperature range from day to night is more extreme and the rainfall diminishes.

In North Eastern Brazil, for example, an area which was once rainforest is now an arid wasteland. If more than 60 per cent of a forest is cleared, it may cause irreversible changes in the climate of the whole region. This could turn the region into an unproductive desert.

Forests and biodiversity

One of the most characteristic features of tropical forests is the enormous diversity of species they contain. In Britain, a forest or wood may consist of only one or two species of tree such as oak, ash, beech or pine. In tropical forests there are many more species and they are widely dispersed throughout the habitat. It follows that there is also a wide diversity of animals which live in such habitats. In fact, it has been estimated that half of the world's 10 million species live in tropical forests.

Destruction of tropical forest, therefore, destroys a large number of different species, driving many of them to the verge of extinction, and also drives out the indigenous populations of humans. In addition, we may be depriving ourselves of many valuable sources

of chemical compounds which the plants and animals produce. The US National Cancer Institute has identified 3000 plants which have products active against cancer cells and 70 per cent of them come from the rainforest (Fig. 15).

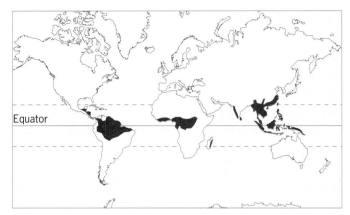

Figure 15 The world's rainforests

Agriculture and the soil

Soil erosion

Bad methods of agriculture lead to soil erosion. This means that the soil is blown away by the wind, or washed away by rain water. Erosion may occur for a number of reasons.

Deforestation

The soil cover on steep slopes is usually fairly thin but can support the growth of trees. If the forests are cut down to make way for agriculture, the soil is no longer protected by a leafy canopy from the driving rain. Consequently, some of the soil is washed away eventually reaching streams and rivers (Figs 14 and 17).

Bad farming methods

If land is ploughed year after year and treated only with chemical fertilizers, the soil's structure may be destroyed and it becomes dry and sandy. In strong winds it can be blown away as dust (Fig. 16),

Figure 16 Topsoil blowing in the wind. A dry, sandy soil can easily be eroded by the wind.

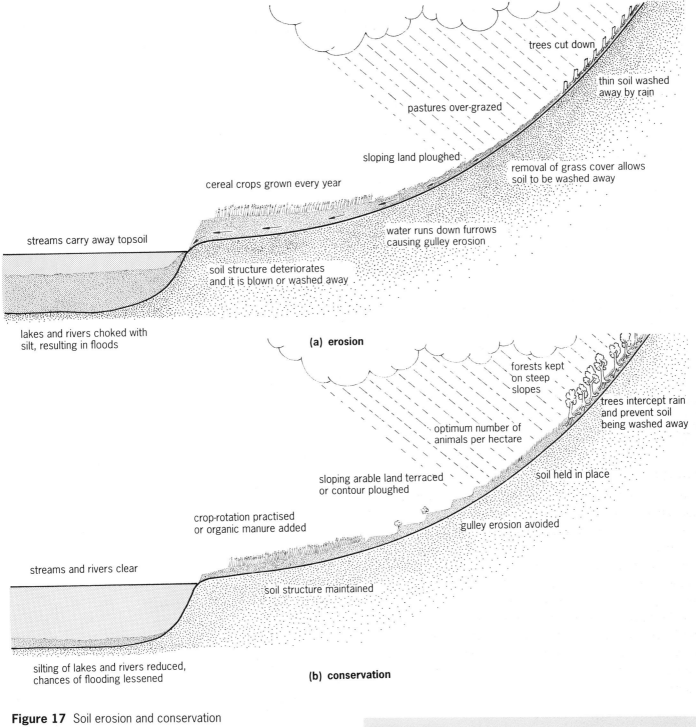

trees cut down

thin soil washed
away by rain

pastures over-grazed

removal of grass cover allows
soil to be washed away

sloping land ploughed

cereal crops grown every year

water runs down furrows
causing gulley erosion

streams carry away topsoil

soil structure deteriorates
and it is blown or washed away

lakes and rivers choked with
silt, resulting in floods

(a) erosion

forests kept
on steep
slopes

trees intercept rain
and prevent soil
being washed away

optimum number of
animals per hectare

sloping arable land terraced
or contour ploughed

soil held in place

crop-rotation practised
or organic manure added

gulley erosion avoided

streams and rivers clear

soil structure maintained

silting of lakes and rivers reduced,
chances of flooding lessened

(b) conservation

Figure 17 Soil erosion and conservation

leading to the formation of 'dust bowls', as in central USA in the 1930s, and even to deserts.

Overgrazing

If too many animals are kept on a pasture, they eat the grass down almost to the roots, and their hooves trample the surface soil into a hard layer. As a result, the rain water will not penetrate the soil and so it runs off the surface, carrying the soil with it.

QUESTIONS

6 What pressures lead to destruction of tropical forest? Give three important reasons for trying to preserve tropical forests.

7 In what ways might trees protect the soil on a hillside from being washed away by the rain?

8 If a farmer ploughs a steeply sloping field, in what direction should the furrows run to help cut down soil erosion?

9 The graph in Fig. 18 shows the change in the numbers of mites and springtails in the soil after treating it with an insecticide. Mites eat springtails. Suggest an explanation for the changes in numbers over the 16-month period.

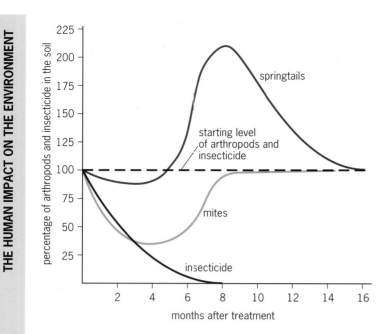

Figure 18 The effect of insecticide on some soil organisms

Use of pesticides

The effect of insecticides on food webs was described on p. 245. When insecticides get into the soil, they kill the insect pests but they also kill other organisms. The effects of this on the soil's fertility are not very clear. An insecticide called **aldrin** was found to reduce the number of species of soil animals in a pasture to half the original number. Ploughing up a pasture also reduces the number of species to the same extent, so the harm done by the insecticide is not obvious.

Water pollution

Human activity sometimes pollutes streams, rivers (Fig. 19), lakes and even coastal waters. This affects the living organisms in the water and sometimes poisons humans or infects them with disease.

Figure 19 River pollution. The river is badly polluted by the effluent from a paper mill.

Sewage

Diseases like typhoid and cholera are caused by certain bacteria when they get into the human intestine. The faeces passed by people suffering from these diseases will contain the harmful bacteria. If the bacteria get into drinking water they may spread the disease to hundreds of other people. For this reason, among others, untreated sewage must not be emptied into rivers. It is treated at the sewage works so that all the solids are removed and the water discharged into rivers is free from harmful bacteria and poisonous chemicals (but see 'Eutrophication' on p. 246).

Eutrophication

When nitrates and phosphates from farmland and sewage escape into water they cause excessive growth of microscopic green plants. This may result in a serious oxygen shortage in the water as explained on p. 246.

Chemical pollution

Many industrial processes produce poisonous waste products. Electroplating, for example, produces waste containing copper and cyanide. If these chemicals are released into rivers they poison the animals and plants and could poison humans who drink the water. It is estimated that the River Trent receives 850 tonnes of zinc, 4000 tonnes of nickel and 300 tonnes of copper each year from industrial processes (p. 256).

In 1971, 45 people in Minamata Bay in Japan died and 120 were seriously ill as a result of mercury poisoning. It was found that a factory had been discharging a compound of mercury into the bay as part of its waste. Although the mercury concentration in the sea was very low, its concentration was increased as it passed through the food chain (see p. 245). By the time it reached the people of Minamata Bay, in the fish and other sea food which formed a large part of their diet, it was concentrated enough to cause brain damage, deformity and death.

High levels of mercury have also been detected in the Baltic Sea and in the Great Lakes of North America.

Oil pollution of the sea has become a familiar event. In 1989, a tanker called the *Exxon Valdez* ran onto Bligh Reef in Prince William Sound, Alaska, and 11 million gallons of crude oil spilled into the sea. Around 400 000 sea birds were killed by the oil (Fig. 20) and the populations of killer whales, sea otters and harbour seals among others, were badly affected. The hot water high pressure hosing techniques and chemicals used to clean up the shoreline, killed many more birds and sea creatures living on the coast. Since 1989, there have continued to be major spillages of crude oil from tankers and off-shore oil wells.

QUESTIONS

10 What are the possible dangers of dumping and burying poisonous chemicals on the land?

11 Before most water leaves the waterworks, it is exposed for some time to the poisonous gas, chlorine. What do you think is the point of this?

Figure 20 Oil pollution. Oiled sea birds like these long-tailed ducks cannot fly to reach their feeding grounds. They also poison themselves by trying to clean the oil from their feathers.

Air pollution

Some factories and all motor vehicles release poisonous substances into the air. Factories produce smoke and sulphur dioxide; cars produce lead compounds, carbon monoxide and the oxides of nitrogen which lead to smog (Fig. 21).

Figure 21 Photochemical 'smog' over Frankfurt

Smoke

This consists mainly of tiny particles of carbon and tar which come from burning coal either in power stations or in the home. The tarry drops contain chemicals which may cause cancer. When the carbon particles settle, they blacken buildings and damage the leaves of trees. Smoke in the atmosphere cuts down the amount of sunlight reaching the ground. For example, since the Clean Air Act of 1956, London has received 70 per cent more sunshine in December.

Particulates

Although smoke has been largely eliminated from our towns, vehicle exhaust gases (particularly from diesels), contain microscopic particles coated with hydrocarbons. These particulates are thought to be a cause of about 10 000 deaths per year, particularly of people already suffering from chronic lung diseases such as emphysema and bronchitis.

Sulphur dioxide and oxides of nitrogen

Coal and oil contain sulphur. When these fuels are burned, they release sulphur dioxide (SO_2) into the air. Although the tall chimneys of factories (Fig. 22) send smoke and sulphur dioxide high into the air, the sulphur dioxide dissolves in rain water and forms an acid. When this acid falls on buildings, it slowly dissolves the limestone and mortar. When it falls on plants, it reduces their growth and damages their leaves.

Figure 22 Air pollution by industry. Tall chimneys keep pollution away from the immediate surroundings but the atmosphere is still polluted.

This form of pollution has been going on for many years and is getting worse. In North America, Scandinavia and Scotland, forests are being destroyed (Fig. 23) and fish are dying in lakes, at least partly as a result of **'acid rain'**.

Figure 23 Effects of acid rain on conifers in the Black Forest, Germany

251

Oxides of nitrogen from power stations and vehicle exhausts also contribute to atmospheric pollution and acid rain. The nitrogen oxides dissolve in rain drops and form nitric acid.

Oxides of nitrogen also take part in reactions with other atmospheric pollutants and produce ozone. It may be the ozone and the nitrogen oxides which are largely responsible for the damage observed in forests.

One effect of acid rain is that it dissolves out the aluminium salts in the soil. These salts eventually reach toxic levels in streams and lakes.

There is still some argument about the source of the acid gases which produce acid rain. For example, a large proportion of the sulphur dioxide in the atmosphere comes from the natural activities of certain marine algae. These microscopic 'plants' produce the gas, dimethylsulphide, which is oxidized to sulphur dioxide in the air.

Nevertheless, there is considerable circumstantial evidence that industrial activities in Britain, America and Central and Eastern Europe add large amounts of extra sulphur dioxide and nitrogen oxides to the atmosphere (Fig. 24).

Smog

This is a thin fog which occurs in cities in certain climatic conditions (see Fig. 21). Smog is irritating to the eyes and lungs and also damages plants. It is produced when sunlight and ozone (O_3) in the atmosphere, act on the oxides of nitrogen and unburnt hydrocarbons released from vehicle exhausts. This type of smog is called 'photochemical smog' to distinguish it from the smoke plus fog that used to afflict British cities.

Carbon monoxide

This gas is also a product of combustion in the engines of cars and trucks. When inhaled, carbon monoxide combines with haemoglobin in the blood to form a fairly stable compound, carboxyhaemoglobin. The formation of carboxyhaemoglobin reduces the oxygen-carrying capacity of the blood and this can be harmful, particularly in people with heart disease or anaemia.

A smoker is likely to inhale far more carbon monoxide from cigarettes than from the atmosphere. Nevertheless, the carbon monoxide levels produced by heavy traffic in towns can be harmful.

Chlorofluorocarbons (CFCs)

These are gases which readily liquefy when compressed. This makes them useful as refrigerants, propellants in aerosol cans and in plastic foams. Chlorofluorocarbons are very stable and accumulate in the atmosphere, where they react with ozone (O_3).

Ozone is present throughout the atmosphere but reaches a peak at about 25 km, where it forms what is called the **'ozone layer'**. This layer filters out much of the ultraviolet radiation in sunlight.

The chlorine from CFCs reacts with ozone and reduces its concentration in the ozone layer. As a result, more ultraviolet (UV) radiation reaches the Earth's surface. Higher levels of UV radiation can lead to an increased incidence of skin cancer. It can also affect crops, damage marine plankton and even distort weather patterns.

The reactions involved are very complex. There are also natural processes which destroy or generate ozone.

CFCs are gradually being replaced with, so-called, 'ozone-friendly' chemicals, though some of these are still active against the ozone layer (p. 256).

Lead

Compounds of lead are mixed with petrol to improve the performance of motor cars. The lead is expelled with the exhaust gases into the air. In some areas of heavy traffic it may reach levels which are dangerous and may cause damage to the brain in children.

Although there are other sources of lead pollution, such as some canned food, or water from lead pipes, the main source of lead entering the body is leaded petrol.

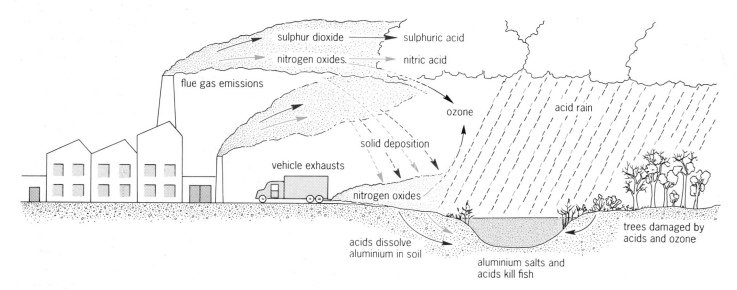

Figure 24 Unnaturally acid rain in Britain. The pollution comes from British factories, power stations, homes and vehicles. Most emissions start as dry gases and are converted slowly to dilute sulphuric and nitric acids.

Laws have been passed to reduce the level of lead in petrol and the results of such legislation in America are shown in Fig. 25. In 1985 in Britain, the lead content in petrol was reduced from 0.4 to 0.15 grams per litre by law and 'lead-free' petrol is now widely available. Unfortunately, some of the additives now used in unleaded petrol produce equally damaging pollutants.

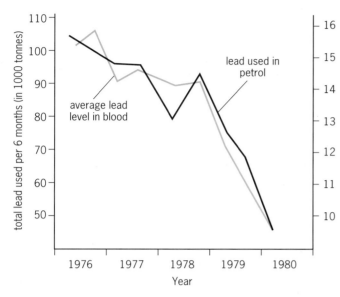

Figure 25 The effect of reducing lead in petrol. In 1975 the US government began to phase out the use of lead in petrol. This was subsequently followed by a fall in the levels of lead in people's blood. This suggests (but does not prove) a close connection between lead in exhaust fumes and the lead in the body.

The greenhouse effect and possible global warming

The Earth's surface receives and absorbs radiant heat from the sun. It re-radiates some of this heat back into space. The sun's radiation is mainly in the form of short-wavelength energy and penetrates our atmosphere easily. The energy radiated back from the Earth is in the form of long wavelengths (infrared or IR), much of which is absorbed by the atmosphere. The atmosphere acts like the glass in a greenhouse. It lets in light and heat from the sun but reduces the amount of heat which escapes (Fig. 26).

If it were not for this 'greenhouse effect' of the atmosphere, the Earth's surface would probably be at −18 °C. The 'greenhouse effect', therefore, is entirely natural and desirable.

Not all the atmospheric gases are equally effective at absorbing IR radiation. Oxygen and nitrogen, for example, absorb little or none.

The gases which absorb most IR radiation, in order of maximum absorption are water vapour, carbon dioxide, methane and atmospheric pollutants such as oxides of nitrogen and CFCs. Apart from water vapour, these gases are in very low concentrations in the atmosphere, but some of them are strong absorbers of IR radiation. It is assumed that if the concentration of any of these gases were to increase, the greenhouse effect would be enhanced and the Earth would get warmer.

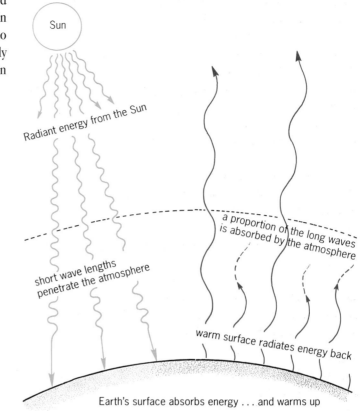

Figure 26 The short-wave energy (ultraviolet, visible light, short-wave infrared) from the Sun can get through the Earth's atmosphere and warms up the Earth's surface. The warm surface re-radiates the energy but in the form of long-wave infrared (radiant heat). About 30% of this energy is absorbed by the 'greenhouse gases' in the atmosphere. These are water vapour, carbon dioxide, methane and chloro-fluorocarbons (CFCs) mainly. Clouds reflect and absorb energy from both sources.

In recent years, attention has focused principally on carbon dioxide. If you look back at the 'carbon cycle' on p. 236, you will see that the natural processes of photosynthesis, respiration and decay would be expected to keep the carbon dioxide concentration at a steady level. However, since the Industrial Revolution, we have been burning the 'fossil fuels' derived from coal and petroleum and releasing extra carbon dioxide into the atmosphere. As a result, the concentration of carbon dioxide has increased from 0.29 to 0.35 per cent since 1860. It is likely to go on increasing as we burn more and more fossil fuel. Could this result in an increase in the Earth's temperature, i.e. global warming?

The answer is that we do not know. There are many computer models of the possible effects but they depend on the very complex and uncertain interactions of many variables.

Changes in climate might increase cloud cover and this might reduce the heat reaching the Earth from the sun. Oceanic plankton absorb a great deal of carbon dioxide. Will their rate of absorption increase or will a warmer ocean absorb less of the gas? An increase in carbon dioxide should, theoretically, result in increased rates of photosynthesis, bringing the system back into balance.

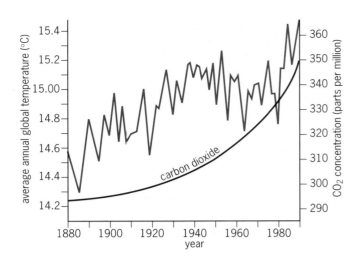

Figure 27 Annual average global temperatures since 1880

12 To what extent do tall chimneys on factories reduce atmospheric pollution?
13 What are thought to be the main causes of 'acid rain'?
14 If compounds of lead and mercury get into the body, they are excreted only very slowly. Why do you think this makes them dangerous poisons even when they are in low concentrations in the air or the water?
15 It costs money to prevent harmful chemicals escaping into the air from factories and cars. The effects of pollution also cost a great deal of money. List some of the ways in which the effects of pollution **a** affect our health and **b** cost us money.

None of these possibilities is known for certain. The worst scenario is that the climate and rainfall distribution will change, and disrupt the present pattern of world agriculture; the oceans will expand and the polar ice-caps will melt causing a rise in sea level; extremes of weather may produce droughts and food shortages.

In fact, it is not possible to produce figures which show that the global temperature is rising. An average of temperature records from round the world suggests that, since 1860, there has been a rise of 0.5–0.7 °C, most of it in the last 10 years, but this is too short a period to draw any conclusion about long-term trends (Fig. 27). There is evidence that there have been far more extreme fluctuations in the distant past. If the warming trend continues, however, it could produce a rise in sea level of between 0.2 and 1.5 metres in the next 50–100 years.

Despite the uncertainties, many scientists think that there is a threat of global warming. Governments are sufficiently concerned to be getting together to discuss ways of reducing the output of carbon dioxide and other greenhouse gases such as methane, CFCs and nitrogen oxides. Current estimates are that the output of greenhouse gases will have to be reduced by 50 per cent over the next 50 years if the atmospheric temperature is to settle down at 2 °C warmer than it is now.

If the whole concept of global warming turns out to be ill-conceived, it is still worth taking certain precautions which have value for other reasons. It is a good idea to save energy and reduce pollution by burning less fuel and to reduce output of pollutants which are potential greenhouse gases.

The cost of cleaning up

Most of the forms of pollution described in this chapter could be prevented provided we were prepared to pay the cost of the necessary measures. For example, removal of sulphur dioxide from the waste gases of power stations might increase our electricity bills by 5 per cent. It is probably essential to bear these extra costs if we are to preserve our environment. Furthermore, when the costs of reducing pollution are compared with the costs of environmental damage and human ill-health, the difference may not be all that great.

Checklist

- The plants and animals in a food web are so interdependent that even a small change in the numbers of one group has a far-reaching effect on all the others.

- Hunting activities and farming upset the natural balance between other living organisms.

- Pesticides kill insects, weeds and fungi that could destroy our crops.

- Pesticides help to increase agricultural production but they kill other organisms as well as pests.

- A pesticide or pollutant which starts off at a low, safe level can become dangerously concentrated as it passes along a food chain.

- Eutrophication of lakes and rivers results in the excessive growth of algae followed by an oxygen shortage when the algae die and decay.

- Soil erosion results from removal of trees from sloping land, use of only chemical fertilizers on ploughed land and putting too many animals on pasture land.

- The conversion of tropical forest to agricultural land usually results in failure because forest soils are poor in nutrients.

- Removal of forests can lead to erosion, silting-up of lakes and rivers and to flooding.

- We pollute our lakes and rivers with industrial waste and sewage effluent.

- We pollute the sea with crude oil and factory wastes.

- We pollute the air with smoke, sulphur dioxide and nitrogen oxides from factories, and lead and nitrogen oxides from motor vehicles.

- The acid rain resulting from air pollution leads to poisoning of lakes and possibly destruction of trees.

- The extra carbon dioxide from fossil fuels might lead to global warming.

26 PROTECTING THE ENVIRONMENT

Reduction of pollution Clean Air Acts, oxides of nitrogen and sulphur, lead, water pollution, sewage disposal. **Waste disposal** Household waste, industrial waste.	**Recycling of materials** Non-renewable resources, renewable resources, recycling. **Biological control** Predators, parasites and pathogens. Pheromones. Sterilization.

Reduction of pollution

Pollution can be reduced voluntarily or by passing laws (legislation) which restrict the emission of pollutants. Pressure of public opinion has sometimes been effective in persuading a factory voluntarily to cut down its emission of polluting gases or liquid effluent. However, to reduce pollution, on a national or global scale, it is necessary to introduce legislation, i.e. pass laws which set limits on the amount of pollutants that may be released into the environment. The laws include penalties that may be applied if these levels are exceeded. One example of legislation is the Clean Air Act.

Control of air pollution

The Clean Air Acts of 1956 and 1968
These Acts designated certain city areas as 'smokeless zones'. The use of coal for domestic heating was prohibited and factories were not allowed to emit black smoke. This was effective in abolishing dense fogs in cities but did not stop the discharge of sulphur dioxide and nitrogen oxides in the country as a whole (Fig. 1). The effect of legislation on lead pollution can be seen in Fig. 25 on p. 253.

Reduction of acid gases
The concern over the damaging effects of acid rain has led many countries to press for regulations to reduce emissions of sulphur dioxide and nitrogen oxides.

Reduction of sulphur dioxide can be achieved either by fitting desulphurization plants to power stations or by changing the fuel or the way it is burnt. In 1986, Britain decided to fit desulphurization plants to three of its major power stations, but recently it has agreed to a United Nations protocol to reduce sulphur dioxide emissions to 50 per cent of 1980 levels by the year 2000, and to 20 per cent by 2010. This will be achieved largely by changing from coal-fired, to gas-fired power stations.

Reduction of vehicle emissions
Oxides of nitrogen come, almost equally, from industry and from motor vehicles (Fig. 2). Flue gases from industry can be treated to

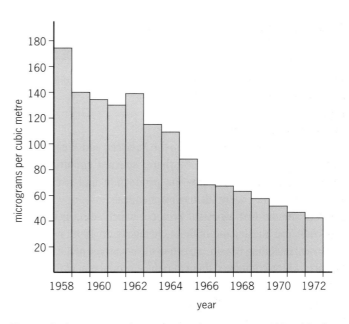

Figure 1 Average smoke concentration near ground level in the United Kingdom 1958–72

remove most of the nitrogen oxides. Vehicles can have **catalytic converters** fitted to their exhaust systems. These converters remove most of the nitrogen oxides, carbon monoxide and unburned hydrocarbons. They add £200–600 to the cost of a car and will work only if lead-free petrol is used, because lead blocks the action of the catalyst.

Another solution is to redesign car engines to burn petrol at lower temperatures ('**lean burn**' engines). These emit less nitrogen oxide but just as much carbon monoxide and hydrocarbons as normal engines.

In the long term, it may be possible to use fuels such as alcohol or hydrogen which do not produce so many pollutants.

The European Union has set limits on exhaust emissions. From 1989, new cars over 2 litres had to have catalytic converters and from 1993, smaller cars had to fit them as well.

In 1995 new regulations are expected to cut emission of particulates by 75 per cent and nitrogen oxides by 50 per cent.

These reductions will have less effect if the volume of traffic continues to increase. Significant reduction of pollutants is more likely if the number of vehicles is stabilized and road freight is reduced.

255

Figure 2 Pollutants from vehicles. The car is being monitored for its output of polluting gases.

Figure 3 River pollution. Untreated industrial effluent being discharged into a tributary of the Mersey.

Protecting the ozone layer

The appearance of 'ozone holes' in the Antarctic and Arctic and the thinning of the ozone layer elsewhere, spurred countries to get together and agree to reduce the production and use of CFCs (p. 252) and other ozone-damaging chemicals.

1988 saw the first Montreal protocol which set targets for the reduction and phasing out of these chemicals. In 1990, nearly 100 countries, including Britain, agreed to the next stage of the Montreal protocol which committed them to reduce production of CFCs by 85 per cent in 1994 and phase them out completely by 1996. The European Union has advanced the phase-out date to 1995.

Control of water pollution

The 1973 Water Act put the control of water supplies and sewage treatment under ten regional water authorities. In 1989, when these were 'privatized', the National Rivers Authority (NRA) was set up to oversee their activities. In 1996 the function of the NRA was taken over by the Environment Agency.

The Environment Agency is required to keep a check on the quality of water reaching rivers, estuaries and bathing beaches. It exercises control over conservation of water, pollution, drainage, flood protection, inland fishing and navigation.

Responsibility for the quality of drinking water, however, rests with the Department of the Environment.

Under current laws it is an offence to discharge water that has been used for any industrial purpose, into a sewer or a river without the consent of the regional water authority.

In practice, this may reduce pollution but does not stop it (Fig.3.). The industries, the water authority and the Environment Agency agree what levels of pollution are acceptable (**consent levels**). In the past, these levels have been set too high. Even when they were exceeded, prosecutions were rare and any fines were small. It remains to be seen whether 'privatization' and the Environment Agency will improve matters.

The Thames Water Authority has been successful in cleaning up the Thames, to the extent that Atlantic salmon have returned to the river after an absence of 60 years. However, there are still 2800 km of other rivers classified as unfit for fish and in 1991 the NRA reported that there had been a net deterioration of water quality in rivers, canals and estuaries since 1985.

Sewage disposal

Sewage contains bacteria from the human intestine which can be harmful (see p. 352). These bacteria must be destroyed in order to prevent the spread of intestinal diseases. Sewage also contains substances from household wastes (such as soap and detergent) and chemicals from factories. These too must be removed before the sewage effluent is released into rivers. Rain water from the streets is also combined with the sewage.

Many coastal towns pour untreated sewage into the sea where it is thought to become diluted enough to be harmless. In practice, winds and tides sometimes carry it back to the shore, and it may become a health risk.

Inland towns have to make their sewage harmless in a sewage treatment plant before discharging the effluent into rivers. A sewage works removes solid and liquid waste from the sewage, so that the water leaving the works is safe to drink.

There are two main methods of large-scale sewage disposal. Most large towns (or groups of towns) use the activated sludge process (Figs 4 and 5). Most small towns use the percolating filter process.

The activated sludge process

a Screening The sewage entering the sewage works is first 'screened'. That is, it is made to flow through a metal grid which removes the solids like rags, plastics, wood and so forth. The 'screenings' are raked off and disposed of – by incineration, for example.

Figure 4 Sewage treatment—activated sludge

Figure 5 Sewage treatment—activated sludge method. In the foreground are the rectangular aeration tanks. Behind these are the circular settlement tanks. In the background are the sludge digesters.

b Grit The sewage next flows slowly through long channels. As it flows, grit and sand from roads and gardens settle down to the bottom and are removed from time to time. The grit is washed and used for landfill.

c First settling tanks The liquid continues slowly through another series of tanks. Here about 40 per cent of the organic matter settles out as crude sludge. The rest of the organic matter is in the form of tiny suspended particles which pass, with the liquid, to the aeration tanks.

The semi-liquid sludge from the bottom of the tank is pumped to the sludge digestion plant (see **f** below).

d Aeration tanks Oxygen is added to the sewage liquid, either by stirring it or by bubbling compressed air through it. Aerobic bacteria and protozoa (p. 284) grow and reproduce rapidly in these conditions.

These micro-organisms cause the suspended organic particles to clump together (flocculate). The bacteria in the clumps (flocs) produce extra-cellular enzymes (p. 15), which digest the organic solids. The soluble products are absorbed by the bacteria and used for energy and growth.

The bacteria also absorb the dissolved substances in the sewage and use them for their metabolism. Some bacteria turn urea (p. 139) into ammonia. Others change ammonia to nitrates. Denitrifying bacteria change nitrates to nitrogen gas. The bacteria get energy from these chemical changes. They also absorb phosphates for use in their energy-transfer systems (see, for example, 'ATP', p. 21). The protozoa help in the flocculation process and they also eat the bacteria.

In this way, the suspended solids and dissolved substances in sewage are converted to nitrogen, carbon dioxide (from respiration) and the protoplasm of the bacteria and protozoa, leaving fairly pure water.

e Second settling tanks In these tanks, the micro-organisms settle out. They form a fine sludge, which is returned to the aeration tanks to keep a high population of organisms there. This is the 'activated sludge' from which the process gets its name (Fig. 6). Although the sewage stays in the aeration tanks for only six to eight hours, the recycling of activated sludge enables the micro-organisms to act on it for as much as 20–30 days.

When all the sludge has settled, the water is pure enough to put into a river.

Figure 6 Protozoa in activated sludge (×150). These single-celled organisms ingest bacteria in the liquid sewage.

257

f Sludge digesters The crude sludge from the first settling tanks is pumped into large containers, where anaerobic bacteria (p. 300) 'digest' it. During this digestion methane gas is produced. The methane is used as fuel to drive the compressed air pumps and other machinery at the sewage works.

g Sludge disposal Provided it is free from harmful bacteria and poisonous metals, much of the dried sludge can be used as an agricultural fertilizer. Some is used as landfill and, at present, about a quarter of it is dumped at sea. After 1998 the dumping will stop and the sludge will be incinerated.

Percolating filter

Stages **a** to **c** are much the same as for the activated sludge method.

d Clinker beds Instead of the liquid sewage passing into aeration tanks, it is sprayed over beds of clinker (Fig. 7). You have probably seen these circular beds with the sprinkler arms going round and round over them. In the aerobic conditions of the clinker bed, bacteria, protozoa and algae grow, using the dissolved substances and suspended particles in the sewage.

The purified liquid is released into the river. The film of micro-organisms on the clinker bed falls away from time to time and collects as sludge at the bottom of the filter.

There are no sludge digesters because the process uses very little energy. The water flow and the sprinklers are driven by the pressure of the sewage.

Figure 7 Sewage treatment—percolating filter. Micro-organisms in the filter beds remove bacteria and organic matter from the effluent.

Biodegradability

It is important that factory wastes should contain no chemicals that might kill the micro-organisms in the activated sludge or the filter beds. And it is also important that the bacteria and protozoa should

be able to digest any chemicals, such as detergents. That is, the chemicals should be **biodegradable** (p. 259). If they are not biodegradable, the chemicals will pass unchanged into the river.

Eutrophication

Although sewage effluent is 'pure' in that it contains no poisonous chemicals or harmful bacteria, it may still retain enough nitrates and phosphates to cause problems of eutrophication (p. 246). Overloaded sewage works are the main source of river pollution.

QUESTIONS

1 Why have the Clean Air Acts not prevented atmospheric pollution?
2 How can **a** sulphur dioxide emissions, **b** nitrogen oxide emissions be reduced?
3 In what ways does sewage treatment **a** reduce pollution, **b** cause pollution?
4 In the long run, who pays the cost of reducing pollution?
5 What part do micro-organisms (bacteria and protozoa) play in sewage treatment?

▌ *Waste disposal*

Household waste

This consists of things like food residues, paper, fabrics, bottles and cans. Some householders sort these materials, take the waste glass to bottle banks and pass on the waste paper to collectors. The local authority (for example, the district council) is responsible for the collection and disposal of the household waste that ends up in the dustbin.

The waste may simply be taken to a landfill site (Fig. 8), such as a disused gravel pit, tipped in, left to settle down to a compact mass and then covered with soil. Sometimes it is partially ground up and compressed before dumping. Sometimes the glass, metals and so

Figure 8 Disposal of domestic waste by landfill. In this case, the rubbish is being used to 'reclaim' marshland, which may well have been a valuable wildlife habitat.

on are sorted out before the residue is dumped. Once the landfill site is full and the rubbish has compacted and been covered, the site may be reclaimed for agriculture or recreational use.

Landfill is usually the cheapest method of disposal, if the landfill sites are not too far away.

The organic materials, such as vegetables and paper, are decomposed by bacteria in the tip and they produce methane. If pipes are installed in the rubbish tip, this methane can be collected and used as fuel. If the methane is not piped off in this way, it may accumulate and seep underground to reach any buildings near the landfill site. In some cases it has caused explosions.

If the rubbish is not dumped, it may be burned in incinerators. In some cities the heat from incineration is used for district heating schemes (up to 7000 houses, shops and public buildings). But only four out of about 50 rubbish incinerators recover the energy in this way. Incineration is expensive if the heat is not used.

Some local authorities have set up installations which sort the rubbish and re-use nearly all of it (Fig. 9). Tin cans, other metals and glass are sorted mechanically from the rubbish and sent for recycling. The organic wastes are placed in fermenters where they decompose and produce methane. The methane is used as fuel to run the plant. The plastic waste may be recycled or it may be compacted with paper and rags to make fuel pellets.

Organic waste is derived from material which was once alive, e.g. paper (wood pulp), cotton and wool cloth, waste food and garden refuse. These materials can be attacked by bacteria which break them down to methane, carbon dioxide, water and soluble salts. Such materials are said to be **biodegradable**; they are 'degraded' to harmless compounds by the biological activity of micro-organisms.

Substances such as glass, metal and most plastic are **non-biodegradable**. There are no bacteria which can use them as a source of food. Work is in progress to produce plastics which contain substances, such as starch, which are partially or wholly biodegradable.

Industrial waste

The least harmful waste products from industry may be mixed with household refuse and dumped in the same landfill sites. This is acceptable so long as the types and quantities of industrial waste are carefully controlled.

More hazardous waste may be deposited in separate sites, dumped at sea or burned in incinerators (Fig. 10).

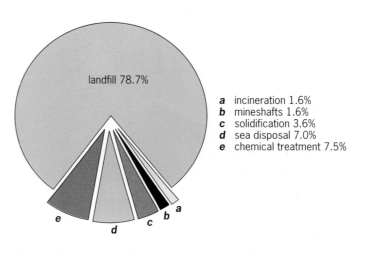

landfill 78.7%

a incineration 1.6%
b mineshafts 1.6%
c solidification 3.6%
d sea disposal 7.0%
e chemical treatment 7.5%

Figure 10 Disposal of hazardous waste in 1985. At the moment landfill is the cheapest method of disposal, but not always the safest.

Figure 9 Recycling domestic waste. This is an experimental plant for sorting and recycling domestic waste. Tin cans and other metals are removed and recycled; paper and plastic are pulverized and compressed into fuel pellets.

Landfill

Many industrial wastes are poisonous or harmful in other ways, so the landfill tips must be sited with great care. They must not be in a place where substances leaking from the tip can find their way into ground water (the water beneath the soil) which might enter our water supplies. The tips must also be well fenced and manned so that no one dumps materials without permission and people cannot wander in.

Staff must make sure that only substances for which the tip is licensed are dumped and that reactive substances are not mixed (Fig. 11).

Figure 11 Well-regulated disposal. The lorry driver must check in at the control gate. His documents are checked to see what waste products he is carrying and he will be directed to the correct disposal area.

Britain is unusual in that most of the tips are run by private companies, who compete with each other. The tips are licensed and inspected by the local authority. But the system is open to mismanagement and misuse. The Hazardous Wastes Inspectorate in 1985 was very critical of several badly managed sites. Competing waste disposal firms had 'cut corners', and the local authorities had not inspected and supervised the sites properly.

Landfill for industrial waste is becoming increasingly unpopular because of the dangers of poisonous chemicals seeping into ground water. Some waste disposal firms treat the waste to make it less harmful before dumping. Other methods of disposal are being sought. But while landfill is cheap and poorly regulated, firms will probably find it does not pay them to use other methods.

Incineration

Industrial wastes may also be incinerated. But incineration plants are not very popular with the people who live near them. Incinerator ships overcome this problem, to some extent, by burning the wastes out at sea.

Incinerating some industrial wastes produces poisonous gases unless they are burned at high temperatures, or treated first to neutralize the acid content.

All these disposal methods have their disadvantages. To ensure safe disposal and to avoid pollution we will have to spend more on control measures and improved methods. We can also try to produce less waste by recycling materials. This will also help to conserve the sources of these materials.

Recycling of materials

Non-renewable resources

Mineral resources such as coal, oil and metals cannot be renewed once they are used up. It is difficult to work out how long these resources may last. In the 1970s, predictions were made of the number of years certain non-renewable sources would last. The following are a selection from the list:

World resources	Years' supply
oil and gas	50
coal	300
aluminium	100
iron	360
lead	26
tin	17

The predictions assume that no new deposits will be found and that our rate of use will not change. In fact, 20 years later, the non-renewable resources seem not to have diminished significantly. It is estimated, for example, that the known reserves of oil and gas are greater than they were in 1970. Even so, the resources are limited and, presumably, will run out eventually or become too expensive to recover. We will need to find other sources of energy, and recycle as much manufactured material as possible.

Renewable resources

These are either materials derived from plants, or sources of energy that do not come from fossil fuels or uranium (hydro-electric power, for example).

By photosynthesis (p. 37), plants can make a wide variety of substances from carbon dioxide, water and mineral salts, which are recycled, rather than being used up. As well as food, we can obtain the following materials from plants:

- wood, and paper made from it,
- sugar, which can be fermented to alcohol,
- alcohol, which can be used as a fuel or for making other chemicals,
- oils, such as sunflower oil, palm oil, which may become important as fuels one day.

At the moment, fuel from plants is more expensive than fuel from coal or petroleum.

Renewable sources of energy (other than plant products) are:

- hydroelectric power,
- wind power (Fig. 12),
- tidal power,
- wave power.

Figure 12 Wind generators in California. On otherwise unproductive land, these generators can make a small but useful input to the electricity supply.

Figure 13 Bottle bank. Recycling glass can save on raw materials. Recycling cans saves on materials and energy.

Apart from hydroelectric power, none of these is yet adding much to our energy supplies. Some of them are still in the experimental stage. All of them damage the environment to some extent.

Recycling

As minerals and other resources become scarcer, they also become more expensive. It then pays to use them more than once. The recycling of materials may also reduce the amount of energy used in manufacturing. In turn this helps to conserve fuels and reduce pollution.

For example, producing aluminium alloys from scrap uses only 5 per cent of the energy that would be needed to make them from aluminium ores. In 1986, Britain used 31 700 tonnes of aluminium scrap.

We also recycle 56 per cent of the lead used in Britain. This seems quite good until you realize that it also means that 44 per cent of this poisonous substance enters the environment.

Manufacturing glass bottles uses about three times more energy than if they were collected, sorted, cleaned and re-used. Recycling the glass from bottles does not save energy but does reduce the demand for sand used in glass manufacture. In 1985, 200 000 tonnes of glass were recycled (Fig. 13).

Waste paper can be pulped and used again, mainly for making paper and cardboard. Newspapers are de-inked and used again for newsprint. One tonne of waste paper is equivalent to perhaps 17 trees. (Paper is made from wood pulp.) So collecting waste paper may help to cut our import bill for timber and spare a few more hectares of moorland from the spread of commercial forestry.

By 1998, the European Union expects 50 per cent of packaging to be recovered (i.e. used more than once) and 25 per cent to be recycled.

QUESTIONS

6 Explain why some renewable energy sources depend on photosynthesis.
7 In what ways does the recycling of waste help to save energy and conserve the environment?
8 Explain why some of the alternative and renewable energy resources are less likely to cause pollution than coal and oil.

Biological control

An alternative to using chemicals to control pests is to limit their numbers by using predators, parasites or pathogens (micro-organisms which cause disease, p. 300).

More than 100 species of pests are now controlled, at least partially, by one of these methods. Biological control can be effective, economical and, if properly tried and tested, does not harm the environment.

Predators

Barn owls are used successfully in Malaysia to keep down the rat population in oil palm plantations. The oil palm fruit produces an oil which is used for making margarine, cooking fat and soap. Rats eat and damage the fruits on the trees or in storage.

Barn owls were introduced and encouraged to breed by providing nest boxes on poles. A single family of owls can catch and eat about 1300 rats per year and this programme of biological control is proving to be very effective (Fig. 14 and 15).

Biological control does not eliminate pests but does keep their numbers low so that damage to the crop is minimal. The outcome of a theoretically ideal programme of biological control is shown in Fig. 15.

261

(a)

(b)

Figure 14 (a) Oil palm plantation in Malaysia. Barn owls keep the rat population down. (b) Barn owl returning to nest with prey.

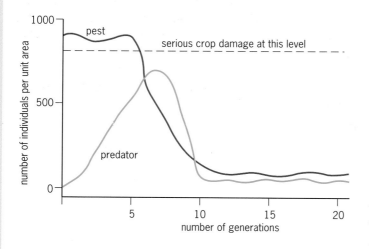

Figure 15 A theoretically ideal programme of biological control of a crop pest

In the 1820s, the prickly pear cactus (*Opuntia*) from South America was introduced into Queensland, Australia, as an ornamental plant. Some of the plants escaped from cultivation and spread until they covered about 25 million hectares of farmland (Fig. 16). Biologists introduced an Argentinian moth, *Cactoblastis cactorum*, whose larvae ate the leaves of the prickly pear. In this way, the invasion of the cactus was brought under control. There is now a balance between the moth and a tolerable level of the cactus.

This story illustrates the danger of introducing and releasing alien species. Introduced species can often thrive in their new environment and have no natural predators to keep them in check.

Figure 16 The larvae of the moth *Cactoblastis cactorum* eat the leaves of the prickly pear cactus

If the cactus moth was a success story for Queensland, the cane toad is a disaster. In the 1930s, the toad was introduced from Hawaii to try and control the cane beetle and its grub which were causing great damage in the sugar-cane crops of Queensland. Unfortunately it was not realized that one of the cane beetle species stayed in the foliage and never descended to the ground where the toads could reach it. The other species was on the ground only when the crop was just starting to grow and no toads were present. The toads, however, are thriving and eating everything except the cane beetles. Although there are potential predators, the toads are poisonous and kill any animal which eats them or even tries to eat them. Unless they can be controlled, it is likely that they will do great harm to the wildlife population of North Eastern Australia.

Before an alien species is used for biological control, it is essential to conduct thorough trials to make sure that the new predator will attack only the target organism and not turn its attention to other creatures when its original prey is reduced.

In 1951, ladybirds were introduced to Bermuda in the hope that they would attack the scale insects that were damaging cedar trees. However, the ladybirds were rapidly eaten by lizards which had been introduced in 1905 in an attempt to control a fruit fly pest.

Parasites

The greenhouse whitefly can be controlled by introducing a parasitic wasp into the greenhouse. The whitefly causes damage to tomatoes and cucumbers. Growers can buy pupae (p. 326) of a tiny South American wasp (*Encarsia formosa*). The pupae hatch out into wasps which lay eggs in the larvae (grubs) of the whitefly. These eggs hatch into tiny maggots which devour the larva from the inside, eventually killing it (Fig. 17).

(a)

(b)

Figure 17 The parasitic wasp, *Encarsia formosa*, is used to control greenhouse whitefly. (a) shows a whitefly and the pupal cases. The black ones have been parasitised. (b) shows a wasp which has hatched from the whitefly pupal case.

Pathogens

Bacteria, viruses and fungi can all be used to control pests, though many of these are only at the experimental stage. Trials are being conducted with fungi to control soil nematodes (roundworms), aphids and bracken. Viruses have been tried in the battle against the gipsy moth and the cotton bollworm in North America.

The bacterium, *Bacillus thuringiensis*, is being successfully used against a wide range of insect pests. At some stage in its life cycle, the bacterium produces a toxic protein which kills insect larvae when they ingest it. The bacterium and its protein have been shown to be harmless to humans and other vertebrates. One strain of the bacterium is effective against the larvae (caterpillars) of moths and butterflies. Another strain kills larvae of blackfly (blackfly transmit river blindness in Africa), and may be used against mosquitoes which transmit malaria.

The commercial product from *Bacillus thuringiensis* contains the toxic protein and not the living organisms.

Figure 18 Sugar beet. The crop on the left has been damaged by 'cutworms' (moth larvae which live in the soil). The crop on the right has been treated with *Bacillus thuringiensis* toxin which kills the cutworms.

Pheromones

Many insects have an acute sense of smell and communicate with each other by releasing chemicals called **pheromones**. Pheromones influence the behaviour of the insects. For example, the males of many insect species are attracted to the female by pheromones which she releases. By setting up traps baited with pheromones and treated with insecticide, insect pests can be lured to their deaths, e.g. the codling moth, whose larvae (maggots) burrow inside apples. One advantage of this method is that a particular pheromone is specific for only the target insect. So other species are unharmed.

Pheromone traps (Fig. 19) do not directly control pests but monitor their numbers so that spraying can be restricted to periods when it will be most effective.

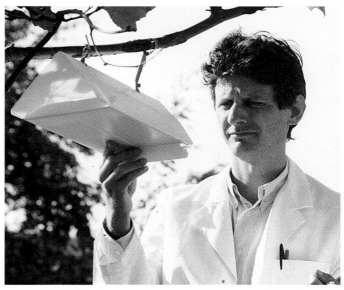

Figure 19 Inspecting a pheromone trap. The insects are trapped on a sticky substance inside the trap so their numbers can be estimated.

Sterilization

In some cases, the release of sterile male insects into the pest population can effectively control the pest.

In the USA, the screw-worm fly, which attacks cattle, has almost been eradicated by this method. Pupae are sterilized by radioactive emissions and released into the environment. When they hatch, the sterile males outnumber the normal males and mate with the females who then fail to produce eggs. Since the females mate only once in their lifetime the fly population falls rapidly.

Integrated pest control

This implies the use of a variety of methods and appropriate strategies of cultivation. It is more subtle than simply dosing the crops with poisons but needs a careful study of the pest and its relation to the crop.

Crops with a high degree of natural resistance are selected in the first place. Then it is important to inspect the crop regularly and estimate whether the level of infestation is likely to reach serious proportions. If action needs to be taken, the best combination of chemical and biological methods is chosen. The life cycle of the pest must be known so that it can be attacked when it is most vulnerable.

QUESTIONS

9 Why are introduced species likely to cause problems in a new environment?

10 What are the desirable characteristics of a predator or parasite introduced to control a pest species?

11 What is the difference between 'eradication' and 'control' of a pest species?

Checklist

- Sulphur dioxide emissions can be reduced by removing sulphur compounds from coal.
- Sulphur dioxide and nitrogen oxides can be removed from flue gases.
- Nitrogen dioxide, carbon monoxide and hydrocarbons can be removed from vehicle exhausts by fitting catalytic converters.
- Removal of lead from petrol reduces lead pollution of the air.
- The EU is pressing for regulations to reduce discharge of all these pollutants.
- Water pollution of rivers is regulated by the National Rivers Authority.
- Household and industrial waste is disposed of mainly by landfill and incineration.
- Raw materials and energy can be saved by recycling processes.
- Alternative sources of energy are less polluting than fossil fuels.
- No form of energy production can entirely avoid causing some environmental damage or problems of waste disposal.
- Pesticide use can be reduced by using biological control methods.
- Introduction of alien species for biological control must be done with caution.

27 ECOSYSTEMS

Definitions
Environment, habitat, community, ecosystem, competition, population.

Artificial and managed ecosystems
Arable farms, fish farms, fishing.

A freshwater ecosystem
Physical aspects, plant and animal communities, adaptations.

Ecology is the study of living organisms in relation to their natural environment, as distinct from in the laboratory. This does not mean that laboratory work is ruled out, but its object is always to explain how the organism survives, how it relates to other organisms and why it is successful in its particular environment.

The following are some of the terms used in any discussion of ecology.

Environment

This means everything in the surroundings of an organism that could possibly influence it. The environment of a tadpole consists of water. The temperature of the water will influence the tadpole's rate of growth and activity. The watery environment contains plants and animals on which the tadpole will feed, but it also contains fish and insects which may eat the tadpole. The water contains dissolved oxygen which the tadpole breathes by means of its gills. The water, the oxygen, the food and the predators are all part of the tadpole's environment.

Habitat

A habitat is where an organism lives, i.e. where it obtains its food and shelter, and where it reproduces. The habitat of a limpet is a rocky shore. The environment includes air, sea water and sunlight but the habitat is the shore. The habitat of the tapeworm is the intestine of a mammal. Its environment, however, is the warm digested food and digestive juices of its host. The habitat of an aphid may be a bean plant, but its environment will include sun, wind, rain, ladybirds, ants and bacteria.

Population

In biology, this term always refers to a single species. A biologist might refer to the population of sparrows in a farmyard or the population of carp in a lake. In each case this would mean the total numbers of sparrows or the total numbers of carp in the stated area.

Community

A community is made up of all the plants and animals living in a habitat. In the soil there is a community of organisms which includes earthworms, springtails and other insects, mites, fungi and bacteria. In a lake, the animal community will include fish, insects, crustacea, molluscs and protozoa.

The plant community will consist of rooted plants with submerged leaves, rooted plants with floating leaves, reed-like plants growing at the lake margin, plants floating freely on the surface, filamentous algae and single-celled algae (p. 295) in the surface waters.

Ecosystem

The community of organisms in a habitat, plus the non-living part of the environment (air, water, soil, light, etc.) make up an ecosystem. A lake is an ecosystem which consists of the plant and animal communities mentioned above, and the water, minerals, dissolved oxygen, soil and sunlight on which they depend. An ecosystem is self-supporting (Fig. 1).

$$\left.\begin{array}{l}\text{individuals} \\ \text{of the same} \\ \text{species}\end{array}\right\} = \text{POPULATION} \left.\begin{array}{l} + \\ \text{populations} \\ \text{of other} \\ \text{species}\end{array}\right\} = \text{COMMUNITY} \left.\begin{array}{l}\text{non-living} \\ \text{part of} \\ \text{environment} \\ + \end{array}\right\} = \text{ECOSYSTEM}$$

In a woodland ecosystem, the plants absorb light and rain water for photosynthesis, the animals feed on the plants and on each other. The dead remains of animals and plants, acted upon by fungi and bacteria, return nutrients to the soil.

Lakes and ponds are clear examples of ecosystems. Sunlight, water and minerals allow the plants to grow and support animal life. The recycling of materials from the dead organisms maintains the supply of nutrients.

All ecosystems will contain producers, consumers and decomposers as described on p. 233.

It is quite reasonable to regard a single oak tree as an ecosystem. The tree grows as a result of photosynthesis. Lichens may grow on its branches; ivy may grow up its trunk, squirrels will eat its acorns, and aphids and caterpillars feed on its leaves. Birds eat the caterpillars and may nest in the branches. The tree, the community

Figure 1 The 'Ecosphere'. The 5-inch globe contains sea water, bacteria, algae, snails and a few Pacific shrimps. Given a source of light it is a self-supporting system and survives for several years (at least). The shrimps live for up to 7 years but few reproduce.

Figure 2 Intraspecific competition in red deer. The 'pushing contest' will determine who has access to the females.

of plants and animals which live on it, and the air, water and sunlight which nourish it, all form an ecosystem.

On the other hand, the whole of that part of the Earth's surface which contains living organisms (called the **biosphere**), may be regarded as one vast ecosystem.

No new material (in significant amounts) enters the Earth's ecosystem from space and there is no significant loss of materials. The whole system depends on a constant input of energy from the sun (p. 37) and recycling of the chemical elements (p. 235).

Competition

Living organisms compete with each other for resources such as food, light, rooting space and breeding partners. As explained on p. 222, all organisms produce more offspring than can possibly survive, so competition is unavoidable.

Intraspecific competition takes place between members of the same species for territory, food, nesting sites or mates (Fig. 2).

Interspecific competition occurs between different species. On page p. 277 there is a description of how competition between two species of *Paramecium* leads to a decline in the population of one of them.

So, a *population* of carp forms part of the animal *community* living in a *habitat* called a lake. The communities in this habitat, together with their watery *environment*, make up a self-supporting *ecosystem*.

A carp is a *secondary consumer* at the top of a *food chain*, where it is in *competition* with other species of fish for food and with other carp for food and mates.

QUESTIONS

1 What communities might be present in an area of woodland?
2 a What is the habitat of an earthworm?
  b What makes up the environment of the earthworm?
3 Name some of the producers, consumers and decomposers that might be present in a grassland ecosystem.
4 Plants make their food by photosynthesis, so in what ways can they be said to 'compete for food'?

Artificial and managed ecosystems

There are few ecosystems that are unaffected by human activity. Sometimes the effects are the product of carelessness or unwitting damage, and sometimes they are due to exploitation of the ecosystem for food, sport or leisure activities.

Artificial ecosystems

Arable farms and market gardens are artificial ecosystems. The plants are deliberately introduced by the farmer and grow in dense stands of a single species. They are usually specifically bred for their food value or commercial value to humans. Such plants would be unlikely to survive for long in a natural environment.

Attempts are made to abolish food chains and food webs in the artificial ecosystem. Competing plants are excluded with the aid of herbicides; fungi are kept at bay with fungicides and insects are eliminated, as far as possible, with insecticides. Animals, other than insects, usually find the ecosystem inhospitable. Such ecosystems could not be maintained without constant human intervention (Fig. 3).

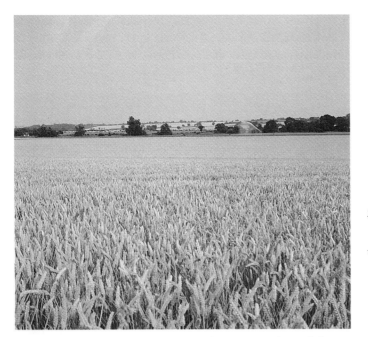

Figure 3 An artificial ecosystem—a wheat monoculture. Only wheat is allowed to grow. All competition is removed.

The soil in these situations is, to some extent, still a natural ecosystem but it is modified by constant ploughing and addition of fertilizers.

Ecologically, artificial ecosystems are not very exciting but they are essential to our survival. We would starve without them. Unless carefully managed, however, they may do long-term damage to their own and other environments. The potential harm from agricultural practices is outlined on pp. 242–50.

Fish farming

One example of an artificial, managed ecosystem is the fish farm. Fish farming has been a growth industry over the last 20 years (Figs 4 and 5). When well managed, it can provide valuable protein at a reduced cost without damage to the environment. If poorly managed it can cause pollution and disturb natural communities.

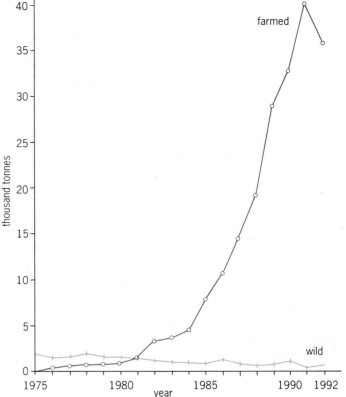

Figure 5 Farmed salmon production and catch of wild salmon in Scotland. Due to difficult market conditions, the farmed production is now stabilizing at about 40 000 tonnes per annum.

Inland fish farms usually raise trout or young salmon using fresh water taken from a river and led through a series of ponds. In some cases, up to half the river's flow can be diverted through the ponds. The water returned to the river is short of oxygen, as a result of the respiration of so many fish, and will contain their faeces and any uneaten food. This can lead to pollution and eutrophication as described on p. 246.

The harmful effects can be reduced by letting the effluent flow through settling tanks where the organic debris precipitates out, before the water returns to the river. By keeping the stocks of fish at relatively low densities and carefully controlling the feeding regime, the amount of harmful effluent can be reduced.

Salmon are farmed mainly in floating cages sited in estuaries (Fig. 6). As with trout farming, this profitable industry can be run with a minimum of environmental disturbance. The risks are similar to those described for trout farms. The faeces and excess food may settle on the sea bed below the cages and destroy the local ecosystem. This can be reduced by moving the cages, though it still takes up to 18 months for the ecosystem to recover. It is more likely, today, that cages are not moved but sited in areas where water currents disperse the debris and feeding regimes are carefully controlled.

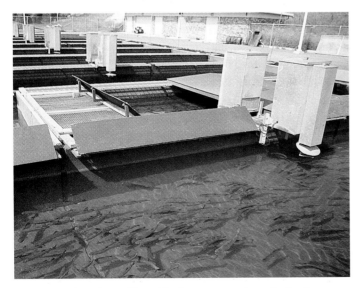

Figure 4 Trout farm. The white boxes are automatic feeders.

Figure 6 Salmon-rearing cages in Scotland

Another worry concerns the spread of disease. As with any artificial ecosystem where a single species is grown or reared at a high density, it is easy for disease to spread rapidly from one individual to the next. If diseased individuals escape from the cages, they could spread disease to the wild stock. Also, the fish farmers use antibiotics to control the diseases. This may lead to the pathogens developing immunity and also puts antibiotics into the human food chain.

One of the main threats to salmon farms is the spread of fish lice. Farmers use a chemical called 'dichlorvos' to control these parasites. Dichlorvos is very toxic to marine crustacea and molluscs. Less harmful is biological control in the form of small fish, wrasse, which live with the salmon and pick the lice off their bodies.

A long-term concern is the possibility of escaped fish interbreeding with the wild stock. The characteristics inbred into the farmed fish are unlikely to be beneficial in the natural environment, so the entire wild population may be weakened.

Managed ecosystems

Some forms of animal husbandry employ managed, rather than artificial ecosystems. Sheep or cattle are allowed to feed on grassland. The animals' grazing prevents the growth of tree seedlings and so maintains a grassland ecosystem. The 'management' may be no more than cutting the hedges, digging ditches and adding fertilizer.

It is not possible to manage every feature of a natural ecosystem. Attempts may be made to control the population of animals which are taken for food or 'sport' by controlling the numbers caught, or by artificially 'stocking' rivers with salmon or trout. A plant community may be partially managed to favour particular creatures. On a grouse moor for example, the heather is regularly burned so that grouse can feed on the young heather shoots which sprout up after the burning. The fire kills seeds, shallow-rooted plants and tree seedlings and helps maintain an ecosystem dominated by heather, which suits the grouse.

Nature reserves, national parks and SSSIs (Sites of Special Scientific Interest) are managed ecosystems which may incorporate

artificial and natural ecosystems. In many wildfowl reserves, shallow lagoons are made and the water level controlled in order to provide an environment that suits ducks, geese and wading birds (Fig. 7).

Figure 7 A managed ecosystem. The Wildfowl & Wetlands Trust manages this site for the benefit of waterfowl and waders.

Fishing—an exploited ecosystem

Although the bulk of our food comes from the artificial ecosystems of farms and market gardens, we still rely on 'hunting' for most of our supplies of fish.

When animals are taken from a natural ecosystem, it is important that the population is not driven to extinction or to levels where it is no longer profitable to hunt. Large-scale hunting for fish and whales in the marine ecosystem has a very poor record in this respect.

If the number of fish removed from a population exceeds the number of young fish reaching maturity, then the population will decline (Fig. 8). At first, the catch size remains the same but it takes longer to catch it. Then the catch starts to contain a greater number of small fish so that the return per day at sea goes down even more. Eventually the stocks are so depleted that it is no longer economical to exploit them. The costs of the boats, the fuel and the wages of the crew exceed the value of the catch. Men are laid off, boats lie rusting in the harbour and the economy of the fishing community and those who depend on it is destroyed. Over-fishing has severely reduced stocks of many fish species; herring in the North Sea, halibut in the Pacific and anchovies off the Peruvian coast, for example. In 1965, 1.3 million tonnes of herring were caught in the North Sea. By 1977 the catch had diminished to 44 000 tonnes, i.e. about 3 per cent of the 1965 catch.

Similarly, whaling has reduced the population of many whale species to levels which give cause for concern.

Figure 8 Landings of North Sea cod since 1972

The only way to avoid reducing the stocks to uneconomic levels, or even extinction, is to try and manage the industry. One way is to aim for the **maximum sustainable yield** by restricting the mass of a species that any one nation may catch, i.e. a 'quota' is set. The maximum sustainable yield is the greatest mass of fish that may be removed without reducing the population over the years.

It is very difficult to estimate the size of fish populations. Furthermore, the population fluctuates from year to year. Consequently, it is difficult to put a figure on the maximum sustainable yield. It has usually been set too high, resulting in falling catches and stock depletion. For plaice, the North Sea quota is 112 000 tonnes per year, but this figure is reduced if the stock is estimated to be falling.

Another management strategy is to fix the mesh size of nets (Fig. 9) so that immature fish are left to breed, and only mature fish above a certain size are caught. This is difficult to achieve if a mixture of fish is to be caught.

Fixed effort harvesting is another management system. The 'effort' is determined by the size of the fishing fleet and the number of days at sea. For example, the Pacific halibut fishery is regulated by having 'closed seasons' when no fishing is allowed. The advantage of this system is that the catch size is related to the population size.

All these systems have their disadvantages and their opponents. They all need policing if they are to be effective and this inevitably adds to the cost of fishing. This, however, is better than losing the fish stocks or the fishing industry.

QUESTIONS

5 In what ways does a managed ecosystem differ from an artificial ecosystem? How would you classify **a** a plantation of Sitka spruce, **b** a hay meadow?

6 Why are artificial ecosystems more vulnerable to pests and diseases than natural ecosystems?

7 Why are artificial ecosystems, in general, essential for human survival?

Figure 9 Checking the mesh size of a trawl net

A freshwater ecosystem

Ponds, lakes and rivers form clearly defined examples of ecosystems. Some of the properties of a pond can be reproduced by setting up a balanced aquarium in the laboratory.

Physical aspects of a freshwater environment

Density
Water is far more dense than air. It offers resistance to moving animals but it also physically supports the animals and plants. Plants and animals living in rivers and streams must be able to withstand or avoid the force of the flowing water.

Surface tension
The surface of water behaves as if it had a thin elastic skin on it. There is, in fact, no skin; the effect is due to the attraction of water molecules at the surface, downwards into the bulk of the liquid. Nevertheless, the surface 'film' is used by animals above and below the surface.

Pond skaters (see Fig. 19(a)) can walk over the surface film without getting wet; pond snails and flatworms crawl along under the surface film; insects such as water boatmen (see Fig. 23(b)) and mosquito larvae (see Fig. 21) can hang from it while they take in air at the surface.

Temperature
Water can absorb a good deal of heat from the sun without its temperature rising much. Similarly, when water loses heat, its temperature does not fall much. A very small or shallow pond might heat up during the day and cool down at night but in most fresh water habitats, the temperature remains fairly constant (Fig. 10)

Figure 10 Daily temperature change in air and water in a small tropical lake

When the temperature of water rises, it can hold less dissolved oxygen. Thus, a rise in temperature, well below the lethal level, might impair the breathing of some aquatic animals.

Water also has an unusual property. Like most liquids, water contracts when it cools, but at temperatures below 4 °C, it expands. becomes less dense and rises. So, ice will form only at the surface of a pond. Water at lower levels will not cool below 4 °C.

Light

Except for places where trees and shrubs grow on the banks, the surface water will receive a high light intensity. The small particles suspended in the water absorb the light so that at a depth of a metre or two, there may not be enough light to allow plants to photosynthesize and grow. The changes in light intensity and other conditions are shown in Fig. 11.

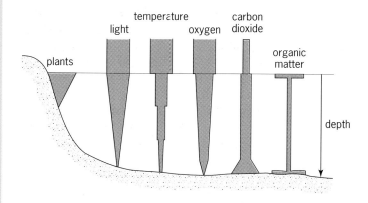

Figure 11 How conditions change with depth. The width of each column is proportional to the factors named.

Oxygen

Although water is H_2O, the oxygen in its molecule is not available for respiration. The oxygen that plants and animals use for respiration is dissolved in the water. It comes from the plants'

photosynthesis during the day and also diffuses continuously through the water surface from the air.

There is much less oxygen in water than there is in air. At 0 °C, 100 cm³ water can hold only about 1 cm³ dissolved oxygen (100 cm³ air contains about 20 cm³ oxygen). This means that a stationary animal or plant in still water quickly uses all the oxygen from the water immediately around it. Anything, such as eutrophication (p. 246) which reduces this small oxygen concentration in fresh water puts the animals at risk of suffocation. Any event which breaks up the surface of the water, e.g. waterfalls or breaking waves, helps to introduce more oxygen from the air.

Minerals

The water flowing in a river or into a lake brings a supply of minerals. Water draining from intensively farmed land may even bring too many minerals (see p. 246). In enclosed lakes and ponds, the supply of minerals will be maintained by normal recycling processes (p. 235).

Substrate

This is the mud or silt on the bottom. It allows plants to root and animals to burrow in it. If it contains a lot of decaying organic matter, it may be very short of oxygen. This is because the bacteria which break down the organic matter use up oxygen in their respiration. Streams which flow into a lake may bring down deposits of silt which collect on the bottom and gradually fill the lake.

QUESTIONS

8 What processes might **a** increase, **b** decrease the amount of oxygen dissolved in water?
9 What factors are likely to limit the growth of plants in a lake?
10 What properties of water make it a good environment for living organisms?

The plant community

The surface waters

The surface waters of ponds, lakes and rivers contain the microscopic algae which form the phytoplankton (p. 233). There will be diatoms and blue-green algae, and in small ponds there may be filamentous algae. The high light intensity of these surface waters allows rapid photosynthesis.

Floating freely on the surface of still waters in lakes and ponds are plants such as duckweed (*Lemna*). Since these plants receive direct sunlight and can reproduce rapidly by vegetative propagation (p. 78) they may spread to cover a large surface of the water and restrict the light reaching the submerged plants. The roots and lower parts of the leaves contain air spaces which enable the plants to float, and the waxy cuticle on the leaves repels water if the leaves are temporarily submerged by waves.

Submerged plants

Plants such as Canadian pondweed (*Elodea canadensis*, Fig. 12) are rooted in the substrate and can only grow where the water is shallow enough or clear enough to allow plenty of light to penetrate. The many leaves of *E. canadensis* and the thin, branching leaves of hornwort (*Ceratophyllum*, Fig. 13) and the river crowfoot (*Ranunculus fluitans*, Fig. 14) present a large surface area to the water. This probably helps to speed up gaseous exchange between the plant and the water. In *Ceratophyllum*, the cuticle is very thin and the epidermal cells, unlike most land plants, contain chloroplasts.

Figure 12 *Elodea canadensis*; Canadian pondweed

Figure 13 *Ceratophyllum*; hornwort

Figure 14 *Ranunculus fluitans*; river crowfoot. The thin flexible leaves stream out in the water current.

Inside the leaves and stems there are air spaces which keep the plant shoots or leaves floating as near to the surface as possible.

The thin leaves of plants such as river crowfoot also offer very little resistance to water flow. They stream out with the current and are not likely to be pulled off.

Because water is so much denser than air, it buoys up and supports the plants submerged in it. Consequently the plants do not need as much strengthening tissue in their stems and leaves as do land plants. For this reason aquatic plants removed from their environment are limp and floppy.

Plants with leaves floating at the surface

The leaves of plants such as the water lily (*Nymphaea*) receive direct sunlight. Their stomata are on the upper surface and so exchange gases directly with the atmosphere rather than with the water. The upper surface of the leaf has a waxy cuticle which allows wave splashes and rain water to run off.

There is very little oxygen in the mud at the bottom of lakes and rivers and it is not always clear how plant roots can respire in these conditions. In some cases, such as in the waterlily, there are air spaces running from leaves to roots, which would allow diffusion of oxygen (Fig. 15).

Figure 15 *Nymphaea* (water lily) leaf stalk, transverse section (×12). The large air spaces may allow diffusion of carbon dioxide and oxygen between the leaves and roots.

The flowers of these plants and of those in the next group are brought above the surface and pollinated probably by insects or the wind.

Growing on the underwater parts of all these plants there is a community, called the **periphyton**, of small organisms such as single-celled creatures, filamentous algae (p. 295) and blue-green algae. These are grazed by fish, tadpoles, pond snails and insect larvae.

Submerged and aerial leaves

In plants such as water crowfoot (*Ranunculus aquatilis*, Fig. 16) and arrow head (*Sagittaria sagittifolia*) the submerged leaves

Figure 16 *Ranunculus aquatilis*; water crowfoot. Notice the difference between aerial and submerged leaves.

271

show the adaptations described for an aquatic environment, while the aerial leaves are characteristic of land plants.

Plants growing at the edge

If the banks are shelving, there is usually a gradual change-over from water plants to land plants (Figs 17 and 18).

The roots and lower stems of plants such as the reed mace (*Typha latifolia*) or reed (*Phragmites communis*) are totally submerged in shallow water. The abundant air spaces in the lower stem permit oxygen to diffuse into the roots. The types of plants represented in the change from aquatic to terrestrial vegetation probably reflect the extent to which they can tolerate having their roots submerged and deprived of oxygen.

Figure 17 Zonation of vegetation at a lake side. The succession does not correspond exactly to Fig. 18 but the transition from submerged plants (*Nuphar*) to swamp scrub can be seen.

The animal community

There are a great many different species of animals living in freshwater habitats and it is possible to mention only a few representative examples here.

Surface film

Pond skaters (Fig. 19(a)) are insects which move about on the surface film of static or sluggish water. The water-repelling bristles on the tips of their legs prevent them breaking through the surface film and so the pond skater can glide over the water surface by brisk movements of the middle pair of legs. Most species feed on insects and crustacea at the surface, by sucking the fluids from their bodies through the piercing mouth parts. Tiny springtails (Fig. 19(b)) also live on the surface film. They are simple, wingless insects and are thought to feed on bacteria and algae in the surface film.

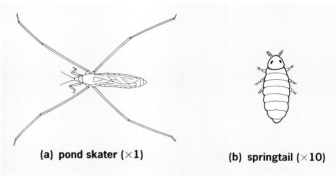

(a) pond skater (×1)   **(b) springtail (×10)**

Figure 19 Animals of the surface film

Plankton

In the top few centimetres of water, there is usually a dense population of zooplankton (p. 233). This includes protozoans such as *Paramecium* (p. 329) and small crustacea such as water fleas and *Cyclops* (Figs 20(a) and (b)). *Paramecium* feeds on the microscopic plants and bacteria in the phytoplankton. The water fleas have a filter-feeding mechanism. They draw a current of water between the two halves of their body covering and filter out the edible plankton.

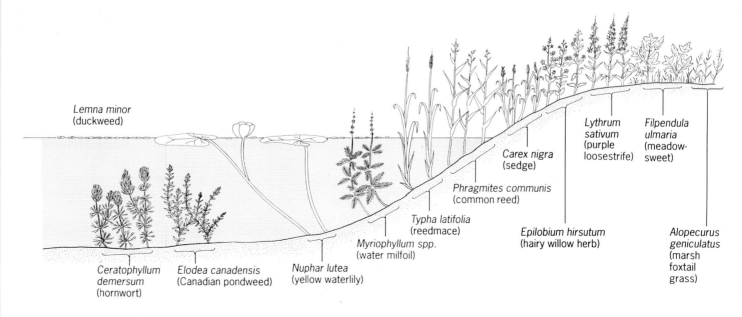

Lemna minor (duckweed)

Lythrum sativum (purple loosestrife)

Filipendula ulmaria (meadow-sweet)

Carex nigra (sedge)

Phragmites communis (common reed)

Typha latifolia (reedmace)

Myriophyllum spp. (water milfoil)

Epilobium hirsutum (hairy willow herb)

Alopecurus geniculatus (marsh foxtail grass)

Ceratophyllum demersum (hornwort)

Elodea canadensis (Canadian pondweed)

Nuphar lutea (yellow waterlily)

Figure 18 Transition from water to land. This is just one example; there are many other sequences and combinations of plants.

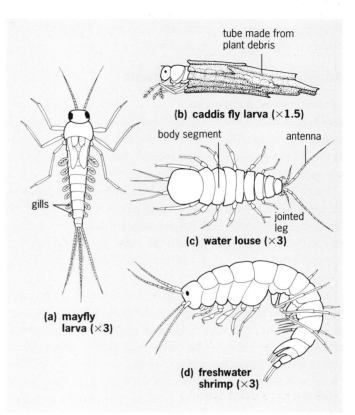

Figure 20 Plankton

Surface feeders

Mosquito larvae hang from the surface film and filter out plankton by the flicking movements of their 'mouth brushes' (Fig. 21(a)). Some species of flatworm may be found gliding along under the surface film by means of their cilia. They are probably feeding on insects and crustacea in the surface film. Flatworms and pond snails are also to be found on the underside of floating leaves. The snails feed by rasping off the algae covering the leaf, with their 'tongues'. The phantom midge larva (*Chaoborus*, Fig. 21(b)) floats at the surface where it captures its prey.

Figure 21 Surface feeders

Bottom-living animals

Dragonflies and mayflies have larvae (Fig. 22(a)) which live in fresh water. When the larvae reach full size, they emerge from the water and change into the adult form. There are many different types of mayfly and their larvae may be adapted to burrowing in mud, clinging to stones in fast streams, or swimming about fairly freely in sluggish waters. Caddis fly larvae (Fig. 22(b)) also live in the bottom waters of well-aerated streams and lakes. They make themselves a tubular casing of pebbles, sand or vegetation and some species spin a silk net for catching their food. Also living on the substrate or in the vegetation is the water louse, *Asellus* (Fig. 22(c)). This appears to be a scavenger in its feeding. Another crustacean, the freshwater shrimp, *Gammarus* (Fig. 22(d)) prefers well-oxygenated water.

Freshwater mussels (*Anodonta cygnea*) and pea shells (*Pisidium amnicum*) are molluscs (p. 292) which live in the mud

at the bottom of lakes and rivers. Inside their two shells are net-like gills, covered with cilia. The beating of the cilia draws water into the shells and the gills filter out small organisms from the water. These organisms are then trapped in sticky mucus and swallowed by the clam.

Figure 22 Bottom-living animals

Free-swimming animals

Water beetles (Fig. 23(a)) and water boatmen move about freely in ponds and lakes though they do cling to water weed to stop themselves floating to the surface when they stop swimming. There are many different species of water beetle which may be either

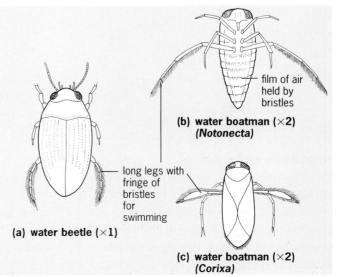

Figure 23 Free-swimming animals

carnivorous or plant eaters. There are two common families of water boatmen, *Notonecta* (Fig. 23(b)), which swims on its back, and *Corixa* (Fig. 23(c)), which swims the right way up. They have short, tubular mouth parts which they use to suck up particles of plant and animal debris.

Fish such as the minnow, stickleback and roach may be found swimming in any part of ponds, lakes and rivers though they have preferred areas for obtaining their food.

Adaptations of aquatic animals

Adaptations for movement

Rapid movement through a dense medium like water requires a streamlined shape. This is seen in most fish and in some mayfly larvae. The latter fold their legs and gills flat against their bodies and swim with rapid wriggling movements.

Water beetles and water boatmen have legs which are adapted for propulsion through water. The last pair of legs is particularly long, slightly flattened and fringed with bristles. The long legs enable the insect to 'row' through the water, the bristles offering maximum resistance during the driving stroke and very little in the recovery stroke.

Some of the aquatic animals have adaptations to make them buoyant. Thus they do not sink to the bottom when they stop swimming. Fish have a swim bladder (p. 331); the transparent phantom midge larvae (*Chaoborus*, Fig. 21(b)) which floats, almost invisible, near the surface has two pairs of air sacs. The abundant bristles on the antennae of the water flea offer a lot of water resistance and so help it to swim, but they also slow down its rate of sinking.

In swiftly moving streams and rivers, some animals may have adaptations which prevent them being swept away in the current. *Gammarus* and some mayfly larvae are flattened and so offer little resistance to the water flow. Leeches (Fig. 24) have suckers which attach firmly to their host animal but also prevent them being carried off in water currents.

Figure 24 Leech (×1)

Adaptations for breathing

There are two options open to freshwater animals; either they extract the dissolved air from the water or they go to the surface and breathe air directly from the atmosphere. Those which use the dissolved oxygen usually have some form of gills which present a large surface to the water. They also have some method of changing the water in contact with the gills. The breathing method of fish is

described on p. 315. The gills of the mayfly larvae (Fig. 22(a)) have air tubes running into them. A rhythmic flicking movement of the gills keeps fresh supplies of water moving past them. The caddis larva has gills and forces a stream of water over them by undulations of the body inside the tube.

Tubifex worms (Fig. 25(a)) live in the mud at the bottom of ponds where the oxygen concentration is usually low. They build tubes in the mud and their bodies project from them and wave about, so renewing the supply of water in contact with them. The lower the oxygen concentration, the more of their body protrudes from the tube. Their blood contains haemoglobin which helps to absorb what little oxygen there is. Some species of the midge *Chironomus* have larvae (Fig. 25(b)) which, unlike most insects, have haemoglobin in their blood. These, too, live in tubes in poorly oxygenated mud.

(a) *Tubifex* worms (×2)

(b) *Chironomus* larva (×3)

Figure 25 Animals of the mud

Mosquito larvae (Fig. 21(a)) and pupae hang by a tube ('siphon') from the surface film (p. 273). This siphon allows a gaseous exchange to take place between the air in the breathing system and the atmosphere. Water beetles carry a store of air under their wing cases and rise to the surface from time to time to replenish the supply. Water boatmen carry a film of air trapped in bristles covering the surface of their bodies. Pond snails have a simple 'lung' under their shells. They fill the lung with air at the water surface at intervals.

Interrelationships

Ponds and lakes may be self-sufficient ecosystems with no new material being brought into them. The oxygen produced by the photosynthesis of green plants in daylight replaces the oxygen used up by the plants and animals. The carbon dioxide produced by plant and animal respiration is used by plants during their daytime photosynthesis.

The dead remains of plants and animals fall to the bottom of the lake or river and decay as a result of bacterial action; the nitrates and phosphates so released into the water are used by the plants for their growth (p. 45).

The plants and animals of fresh water form a complex food web (p. 234). For example, protozoa eat single-celled plants; water fleas eat protozoa; fish and *Hydra* (p. 321) eat water fleas; big fish eat the smaller fish, and herons may eat some of the big fish (Fig. 26).

Although suggestions have been made in the text about the food of the pond animals, there is a great deal still to be learned about what they eat. It is not always easy to find out exactly what some of the small animals are eating. Only by careful observation and by examining the gut contents is it possible to decide what is the main food source of an animal.

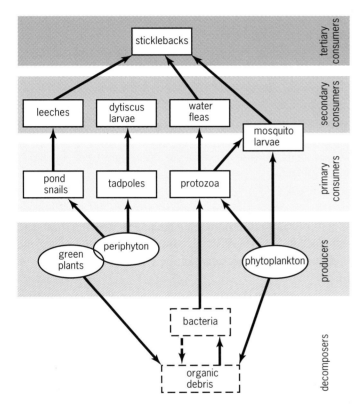

Figure 26 Part of a freshwater food web. Many other organisms could be added. Sticklebacks are themselves eaten by perch and pike. The animals also contribute to the organic debris.

Checklist

- A habitat is where an organism lives, feeds and breeds.
- A community is all the organisms living in a habitat.
- An ecosystem is a self-supporting community plus the physical features of its environment.
- A population is the number of a given species in a defined habitat.
- There is competition within and between species for food, light, space and mates.
- Arable farms, market gardens and fish farms are examples of artificial ecosystems.
- Pastures, nature reserves and grouse moors are managed ecosystems.
- A freshwater environment is affected by light, temperature, oxygen, pH and minerals.
- A plant community in fresh water consists of floating plants, submerged plants, rooted plants with floating leaves and rooted plants with leaves and flowers above water.
- The numerous, thin or finely divided leaves of submerged plants help speed up gaseous exchange.
- Freshwater animals may live on or just under the surface film, in the mud at the bottom, on the leaves of aquatic plants, on or under stones or swimming freely in the water.
- The activities of decomposers (mainly bacteria) in the mud, may lower the oxygen concentration.
- Aquatic animals obtain oxygen either from the air at the surface or from dissolved oxygen in the water.
- Many aquatic animals have adaptations which make them buoyant or enable them to move effectively in water.

QUESTIONS

11 Give one example of an animal and one of a plant which is adapted in a way that prevents it being swept away in a flow of water. Say what the adaptive feature is.

12 In what two ways do aquatic animals obtain their oxygen? Give an example of each.

13 Give three examples of animals which exploit the surface tension of water and say how they use it.

14 What disadvantages might there be for plants and animals living at the bottom of a pond?

28 POPULATIONS

Population changes

On p. 265 it was explained that a biological population is defined as the total number of individuals of any one species in a particular habitat. Such a population will not necessarily be evenly spread throughout the habitat, nor will its numbers remain steady. The population will also be made up of a wide variety of individuals: adults (male and female), juveniles, larvae, eggs or seeds, for example. In studying populations, these variables often have to be simplified.

Population growth

In the simplest case, where a single species is allowed to grow in laboratory conditions, the population develops more or less as shown in Fig. 1.

Figure 1 The sigmoid curve (*Paramecium caudatum*). This is the characteristic growth pattern of a population when food is abundant at first and there are no other factors limiting growth and reproduction.

The population might be of yeast cells (p. 304) growing in a sugar solution, flour beetles in wholemeal flour or weevils in a grain store. The experiment illustrated by Fig. 1 uses a single-celled organism called *Paramecium* (see p. 329), which reproduces by dividing into two (binary fission). The **sigmoid** (S-shaped) form of the graph can be explained as follows:

A The population increase is exponential or logarithmic, i.e. it does not increase 2–4–6–8, etc. but 2–4–8–16–32, etc. One *Paramecium* divides into two, the two offspring each divide, producing 4, the 4 divide into 8 cells, and so on. In other words, the population doubles at each generation. In 10 generations it would reach 1024. When a population of 4 organisms doubles, it is not likely to strain the resources of the habitat, but when a population of 1024 doubles, there is likely to be considerable competition for food and space.

B The population continues to grow but at a steady rate. This may be because the food resources are limiting the rate of growth and reproduction: or the effect of crowding may itself reduce the reproduction rate. Also, some of the mature organisms may be dying.

C At this point the population ceases to grow. The **reproduction rate** equals the **mortality rate** (death rate). The number of offspring produced will still be greater than the number of adults which die, but fewer of these offspring will live long enough to reproduce themselves.

After this stage, the population may start to decline. This can happen because the food supply is insufficient, waste products contaminate the habitat or disease spreads through the dense population.

QUESTIONS

1 In Fig. 1, how many days does it take for the mortality rate to equal the replacement rate?
2 From the graph in Fig. 1, what is the approximate increase in the population of *Paramecium* **a** between day 0 and day 2, **b** between day 2 and day 4, and **c** between day 8 and day 10?

Limits to population growth

The sigmoid curve is a very simplified model of population growth. Few organisms occupy a habitat on their own, and the conditions in a natural habitat will be changing all the time. The steady state of the population in part *C* of the sigmoid curve is rarely reached in nature. In fact, the population is unlikely to reach its maximum theoretical level because of the many factors limiting its growth

If, in the laboratory, two species of *Paramecium (P. aurelia* and *P. caudatum*) are placed in an aquarium tank, the population growth of *P. aurelia* follows the sigmoid curve but the population of *P. caudatum* soon declines to zero because *P. aurelia* takes up food more rapidly than *P. caudatum* (Fig. 2).

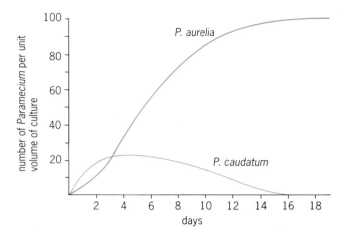

Figure 2 The effect of competition. *Paramecium aurelia* and *P. caudatum* eat the same food but *P. aurelia* can capture and ingest it faster than *P. caudatum*.

This example of competition for food is only one of many factors in a natural environment which will limit a population or cause it to change.

Plant populations will be affected by **abiotic** (non-biological) factors such as rainfall, temperature and light intensity. The population of small annual plants may be greatly reduced by a period of drought; a severe winter can affect the numbers of more hardy perennial plants. **Biotic** (biological) factors affecting plants include their leaves being eaten by browsing and grazing animals or by caterpillars and other insects, and the spread of fungus diseases.

Animal populations, too, will be limited by abiotic factors such as seasonal changes. A cold winter can severely reduce the populations of small birds. However, animal populations are also greatly affected by biotic factors such as the availability of food, competition ior nest sites (Fig. 3), predation (i.e. being eaten by other animals), parasitism and diseases.

The size of an animal population will also be affected by the numbers of animals entering from other localities (immigration) or leaving the population (emigration).

The effect of predation in a laboratory experiment using *P. aurelia* and yeast (*Paramecium* eats yeast cells) is shown in Fig. 4. It can be seen that as the population of *P. aurelia* increases, the yeast population declines. When the population of *P. aurelia* is low, the yeast population recovers.

In this simple experimental situation, it is reasonable to deduce that when the population of *P. aurelia* is high, yeast cells are eaten faster than they can reproduce. When the yeast population falls, there is insufficient food to maintain the reproduction rate of *P. aurelia*.

Figure 3 Colony of nesting gannets. Availability of suitable nest sites is one of the factors which limits the population.

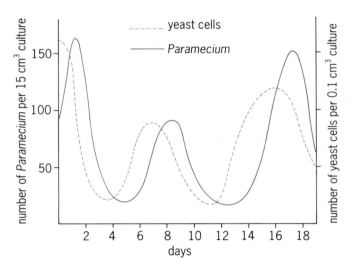

Figure 4 Effects of predation. Yeast cells multiply rapidly in a culture solution containing sugar, but *Paramecium* eats the yeast cells and multiplies at their expense.

In a natural environment, it is rarely possible to say whether the fluctuations observed in a population are mainly due to one particular factor because there are so many factors at work. In some cases, however, the key factors can be identified as mainly responsible for limiting the population.

A classical example of predator–prey relationships comes from an analysis of the fluctuating populations of lynxes and snowshoe hares in Canada. The figures are derived from the numbers of skins sold by trappers to the Hudson's Bay Company between 1845 and 1945.

The lynx preys on the snowshoe hare, and the most likely explanation of the graph in Fig. 5 is that an increase in the hare population allowed the predators to increase. Eventually the increasing numbers of lynxes caused a reduction in the hare population.

However, seasonal or other changes affecting one or both of the animals could not be ruled out.

277

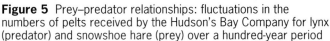

snowshoe hare   lynx

Figure 5 Prey–predator relationships: fluctuations in the numbers of pelts received by the Hudson's Bay Company for lynx (predator) and snowshoe hare (prey) over a hundred-year period

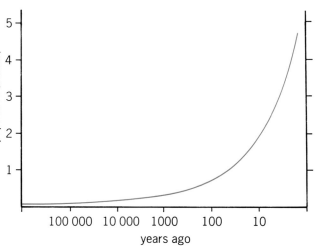

Figure 6 World population growth. The time scale (horizontal axis) is logarithmic. The right-hand space (0–10) represents only 10 years, but the left-hand space (100 000–1 million) represents 900 000 years. The greatest population growth has taken place in the last 300 years.

QUESTIONS

3 In section *B* of the graph in Fig. 1, what is the approximate reproduction rate of *Paramecium* (i.e. the number of new individuals per day)?

4 In 1937, 2 male and 6 female pheasants were introduced to an island off the NW coast of America. There were no other pheasants and no natural predators. The population for the next 6 years increased as follows:

1937 –   24   1940 –   563
1938 –   65   1941 – 1122
1939 – 253   1942 – 1611

Plot a graph of these figures and say whether it corresponds to any part of the sigmoid curve.

5 In Fig. 2, which part of the curve approximately represents the exponential growth of the *P. aurelia* population? Give the answer in days.

6 What forms of competition might limit the population of sticklebacks in a pond?

7 Suggest **a** some abiotic factors, **b** some biotic factors that might prevent an increase in the population of sparrows in a farmyard.

Human population

In AD 1000, the world population was probably about 300 million. In the early 19th century it rose to 1000 million (1 billion), and by 1984 it had reached 4.7 billion. The United Nations expect this number to increase to 6.1 billion by the year 2000. And the population might stabilize at 10 billion by 2100—about twice its present level. The graph in Fig. 6 shows that the greatest population surge has taken place in the last 300 years.

Population growth

About 20 years ago, the human population was increasing at the rate of 2 per cent a year. This may not sound very much. But it means that the world population was doubling every 35 years. This doubles the demand for food, space and other resources. Recently, the growth rate has slowed to 1.7 per cent. But it is not the same everywhere. Kenya's population is growing by 3.9 per cent each year, but Western Europe's grows at only 0.1 per cent.

Traditionally, it is assumed that population growth is limited by famine, disease or war. These factors are affecting local populations in some parts of the world today but they are unlikely to have a limiting effect on the rate of overall population growth.

Diseases such as malaria (spread by mosquitoes), and sleeping sickness (spread by tsetse flies) have for many years limited the spread of people into areas where these insects carry the infections. Diseases such as bubonic plague and influenza have checked population growth from time to time.

Factors affecting population growth

If a population is to grow, the birth rate must be higher than the death rate. But if the birth rate is higher than the death rate, the population does not automatically grow.

Suppose a population of 1000 people produces 100 babies each year but only 50 people die each year. This means that 50 new individuals are added to the population each year and the population will double in 20 years (or less if the new individuals start reproducing at 16) (Fig. 7).

However, if half the babies die before reaching their first birthday, the population will remain stable. Populations in the developing world are growing, not because of an increase in the number of babies born per family, but because more babies are surviving to reach reproductive age. Infant mortality is falling and more people are living longer. That is, **life expectancy** is increasing.

278

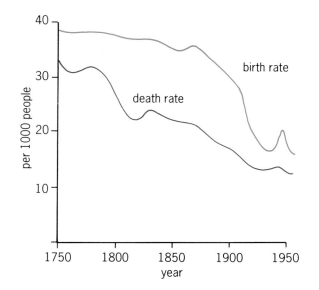

Figure 7 Birth and death rates in England and Wales from 1750 to 1950. Although the birth rate fell during this period, so did the death rate. As a result, the population continued to grow. Note the 'baby boom' after the Second World War.

Increase in life expectancy

The life expectancy is the average age to which a newborn baby can be expected to live. In Europe between 1830 and 1900 the life expectancy was 40–50 years. Between 1900 and 1950 it rose to 65 and it now stands at 72–3 years.

In the developing world, the life expectancy has increased since 1950 from 40 years to 58 years.

These figures are averages. They do not mean, for example, that everyone in the developing world will live to the age of 58. In the developing world, 40 per cent of the deaths are of children younger than 5 years and only 25–30 per cent are deaths of people over 60. In Europe, only 5–20 per cent of deaths are those of children below the age of 5, but 70–80 per cent are of people over 60.

An increase in the number of people over the age of 60 does not change the rate of population growth much, because these people are past child-bearing age. On the other hand, if the death rate among children falls and the extra children survive to reproduce, the population will continue to grow. This is the main reason for the rapid population growth in the developing world since 1950.

Causes of the reduction in death rate

The causes are not always easy to identify and vary from one community to the next. In 19th-century Europe, agricultural development and economic expansion led to improvements in nutrition, housing and sanitation, and to clean water supplies. These improvements reduced the incidence of infectious diseases in the general population. And better-fed children could resist these infections when they did meet them. The drop in deaths from infectious diseases probably accounted for three-quarters of the total fall in deaths.

The social changes probably affected the population growth more than did the discovery of new drugs or improved medical techniques. Because of these techniques, particularly immuniza-

tion, diphtheria, tuberculosis and polio are now rare. And by 1977 smallpox had been wiped out by the World Health Organization's vaccination campaign.

In the developing world, sanitation, clean water supplies and nutrition are improving slowly. The surge in the population since 1950 is likely to be at least 50 per cent due to modern drugs, vaccines and insecticides.

Stability and growth

Up to 300 years ago, the world population was relatively stable. Fertility (the birth rate) was high and so was the mortality rate (death rate). Probably less than half the children born lived to have children of their own. Many died in their first year (infant mortality), and many mothers died during childbirth.

No one saw any point in reducing the birth rate. If you had a lot of children, you had more help on your land and a better chance that some of them would live long enough to care for you in your old age.

In the past 300 years, the mortality rate has fallen but the birth rate has not gone down to the same extent. As a result the population has expanded rapidly.

In 18th-century Europe, the **fertility rate** was about 5. This means that, on average, each woman would have five children. When the death rate fell, the fertility rate lagged behind so that the population increased. However, the fertility rate has now fallen to somewhere between 1.4 and 2.6, and the European population is more or less stable.

A fall in the fertility rate means that young people will form a smaller proportion of the population. There will also be an increasing proportion of old people for the younger generation to look after. In Britain it is estimated that between 1981 and 1991, the number of people aged 75–84 increased by 16 per cent. The number of those over 85, increased by about 46 per cent (Fig. 8).

In the developing world, the fertility rate has dropped from about 5.8 to 4.7. But this is still very much higher than the mortality rate. An average fertility rate of 2.1 is necessary to keep the population stable.

As a community grows wealthier, the birth rate goes down. There are believed to be four reasons.

Longer and better education

Marriage is postponed and a better-educated couple will have learned about methods of family limitation.

Better living conditions

Once people realize that half their offspring are not going to die from disease or malnutrition, family sizes fall.

Agriculture and cities

Modern agriculture is no longer labour intensive. Farmers do not need large families to help out on the land. City dwellers do not depend on their offspring to help raise crops or herd animals.

Application of family planning methods

Either natural methods of birth control or the use of contraceptives is much more common.

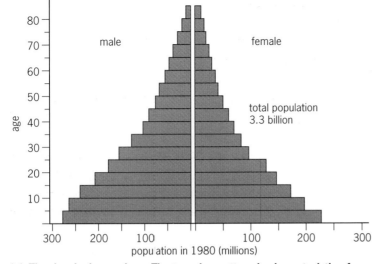

(a) **The developing regions. The tapering pattern is characteristic of a population with a high birth rate and a low average life expectancy. The bulk of the population is under 25.**

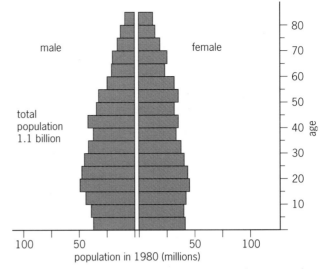

(b) **The developed world. The almost rectangular pattern is characteristic of an industrialized society, with a steady birth rate and a life expectancy of about 70. (The horizontal scale is not the same as in a).**

Figure 8 Age distribution of population in 1980

It takes many years for social improvements to produce a fall in birth rate. Some countries are trying to speed up the process by encouraging couples to limit their family size (Fig. 9), or by penalizing families who have too many children.

Figure 9 Family planning. A health worker in Bangladesh explains the use of a condom.

Meanwhile the population goes on growing. The United Nations expect that the birth rate and death rate will not be in balance until the year 2100. By that time the world population may have reached 10 billion, assuming that the world supply of food will be able to feed this population.

In the past few decades, the world has produced enough food to feed, in theory, all the extra people. But the extra food and the extra people are not always in the same place. As a result, 72 per cent of the world's population has a diet which lacks energy, as well as other nutrients.

Every year between 1965 and 1975, food production in the developed nations rose by 2.8 per cent, while the population rose by 0.7 per cent. In the developing nations during the same period, food production rose by only 1.5 per cent each year, while the annual population rise was 2.4 per cent.

The western world can produce more food than its people can consume. Meanwhile people in the drier regions of Africa face famine due to drought and population pressure on the environment. Even if the food could be taken to the developing world, people there are often too poor to buy it. Ideally, each region needs to grow more food or reduce its population until the community is self-supporting. Some countries grow tobacco, cotton, tea and coffee (cash crops) in order to obtain foreign currency for imports from the western world. This is fine, so long as they can also feed their people. But when food is scarce, people cannot live on the cash crops.

Population pressures

More people, more agriculture and more industrialization will put still more pressure on the environment unless we are very watchful (see Chapters 25 and 26). If we damage the ozone layer (p. 252), increase atmospheric carbon dioxide (p. 253), release radioactive products or allow farmland to erode, we may meet with additional limits to population growth.

QUESTIONS

8 Look at the graph in Fig. 7.
  a When did the post-war 'baby-boom' occur?
  b What was the growth rate of the population in 1800?

9 Which of the following causes of death are likely to have most effect on the growth rate of a population: smallpox, tuberculosis, heart disease, polio, strokes, measles?
Give reasons for your answer.

10 Suggest some reasons why the birth rate tends to fall as a country becomes wealthier.

11 Give examples of the kind of demands that an increasing population makes on the environment. In what ways can these demands lead to environmental damage?

12 If there are 12 000 live births in a population of 400 000 in one year, what is the birth rate?

13 Try to explain why, on average, couples need to have just over 2 children if the population is to remain stable.

14 Study Fig. 8(b) and then comment on **a** the relative number of boy and girl babies, **b** the relative number of men and women of reproductive age (20–40), and **c** the relative numbers of the over-70s.

Checklist

- A population is the total number of individuals of a species in a habitat.

- The growth of plant populations is limited by competition for, e.g. light, water, minerals, rooting space and the abundance of herbivores.

- The growth of animal populations is limited by competition for food, water, breeding space and the abundance of predators.

- The world population is growing at the rate of 1.7 per cent each year. At this rate, the population more than doubles every 50 years.

- The rate of increase is slowing down and the population may stabilize at 10 billion by the year 2100.

- A population grows when the birth rate exceeds the death rate, provided the offspring live to reproduce.

- In the developed countries, the birth rate and the death rate are now about the same.

- In the developing countries, the birth rate exceeds the death rate and their populations are growing. This is not because more babies are born, but because more of them survive.

- The increased survival rate may be due to improved social conditions, such as clean water, efficient sewage disposal, better nutrition and better housing.

- It is also the result of vaccination, new drugs and improved medical services.

- As a population becomes more wealthy, its birth rate tends to fall.

- Food production in the developed countries has increased faster than the population growth.

- Food production in the developing countries has not kept pace with population growth.

EXAMINATION QUESTIONS

Section 5 Organisms and their Environment

Do not write on this page. Where necessary copy drawings, tables or sentences.

1 Grass absorbs light energy and produces food. Clover also makes its food by photosynthesis. Grass is eaten by herbivorous animals such as rabbits and field voles. Field voles and the caterpillars of some butterflies eat other plants including clover. Insectivorous birds such as robins and blue tits may eat the caterpillars. Rabbits, voles and robins are animals which provide food for carnivores such as the fox and kestrel. Foxes eat rabbits and voles. Voles and robins form part of the diet of kestrels. Draw a diagram of a food web based on the feeding relationships described above. (10)

[M]

2 The diagrams below show an animal which lives in fast-flowing streams. Using ONLY the information shown in the diagrams, explain THREE ways in which the animal is adapted for life in running water. (6)

[N]

3 In the 1970s an insect, the plant hopper, destroyed millions of tonnes of rice in Indonesia. Between 1980 and 1986, the government controlled the problem by spending 35 million dollars a year on pesticides. After this time, the pest was destroying almost as much rice as in the 1970s. By then, the insects had become resistant to the pesticide. The pesticide was killing insects and spiders which normally feed on the plant hopper.
(a) What is a pesticide? (1)
(b) Give two reasons why pesticides have not solved the problems outlined above. (2)
(c) Suggest how you could save $35 million per year for the Indonesian government by solving the problem outlined above. (1)

[W]

4 The diagram shows the action of a percolating filter at a sewage treatment works. The sewage effluent is sprayed through the air on to a bed of stones. The stones are covered with micro-organisms (microbes).

(a) Why is it important that the effluent and the filter bed should have a good supply of air? (2)
(b) How do micro-organisms (microbes) on the stones remove organic matter from the sewage effluent. (4)

[S]

5 The diagram shows part of the nitrogen cycle.

(a) Draw FIVE arrows on the diagram, joining the boxes, to show how nitrogen is recycled. (5)
(b) Discuss the benefits and drawbacks of cultivating crops using artificial fertilisers instead of organic manures and composts. (6)

[M]

6 (a) In the boxes below, write the names of THREE organisms which form a food chain in an ecosystem you have studied. (3)
(b) Which organism in your food chain is a carnivore? (1)
(c) What types of organisms are producers in food chains? (1)

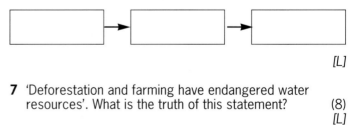

[L]

7 'Deforestation and farming have endangered water resources'. What is the truth of this statement? (8)

[L]

8 All the energy which passes along a food chain comes originally from:
A Oxygen   **C** Glucose
B Carbon dioxide   **D** Light

[N]

Diversity of Organisms

29 CLASSIFICATION

You do not need to be a biologist to realize that there are millions of different organisms living on the Earth, but it takes a biologist to sort them out into some kind of meaningful order, i.e. to **classify** them.

There are many possible ways of classifying organisms. You could group all aquatic organisms together or put all black and white creatures into the same group. However, these do not make very meaningful groups; a seaweed and a porpoise are both aquatic organisms; a magpie and a zebra are both black and white but neither of these pairs has much in common apart from being living organisms and the latter two being animals.

A biologist looks for important features which are shared by as large a group as possible. In some cases it is easy. Birds all have wings, beaks and feathers; there is rarely any doubt about whether a creature is a bird or not. In other cases it is not so easy. As a result, biologists change their ideas from time to time about how living things should be grouped. New groupings are suggested and old ones abandoned.

Kingdoms

The largest group of organisms recognized by biologists is the kingdom. But how many kingdoms should there be? Most biologists used to favour the adoption of two kingdoms, namely **Plants** and **Animals**. This, however, caused problems in trying to classify fungi, bacteria and single-celled organisms which do not fit obviously into either kingdom. A scheme now favoured by many, but by no means all, biologists is the Whittaker 5-kingdom scheme comprising the **Monera**, **Protoctista**, **Fungi**, **Plants** and **Animals**.

Kingdom Monera

These are the bacteria (p. 299) and the blue-green algae. They consist of single cells but differ from other single-celled organisms because their chromosomes are not organized into a nucleus.

Kingdom Protoctista

These are single-celled (unicellular) organisms which have their chromosomes enclosed in a nuclear membrane to form a nucleus.

Some of them, e.g. *Euglena* (Fig. 1), possess chloroplasts and make their food by photosynthesis. These protoctista are often referred to as unicellular 'plants' or **protophyta**.

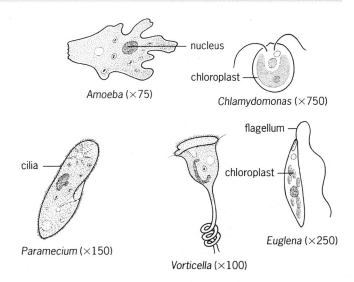

Figure 1 Protoctista. *Chlamydomonas* and *Euglena* have chloroplasts and can photosynthesize. The others are protozoa and ingest solid food.

Organisms such as *Amoeba* and *Paramecium* (Fig. 1), take in and digest solid food and thus resemble animals in their feeding. They may be called unicellular 'animals' or **protozoa**.

Amoeba (Fig. 1) is a protozoan which moves by a flowing movement of its cytoplasm. It feeds by picking up bacteria and other microscopic organisms as it goes. *Vorticella* has a contractile stalk and feeds by creating a current of water with its cilia (p. 7). The current brings particles of food to the cell. *Euglena* and *Chlamydomonas* have chloroplasts in their cells and feed, like plants, by photosynthesis.

Kingdom Fungi

Most fungi are made up of thread-like hyphae (p. 302), rather than cells, and there are many nuclei distributed throughout the cytoplasm in their hyphae.

The fungi include fairly familiar organisms such as mushrooms, toadstools, puffballs, and the bracket fungi that grow on tree-trunks. There are also the less obvious, but very important, mould fungi which grow on stale bread, cheese, fruit or other food. Many of the mould fungi live in the soil or in dead wood (Fig. 2). The yeasts are single-celled fungi similar to the moulds in some respects.

Figure 2 Toadstools. They are getting their food from the rotting tree stump.

Some fungal species are parasites and live in other organisms, particularly plants, where they cause diseases which can affect crop plants (Fig. 3).

Figure 3 Mildew on wheat. Most of the hyphae are inside the leaves, digesting the cells, but some grow out and produce the powdery spores seen here.

Kingdom Plants

These are made up of many cells – they are multicellular. Plant cells have an outside wall made of cellulose. Many of the cells in plant leaves and stems contain chloroplasts with photosynthetic pigments, e.g. chlorophyll. Plants make their food by photosynthesis.

Kingdom Animals

Animals are multicellular organisms whose cells have no cell walls or chloroplasts. Most animals ingest solid food and digest it internally.

It is still not easy to fit all organisms into this scheme. For example, many protoctista with chlorophyll (the protophyta) show important resemblances to some members of the algae, but the algae are classified into the plant kingdom. The viruses are not included in the scheme because, in many respects, viruses are not independent living organisms.

This kind of problem will always occur when we try to devise rigid classificatory schemes with distinct boundaries between groups. The process of evolution would hardly be expected to result in a tidy scheme of classification for biologists to use.

An outline classification of plants and animals is given below and illustrated in Figs 4–7. Bacteria and fungi are described more fully on pp. 299–304.

QUESTIONS

1 Which kingdoms contain organisms with **a** many cells, **b** nuclei in their cells, **c** cell walls, **d** hyphae, **e** chloroplasts?
2 State the main differences between animals and plants.
3 If it was decided to use a 2-kingdom classification (plants and animals), in which kingdoms would you place the bacteria, protoctista and fungi? Suggest reasons for your decision.

Species

The smallest natural group of organisms is the species. Robins, blackbirds and sparrows are three different species of bird. Apart from small variations, members of a species are almost identical in their anatomy, physiology and behaviour.

Members of a species also often resemble each other very closely in appearance, unless humans have taken a hand in the breeding programmes. For example, all dogs belong to the same species but there are wide variations in the appearance of different breeds.

One of the main features which determines whether organisms belong to the same species is whether they can successfully breed together. A spaniel and a labrador retriever may look very different but they have no problems in breeding together to produce a litter of 'mongrel' puppies.

Closely related species are grouped into a **genus** (plural **genera**). For example, stoats, weasels and polecats are grouped into the genus *Mustela*.

Binomial nomenclature

Species must be named in such a way that the name is recognized all over the world.

The 'cuckoo flower' and the 'lady's smock' are two common names for the same wild plant. If you are not aware that these are alternative names this could lead to confusion. If the botanical name, *Cardamine pratensis*, is used, however, there is no chance of error. The Latin form of the name allows it to be used in all the countries of the world irrespective of language barriers.

Binomial means 'two names'; the first name gives the genus and the second gives the species. For example, the stoat and weasel are both in the genus *Mustela* but they are different species; the stoat is *Mustela erminea* and the weasel is *Mustela nivalis*.

The name of the genus (generic name) is always given a capital letter and the name of the species (specific name) always starts with a small letter.

Frequently, the specific name is descriptive, e.g. *hirsutum* = hairy, *aquatilis* = living in water, *bulbosus* = having a bulb.

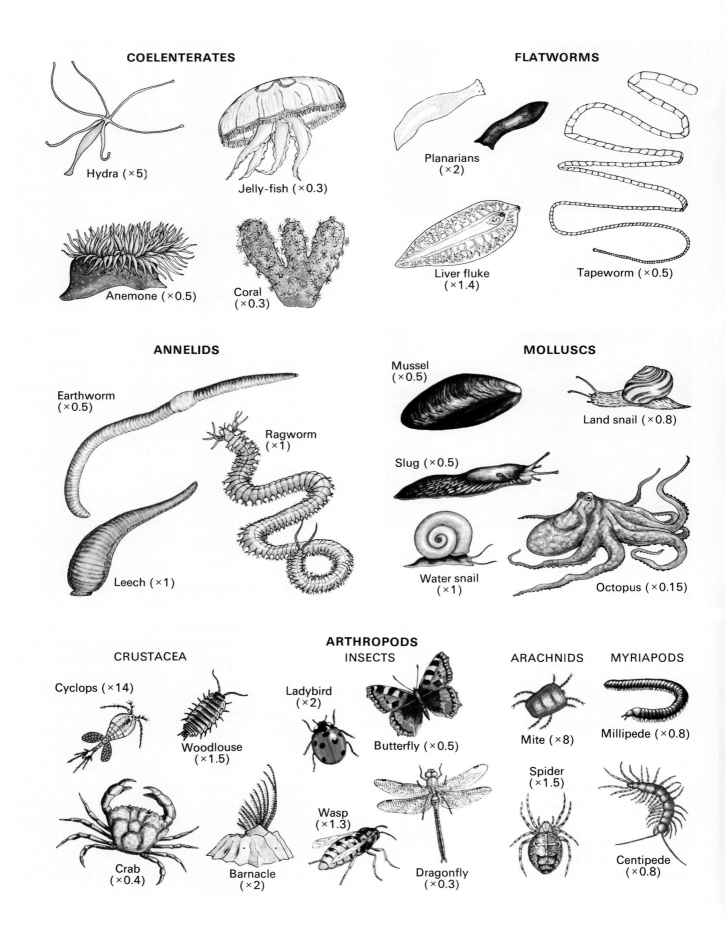

COELENTERATES

Hydra (×5)

Jelly-fish (×0.3)

Anemone (×0.5)

Coral (×0.3)

FLATWORMS

Planarians (×2)

Liver fluke (×1.4)

Tapeworm (×0.5)

ANNELIDS

Earthworm (×0.5)

Ragworm (×1)

Leech (×1)

MOLLUSCS

Mussel (×0.5)

Land snail (×0.8)

Slug (×0.5)

Water snail (×1)

Octopus (×0.15)

ARTHROPODS

CRUSTACEA

Cyclops (×14)

Woodlouse (×1.5)

Crab (×0.4)

Barnacle (×2)

INSECTS

Ladybird (×2)

Butterfly (×0.5)

Wasp (×1.3)

Dragonfly (×0.3)

ARACHNIDS

Mite (×8)

Spider (×1.5)

MYRIAPODS

Millipede (×0.8)

Centipede (×0.8)

Figure 4 The animal kingdom; 5 invertebrate phyla

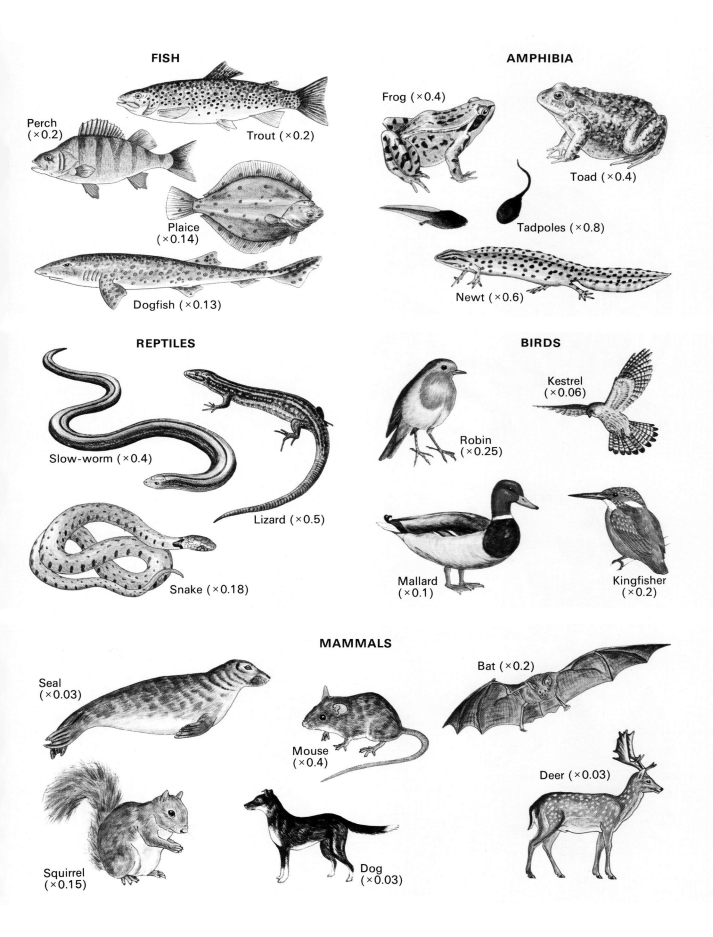

FISH

Perch (×0.2)

Trout (×0.2)

Plaice (×0.14)

Dogfish (×0.13)

AMPHIBIA

Frog (×0.4)

Toad (×0.4)

Tadpoles (×0.8)

Newt (×0.6)

REPTILES

Slow-worm (×0.4)

Lizard (×0.5)

Snake (×0.18)

BIRDS

Kestrel (×0.06)

Robin (×0.25)

Mallard (×0.1)

Kingfisher (×0.2)

MAMMALS

Seal (×0.03)

Mouse (×0.4)

Bat (×0.2)

Deer (×0.03)

Squirrel (×0.15)

Dog (×0.03)

Figure 5 The animal kingdom; the vertebrate classes

ALGAE

Sea lettuce
(×0.1)

Laminaria
(×0.15)

Dulse
(×0.3)

Bladder wrack
(×0.3)

BRYOPHYTES

(*a*) LIVERWORTS

Pellia (×2)

Lophocolea
(×3)

Marchantia (×1.5)

FERNS

Bracken (×0.1)

Spleenwort (×0.5)

Hart's tongue
(×0.3)

Male fern
(×0.1)

Polypody (×0.3)

(*b*) MOSSES

Funaria (×1)

Hypnum (×1.5)

Sphagnum
(×0.8)

Polytrichum (×0.75)

Figure 6 The plant kingdom; plants that do not bear seeds

CONIFERS

FLOWERING PLANTS

(a) MONOCOTYLEDONS

Pine
(×0.004)

Spruce
(×0.004)

Cypress
(×0.005)

Cedar
(×0.0035)

Meadow grass
(×0.6)

Iris (×0.3)

Cocksfoot
(×0.4)

Daffodil
(×0.3)

(b) DICOTYLEDONS

(i) TREES

(ii) SHRUBS

Broom (×0.03)

Horse chestnut (×0.002)

(iii) HERBS

Forget-me-not (×0.5)

Buttercup
(×0.5)

Poppy
(×0.4)

Figure 7 The plant kingdom; seed-bearing plants

289

Animal kingdom

(Only 8 groups out of 23 are listed here.)

1 {
Coelenterates (sea anemones, jellyfish)
Flatworms
Nematode worms
Annelids (segmented worms)
Arthropods
    CLASS
    Crustacea (crabs, shrimps, water fleas)
    Insects
    Arachnids (spiders and mites)
    Myriapods (centipedes and millipedes)
Molluscs (snails, slugs, mussels, octopuses)
Echinoderms (starfish, sea urchins)
}

Vertebrates
    CLASS
    Fish
    Amphibia (frogs, toads, newts)
    Reptiles (lizards, snakes, turtles)
    Birds
    Mammals
    (Only 4 subgroups out of about 26 are listed.)
        Insectivores
        Carnivores
        Rodents
        Primates

[1] All the organisms which do not have a vertebral column are often referred to as invertebrates. Invertebrates are not a natural group, but the term is convenient to use.

Annelids

Annelids are worms (see Fig. 4 on p. 286). Most of them have elongated, cylindrical bodies which are divided into segments. All the segments have identical sets of organs, though those at the front end may have specialized structures. Some organs, e.g. the alimentary canal, the nerve cord and the main blood vessels run the whole length of the body.

In most annelids, each segment carries a number of bristles, called **chaetae** which help in locomotion.

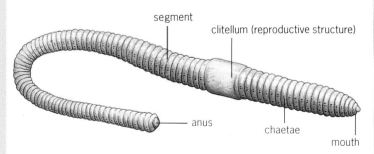

Figure 8 Earthworm

Earthworms are annelids (Fig. 8), but there are many more annelid species living in fresh water and the sea. Lugworms, bristle-worms and ragworms, for example, are annelids which burrow in the sand on the sea shore. *Tubifex* (Fig. 25, p. 274) is a freshwater annelid living in the mud at the bottom of ponds.

Arthropods

The arthropods include the crustacea, insects, centipedes and spiders (Fig. 4 on p. 286). The name arthropod means jointed limbs, and this is a feature common to them all. They also have a hard, firm external skeleton, called a **cuticle**, which encloses their bodies. Their bodies are segmented and, between the segments, there are flexible joints which permit movement. In most arthropods, the segments are grouped together to form distinct regions, e.g. the head, thorax and abdomen.

Crustacea

Marine crustacea are crabs, prawns, lobsters, shrimps and barnacles. Freshwater crustacea are water fleas, *Cyclops*, the freshwater shrimp (*Gammarus*) and the water louse (*Asellus*). Woodlice are land-dwelling crustacea. Some of these crustacea are illustrated on pp. 273 and 286.

Like all arthropods, crustacea have an exoskeleton and jointed legs. They also have two pairs of antennae which are sensitive to touch and to chemicals, and they have **compound eyes** (Fig. 9). Compound eyes are made up of tens or hundreds of separate lenses with light-sensitive cells beneath. They are able to form a crude image and are very sensitive to movement.

Typically, crustacea have a pair of jointed limbs on each segment of the body, but those on the head segments are modified to form antennae or specialized mouth parts for feeding (Fig. 10).

Figure 9 Compound eyes of a horsefly (×20)

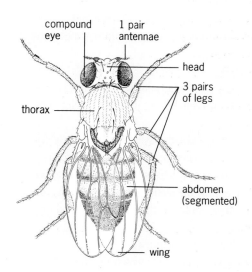

Figure 10 Lobster (×0.2)

Insects

The insects form a very large class of arthropods. Bees, butterflies, mosquitoes, houseflies, earwigs, greenfly and beetles are just a few of the subgroups in this class.

Insects have segmented bodies with a firm exoskeleton, three pairs of jointed legs, compound eyes and, typically, two pairs of wings. The segments are grouped into distinct head, thorax and abdomen regions (Fig. 11).

Insects differ from crustacea in having wings, only one pair of antennae and only three pairs of legs. There are no limbs on the abdominal segments.

The insects have very successfully colonized the land. One reason for their success is the relative impermeability of their cuticles which prevents desiccation even in very hot, dry climates.

Arachnids

These are the spiders, scorpions, mites and ticks. Their bodies are divided into two regions, the cephalothorax and abdomen (Fig. 12). They have four pairs of limbs on the cephalothorax but no antennae. In place of antennae there is a pair of *chelicerae* which are used to hold prey. In spiders the chelicerae pierce the prey and paralyse it with a poison secreted by a gland at their base. There are usually several pairs of simple eyes.

Figure 11 External features of an insect (fruit fly (x10)). Flies, midges and mosquitoes have only one pair of wings.

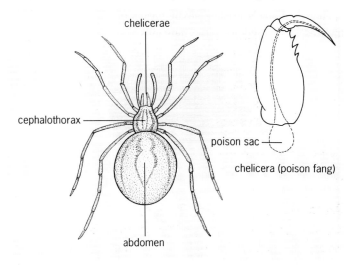

Figure 12 Spider (×2.5)

Myriapods

These are millipedes and centipedes. They have a head and a segmented body which is not obviously divided into thorax and abdomen. There is a pair of legs on each body segment but in the

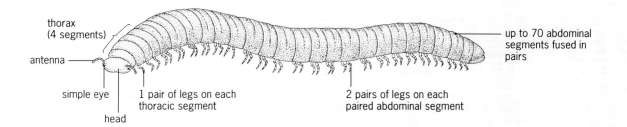

Figure 13 Millipede (×2.5)

millipede, the abdominal segments are fused in pairs and it looks as if it has two pairs of legs per segment (Fig. 13).

As the myriapod grows, additional segments are formed. The myriapods have one pair of antennae and simple eyes. Centipedes are carnivorous; millipedes feed on vegetable matter.

Molluscs

Molluscs include snails, whelks, slugs, mussels, oysters and (perhaps surprisingly) squids and octopuses (see Fig. 4 on p. 286).

Many of the molluscs have a shell. In snails, the shell is usually a coiled, tubular structure. In mussels and clams (the bivalves), the shell consists of two halves which can be partially open or tightly closed. In squids the shell is a plate-like structure enclosed in the body. In other molluscs, the shell is reduced or absent, e.g. slugs, octopuses.

All molluscs have a muscular **foot**. In the snails and slugs it forms a flattened structure which protrudes from the shell during locomotion (Fig. 14). In bivalves, the foot can protrude from between the halves of the shell and burrow in the sand (e.g. cockles). In the squids and octopuses, the foot has become the array of tentacles.

Slugs and snails breathe air by means of a simple 'lung'. The bivalves, squids and octopuses have gills.

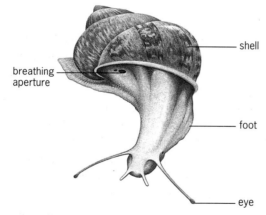

Figure 14 Snail (x1)

Vertebrates

Vertebrates are animals which have a vertebral column. The vertebral column is sometimes called the spinal column or just spine and consists of a chain of cylindrical bones (vertebrae) joined end to end (p. 162).

Each vertebra carries an arch of bone on its dorsal (upper) surface. This arch protects the spinal cord (p. 182), which runs most of the length of the vertebral column. The front end of the spinal cord is expanded to form a brain which is enclosed and protected by the skull.

The skull carries a pair of jaws which, in most vertebrates, have rows of teeth.

The five classes of vertebrates are fish, amphibia, reptiles, birds and mammals.

Body temperature

Fish, amphibia and reptiles are often referred to as 'cold-blooded'. This is a misleading term. A fish in a tropical lagoon, or a lizard basking in the sun will have warm blood. The point is that these animals have a variable body temperature which, to some extent, depends on the temperature of their surroundings. Reptiles, for example, may control their temperature by moving into sunlight or retreating into shade but there is no internal regulatory mechanism.

So-called 'warm-blooded' animals, for the most part, have a body temperature higher than that of their surroundings. The main difference, however, is that these temperatures are kept more or less constant despite any variations in external temperature. There are internal regulatory mechanisms (p. 148), which keep the body temperature within narrow limits.

It is better to use the terms **poikilothermic** (variable temperature) and **homoiothermic** (constant temperature).

The advantage of homoiothermy is that an animal's activity is not dependent on the surrounding temperature. A lizard may become sluggish if the surrounding temperature falls. This could be a disadvantage if the lizard is being pursued by a homoiothermic predator whose speed and reactions are not affected by low temperatures.

QUESTIONS

4 What position do **a** earthworms, **b** snails occupy in their respective food webs?

5 An earthworm can withdraw into its burrow; a snail can retract into its shell. What is the likely benefit of this behaviour to these two organisms?

6 Why do you think poikilothermic animals are slowed down by low temperatures (see p. 15)?

Fish

Fish are poikilothermic vertebrates. Many of them have a smooth, streamlined shape which offers minimal resistance to the water through which they move (Fig. 15). Their bodies are covered with overlapping scales and they have fins which play a part in movement (p. 330).

Fish breathe by means of filamentous gills which are protected by a bony plate, called the **operculum** (p. 316).

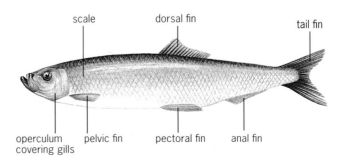

Figure 15 Herring (*Clupea*) (x0.3)

Cartilaginous fish such as dogfish, sharks and rays differ from the 'bony' fish described above in having spiny scales and separate gill openings.

Fish reproduce sexually but fertilization usually takes place externally, i.e. the female lays eggs and the male sheds sperms on them after they are laid.

Amphibia

Amphibia are poikilothermic vertebrates with four limbs and no scales. The class includes frogs, toads and newts. The name, amphibian, means double life and refers to the fact that the organism spends part of its life in water and part on the land. In fact, most frogs, toads and newts spend much of their time on the land, albeit in moist situations, and return to ponds, etc. only to lay eggs.

The external features of the common frog are shown in Fig. 16; Fig. 5 on p. 287 shows the toad and the newt.

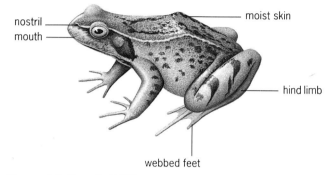

Figure 16 *Rana* (x 0.75)

The toad's skin is drier than that of the frog and it has glands which can exude an unpleasant-tasting chemical which discourages predators. Newts differ from frogs and toads in having a tail. All three groups are carnivorous.

Amphibians have four limbs. In frogs and toads, the hind feet have a web of skin between the toes. This offers a large surface area to thrust against the water when the animal is swimming. Newts swim by a wriggling, fish-like movement of their bodies and make less use of their limbs for swimming.

Amphibia have moist skins with a good supply of capillaries which can exchange oxygen and carbon dioxide with the air or water. They also have lungs which can be inflated by a kind of swallowing action. They do not have a diaphragm or ribs.

Frogs and toads migrate to ponds where the males and females pair up. The male climbs on the female's back and grips firmly with his front legs (Fig. 5, p. 319). When the female lays eggs, the male simultaneously releases sperms over them. Fertilization, therefore, is external even though the frogs are in close contact for the event.

Reptiles

Reptiles are land-living vertebrates. Their skins are dry and the outer layer of epidermis forms a pattern of scales. This dry, scaly skin resists water loss. Also the eggs of most species have a tough, parchment-like shell. Reptiles, therefore, are not restricted to damp habitats, nor do they need water in which to breed.

The reptiles are poikilothermic but they can regulate their temperature to some extent. They do this by basking in the sun until their bodies warm up. This enables them to move about rapidly in pursuit of insects and other prey.

The reptiles include lizards, snakes, turtles, tortoises and crocodiles. In Britain we have only three species of lizard and three species of snake (Fig. 17 and Fig. 5 on p. 287).

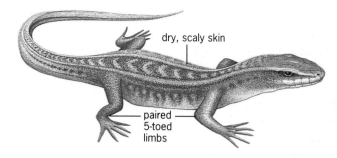

Figure 17 *Lacerta* (x 1.5)

Apart from the snakes, reptiles have four limbs, each with five toes. Some species of snake still retain the vestiges of limbs and girdles.

Male and female reptiles mate, and sperms are passed into the female's body. The eggs are, therefore, fertilized internally before being laid. In some species, the female retains the eggs in the body until they are ready to hatch.

Birds

Birds are homoiothermic vertebrates.

The vertebral column in the neck is flexible but the rest of the vertebrae are fused to form a rigid structure. This is probably an adaptation to flight, as the powerful wing muscles need a rigid air-frame to work against.

The epidermis over most of the body, produces a covering of feathers but, on the legs and toes, the epidermis forms scales. The feathers are of several kinds. The fluffy down feathers form an insulating layer close to the skin; the contour feathers cover the body and give the bird its shape and coloration; the large quill feathers on the wing are essential for flight.

Birds have four limbs, but the forelimbs are modified to form wings. The feet have four toes with claws which help the bird to perch, scratch for seeds or capture prey, according to the species.

The upper and lower jaws are extended to form a beak which is used for feeding in various ways (Fig. 18).

In birds, fertilization is internal and the female lays hard-shelled eggs in a nest where she incubates them.

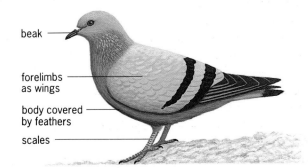

Figure 18 The main features of a bird

Mammals

Mammals are homoiothermic vertebrates with four limbs. They differ from birds in having hair rather than feathers. Unlike the other vertebrates they have a diaphragm which plays a part in breathing (p. 132). They also have mammary glands and suckle their young on milk.

A sample of mammals is shown in Fig. 5 on p. 287 and Fig. 19 illustrates some of the mammalian features.

Figure 19 Mammalian features. The furry coat, the external ear pinnae and the facial whiskers (vibrissae) are visible mammalian features in this gerbil.

Humans are mammals and most of the physiology described in Section 3 applies to all mammals.

Mammals give birth to fully formed young instead of laying eggs. The eggs are fertilized internally (see p. 153) and undergo a period of development in the uterus.

The young may be blind and helpless at first, e.g. dogs and cats, or they may be able to stand up and move about soon after birth, e.g. sheep and cows. In either case, the youngster's first food is the milk which it sucks from the mother's teats. The milk is made in the mammary glands and contains all the nutriments that the offspring need for the first few weeks or months, depending on the species.

As the youngsters get older, they start to feed on the same food as the parents. In the case of carnivores, the parents bring the food to the young until they are able to fend for themselves.

QUESTIONS

7 Which vertebrate classes **a** are warm-blooded, **b** have four legs, **c** lay eggs, **d** have internal fertilization and **e** have some degree of parental care?

8 Draw up a key (see p. 297) which a biologist could use to place an animal in its correct group.

Plant kingdom

DIVISION

Red algae
Brown algae } seaweeds and filamentous forms; mostly aquatic
Green algae

Bryophytes (no specialized conducting tissue)

    CLASS
    Liverworts
    Mosses

Vascular plants (well-developed xylem and phloem)

    CLASS
    Ferns
[1] { Conifers (seeds not enclosed in fruits)
     Flowering plants (seeds enclosed in fruits)
        SUBCLASS
        Monocotyledons[2] (grasses, lilies)
        Dicotyledons (trees, shrubs, herbaceous plants)
           FAMILY
           e.g. Ranunculaceae (one of about 70 families)
             GENUS
             e.g. *Ranunculus*
                SPECIES
                e.g. *Ranunculus bulbosus*
                (bulbous buttercup)

[1] Sometimes called, collectively, 'seed-bearing plants'.
[2] The monocotyledons (monocots for short), are flowering plants which have only one cotyledon in their seeds (p. 83). Most, but not all, monocots also have long, narrow leaves (e.g. grasses, daffodils, bluebells) with parallel leaf veins (Fig. 20).

The dicotyledons (dicots for short), have two cotyledons in their seeds (p. 83). Their leaves are usually broad and the leaf veins form a branching network.

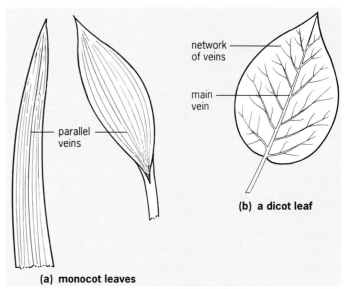

(a) monocot leaves

(b) a dicot leaf

Figure 20 Leaf types in flowering plants

Algae

The algae are plants which range from very simple filaments, consisting of a single row of identical cells, to large seaweeds. The Protoctista with chloroplasts (the protophyta) may also, with good reason, be classified as algae.

Most members of the algae have cellulose cell walls and chloroplasts. The chloroplasts contain chlorophyll, but other pigments may be present in brown and red algae.

The algae, for the most part, grow in water and have little need of specialized supporting or water-conducting tissue. Compared with vascular plants, therefore, their cells and tissues are relatively unspecialized. The aquatic algae cannot survive prolonged desiccation when out of water.

In open water, it is the single-celled algae which form the basis of the food web. These algae, particularly the forms called **diatoms**, are abundant in the surface waters of ponds, lakes and the sea. Usually they are invisible to the naked eye, but occasionally, in a pond or neglected swimming-pool, the unicellular plants are so numerous that the water looks green.

Bryophytes

The bryophytes include the liverworts and mosses.

Liverworts

Liverworts are small, flat, green plants, usually growing close together in very moist, shady places such as in the mouths of caves or on a stream bank near the water-line. They look like small, over-lapping strips of dark green leaf stuck to the rock or the bank (Fig. 21).

Figure 21 A colony of liverworts. The small black knobs on the ends of the green stalks are the spore capsules.

Liverworts have no stems or roots but single-celled **rhizoids** (like root hairs, p. 59) grow from their lower surface, anchor them in the ground and absorb water.

Mosses

Mosses are simple but successful land plants. Each plant is quite small, on average about 1–5 cm long, but they usually grow in dense tufts which probably helps to support them and to conserve moisture.

Each plant has a slender stem with numerous small leaves arising from it (Fig. 22). Sometimes the stem is horizontal and has many branches. Most moss leaves are only one cell thick, so there is no specialization of cells like there is in the leaves of flowering plants. In the stem, the innermost cells conduct water and food, but apart from being longer than other cells, they are not particularly specialized in their structure. The outer cells have thick walls and probably give the stem strength.

Figure 22 Moss plants growing in a wood. Some of the plants have produced spore capsules.

As in liverworts, there are no proper roots. The rhizoids may consist of several cells and grow from the base of the stem into the soil.

Four different species of moss are illustrated in Fig. 6 on p. 288.

Mosses and liverworts reproduce sexually but do not form seeds. The male gametes have flagella and swim in the film of water on the surface of the plant in order to reach the female gamete. When the gametes fuse the zygote grows into a spore capsule on a long stalk. The spore capsules eventually break open and release thousands of single-celled **spores**. Each of these spores can grow into a new moss or liverwort plant.

The rhizoids of liverworts and mosses penetrate only into the top few millimetres of the soil. The plants lack a waxy cuticle and so are unable to resist desiccation. The male gametes have to reach the female gamete by swimming in water. These characteristics restrict most bryophytes to damp, shady habitats.

Ferns

Ferns are land plants, usually much larger than mosses, with more highly developed structures. Their stems, leaves and roots are very similar to those of the flowering plants.

The stem is usually entirely below ground and takes the form of a **rhizome** (p. 78). In bracken, the rhizome grows horizontally below ground, sending up leaves at intervals. The roots which grow from the rhizome are called adventitious roots. This is the name given to any roots which grow directly from the stem rather than from other roots.

The stem and leaves have sieve tubes and water-conducting cells similar to those in the xylem and phloem of a flowering plant (p. 56). For this reason, the ferns and seed-bearing plants are sometimes referred to as vascular plants, because they all have vascular bundles or vascular tissue. Ferns also have multicellular roots with vascular tissue.

The leaves of ferns vary from one species to another (Fig. 23 and Fig. 6 on p. 288), but they are all much larger than those of mosses and are several cells thick. Most of them have an upper and lower epidermis, a layer of palisade cells and a spongy mesophyll similar to the leaves of a flowering plant.

Figure 23 Young fern leaf. Ferns do not form buds like those of the flowering plants. The midrib and leaflets of the young are tightly coiled and unwind as the leaf grows.

Ferns produce gametes but no seeds. The zygote gives rise to the fern plant, which then produces single-celled spores from numerous **sporangia** (spore capsules) on its leaves. The sporangia are formed on the lower side of the leaf but their position depends on the species of fern. The sporangia are usually arranged in compact groups (Fig. 24).

Figure 24 Polypody fern. Each brown patch on the underside of the leaf is made up of many sporangia.

Conifers

Conifers are trees or shrubs often with needle-like leaves (Fig. 25). They reproduce by seeds rather than spores but the seeds are formed in cones and not in flowers and are not enclosed in an ovary.

Figure 25 Pine needles and male cones. The needle-like shape of the leaves helps to reduce the rate of transpiration (p. 63) in windy conditions.

There are male and female cones on the same plant. The male cones produce pollen which is light and powdery and falls, or is blown, onto the female cones (Fig. 26). Familiar conifers are the pine, larch, spruce and cypress.

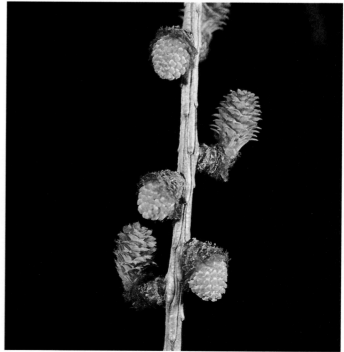

Figure 26 Larch cones. The green cones are male and are producing pollen. The pink female cones face upwards and will produce the seeds.

Flowering plants

Flowering plants reproduce by seeds which are formed in flowers. The seeds are enclosed in an ovary. The structure, physiology and life cycle of flowering plants are described on pp. 52–85.

> ### QUESTIONS
>
> **9** Try to classify the following organisms: beetle, sparrow, weasel, gorilla, pine tree, buttercup, moss. e.g. Butterfly: Kingdom, animal; Group, arthropod; Class, insect.
> **10** The white deadnettle is *Lamium album*; the red deadnettle is *Lamium purpureum*. Would you expect these two plants to cross-pollinate successfully?
> **11** If a heath fire destroys all the above-ground vegetation, the bracken will still grow well next season. Suggest why this should be so.
> **12** Which groups of plants are sometimes called vascular plants and why are they so called?

Keys for identification

Once you know the main characteristics of a group, it is possible to draw up a systematic plan for identifying an unfamiliar organism. One such plan is shown in Fig. 27.

An alternative form of key is the **dichotomous key**. Dichotomous means two branches, so you are confronted with two possibilities at each stage.

DICHOTOMOUS KEY FOR VERTEBRATE CLASSES

1 { Poikilothermic 2
{ Homoiothermic 4

2 { Has fins but no limbs......................... **Fish**
{ Has 4 limbs 3

3 { Has no scales on body........................ **Amphibian**
{ Has scales .... **Reptile**

4 { Has feathers .... **Bird**
{ Has fur .... **Mammal**

Above is an example of a dichotomous key that could be used to place an unknown vertebrate in the correct class. Item 1 gives you a choice between 2 alternatives. If the animal is poikilothermic you move to item 2 and make a further choice. If it is homoiothermic you move to item 4 for your next choice.

The same technique may be used for assigning an organism to its class, genus or species. However, the important features may not always be easy to see and you have to make use of less fundamental characteristics.

Fig. 28 is a key for assigning pond organisms to their correct group or subgroup. It is not based on the fundamental characteristics of the group, e.g. 'not more than 2 cm long' is not a feature of flatworms, but it is a useful guide to those flatworms that live in ponds. A key such as this is sometimes called an artificial key.

KEY FOR SOME FRESHWATER INVERTEBRATES

1 { Body flattened; not more than about 2 cm long; gliding movement .... **Flatworm**
{ Body not flattened 2

2 { Body with shell........................... **Mollusc** (e.g. snail)
{ Body with no shell 3

3 { Body narrow, cylindrical and segmented 4
{ Body not narrow and cylindrical 5

4 { Obvious head; appendages or mouth parts on front segments .... Insect larva (e.g. *Chironomus*)
{ No obvious head; no appendages or mouth parts.. **Annelid** (e.g. *Tubifex*)

5 { Head appendages used for locomotion Water fleas or *Cyclops*
{ Other appendages used for locomotion 6

6 { Three pairs of legs............................. **Insect** (e.g. water boatman)
{ More than three pairs of legs............. **Crustacean** (e.g. *Gammarus*)

Figure 28 Key for pond organisms

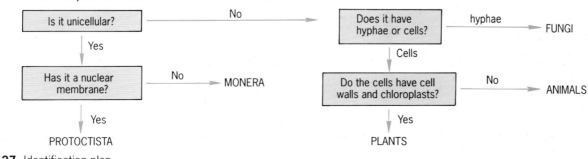

Figure 27 Identification plan

13 Fig. 1 on p. 284 shows some protoctista. Using only the features shown in the drawings, construct a dichotomous key that could be used to identify these organisms.

Checklist

- BACTERIA are microscopic organisms; they have no proper nucleus.

- PROTOCTISTA are single-celled organisms containing a nucleus.

- FUNGI are made up of thread-like hyphae. They reproduce by spores.

- PLANTS make their food by photosynthesis.

    a Algae are simple plants with few special organs. Most of them live in water.

    b Liverworts and mosses are land plants. Mosses have stems and leaves.

    c Ferns have well-developed stems, leaves and roots. They reproduce by spores.

    d Seed-bearing plants reproduce by seeds.

        i Conifers have no flowers; their seeds are not enclosed in an ovary.

        ii Flowering plants have flowers; their seeds are in an ovary which forms a fruit. Monocots have one cotyledon in the seed; dicots have two cotyledons in the seed.

- ANIMALS get their food by eating plants or other animals.

    a Coelenterates live mostly in the sea. They have tentacles and sting cells.

    b Flatworms have flat bodies; some have cilia. They live in fresh water or are parasitic.

    c Annelids have tubular bodies divided into segments.

    d Molluscs have an external shell and a creeping or holding 'foot'.

    e Arthropods have a hard exoskeleton and jointed legs.

        i Crustacea mostly live in water and have more than three pairs of legs.

        ii Insects mostly live on land and have wings and three pairs of legs.

        iii Arachnids have four pairs of legs and poisonous mouth parts.

        iv Myriapods have many pairs of legs.

    f Vertebrates have a spinal column and skull.

        i Fish have gills, fins and scales.

        ii Amphibia can breathe in air or in water.

        iii Reptiles are land animals; they lay eggs with leathery shells.

        iv Birds have feathers, beaks and wings; they are 'warm-blooded'.

        v Mammals have fur, and suckle their young; the young develop inside the mother.

30 MICRO-ORGANISMS

The term 'micro-organism' includes viruses, bacteria, protoctista and some fungi and algae. Most of these organisms are, indeed, microscopic.

Bacteria and protoctista are single-celled organisms. Fungi and algae are multicellular. (That means, their bodies are made up of many cells.) Viruses do not have a cellular structure.

'Micro-organisms' is a convenient term by which to refer to a wide variety of fairly simple organisms. But the word is not used in classifications, in the way we use words like 'mammal' or 'vertebrate'.

Bacteria

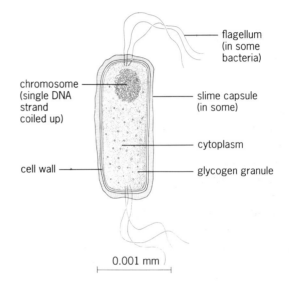

Figure 1 Generalized diagram of a bacterium

flagellum (in some bacteria)
chromosome (single DNA strand coiled up)
slime capsule (in some)
cytoplasm
cell wall
glycogen granule
0.001 mm

Bacterial structure

Bacteria (singular = bacterium) are very small organisms consisting of single cells rarely more than 0.01 mm in length. They can be seen only with the higher powers of the microscope.

Their cell walls are made, not of cellulose, but of a complex mixture of proteins, sugars and lipids. Some bacteria have a **slime capsule** outside their cell wall. Inside the cell wall is the cytoplasm which may contain granules of glycogen, lipid and other food reserves (Fig. 1).

Each bacterial cell contains a single chromosome, consisting of a circular strand of DNA (p. 202). The chromosome is not enclosed in a nuclear membrane but is coiled up to occupy part of the cell (Fig. 2).

Individual bacteria may be spherical, rod-shaped or spiral and some have filaments, called **flagella**, projecting from them. The flagella can flick, and so move the bacterial cell about.

Bacterial physiology

Nutrition

There are a few species of bacteria which contain a photosynthetic pigment like chlorophyll, and can build up their food by photosynthesis. Most bacteria, however, live in or on their food. They

Figure 2 Longitudinal section through a bacterium (× 27,000). The light areas are coiled DNA strands. There are three of them because the bacterium is about to divide twice (see Fig. 3).

299

produce and release enzymes which digest the food outside the cell. The liquid products of digestion are then absorbed back into the bacterial cell.

Respiration

The bacteria which need oxygen for their respiration are called **aerobic bacteria**. Those which do not need oxygen for respiration are called **anaerobic bacteria**. The bacteria used in the filter beds of sewage plants (p. 257) are aerobic, but those used to digest sewage sludge and produce methane, are anaerobic.

Reproduction

Bacteria reproduce by cell division or **fission**. Any bacterial cell can divide into two and each daughter cell becomes an independent bacterium (Fig. 3). In some cases, this cell division can take place every 20 minutes so that, in a very short time, a large colony of bacteria can be produced. This is one reason why a small number of bacteria can seriously contaminate our food products.

This kind of reproduction, without the formation of gametes, is called **asexual reproduction** (p. 321).

(a) bacterial cell

(b) chromosome replicates

(c) cell divides

(d) each cell divides again

Figure 3 Bacterium reproducing. This is asexual reproduction by cell division.

Effect of heat

Bacteria, like any other living organisms, are killed by high temperatures. The process of cooking destroys any bacteria in food, provided high enough temperatures are used. If drinking water is boiled, any bacteria present are killed.

However, some bacteria can produce spores which are resistant to heat. When the cooked food or boiled water cools down, the spores germinate to produce new colonies of bacteria, particularly if the food is left in a warm place for many hours. For this reason, cooked food should be eaten at once or immediately refrigerated. (Refrigeration slows down bacterial growth and reproduction.) After refrigeration, food should not merely be warmed up but either eaten cold or heated to a temperature high enough to kill any bacteria that have grown, i.e. to 90 °C or more (p. 101).

Useful and harmful bacteria

When people talk about bacteria, they are usually thinking about those which cause disease or spoil our food. In fact, only a tiny minority of bacteria are harmful. Most of them are harmless or extremely useful.

Bacteria which feed saprotrophically (p. 309), bring about decay. They secrete enzymes into dead organic matter and liquefy it. This may be a nuisance if the organic matter is our food but, in most cases, it consists of the excreta and dead bodies of organisms. If it were not for the activities of the decay bacteria (and fungi), we should be buried in ever-increasing layers of dead vegetation and animal bodies.

The decay bacteria also release essential elements from the dead remains. For example, proteins are broken down to ammonia and the ammonia is turned into nitrates by nitrifying bacteria (p. 237). The nitrates are taken up from the soil by plants, which use them to build up their proteins. In a similar way, sulphur, phosphorus, iron, magnesium and all the elements essential to living organisms are recycled in the course of bacterial decomposition.

Humans exploit bacterial physiology in the course of **biotechnology** (p. 340).

The bacteria which cause disease are parasites (p. 309). They live in the cells of plants or animals and feed on the cytoplasm. Parasitic organisms which cause disease are called **pathogens** (Fig. 4). The organism in which they live and reproduce is called the **host**.

Pathogenic bacteria (p. 347) may cause diseases because of the damage they do to the host's cells, but most bacteria also produce poisonous waste products called **toxins**. The toxin produced by the *Clostridium* bacteria (which cause tetanus) is so poisonous that as little as 0.00023 g is fatal.

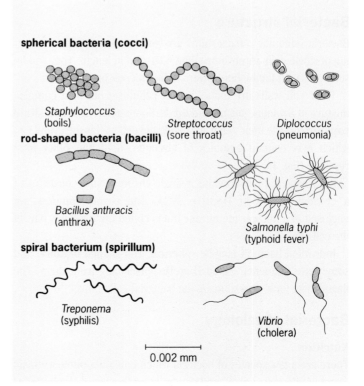

spherical bacteria (cocci)

Staphylococcus (boils)

Streptococcus (sore throat)

Diplococcus (pneumonia)

rod-shaped bacteria (bacilli)

Bacillus anthracis (anthrax)

Salmonella typhi (typhoid fever)

spiral bacterium (spirillum)

Treponema (syphilis)

Vibrio (cholera)

0.002 mm

Figure 4 Some pathogenic bacteria

Viruses

Most viruses are very much smaller than bacteria and can be seen only with the electron microscope at magnifications of about ×30 000.

Virus structure

There are many different types of virus and they vary in their shape and structure. All viruses, however, have a central core of RNA or DNA (p. 202) surrounded by a protein coat. Viruses have no nucleus, cytoplasm, cell organelles, or cell membrane, though some forms have a membrane outside their protein coats.

Virus particles, therefore, are not cells. They do not feed, respire, excrete or grow and it is debatable whether they can be classed as living organisms. Viruses do reproduce, but only inside the cells of living organisms, using materials provided by the host cell.

A generalized virus particle is shown in Fig. 5. The nucleic acid core is a coiled single strand of RNA. The coat is made up of regularly packed protein units called **capsomeres** each containing many protein molecules. The protein coat is called a **capsid**. Outside the capsid, in the influenza and some other viruses, is an envelope which is probably derived from the cell membrane of the host cell (Fig. 6).

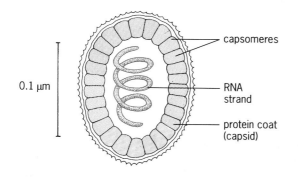

0.1 µm

capsomeres

RNA strand

protein coat (capsid)

Figure 5 Generalized structure of a virus

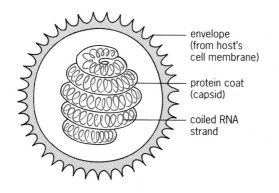

envelope (from host's cell membrane)

protein coat (capsid)

coiled RNA strand

Figure 6 Structure of the influenza virus

Multiplication of viruses

Viruses can survive outside the host cell, but in order to reproduce they must penetrate into a living cell. How they do this, in many cases, is not known for certain. In most instances, the virus particle first sticks to the cell membrane. It may then 'inject' its DNA or RNA into the cell's cytoplasm or the whole virus may be taken in by a kind of endocytosis (p. 29).

Once inside the host cell, the virus is 'uncoated', i.e. its capsid is dispersed, exposing its DNA or RNA. The DNA or RNA then takes over the host cell's physiology. It arrests the normal syntheses in the cell and makes the cell produce new viral DNA or RNA and new capsomeres. The nucleic acid and the capsomeres are assembled in the cell to make new virus particles which escape from the cell (Fig. 7).

(a) The virus sticks to the membrane of a suitable host cell.

(b) The virus enters the cell, the protein coat breaks down and releases the DNA or RNA.

(c) The viral DNA or RNA replicates and directs the host cell to make new protein coats.

(d) The new viruses escape from the host cell.

Figure 7 Reproduction of a virus

The cell may be destroyed in this process or the viruses may escape, wrapping themselves in pieces of the host's cell membrane as they do so. These activities give rise to the signs and symptoms of disease.

Many viruses cause diseases in plants and animals. Human virus diseases include the common cold, poliomyelitis, measles, mumps, chickenpox, herpes, rubella, influenza and AIDS (p. 349).

Tobacco mosaic virus affects tomato plants as well as tobacco, but the virus which causes the striking patterns (called 'breaks') in tulip petals does not seem to harm the plant.

Viruses may be used as vectors to 'deliver' recombinant DNA in genetic engineering (p. 206).

1 Which of the following structures are present in both bacterial cells and plant cells: cytoplasm, cellulose, DNA, cell wall, nucleus, chromosome, vacuole, glycogen?

2 If 5 bacteria landed in some food and reproduced at the maximum possible rate, what would be the population of bacteria after 4 hours?

3 **a** Why is a virus particle not considered to be a cell?
b Why are viruses not easy to classify as living organisms?

4 Why do you think that commercial growers of tomatoes forbid smoking in their greenhouses?

5 How does the reproduction of a virus differ from that of a bacterium?

Fungi

The kingdom of the fungi includes fairly familiar organisms such as mushrooms, toadstools, puffballs, and the bracket fungi that grow on tree-trunks. There are also the less obvious, but very important, mould fungi which grow on stale bread, cheese, fruit or other food. Many of the mould fungi live in the soil or in dead wood. The yeasts are single-celled fungi similar to the moulds in some respects.

Some fungal species are parasites and live in other organisms, particularly plants, where they cause diseases which can affect crop plants.

Structure

Many fungi are not made up of cells but of microscopic threads called **hyphae**. The branching hyphae spread through the material on which the fungus is growing, and absorb food from it. The network of hyphae that grow over or through the food material is called the **mycelium** (Fig. 8).

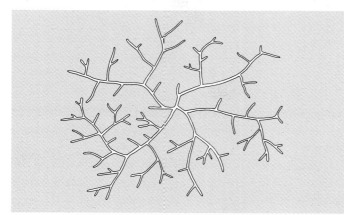

Figure 8 The branching hyphae form a mycelium

Mushrooms and toadstools are the reproductive structures, 'fruiting bodies', of an extensive mycelium that spreads through the soil or the dead wood on which the fungus is growing.

The hyphae are like microscopic tubes lined with cytoplasm. In the centre of the older hyphae there is a vacuole, and the cytoplasm contains organelles and inclusions (p. 3). The inclusions may be lipid droplets or granules of glycogen but, unlike plants, there are no chloroplasts or starch grains (Fig. 9).

The hyphal wall may contain cellulose or chitin or both, according to the species. Chitin is similar to cellulose but the chitin molecule contains nitrogen atoms.

In some species of fungi, there are incomplete cross-walls dividing the hyphae into cell-like regions, but the cytoplasm is free to flow through large pores in these walls. In the species which do have cross-walls (septa), there may be one, two or more nuclei in each compartment. In the species without septa, the nuclei are distributed throughout the cytoplasm.

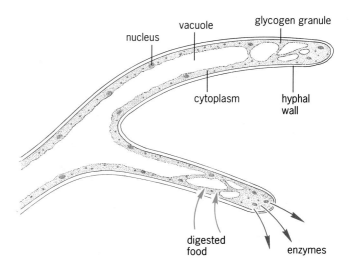

Figure 9 Structure of a fungal hyphae

Figure 10 Toadstools growing on a fallen tree. The toadstools are the reproductive structures which produce spores. The feeding hyphae are inside the tree, digesting the wood.

Nutrition

Saprotrophic fungi

Most fungi are saprotrophs (p. 309) living on dead organic matter (Fig. 10).

The hyphae secrete enzymes into the organic material and digest it to liquid products. The digested products are then absorbed back into the hyphae and used for energy or for the production of new protoplasm or hyphal walls.

The type of enzymes produced by the hyphae will depend on the species of fungus and the material it is growing on. Those species which produce a **cellulase**, i.e. an enzyme which digests cellulose, are important in helping to break down plant remains (Fig. 11). Saprotrophic fungi and bacteria are the 'decomposers' in most food webs (p. 234), and are largely responsible for the recycling of essential nutrients in any ecosystem (p. 265).

Figure 11 Saprotrophic bacteria and fungi have started to rot some of the apples. The saprotrophs gained access via wasp damage.

Parasitic fungi

The hyphae of parasitic fungi penetrate the tissues of their host plant and digest the cells and their contents. If the mycelium spreads extensively through the host, it usually causes the death of the plant. The bracket fungus shown in Fig. 12 is the fruiting body of a mycelium that is spreading through the tree and will eventually kill it.

Fungus diseases such as blight, mildews or rusts (Fig. 3, p. 285) are responsible for causing considerable losses to arable farmers, and there is a constant search for new varieties of crop plants which are resistant to fungus disease, and for new chemicals (fungicides) to kill parasitic fungi without harming the host.

A few parasitic fungi cause diseases in animals, including humans. One group of these fungi cause tinea or ringworm. The fungus grows in the epidermis of the skin and causes irritation and inflammation. One form of tinea is athlete's foot, in which the skin between the toes becomes infected.

Tinea is very easily spread by contact with infected towels or clothing, but can usually be cured quickly with a fungicidal ointment.

Figure 12 A parasitic fungus. The 'brackets' are the reproductive structures. The mycelium in the trunk will eventually kill the tree.

Reproduction

Fungi reproduce by releasing microscopic, single-celled **spores**. You can see a cloud of these spores if you squeeze a mature puffball (Fig. 13).

The spores are budded off from the tips of special hyphae. Each spore contains a little cytoplasm and one or more nuclei, depending on the species. The spores are dispersed in air currents or by other methods. When a spore lands on suitable organic matter or on a new host, it germinates to produce a mycelium.

The spores are produced from a single mycelium without the involvement of gametes or any sexual process, and so this is an example of **asexual reproduction** (p. 321).

Most fungi, however, do have a sexual process in their life cycle and this may precede the production of asexual spores.

Figure 13 Puffball dispersing spores. When a rain drop hits the ripe puffball, a cloud of spores is ejected.

Penicillium

Penicillium is a genus of mould fungi that grow on decaying vegetable matter, damp leather and citrus fruits. The mycelium grows over the food, digesting it and absorbing nutrients. Vertical hyphae grow from the mycelium and, at their tips, produce chains of spores (Figs 14 and 15). These give the colony a blue-green colour and a powdery appearance (Fig. 13, p. 239).

The spores are dispersed by air currents and, if they reach a suitable substrate, grow into a new mycelium.

There are many species of *Penicillium*. *P. notatum* was the species which led to the discovery of penicillin when Alexander Fleming, in 1928, observed that the mould suppressed bacterial growth in one of his culture plates. *P. camemberti* and *P. roqueforti* are used in the preparation of cheeses. *P. griseofulvin* is used to produce an antibiotic effective against fungal skin diseases such as athlete's foot.

Figure 14 *Penicillium sp.*

Figure 15 Scanning electron micrograph of *Penicillium* spores

QUESTIONS

6 When something goes mouldy what is actually happening to it?

7 Suggest why bread, wood and leather may go mouldy, while glass and plastic do not.

8 Suggest why toadstools may be found growing in very dark areas of woodland where green plants cannot flourish.

Yeast

The yeasts are a rather unusual family of fungi. Only a few of the several species can form true hyphae. The majority of them consist of separate, spherical cells, which can be seen only under the microscope. They live in situations where sugar is likely to be available, e.g. in the nectar of flowers or on the surface of fruits.

The thin cell wall encloses the cytoplasm, which contains a nucleus and a vacuole. In the cytoplasm are granules of glycogen and other food reserves (Fig. 16(a)).

The cells reproduce by budding. An outgrowth from the cell appears, enlarges and is finally cut off as an independent cell. When budding occurs rapidly, the individuals do not separate at once and, as a result, small groups of attached cells may sometimes be seen (Fig 16(b)).

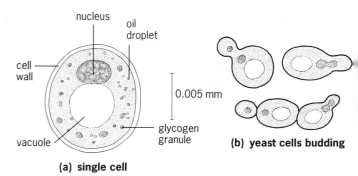

(a) single cell   (b) yeast cells budding

Figure 16 Yeast

Fermentation

Yeasts are of economic importance in promoting alcoholic fermentation. Yeast cells contain many enzymes, some of which can break down sugar into carbon dioxide and alcohol. This chemical change provides energy for the yeast cells to use in their vital processes.

$$C_6H_{12}O_6 \rightarrow 2CO_2 + 2C_2H_5OH + 118 \text{ kJ}$$
$$\text{alcohol}$$

Alcoholic fermentation is a form of anaerobic respiration (p. 21) but, if the yeast is supplied with carbohydrates other than sugar, the yeast needs oxygen to convert these substances to sugar first.

In brewing, barley is allowed to germinate. During germination, the barley grains convert their starch reserves into maltose (p. 13). The germinating barley is then killed by heat and the sugars are dissolved out with water. Yeast is added to this solution and ferments it to carbon dioxide and alcohol.

In making beer, hops are added to give the brew a bitter flavour and the liquid is sealed into casks or bottles so that the carbon dioxide is under pressure. When the pressure is released, the dissolved carbon dioxide escapes from the liquid giving it a 'fizz'.

In making spirits such as whisky, the fermentation is allowed to go on longer and the alcohol is distilled off. Wine is made by extracting the juice from fruit, usually grapes, and allowing the yeasts, which live on the surface of the fruit, to ferment the sugar to alcohol.

If too much oxygen is admitted, or fungi and bacteria are allowed to enter, the alcohol may be oxidized to ethanoic (acetic) acid, i.e. vinegar.

Culturing micro-organisms

Most micro-organisms will grow under controlled conditions in the laboratory. By carefully controlling the composition of the culture medium, its temperature and pH it is possible to encourage the growth of some organisms while suppressing others. For example, bacteria grow readily on an alkaline medium; fungi prefer an acid medium. Most micro-organisms need a culture medium which contains an organic carbon source such as glucose, and some need an organic source of nitrogen (e.g. amino acids) as well. Others will grow if supplied with potassium nitrate as their sole source of nitrogen.

In genetic engineering, culture media containing antibiotics are used to isolate bacteria carrying recombinant DNA (p. 206).

The culture medium can either be a liquid or a solid. If it is a solid, it is based on a jelly made from **agar**. Agar is a jelly derived from seaweed. Bacteria cannot grow on it unless nutrients are added. This makes it possible to be selective as described above.

When fungi and bacteria grow on agar, they reproduce rapidly and form visible colonies (Fig. 17). If the culture plate was inoculated with only a few organisms, each colony will have grown from a single individual. By subculturing from one of these colonies, it is possible to produce a pure strain of the micro-organism.

If bacteria are cultured in a liquid medium, their rapid reproduction soon causes the liquid to take on a cloudy appearance. If this liquid bacterial culture is repeatedly diluted, the concentration of bacteria can be reduced to a point where pouring a little of the culture onto an agar plate will produce isolated colonies, each derived from a single bacterium. If the dilution is known, the population of bacteria in the original culture can be calculated.

Viruses reproduce only in living cells so they cannot be cultured on non-living media. One technique is to inoculate them into living hens' eggs (Fig. 18); another is to use tissue cultures of mammalian cells (p. 9).

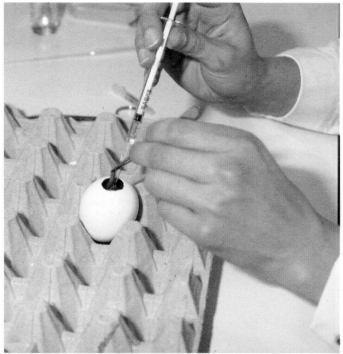

Figure 18 Culturing viruses in hens' eggs

Aseptic techniques

Fungal spores and bacteria are everywhere, in the air, on surfaces, on clothing and the skin. If you want to culture one particular species of micro-organism, it is important to ensure that all other species are excluded. This means using **aseptic techniques**. The glassware and the culture media must be sterilized, and precautions must be taken to avoid allowing bacteria from the air or from the experimenter to contaminate the experiments.

The glassware and media are sterilized by heating them to temperatures which destroy all bacteria and their spores. Contamination is avoided by handling the apparatus in a way which reduces the chances of unwanted bacteria entering the experiments.

Figure 17 Micro-organisms from the air. The nutrient agar was exposed to the air for a few minutes. Subsequently, these bacterial (and one fungus) colonies grew. Each colony has developed from a single micro-organism.

PRACTICAL WORK

1 Culturing bacteria

The glassware must be sterilized by super-heated steam in an autoclave or pressure cooker for 15 minutes at 1 kg/cm^2 so that the unwanted bacteria on the glassware are destroyed.

1.5 g agar is stirred into 100 cm^3 of hot distilled water and 1 g beef extract, 0.2 g yeast extract, 1 g peptone and 0.5 g sodium chloride added. The mixture is sterilized in an autoclave and poured into sterile petri dishes which are covered at once and allowed to cool.

Precautions Culture methods can give rise to very dense colonies of potentially harmful bacteria. **All bacteria should be treated as if they were harmful.** When the petri dishes have been inoculated with the bacterial samples, the lids should be sealed in place with adhesive tape and not removed until the plates have been sterilized again at the end of the experiment. Any colonies which appear must be examined with the lids still on the dishes.

Inoculating the plates If a little cooked potato or other vegetable matter is allowed to rot for a few days in water, bacteria will arrive and grow in the liquid. A drop of this liquid is picked up in a sterile wire loop[1] and streaked lightly across the surface of the agar jelly, lifting the lid of the dish just far enough to admit the wire loop, but not enough to let bacteria from the air fall onto the agar (Fig. 19). The lid is then sealed on the dish with adhesive tape, and the dish is kept, upside-down[2], in an incubator or similar warm place for about 2 days. A 'control' dish is left unopened.

Figure 19 Lift the lid as little as possible

Bacterial colonies should grow on the streaked dish, following the path taken by the loop (Fig. 20). There should be no growth of bacteria in the control dish, showing that the bacteria in the experiment came from the inoculating liquid and not from the glassware or the medium.

All the dishes, still sealed, are then sterilized once more in the autoclave before they are washed up.

2 The effect of antibiotics

The culture medium is made as before. When it is cool, but before it sets, a few drops of a pure bacteria culture[3] are added and thoroughly mixed in with the medium. The medium is then poured into sterile petri dishes as before and allowed to set.

Figure 20 A 'streak plate'. Notice how the bacterial colonies have grown where the plate was streaked.

Using sterile forceps, some discs containing antibiotics are placed on the surface of the agar and the lids sealed on. The dishes are then incubated at about 35 °C for 24 hours.

The bacteria, which are dispersed in the agar, will grow and make the medium look cloudy, but in regions surrounding some of the antibiotic discs, bacterial growth will have been suppressed and the agar will look clear (Fig. 21).

If the bacteria are able to grow right up to the edge of a disc, it means that the antibiotic in that disc is unable to suppress the growth of the species of bacteria used in the trial.

1 The wire loop is sterilized by heating it to redness in a Bunsen flame.
2 If water condenses in the petri dish, it will fall on the lid and not spread over the surface of the agar.
3 e.g. *Escherichia coli* or *Staphylococcus albus* from a reputable supplier.

Figure 21 Testing antibiotics. Antibiotics have diffused out from the discs and suppressed the growth of bacteria.

QUESTION

9 Suggest reasons for the variation in extent of the clear areas round the disc in Fig. 21.

3 Growing mould fungi

Moisten a piece of stale bread, place it in a petri dish base and leave it exposed to the air. After a day, moisten it again if it has dried out, and then cover it with a beaker or small jar. In the humid atmosphere inside the jar, colonies of mould fungi should grow on the bread within a few days.

The experiment can be duplicated by using different food material, e.g. pieces of banana or cooked apple.

4 Culturing fungi

Mould fungi or even small toadstools can be cultured on dishes of agar as described for bacteria on p. 306. Follow the same techniques as described for bacteria, but use vegetable agar in place of nutrient agar. The vegetable agar is made simply by dissolving 2 g agar and 25 cm^3 tomato juice in 100 cm^3 water. The vegetable agar and petri dishes are sterilized in an autoclave. The cool, but still liquid agar is poured into the sterile dishes and allowed to set.

The dishes are then exposed to the air or inoculated with fungus spores or hyphae using a sterile wire loop as shown in Fig. 19. The fungi may be selected from one of the colonies that grew in Experiment 3.

The dishes are incubated for 24 hours at about 35 °C or left at room temperature for 2 or 3 days.

5 Starch digestion by a mould fungus

Prepare starch-agar by dissolving 0.3 g starch powder and 1 g agar in 100 cm^3 water. If the experiment can be examined 1 or 2 days after setting up, there is no need to sterilize the agar or the glassware.

Pour the agar, when cool, into petri dishes, replace the lids and allow the medium to set.

Inoculate the agar with fungal spores or hyphae, by making one or two strokes across it with a sterile wire loop that has been dipped into a mature fungal colony, as described for Experiment 3. Replace the lid and leave the dish for 1 or 2 days.

After this time, the fungus should be seen growing along the lines of the streaks.

Remove the lid, pour iodine solution into the dish to cover the agar and leave for a minute or two. Finally wash away the iodine with water.

Iodine solution will turn starch-agar blue. So, any areas which remain clear have no starch in them and it is reasonable to assume that the fungus has secreted an enzyme into the agar and digested the starch.

A control should be set up at the same time as the experiment, by streaking a starch-agar plate with a wire loop that has been heated to redness in a Bunsen flame. This plate should also be tested with iodine solution.

For experiments on yeast, see pp. 25 and 345.

Checklist

- Bacteria are single cells; they have a cell wall, cytoplasm and a single chromosome.
- Bacteria produce enzymes which digest the surrounding medium.
- Bacteria reproduce by cell division.
- Bacteria are killed by heat but their spores survive.
- Most bacteria are saprotrophs and help to bring about decay.
- Bacterial decay releases essential substances for recycling.
- Viruses are smaller than bacteria and cannot, strictly, be classed as living organisms.
- Each virus particle consists of a DNA or RNA core enclosed in a protein coat.
- Viruses can reproduce only inside a living cell.
- Viruses take over the host cell's physiology and make it produce new virus particles.
- Viruses and bacteria can cause disease.
- Tetanus, pneumonia, typhoid and syphilis are bacterial diseases.
- Influenza, colds, herpes and AIDS are viral diseases.
- Fungi are formed from thread-like hyphae rather than cells.
- The branching hyphae produce a network called a mycelium.
- The hyphae have walls, cytoplasm, nuclei, vacuoles, organelles and inclusions.
- Fungi secrete enzymes into their food and absorb the digested products.
- Saprotrophic fungi digest dead organic matter.
- Parasitic fungi digest living tissues.
- The saprotrophic bacteria are important as decomposers; they release essential nutrients from dead organic matter.
- Parasitic fungi cause plant diseases some of which affect our crops.
- Fungi reproduce asexually by releasing spores which can grow into new mycelia.
- Yeasts are single-celled fungi, which are important in brewing, baking and other forms of biotechnology.
- Yeasts respire anaerobically, causing a fermentation process which produces CO_2 and alcohol.

Characteristics of living organisms

All living organisms, whether they are single-celled, many-celled, plants or animals, do the following things:

Feed   They may take in solid food as animals do, or digest it first and absorb it later like fungi do, or build it up for themselves like plants do.

Breathe   They take in oxygen and give out carbon dioxide. This exchange of gases takes place between the organism and the air or between the organism and water. The oxygen is used for respiration.

Respire   That is, they break down food to obtain energy. Most organisms need oxygen for this.

Excrete   Respiration and other chemical changes in the cells produce waste products such as carbon dioxide. Living organisms expel these substances from their bodies in various ways.

Grow   Bacteria and single-celled creatures increase in size. Many-celled organisms increase the numbers of cells in their bodies, become more complicated and change their shape as well as increasing in size.

Reproduce   Single-celled organisms and bacteria may simply keep dividing into two. Many-celled plants and animals may produce sexually or non-sexually.

Respond   The whole animal or parts of plants respond to stimuli.

Move   Most single-celled creatures and animals move about as a whole. Fungi and plants may make movements with parts of their bodies.

The next five chapters describe some of these characteristics in different living things.

31 FEEDING

All living organisms need food. Some of this food is used to make new tissues for growth and replacement but most of it is used to provide energy (see p. 20). Energy is obtained from food by breaking it down chemically to carbon dioxide and water.

Types of nutrition

There are two principle methods of obtaining food, called **autotrophic** and **heterotrophic** nutrition.

Autotrophic organisms (**autotrophs**) can build up all the organic molecules they need from simple inorganic substances. Green plants are autotrophs. By the process of photosynthesis, they can build up large, complex molecules (carbohydrates, lipids and proteins) from small inorganic molecules (carbon dioxide, water, nitrates, sulphates and other mineral salts, p. 37).

Some bacteria and protoctista are also autotrophs. In any ecosystem, it is the autotrophs which are the producers (p. 233).

Heterotrophic organisms (**heterotrophs**) use ready-made organic compounds as their food source. These organic compounds will have been made, originally, by autotrophs (mainly plants). The heterotrophs digest the organic compounds to simpler substances and absorb the products into their bodies. Animals, fungi and some bacteria and protoctista are heterotrophs.

Animals take in food, in the form of complex organic molecules (carbohydrates, lipids and proteins, p. 11) and, by the process of digestion, break them down into simpler compounds (e.g. glucose and amino acids, p. 106) which can be absorbed.

Fungi and heterotrophic bacteria also digest their food source but do so by a process of external digestion.

In plants and other autotrophs the process of photosynthesis is much the same in all of them. Among the heterotrophs there is a wide variety of methods of obtaining food.

Saprotrophs

These are organisms which feed on dead and decaying matter. Often they release digestive enzymes into their food and absorb the soluble products back into their bodies. Bacteria and fungi are examples of saprotrophs which digest dead wood, rotting vegetation or the humus in the soil (see p. 235).

Parasites

A parasite is an organism which derives its food from another organism, called the **host**, while the host is still alive. **Ectoparasites** live, for most of their life cycle, on the surface of their host: **endoparasites** live inside the host.

A flea is an ectoparasite which lives in the fur or feathers of mammals or birds and sucks blood from the skin. An aphid (greenfly) is a parasite on plants and sucks food from the veins in their leaves or stems. A tapeworm is an endoparasite living in the intestine of a vertebrate host and absorbing the host's digested food. In some cases the host may be weakened by the presence of the parasite, or its metabolism may be upset by the parasite's excretory products. In many cases, however, the host appears to suffer no serious disadvantage. Often, this will depend on how many parasites the host is carrying.

Protoctista

Some protoctista have chloroplasts (p. 40) and make their food in a similar way to plants. *Euglena* (p. 284) is an example. Others, such as *Amoeba*, *Paramecium* and *Vorticella* (p. 284) take in solid food and digest it. The cilia on *Vorticella* and *Paramecium* create water currents which sweep small organisms, such as bacteria, towards the protoctistan. The food is then ingested into the cytoplasm as a food vacuole.

Amoeba has no cilia but ingests its food by 'flowing' round it (Figs 1 and 2) and trapping it in a food vacuole. The cytoplasm surrounding the food vacuole releases enzymes into it. These enzymes digest the food, and the soluble products are absorbed back into the cytoplasm. This method of feeding is similar to that of animals, i.e. ingestion, digestion, absorption.

Animals

Animals feed either on plants or on other animals and there is a great variety of ways in which they obtain their food.

Filter-feeding

Many quite unrelated animals which live in water filter the water and ingest the particles which they extract from it.

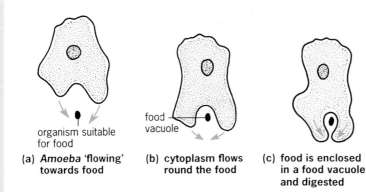

(a) *Amoeba* 'flowing' towards food

(b) cytoplasm flows round the food

(c) food is enclosed in a food vacuole and digested

organism suitable for food

food vacuole

Figure 1 *Amoeba* ingesting food

Figure 2 *Amoeba* feeding

The mosquito larva shown in Fig. 21 on p. 273 flicks its 'mouth brushes' and makes a current of water from which it extracts bacteria and other small organisms.

The mussel (Figs 3 and 4) has large gills covered with cilia. The beat of the cilia draws a current of water in between the shells and forces it through the gills. The gills filter out any food particles and the filtered water is expelled through the siphon. The cilia on the gills also carry the trapped food particles forward to the mouth.

These two creatures stay more or less in the same place and create a water current, but a great many animals swim through the water and filter it as they go.

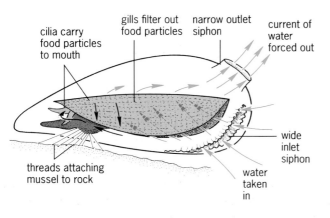

cilia carry food particles to mouth

gills filter out food particles

narrow outlet siphon

current of water forced out

wide inlet siphon

water taken in

threads attaching mussel to rock

Figure 3 Filter-feeding in a mussel

Figure 4 Mussels filter-feeding. The water is going in through the opening fringed with 'tentacles' and coming out of the smooth-edged opening. If anything touches the tentacles, the mussel closes its shells at once. Some of the mussels have barnacles growing on them.

As the herring swims along, it takes in water through its mouth and lets it out through the gill covers (see Fig. 15 on p. 292). This flow of water is partly for gaseous exchange as described on p. 316, but the water is filtered by gill rakers (Fig. 6, p. 316). These form a bony grid between the gills which allows water to pass out but traps small shrimps and other crustacea which are swimming in the surface waters of the sea.

Certain types of whale, enormous though they are, filter-feed in a similar way.

Insects

Insects have a great variety of feeding methods. Grasshoppers, locusts and caterpillars have simple jaws (Fig. 5(a)) with which they bite off pieces of leaf small enough to be ingested.

The butterfly's mouth parts form a long, fine tube which reaches into flowers and sucks the nectar (Fig. 5(b)).

The greenfly also has tubular mouth parts but these are used to pierce through leaves and stems of plants and suck food from the phloem.

The tubular mouthparts of the female mosquito are adapted for piercing the skin and sucking blood from the skin capillaries (Figs. 5(d) and (e)).

The feeding method of the housefly is described on p. 352.

Mammals

Mammals may be classed as **carnivores** (flesh-eaters), **herbivores** (plant-eaters) or **omnivores** (mixed diet).

A carnivore's teeth are used to catch and kill its prey, cut flesh from the carcase, crack bones and reduce the food to sizes suitable for swallowing. The position and shapes of its teeth are adapted to these varied functions as shown in Fig. 6.

A herbivore's teeth are adapted to the single function of collecting vegetation and grinding it up as small as possible. Sheep and cows have no upper incisors but grip the grass between the lower incisors and the gum. The canines are similar in shape and function to the incisors. There is a gap between the canines and the premolars which permits the tongue to manipulate the food between the back teeth as they slide sideways over each other and

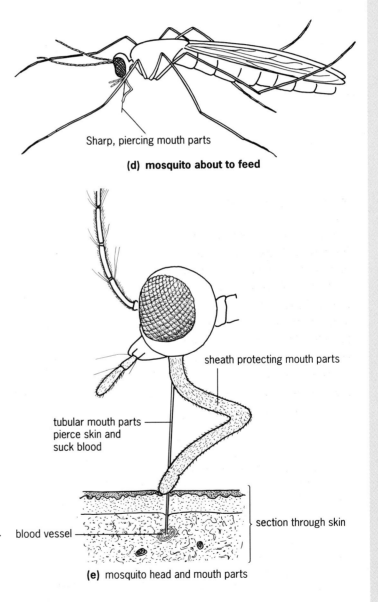

(a) Mouth parts of an insect, such as a grasshopper, which bites off pieces of leaf.

gullet

mouth

jaws cut off pieces of leaf

these mouth parts push the food between the jaws

(b) butterfly feeding
The mouth parts form a tube, like a fine drinking straw, which probes into the flower to suck the nectar.

Sharp, piercing mouth parts

(d) mosquito about to feed

sheath protecting mouth parts

tubular mouth parts pierce skin and suck blood

blood vessel

section through skin

(e) mosquito head and mouth parts

(c) larva of the great diving beetle
This larva lives in fresh water and spears small fish or tadpoles with its jaws. It will change into a water beetle like the one in Fig. 23 on p. 273. This one is breathing air at the surface.

Figure 5 Insects: some feeding methods

grind it up (Fig. 7). The premolars and molars are similar to each other. Throughout the life of the animal they grow as fast as they are worn down by their grinding action.

Many omnivores have teeth similar to those of carnivores but usually with smaller canines and premolars. Pigs and humans are omnivores. Human dentition is shown on p. 168.

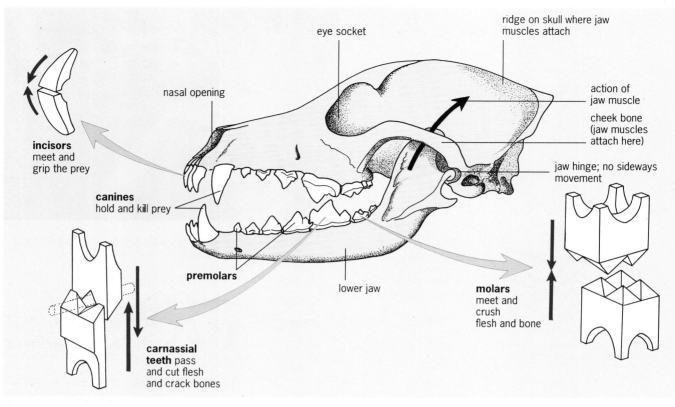

incisors
meet and
grip the prey

nasal opening

eye socket

ridge on skull where jaw
muscles attach

action of
jaw muscle

cheek bone
(jaw muscles
attach here)

canines
hold and kill prey

jaw hinge; no sideways
movement

premolars

lower jaw

molars
meet and
crush
flesh and bone

carnassial
teeth pass
and cut flesh
and crack bones

Figure 6 Dog's skull and action of teeth

hinge joint allows sideways movement

toothless gap

pad of
gum

incisor

molars

premolars

(a) sheep's skull

pad of gum

incisor

these teeth
slide past
each other

premolars

movement
of lower jaw

**(b) section through herbivore's skull to show action
of teeth** (after J. Maynard Smith)

(c) sheep's teeth
The tops of the molars on the left have been worn away to
expose the brown dentine, but those on the right have not
grown far enough out of the jaw to meet the top teeth.

sharp enamel
ridges

the molars slide
over each other
and grind the food

(d) herbivore's molars
The grass cells are torn open between the sharp
ridges of enamel.

Figure 7 Herbivore's teeth

Symbiosis

The term 'symbiosis' strictly refers to two unrelated organisms living more or less permanently together, irrespective of whether one of them is a parasite. More commonly, however, it implies that both organisms derive some benefit from the association.

For example, in the stomachs of cattle and sheep there live large numbers of bacteria. They cause no symptoms of illness and they are thought to be of value to the animal because they help to digest the cellulose in its food. The cow benefits from the relationship because it is better able to digest grass. The bacteria are thought to benefit by having an abundant supply of food, though they are themselves digested when the grass moves along the cow's digestive tract.

The nitrogen-fixing bacteria in root nodules (Fig. 8) provide a further example of symbiosis. The plant benefits from the extra nitrates that the bacteria provide, while the bacteria are protected in the plant's cells and can also use the sugars made by the plant's photosynthesis.

It is often difficult to produce good evidence that one or both organisms are getting some benefit or suffering damage from these relationships. Consequently, it is not always easy to make a distinction between parasitism and symbiosis. The argument is about words rather than biology. It is far more important to try and find out what the relationship between organisms really is rather than to argue about what word to apply to the relationship.

QUESTIONS

1 What type of nutrition is characteristic of **a** an apple tree, **b** a toadstool, **c** a human, **d** a mosquito, **e** a *Streptococcus*?
2 Apart from carbon dioxide, water and salts, what do plants need in order to make their food?
3 In animals, where does digestion and absorption take place?

Checklist

- All living organisms need to obtain food to make new tissue and to provide energy.
- All green plants make their food by photosynthesis.
- Saprophytes feed on decaying organic matter.
- Parasites live in or on another organism and obtain food from it.
- Animals eat plants or other animals. They ingest, digest and absorb their food.
- There is a great variety of ways in which animals obtain their food, such as filter-feeding, chasing and catching prey or chewing up vegetation.
- The teeth and alimentary canals of animals are adapted to deal with the kind of food they eat.

Figure 8 Root nodules of white clover – a leguminous plant

32 BREATHING

In order to get energy from their food, animals and plants have to break it down to carbon dioxide and water by the chemical process called respiration (p. 20). One form of respiration, aerobic respiration, uses oxygen for this process. The oxygen is obtained by gaseous exchange (p. 134) with the air or water around the organism. The same process gets rid of carbon dioxide produced by respiration.

In some animals, gaseous exchange is helped by a process of ventilation (p. 132) but many plants and animals depend on diffusion alone.

Very small organisms, like bacteria and protoctista, have no need of ventilation. The distances over which the oxygen and carbon dioxide have to travel in these creatures are so small that diffusion is fast enough to meet their needs (Fig. 1).

Larger animals need special organs for gaseous exchange and some method of ventilating these organs.

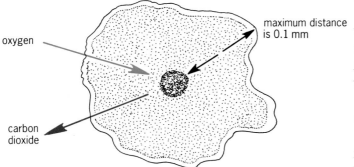

Figure 1 The distance is so small in *Amoeba* that diffusion is a rapid enough process for the cell's needs

Surface area and volume

The ratio of surface area to volume in a small organism is greater than the ratio in a large organism.

As a crude example, imagine an organism that has a volume of 8 cm³ and a surface area of 24 cm² (Fig. 2(a)). The ratio of surface area to volume is 24/8 = 3. A smaller organism, having half the volume (4 cm³) has a surface area of 16 cm². Its surface area to volume ratio is 16/4 = 4 (Fig. 2(b)).

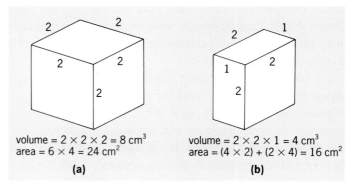

volume = 2 × 2 × 2 = 8 cm³
area = 6 × 4 = 24 cm²
(a)

volume = 2 × 2 × 1 = 4 cm³
area = (4 × 2) + (2 × 4) = 16 cm²
(b)

Figure 2 Surface area and volume

The demand for oxygen depends on the volume of active cytoplasm but the supply of oxygen depends on the exposed surface area. Consequently smaller organisms have the advantage of a relatively large absorbing surface to meet the oxygen requirements of a comparatively small volume of respiring tissue.

In larger organisms, not only is the ratio of surface area to volume low, but also the skin is not very permeable to gases and there is a large distance between the skin and the internal organs. Absorption through the skin plus diffusion through the tissues would not be rapid enough to meet the demands of the living cells.

In these organisms, a blood circulatory system largely replaces diffusion as the means of delivering oxygen to the tissues.

Some comparatively large organisms still rely on their skin as a means of absorbing oxygen. Earthworms and frogs are examples, though the frog has lungs which can supplement the supply of oxygen when necessary (Figs 3 and 4). In both cases, the thin, moist skin is supplied with a network of capillaries which absorb oxygen and deliver it, via the circulatory system, to the rest of the body (Fig. 5). Earthworms are fairly inactive animals, so their oxygen requirements are easily met by the relatively small absorbing surface of the skin.

Active animals have special absorbing surfaces such as lungs or gills. These are internal organs but they still provide a very large absorbing surface. In the gills of a fish, this is achieved by a large number of branched filaments. In mammals, the surface area of the lungs is greatly increased by thousands of tiny air pockets (alveoli, p. 131).

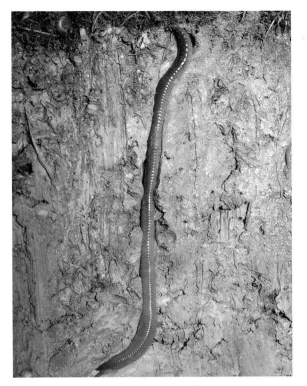

Figure 3 Earthworm in its burrow. The thin, moist skin of the earthworm makes it vulnerable to water loss. It spends most of the day in its burrow where evaporation is at a minimum.

Figure 4 Frog. The skin can exchange gases with either the air or water. Frogs spend most of their life on land but their moist skins restrict them to damp, shady habitats.

All respiratory surfaces have a dense network of capillaries to absorb oxygen and get rid of carbon dioxide. Steep diffusion gradients (p. 29) for oxygen and carbon dioxide are maintained by ventilation mechanisms which constantly exchange the air or water in contact with the respiratory surface.

QUESTIONS

1 What would be the surface area to volume ratio in an organism half the size of the one in Fig. 2(b)?
2 What would be the maximum diffusion distance for oxygen in the *Salmonella typhi* illustrated in Fig. 4 on p. 300?
3 In frogs, gaseous exchange takes place continually through the skin. In what conditions would you expect a frog to use its lungs?
4 Why do large, active animals need special breathing organs and a circulatory system?

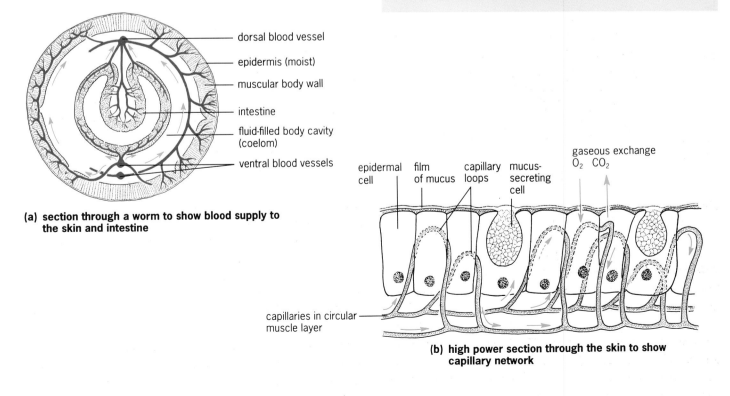

(a) section through a worm to show blood supply to the skin and intestine

(b) high power section through the skin to show capillary network

Figure 5 Gaseous exchange in an earthworm

branches with capillaries in them

tips of filaments greatly magnified

Figure 6 Herring gills. (The gill cover has been cut away to show the gills.)

Fish

Although the water molecule (H_2O) contains oxygen, this cannot be used by aquatic organisms. The oxygen they breathe comes from the atmosphere and is dissolved in the water.

Fish absorb dissolved oxygen from the water by means of gills. The herring has four gills on each side of the head underneath a bony gill cover (Fig. 15 on p. 292). Each gill consists of a curved bar with gill filaments sticking out from one side and gill rakers projecting forwards from the other side (Fig. 6). The gill filaments are branched structures with blood capillaries running in them. The branched filaments present a very large surface to the water for absorbing oxygen into the blood and getting rid of carbon dioxide.

By movements of the floor of the mouth and the gill covers, a fish takes water in through the mouth, and forces it between the gills and out through a gap between the gill cover and its body (Fig. 7). This method of ventilation changes the water in contact with the gills, so bringing fresh supplies of oxygen and carrying away carbon dioxide.

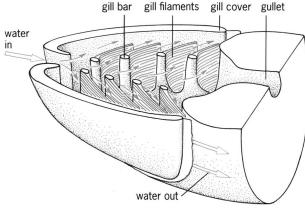

Figure 7 Diagram to show ventilation of gills

The lungs of mammals and gills of fish both show the basic requirements of an efficient structure for gaseous exchange. These structures need a large surface area. The gills achieve this by having hundreds of filaments, with many branches on each filament. In the lungs, it is the millions of alveoli (p. 131) which give a large surface area.

In both gills and lungs, there is a dense network of capillaries which enable rapid exchange of oxygen and carbon dioxide between the blood and the water or air. The capillaries are separated from the water or air by only a thin epithelium (single layer of cells), so that diffusion of the gases is as rapid as possible.

Methods of breathing in other aquatic animals are discussed on p. 274.

Gaseous exchange in flowering plants

Plants do not have special breathing organs. Because they do not move about, like animals, they do not use up oxygen or produce carbon dioxide very rapidly. Their leaves present a very large surface area to the air, and diffusion of oxygen and carbon dioxide through the stomata into the intercellular spaces (p. 55) is fast enough for the respiration going on in their cells. Because most plant leaves are thin, the distances for diffusion are also very short (see also pp. 43 and 55).

In daylight, some of the oxygen produced by photosynthesis is used in respiration and the carbon dioxide released from respiration is used by photosynthesis (see p. 42). In darkness, only respiration is going on and so carbon dioxide passes out of the leaf and oxygen diffuses in.

Plant stems exchange gases with the air through their stomata or structures called lenticels. Roots use oxygen dissolved in the soil water which they absorb.

QUESTIONS

5 In what way might the method of ventilation of a fish's gills be considered more efficient than the ventilation of lungs? (See 'Residual volume' on p. 133.)

6 Ventilation of gills and lungs requires muscular activity. Where does this muscular activity take place in **a** mammals, **b** fish?

7 In daylight there may be no detectable gaseous exchange in a plant leaf. Explain why this might be so.

Figure 8 Scanning electronmicrograph of lower surface of a birch leaf. The epidermis has been broken open, revealing the spongy mesophyll. Notice the extensive air spaces.

Checklist

- Diffusion is rapid enough to supply microscopic organisms with the oxygen they need.

- The ratio of surface area to volume decreases with increasing size of an organism.

- Respiratory organs and blood circulation overcome the disadvantage of this low ratio in large animals.

- All breathing organs have a large surface, thin epithelium and many capillaries.

- Ventilation maintains a favourable diffusion gradient at a respiratory surface.

- Frogs and earthworms breathe through their skin.

- Fish force water over their gills, which absorb the dissolved oxygen.

- Plants exchange carbon dioxide and oxygen with the air by diffusion through their stems, roots and leaves.

33 REPRODUCTION

Sexual reproduction
Plants and animals, external and internal fertilization, parental care.

Asexual reproduction
Animals, fungi, flowering plants.

Sexual and asexual reproduction compared

No organism can live for ever, but part of it lives on in its offspring. Offspring are produced by the process of reproduction. This process may be **sexual** or **asexual** (see below), but in either case it results in the continuation of the species.

Sexual reproduction

The following statements apply equally to plants and animals. Sexual reproduction involves the production of sex cells. These sex cells are called **gametes** and they are made in reproductive organs. The process of cell division which produces the gametes is called **meiosis** and is described on p. 201. In sexual reproduction, the male and female gametes have to come together and **fuse**, that is, their cytoplasm and nuclei join together to form a single cell called a **zygote**. The zygote then grows into a new individual (Fig. 1 on p. 150).

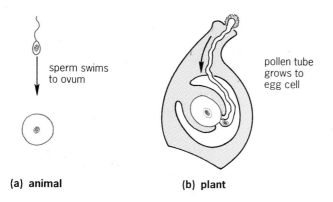

sperm swims to ovum

pollen tube grows to egg cell

(a) animal   **(b) plant**

Figure 1 The male gamete is small and mobile; the female gamete is larger

Sexual reproduction in flowering plants

The male gamete is a cell in the pollen grain. The female gamete is an egg cell in the ovule. The process which brings the male gamete within reach of the female gamete (i.e. from stamen to stigma) is called **pollination**. The pollen grain grows a microscopic tube which carries the male gamete the last few millimetres to reach the female gamete for fertilization. The zygote then grows to form the seed. These processes are all described in more detail on pp. 71–82.

Sexual reproduction in animals

The male gamete is the sperm; the female gamete is the ovum. The release of the sperms and ova is co-ordinated in such a way that they come close to each other. This may involve behaviour patterns or mating or both. The sperms then swim to the ovum and fertilize it. The zygote grows into an embryo. Sexual reproduction in humans is described more fully on pp. 150–161.

In both plants and animals, the male gamete is microscopic in size and mobile (i.e. can move from one place to another). The sperms swim to the ovum and the pollen cell moves down the pollen tube (Fig. 1). The female gametes are always larger than the male gametes and are not mobile. Pollination in the seed-bearing plants and mating in most animals are the processes which bring the male and female gametes close together. Fertilization is the fusion of the gametes, and as a result the zygote undergoes rapid cell division to produce an embryo.

External fertilization

In many animals which live or breed in water, the female sheds her unfertilized eggs into the water and the male releases sperms over them. The eggs are fertilized outside the body of the female and so the process is described as external fertilization. For fertilization to be successful, the eggs and sperms must be released at the same time and close to each other. This usually results from a behaviour pattern in which the male and female are first attracted towards each other and then stimulate each other to produce gametes.

The male stickleback, in the breeding season, develops a red belly and blue eyes (Fig. 2). These colour changes seem to keep

Figure 2 Male three-spined stickleback showing the breeding colour.

318

other males at bay. The male digs a small hollow at the bottom of the pond and roofs it over with pieces of vegetation. When a mature female enters the male's territory, she is attracted by his bright colours. The male's next behaviour is triggered by the sight of her abdomen, swollen with eggs. He swims towards her and then down to the nest. She follows him and enters the nest (Figs 3 and 4). The male taps the female's tail with his snout and this stimulates her to lay eggs. When she leaves the nest, the male enters and sheds sperms on the eggs. The nest and the courtship behaviour ensure that eggs and sperms are shed in the same place and at the same time.

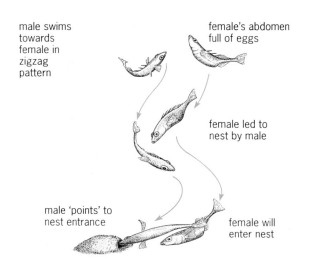

male swims towards female in zigzag pattern

female's abdomen full of eggs

female led to nest by male

male 'points' to nest entrance

female will enter nest

Figure 3 Courtship in the three-spined stickleback

Figure 4 Sticklebacks at the nest. The male's movements in the nest entrance induce the female to enter and lay eggs. Notice her abdomen swollen with eggs.

Frogs and toads, in the spring, migrate to ponds. Here the male climbs on the back of the female and is carried about by her till she lays her eggs (Fig. 5). When the male feels the eggs being laid, he releases his sperms which fertilize the eggs. This is external fertilization, but the male's behaviour in remaining attached to the female makes sure that the sperms are released at the same time as the eggs.

Figure 5 Frogs pairing. The male clings to the female's back and releases his sperm as she lays the eggs.

Parental care

Parental care is the way in which animals look after their offspring. In British frogs and toads there is no parental care; both parents leave the eggs to develop entirely on their own. The male stickleback, on the other hand, guards the developing eggs by driving off intruders, and fans water through the nest with his fins to improve the oxygen supply.

In many birds and mammals, the parental care is even greater (see p. 157). The young are reared in a nest, fed and kept warm usually by both parents (Fig. 6).

Figure 6 Parental care in birds. The song thrush has brought a beakful of worms to feed the chicks in the nest. The bright colour on the inside of their beaks stimulates the parents to put food into them.

Where there is little parental care, there is usually a large number of eggs (Fig. 7). Because the parents do not protect the eggs and young or bring them food, many offspring will die. The large number of eggs, however, ensures that some offspring will survive to maturity. The parental care shown by birds and mammals gives the young a better chance of survival and they have small numbers of offspring.

Figure 7 Large numbers of offspring. One method of survival is to produce a large number of eggs, as in this egg-rope of the perch. Many of the young fish will die but enough will survive to continue the species.

Internal fertilization

In reptiles, birds and mammals, the eggs are fertilized by the male placing sperms inside the body of the female. This is internal fertilization and it requires a behaviour pattern which brings male and female together at the right time so that sperms are released when the eggs are mature. The first step is **pair formation**, which ensures that the male and female stay together at least for the breeding period. There is usually a courtship pattern which brings the pair together in the first place, stimulates them to mate, and keeps them together while the young are reared (Fig. 8). In birds, the male often feeds the female as part of the courtship ritual.

Figure 8 Gannets at nest site. The beak "fencing" ritual helps to maintain the pair bond.

The reproductive organs have to develop in such a way that internal fertilization can occur. This usually means that the male animal has developed a penis which can be inserted into the female's reproductive organs in order to deposit the sperms.

Internal fertilization is essential for animals living on land. External fertilization here is impossible because the eggs and

sperms would dry up when exposed to the air. In reptiles and birds, the egg is prevented from drying up by having a shell, so the sperms have to get to the egg before the shell is put on it. This means that internal fertilization is necessary. In mammals, the egg develops inside the body of the female after internal fertilization.

Other land-dwelling animals such as insects, spiders and snails rely on methods of internal fertilization for their reproduction.

Figure 9 Blackbird's nest with eggs. Compare with Fig. 7. Parental care ensures that a higher proportion of the offspring survive.

QUESTIONS

1 Draw up a table with three columns as shown below. In the first column write **a** male reproductive organs, **b** female reproductive organs, **c** male gamete, **d** female gamete, **e** place where fertilization occurs, **f** zygote grows into.

Now complete the other two columns.

	flowering plants	mammals
a Male reproductive organs		
b female reproductive organs		
c male gamete etc.		

2 Which of the following animals would you expect to have external, and which internal fertilization: butterfly, mussel, trout, sparrow, earthworm? In each case give the reason for your decision.
3 Why do you think courtship behaviour is necessary for the success of both internal and external fertilization?

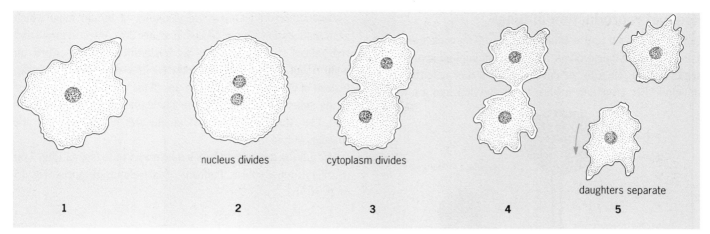

Figure 10 Asexual reproduction in *Amoeba*. This is called 'binary fission'. See 'Bacteria' (p. 300) and 'Yeast' (p. 304) for other examples of binary fission.

Asexual reproduction

Asexual means 'without sex' and this method of reproduction does not involve gametes. In the single-celled protoctista or in bacteria, the cell simply divides into two and each new cell becomes an independent organism (Fig. 10, Fig. 11 and Fig. 3 on p. 300). In more complex organisms, part of the body may grow and develop into a separate individual. For example, a small piece of stem planted in the soil may form roots and grow into a complete plant.

Asexual reproduction in animals

Some species of invertebrate animals are able to reproduce asexually.

Hydra is a small animal, 5–10 mm long, which lives in ponds attached to pondweed. It traps small animals with its tentacles, swallows and digests them. *Hydra* reproduces sexually by releasing its male and female gametes into the water but it also has an asexual method which is shown in Fig. 12.

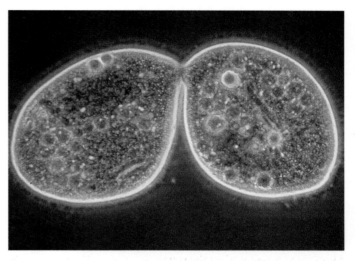

Figure 11 Asexual reproduction. Binary fission in *Paramecium*.

(e)

Figure 12 Asexual reproduction in *Hydra*
(a) A group of cells on the column start dividing rapidly and produce a bulge
(b) The bulge develops tentacles
(c) The daughter hydra pulls itself off the parent
(d) The daughter becomes an independent animal
(e) Hydra with bud

Asexual reproduction in fungi

Fungi have sexual and asexual methods of reproduction. In the asexual method they produce single-celled, haploid spores. These are dispersed, often by air currents and, if they reach a suitable situation, they grow new hyphae which develop into a mycelium.

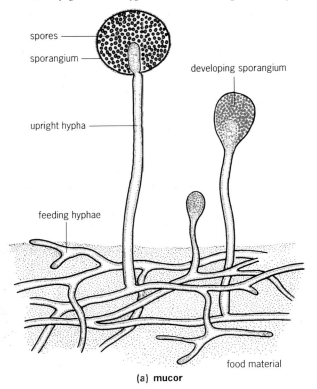

(a) **mucor**

Penicillium and *Mucor* are examples of mould fungi which grow on decaying food or vegetable matter. *Mucor* feeds, grows and reproduces in a similar way to *Penicillium* (as described on p. 304) but the spores are produced in a slightly different way. Instead of chains of spores at the tips of the vertical hyphae, *Mucor* forms spherical sporangia, each containing hundreds of spores (Fig. 13). These are dispersed on the feet of insects or by the splashes of rain drops.

The gills on the underside of a mushroom or toadstool (Fig. 7 on p. 235) produce spores. Puffballs release clouds of spores (Fig. 13 on p. 303).

Asexual reproduction in flowering plants (vegetative propagation)

Although all flowering plants reproduce sexually (that is why they have flowers), many of them also have asexual methods.

In most cases a lateral bud on the stem, instead of producing a flower or a leafy shoot, gives rise to a complete plant which eventually becomes independent of its parent.

Several of these asexual methods (also called 'vegetative propagation') are described on p. 78. An unusual method is shown in Fig. 14.

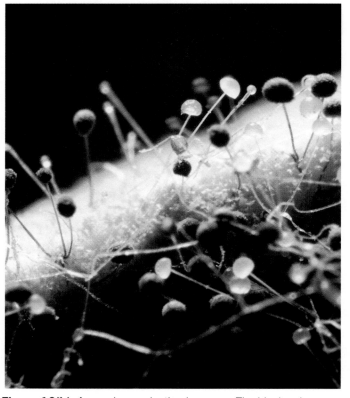

Figure 13(b) Asexual reproduction in mucor. The black spheres are sporangia which have not yet discharged their spores.

Figure 14 Bryophyllum. The plantlets are produced from the leaf margin. When they fall to the soil below, they grow into independent plants.

Sexual and asexual reproduction compared

The relative advantages of asexual and sexual reproduction are discussed on p. 81 in the context of flowering plants. However, the points made there are equally applicable to most forms of sexual and asexual reproduction.

In asexual reproduction, the offspring are usually produced by mitosis (p. 198), mostly in diploid cells. In sexual reproduction the offspring result from the fusion of two haploid cells (gametes) produced by meiosis.

The outcome is that the offspring in asexual reproduction inherit identical sets of genes and there is no variation between them or their parent. This has the advantage of preserving the 'good' characteristics of a successful species from generation to generation. The disadvantage is that there is no variability for natural selection (p. 222) to act on in the process of evolution.

The new combinations of genes which result from meiosis and fertilization (p. 220) produce variety among the offspring. This has the disadvantage that some combinations will produce less successful individuals. On the other hand, there are likely to be some more successful combinations that have greater survival value or produce individuals which can thrive in new or changing environments.

In agriculture and horticulture, asexual reproduction is exploited to preserve desirable qualities in crops; sexual reproduction is exploited to produce new varieties of animals and plants by cross-breeding.

Checklist

- Sexual reproduction involves the male and female sex cells (gametes) joining together.
- The male gamete is small and mobile. The female gamete is larger and not often mobile.
- The male gamete of an animal is a sperm. The male gamete of a flowering plant is the pollen nucleus.
- The female gamete of an animal is an ovum. The female gamete of a flowering plant is an egg cell in an ovule.
- Fish and frogs have external fertilization. The sperms are placed on the eggs after they are laid.
- Reptiles, birds and mammals have internal fertilization. The sperms are placed in the female's body to fertilize the eggs.
- Asexual reproduction does not involve gametes.
- Fungi can reproduce asexually by single-celled spores.
- Many flowering plants reproduce asexually by vegetative propagation.
- Sexual reproduction can produce new varieties of plants and animals.
- Asexual reproduction keeps the characteristics of the organism the same from one generation to the next.

QUESTIONS

4 a In what ways does asexual reproduction in *Mucor* differ from asexual reproduction in flowering plants?
b How does a spore differ from a seed? (See p. 83 for seed structure.)

5 A gardener finds a new and attractive plant produced as a result of a chance mutation (p. 205). Should she attempt to produce more of the same plant by self-pollination or by vegetative propagation? Explain your reasoning.

6 A farmer wants to breed pigs with less fat and longer backs (more lean rashers). Discuss some of the factors he will have to consider in deciding whether this will be cost-effective.

7 Which of the following do not play a part in asexual reproduction: mitosis, gametes, meiosis, cell division, chromosomes, zygote?

8 Revise pp. 78–80 and then say how we exploit the process of asexual reproduction in plants.

9 A fish may lay hundreds of eggs; a bird may lay only five or six. Despite this difference in the number of eggs the numbers of birds and fish do not change much. Suggest reasons for this.

34 GROWTH AND CHANGE

Growth	Metamorphosis in insects
Cell division and enlargement, change in shape.	Complete and incomplete metamorphosis.
	Measurement of growth

Growth

Most living organisms start their lives as a single cell (zygote, p. 150), too small to be seen with the naked eye. This cell divides many times (by mitosis, p. 198) to produce an organism made up of thousands or millions of cells. Not only does this cell division eventually increase the size of the organism, it also makes it far more complicated than it was to start with. The new cells become specialized and form tissues and organs (p. 6). So growth involves an increase in size and mass, and also an increase in complexity.

Cell division on its own does not always produce growth. The fertilized frog's egg in Fig. 1(a) divides into hundreds of cells but the embryo in Fig. 1(g) is no bigger than the single cell from which it came. It consists of hundreds of tinier cells. As the cells continue to divide, the tissues they produce fold or roll up to form organs. They tuck in at the front to form a mouth and alimentary canal. Along the top, they roll up to form a tube which will later become the spinal cord and brain (Fig. 2).

In the early stages of growth, all the cells are able to divide. In the later stages, the dividing cells are restricted to certain regions, e.g. root tips (p. 58), buds (p. 60), Malpighian layer of the skin (p. 146). Specialized cells lose their ability to divide.

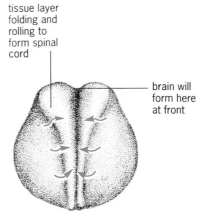

Figure 2 Early frog's embryo seen from above. Soon after stage (g) in Fig. 1 the cells form tissues.

In plants, there are special groups of rapidly dividing cells in the tip of the root or the shoot. Some of these cells, when they stop dividing, start to increase in size because their vacuoles enlarge and stretch the cell walls (Fig. 3). As a result of hundreds of these cells all getting larger, the root grows down into the soil and the shoot grows upwards.

In most organisms, growth is a gradual process, though the rate of growth may vary at different ages. Fig. 4 shows that your increase in weight is rapid for the first three years and slows down in the next two years. Between the ages of 10 and 16 (adolescence), there is an increase in the growth rate, but after the age of 20 you will probably not grow at all.

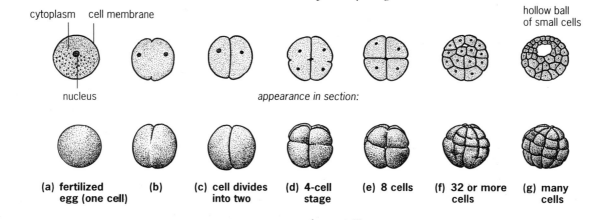

appearance from outside

Figure 1 Cell division in a fertilized frog's egg

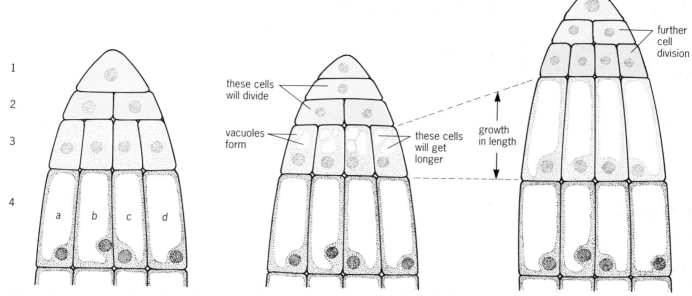

further cell division

these cells will divide

vacuoles form

these cells will get longer

growth in length

Figure 3 Growth by cell division and cell elongation at the tip of a seaweed. The cells divide and expand to make the tip longer and wider. Once cells 4*a*–*d* have formed they cannot divide again but may become specialized, e.g. 4*b* and 4*c* may help carry food up to the growing point.

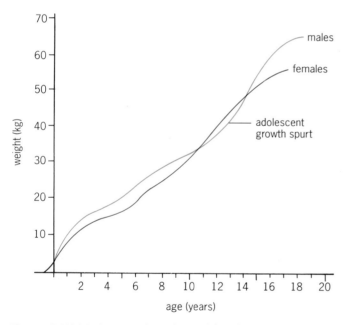

males

females

adolescent growth spurt

Figure 4 Weight increase in males and females

Figure 5 The final moult in a dragonfly

Some fish go on growing all their lives but the growth rate gets slower as they get older.

In the arthropods (p. 290) growth in size takes place in spurts. The hard shell, called the **exoskeleton** or **cuticle**, covering the body of a crustacean like a crab, cannot expand. The crab has to get out of this cuticle before it can grow in size. So the cuticle is partly dissolved and softened from inside, then it splits open, the crab crawls out and allows its body to expand. When the body has expanded, the new cuticle forms and hardens on the outside, and the crab stays this size until the next time it sheds its cuticle.

Although an insect's cuticle is much thinner than the crab's it can not be extended. So insects grow by a series of moults. Fig. 5 shows a dragonfly that has emerged from its last larval cuticle.

Change in shape

Growth involves an increase in size and complexity, but it also results in a change in shape. A tadpole is a very different shape from an adult frog. Also, growth does not take place uniformly in all parts of an organism. In young humans, the head grows relatively little compared with the limbs (Fig. 6).

If our bones were to grow simply by adding cells to the outside, they would become very thick and short (Fig. 7(a)). Fig. 7(b) shows that they grow more rapidly at the ends than they do at the sides. Also, in the process of growth, some parts actually have to be dissolved away and remodelled.

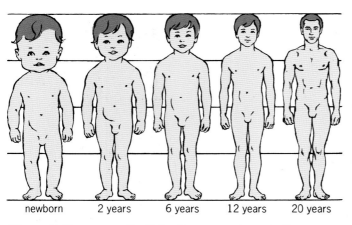

newborn · 2 years · 6 years · 12 years · 20 years

Figure 6 Human growth. All the figures are drawn to the same height to show how the body proportions change with age.

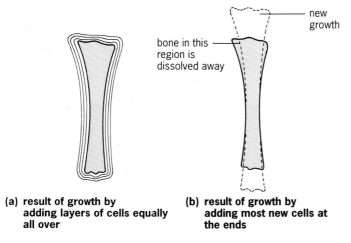

new growth

bone in this region is dissolved away

(a) result of growth by adding layers of cells equally all over

(b) result of growth by adding most new cells at the ends

Figure 7 Growth of limb bones

When a seed germinates to produce a plant, the embryo completely changes its shape and proportions as shown in Fig. 3 on p. 84.

In some organisms, the change in shape at different stages of growth is so extreme that the young form of the creature seems to be totally unlike the adult form. The caterpillar growing into a butterfly or the tadpole into a frog involves drastic changes like this. The process is called **metamorphosis**.

Metamorphosis in insects

Insects shed their cuticles five or six times during growth. After each moult there is an increase in size and sometimes a change in form. The timing and extent of the change depends on whether the insect exhibits complete or incomplete metamorphosis.

Complete metamorphosis

Insects lay eggs which hatch into **larvae**. These larvae are usually quite unlike the adult and are called grubs, maggots or caterpillars according to the species of insect. Generally the larva is the feeding and growing stage, eating voraciously, shedding its cuticle several times and growing rapidly.

When it has reached full size, the larva becomes inactive, neither moving nor feeding, and extensive breakdown and reorganization takes place within its body, giving rise eventually to the adult form. The stage in the insect's life when these changes take place is called the **pupa**. The adults mate and lay eggs.

Butterflies, moths, bees, wasps and flies are examples of insects that undergo complete metamorphosis (Fig. 8).

Incomplete metamorphosis

With insects such as the cockroach, dragonfly or locust, the egg does not hatch into a larva but a **nymph** which, though still very different from the adult, resembles it more closely than a larva does.

The nymph has three pairs of jointed legs, compound eyes and rudimentary wings. At each moult, changes occur which bring it nearer to the adult form. There is no prolonged 'resting' stage as there is in a pupa, though the final moult usually reveals drastic changes that have occurred in the final weeks of the nymph's development (Fig. 9).

In both types of metamorphosis, the larval forms often live in an environment quite different from that of the adult and have different methods of feeding, moving and, in some cases, breathing.

Caterpillars live on vegetation and use their jaws to bite off pieces of leaf. They move by means of three pairs of 'true-legs' and five pairs of 'pro-legs'. Butterflies feed on nectar with their specialized, tubular mouth parts. They have no pro-legs and their principal method of locomotion is by flying.

Some larval forms of insects are aquatic and their methods of breathing differ from those of the adult. Mayfly larvae (Fig. 22a, p. 273) live in fresh water and obtain dissolved oxygen by means of gills on the abdomen. After metamorphosis, the gills are lost and the adult mayfly breathes air.

Measurement of growth

To investigate growth, it is possible to measure increase in length, mass, volume or area. The apparent pattern of growth revealed by these measurements will differ according to which method is used. For example, when a zygote undergoes cell division it does not increase in size or mass; a relatively large single cell simply divides to form a number of smaller cells (Fig. 1).

If growth is defined as increase in the total amount of protoplasm, then the only reliable indicator of growth is the increase of **dry mass** (p. 22). An increase of living mass may be the result of a temporary intake of water by osmosis, which is not necessarily a feature of growth.

However, measuring dry mass involves killing the organism and heating it in an oven at 110 °C until it loses no more weight. This is often neither desirable nor feasible, so one of the other measurements has to be used, but with regard to its possible limitations.

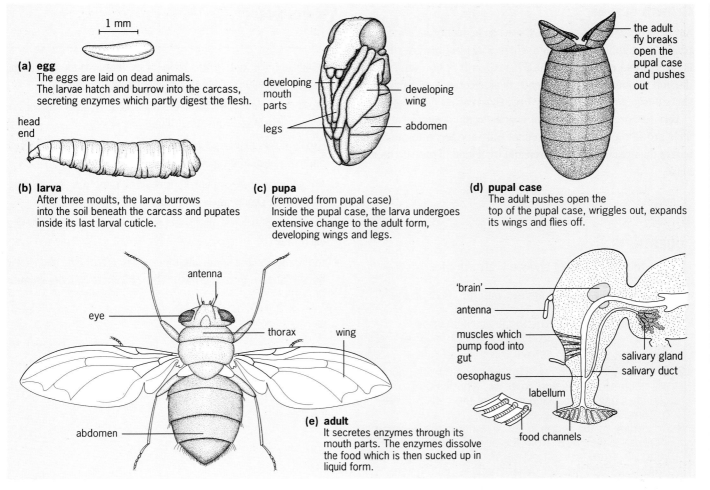

(a) egg
The eggs are laid on dead animals.
The larvae hatch and burrow into the carcass,
secreting enzymes which partly digest the flesh.

head
end

(b) larva
After three moults, the larva burrows
into the soil beneath the carcass and pupates
inside its last larval cuticle.

developing
mouth
parts

developing
wing

legs

abdomen

(c) pupa
(removed from pupal case)
Inside the pupal case, the larva undergoes
extensive change to the adult form,
developing wings and legs.

the adult
fly breaks
open the
pupal case
and pushes
out

(d) pupal case
The adult pushes open the
top of the pupal case, wriggles out, expands
its wings and flies off.

antenna

eye

thorax

wing

abdomen

(e) adult
It secretes enzymes through its
mouth parts. The enzymes dissolve
the food which is then sucked up in
liquid form.

'brain'

antenna

muscles which
pump food into
gut

oesophagus

salivary gland
salivary duct

labellum

food channels

Figure 8 Complete metamorphosis of the blowfly

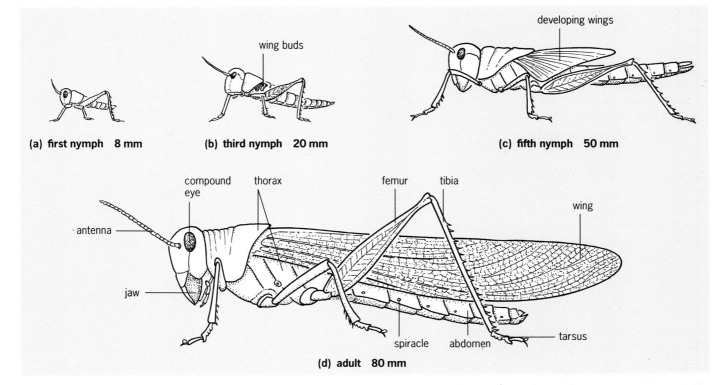

wing buds

developing wings

(a) first nymph 8 mm

(b) third nymph 20 mm

(c) fifth nymph 50 mm

compound
eye

thorax

femur

tibia

wing

antenna

jaw

spiracle

abdomen

tarsus

(d) adult 80 mm

Figure 9 Incomplete metamorphosis of the locust. All stages
feed on green vegetation (mainly grasses and cereals).

Growth curves

Any of the measurements listed above can be plotted against time to produce a graph called a growth curve. The idealized growth curve is the sigmoid curve (Fig. 1, p. 276). Growth is slow at first (cell division without cell enlargement), then becomes rapid and finally slows down and stops when the organism reaches full size.

Very few organisms conform exactly to the sigmoid pattern in their growth, and the growth curve will vary according to which part of the organism is measured and what kind of measurement is made.

QUESTIONS

1 The following figures show the growth of an insect over 200 days.
  a Plot a graph of the figures with the days on the horizontal axis.
  b Say whether you think the graph is sigmoid.

Mass in mg	10	20	40	120	240	400	640	940	950	960
Time in days	20	40	60	80	100	120	140	160	180	200

2 Some of the cells in Fig. 1(g) will become skin, some will become nerves and muscles. What kind of changes must take place in one of the cells for it to become a muscle cell?

3 According to the graph in Fig. 4, what is the average increase in weight **a** from the age of one to five years, **b** from five to ten years?

4 Study Figs 8(b) and (e). What changes have taken place in turning from a maggot into a blowfly?

Checklist

- When animals and plants grow, they increase their size and weight, and their structures become more complicated.
- Most organisms grow from a single cell, by cell division, to become creatures consisting of millions of cells.
- Growth takes place as a result of cell division, cell enlargement and cell specialization.
- Not all parts of an organism grow at the same rate, so the parts of the body change their proportions during growth.
- The arthropods grow in distinct stages, shedding their cuticle at each stage.
- In the insects and some amphibia, there are drastic changes in body structure during growth. This change is called metamorphosis.

35 MOVEMENT AND LOCOMOTION

Protoctista	Vertebrates
Paramecium and *Amoeba*.	Fish, frogs, birds and mammals.
Invertebrates	**Plants**
Earthworms and insects.	

Locomotion refers to the movement of an organism from place to place. **Movement** refers to a change in position of any part of an organism's body, but does not necessarily involve locomotion. Chewing, breathing, heart beat, blinking are all examples of movement in humans.

Most animals and protoctista exhibit locomotion. Even animals such as mussels, barnacles and sea anemones, which remain in one place throughout their lives, have larval stages which are freely mobile.

Mature plants do not exhibit locomotion but the male gametes of algae and bryophytes swim freely during the process of fertilization. Parts of plants do show movement, however. Examples are the tropisms described on p. 88, the movements of insectivorous plants (p.336) and the opening and closing of petals shown on p. 333.

Locomotion gives organisms the advantage of being able to seek food, escape predators, search for mates or move to more congenial conditions. Static (sessile) organisms have to take what comes but they have evolved methods of avoiding some of the hazards. Mussels and barnacles can close their shells when the tide goes out or when predators attack; plants exploit the wind and insects for pollination and their leaves often produce poisonous or unpleasant-tasting compounds which deter predators.

Protoctista

Protoctista such as *Paramecium* are called **ciliates**. They possess cilia over all or part of their unicellular bodies. The cilia flick rhythmically to and fro in a co-ordinated pattern, and propel the ciliate through the water. In the power stroke, each cilium is more or less straight. In the recovery stroke, the cilium returns to its position in such a way as to offer minimal resistance to the water (Fig. 1). Nevertheless, ciliates can reverse the beat of their cilia and move backwards.

Amoeba's method of locomotion is unusual. It has no fixed shape and when a part of its cell membrane bulges out, the cytoplasm flows into it (Fig. 2). If a bulge forms at a different place on the surface, the cytoplasm flows into this one and so *Amoeba* changes direction.

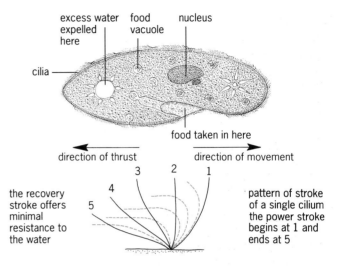

Figure 1 Locomotion in *Paramecium* (a ciliate common in freshwater ponds)

Figure 2 *Amoeba* moving from the left to right. Cytoplasm will flow into one of these 'arms' and the other will be withdrawn.

Invertebrates

There is an enormous variety of methods of locomotion in invertebrate animals. Arthropods have jointed legs for walking, leaping and running, and insects have wings for flying. Free-living flatworms glide over the bottom of ponds, under leaves or even

329

under the surface film using cilia. Slugs and snails have a foot with which they creep over the soil or over plants. Muscular ripples pass down the length of the base of the foot and propel the molluscs forward, usually over a coating of mucus which lubricates the track.

There are many other methods of locomotion, too numerous to list here.

The way in which the limbs of arthropods and insects propel them during walking is shown in Fig. 3.

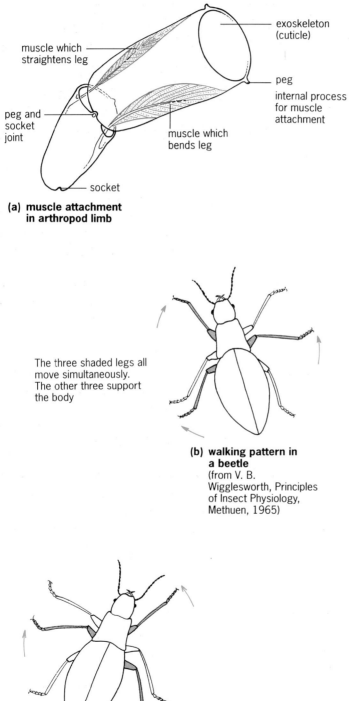

(a) muscle attachment in arthropod limb

The three shaded legs all move simultaneously. The other three support the body

(b) walking pattern in a beetle
(from V. B. Wigglesworth, Principles of Insect Physiology, Methuen, 1965)

Figure 3 Arthropod muscle and movement

Figure 4 Locomotion in the earthworm

How the earthworm effects locomotion without the aid of limbs is shown in Fig. 4. Between the muscular body wall and the intestine is a fluid-filled space called the **coelom** (Fig. 3 (a), p. 315). The body wall contains circular and longitudinal muscle. The inward pressure of the muscle layers against the coelomic fluid maintains the shape of the worm and can also produce movement and locomotion.

When the circular muscles of the front segments contract, it makes them thinner and more elongated so the front end of the worm slides forward. The **chaetae** of the front segments are extended to grip the substratum or the walls of the burrow and the longitudinal muscles start to contract. Behind this region, the next series of segments are elongating and moving forward. In this way, a smooth series of extensions and contractions in successive segments moves the worm forwards.

Vertebrates

Vertebrates move by making their muscles contract and pull on the bones of their skeletons. This process is described more fully on p. 165.

Fish

The external features of a bony fish are shown in Fig. 15 on p. 292. Locomotion is achieved by lateral contractions of the whole of the muscular body with a final thrust by the tail (Fig. 5). The muscles on opposite sides of the body contract alternately to send the tail first to one side and then to the other (Fig. 5(b)). As the tail moves, it pushes sideways and backwards on the water. The sideways movements to left and right cancel each other out, and the backwards thrust of the tail on the water drives the fish forward.

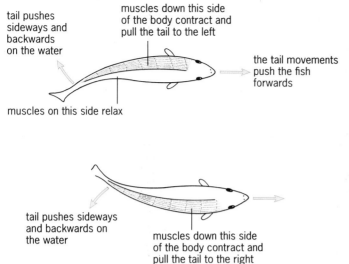

tail pushes sideways and backwards on the water

muscles down this side of the body contract and pull the tail to the left

the tail movements push the fish forwards

muscles on this side relax

tail pushes sideways and backwards on the water

muscles down this side of the body contract and pull the tail to the right

Figure 5 Locomotion in bony fish

The tail fin contributes to the forward propulsion but the other fins, in most fish, serve mainly to steer and stabilize (Fig. 6). The dorsal and ventral fins are called **median fins**; the pectoral and pelvic fins are the **paired fins**.

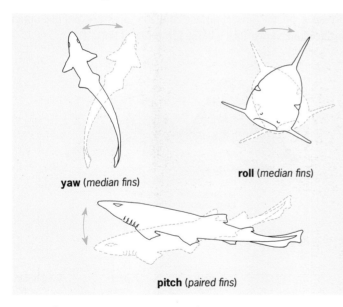

yaw *(median fins)*

roll *(median fins)*

pitch *(paired fins)*

Figure 6 Movements controlled by fins

Figure 7 Frog leaping

The streamlined shape of the fish offers the least possible resistance to forward movement through the water.

The fish may, therefore, be said to be adapted to movement in water by having a streamlined shape, powerful body muscles to move its tail and fins to help it steer.

Many fish also have an air bladder in their bodies which makes them buoyant, so that they don't sink down every time they stop swimming.

Frogs

Frogs' limbs are adapted for locomotion on land and in water. The hind limbs are very long and muscular. The femur, tibia, foot and ankle bones are all elongated and the extensor muscles are large and powerful. When they contract, they straighten the hind legs which thrust against the ground and propel the frog forwards and upwards in its characteristic jump (Figs 7 and 8).

Similar powerful movements of the hind limbs propel it through the water when it swims. The web between the toes provides an increased surface area to thrust against the water.

The front limbs are smaller and the toes are not webbed. They act as shock absorbers when the frog lands from a jump and may help to steer it when swimming (Fig. 8(d)).

(a) at rest; hind legs flexed

(b) hind legs extended, pushing frog forward and upwards

(c) front legs being extended; beginning to draw up the hind legs

(d) front legs being extended to take first shock of landing; hind legs drawn up

Figure 8 Sequence of actions in frog leaping

331

Birds

In mammals, the muscles for locomotion act mainly on the hind limbs. In birds, they act on the forelimbs which are the wings. One pair of muscles pulls the wings down and an antagonistic pair pulls the wings up.

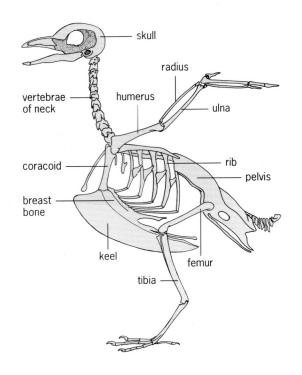

Figure 9 Bird's skeleton

Fig. 9 shows the skeleton of a bird and you will see that the breastbone has a deep keel projecting down. This provides a large surface for the attachment of the powerful flight muscles that act on the wing. The **coracoid** bone acts as a strut between the breastbone and the spine so that when the flight muscles contract, they pull the wing bone down rather than pull the breastbone up. Fig. 10 shows this part of the bird's skeleton as seen from in front. You can see how the flight muscles are attached to the breastbone at one end and the humerus at the other.

When the large flight muscles contract (Fig. 10(a)), they pull the wings down. The small flight muscles, when they contract (Fig. 10(b)), pull the wings up because their tendons run over a groove in each coracoid, like a pulley, and attach to the upper side of the humerus. So the large and small flight muscles are antagonistic to each other. When they contract alternately, they pull the wing up and down.

During the upstroke, the wing is bent at the wrist and so it does not offer much air resistance (Fig. 10(b)). In the downstroke the wing is spread out and pushes downwards on the air. The air resistance causes an upthrust on the wing and so lifts the bird up.

Gliding flight

In gliding flight a bird's wings act as aerofoils (i.e. like aeroplane wings). Air flowing over the curved upper surface moves faster than it does over the lower surface. This effectively produces a high pressure on the lower surface and low pressure on the upper surface and so provides a lifting force (Fig. 11(a)).

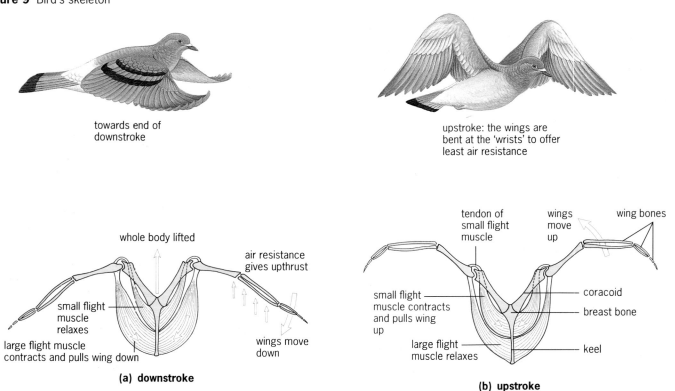

Figure 10 Front view of bird's skeleton to show how the muscles work the wings

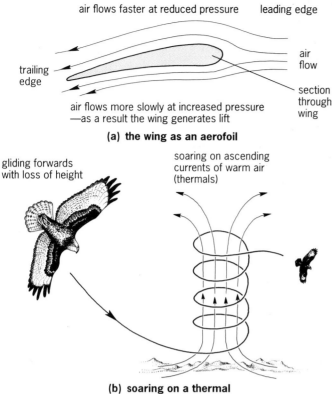

(a) the wing as an aerofoil

air flows faster at reduced pressure

leading edge

trailing edge

air flow

section through wing

air flows more slowly at increased pressure —as a result the wing generates lift

gliding forwards with loss of height

soaring on ascending currents of warm air (thermals)

(b) soaring on a thermal

Figure 11 Gliding flight

With wings outstretched, a bird moves forward and loses height. The lift produced by the air flow over its wings keeps the bird airborne. Most birds glide at some stages in their flight pattern, especially when landing. On upcurrents of air near cliff faces or updraughts caused by thermals, some birds can remain airborne for long periods without flapping their wings (Fig. 11(b)).

Adaptations for flight

The skeleton, limbs and musculature of a bird are modified from the basic vertebrate pattern in the following ways:

a Its front legs are modified to form wings.

b Its flight feathers are light but provide a large air-resistant surface (Fig. 12).

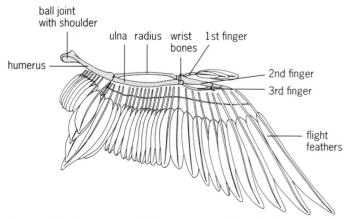

ball joint with shoulder

humerus

ulna   radius   wrist bones   1st finger

2nd finger

3rd finger

flight feathers

Figure 12 Skeleton and flight feathers of a bird's wing

c It has powerful flight muscles.

d The deep keel provides a large surface for attachment of these muscles.

e Its limb bones are hollow and light.

f Its skeleton forms a rigid framework to stand up to the contractions of the flight muscles.

Mammals

Locomotion in most terrestrial mammals is by walking or running in which the four limbs move, usually, in a diagonal pattern, i.e. right forelimb and left hind limb moving at the same time, followed by left forelimb and right hind limb. As the pace quickens, the hind leg is lifted before the diagonal forelimb reaches the ground.

The way the muscles act on the limb bones to produce movement is described for humans on p. 166 and, in general, is applicable to most walking and running mammals.

Mammals' limbs are often adapted to their way of life or to the terrain over which they move, e.g. running, burrowing, climbing (Fig. 13). One example is the horse and its relatives, which are adapted to rapid movement over firm ground such as grassland. In each foot, only the enlarged and extended middle bone is present. The others have been reduced or 'lost' in the course of evolution. Only one toe bone remains, so the horse in effect, walks on the tips of its middle toes. This increases the length of its legs for rapid running.

knee joint

fibula

tibia

ankle joint

middle foot bone

middle toe bones

(a) rabbit

(b) horse

Figure 13 Lower hind limbs of rabbit and horse. The rabbit's leg is fairly characteristic of an unspecialized mammalian limb. The horse's limb shows considerable specialization in the reduction of the number of foot and toe bones. (See also Fig. 15 on p. 227.)

333

Flowering plants

Plants do not show locomotion; they remain in one place for their whole life. They do make movements, however, of parts of their bodies. These movements, such as tropisms (p. 88), are also seen in the opening up of flowers, folding up of leaves at night or climbing plants twining round supports. When these movements are photographed by 'time-lapse' photography and shown speeded up, they reveal plants as active organisms, even though they do not move about as a whole.

Comparatively rapid movements are the 'sleep movements' made by some plants at night or when light intensity falls, e.g.

1 the closing up of the petals of crocuses and tulips
2 the closing of the outer florets of daisies (Fig. 14)
3 the folding of the leaflets of wood sorrel.

(a) daytime:
inflorescences open

(b) evening:
inflorescences closed

Figure 14 'Sleep movements' of daisies

In some cases, the movement is brought about by changes in growth rate. In other cases it is changes in the turgor (p. 32) of groups of cells which bring about the movement.

It is not entirely clear what the advantages of 'sleep movements' are.

Some insectivorous plants make easily visible movements. The sticky tentacles on the leaves of the sundew (Fig. 4, p. 336) can be seen to curl gradually round a trapped insect. The Venus flytrap closes very rapidly when triggered by an insect walking between the leaves (Fig. 15); the leaves of the 'sensitive plant' (*Mimosa pudica*) collapse and droop at once if they are touched.

Figure 15 Venus flytrap with trapped blowfly, which will eventually be digested

QUESTIONS

1 In vertebrates, what systems are directly involved in producing co-ordinated movement?
2 The earthworm is sometimes described as having a 'hydrostatic skeleton'. What do you think this means?
3 In the hind limbs of terrestrial vertebrates, the extensor muscles are usually larger than the flexor muscles. Why do you think this is so?
4 How do birds achieve 'lift' **a** in flapping flight, **b** in gliding flight?

Checklist

- Nearly all living things make movements with parts of their body.
- Protoctista and animals move their whole bodies around (locomotion).
- Some protoctista use cilia to move about.
- Earthworms move by muscular contractions of their body wall acting on their coelomic fluid.
- Arthropods have exoskeletons with muscles attached to the inside of the cuticle.
- A fish is adapted by its shape and muscle arrangement to swim in water.
- Frogs' hind limbs are adapted for jumping and swimming.
- A bird's skeleton and flight muscles are adapted in special ways to produce flight.
- Mammals' limbs may be specialized for running, jumping, digging or climbing.
- Plants make growth movements (tropisms) and move their leaves and flower petals.

36 SENSITIVITY

Sensitivity is the ability of living organisms to respond to stimuli. A **stimulus** is a change in the external or internal environment of an organism.

Animals

An external stimulus might be a touch, a change of light intensity or temperature, a sound or a movement. An internal stimulus might be food arriving in the stomach, a rise in blood pressure or a fall in body temperature.

An organism has the means of detecting these stimuli and responding to them in appropriate ways. In animals, stimuli are detected by structures called **receptors**.

Receptors for external stimuli may be distributed over the surface of an animal (e.g. touch-sensitive bristles on an insect, pressure receptors in the human skin), or they may be concentrated into specialized sensory organs such as the antennae, the eyes or the ears (Figs 1, 2 and 3).

Receptors for internal stimuli may be (a) in the organ itself, e.g. stretch receptors in muscle (p. 182), (b) in a localized part of the system, e.g. blood pressure receptors in the carotid artery, or (c) in the brain, e.g. the receptors in the hypothalamus which monitor the concentration of the blood (p. 188).

The organs which make a response to the stimuli are called **effectors**. An effector could be a muscle which contracts or a gland which produces a secretion.

Figure 2 Gerbils. The whiskers on the face affect nerve endings in the skin. The slightest movement of the whiskers, even those caused by air movements, will cause nerve impulses to be sent to the brain.

Figure 3 Hare. The long ears help to pick up and locate sound vibrations. The eyes at the side of the head give the hare good vision all around.

Figure 1 Grass snake; sense of smell. The flicking tongue picks up chemicals in the air and carries them to a sense organ in the roof of the mouth which 'tastes' them.

335

Plants

Although plants do not respond by moving their whole bodies, parts of them do respond to stimuli. The tropisms described on pp.88–89 are good examples of responses made by the growing parts of shoots or roots to the stimuli of light or gravity. There are no specialized receptors or effectors but the ability to detect and respond to these stimuli may be restricted to certain regions.

Some plants respond to touch. A runner bean shoot that is stimulated by being touched repeatedly on one side for an hour or two will curve towards the stimulus. As a result, the bean shoot twines round a string or a stake as it grows upwards. The insect-eating plant, sundew, has sticky tentacles on its leaves. When the tentacles are touched by an insect walking on the leaf, they curve over the insect, trap it and slowly digest it (Fig. 4).

Figure 4 An insect-eating plant. The sundew plant grows in damp, swampy places and gets part of its food by trapping insects (inset). Once the insect is trapped on the leaf, the sticky tentacles bend towards it, digest it with their enzymes, and then absorb the liquid products of digestion. The insects probably supply the plant with nitrates, which are lacking in the soil.

Co-ordination

For an effective response, there has to be some form of communication between the receptor and the effector.

In animals, this communication is by nerve impulses or hormones. It is described for humans on p. 179.

Plants do not have nervous systems but evidence is accumulating which suggests that electrical signals may be carried from one part of the plant to another. The main system of communication in plants seems to be by means of chemicals called **plant growth substances** or sometimes 'plant hormones'. These substances may be produced in one part of the plant and conducted to another part through the phloem or from cell to cell. Despite their name, these substances control many processes other than growth. Different substances may promote leaf fall, initiate flowering, break dormancy in seeds (p. 87), inhibit growth of buds, stimulate root formation or bring about ripening of fruit.

A simple 'model' of control by a plant growth substance is often offered as an explanation of tropisms. An example is given in Fig. 5. The growth substance (in this case called **auxin**) is thought to accumulate on the lower side of the root tip and inhibit cell extension, so producing a downward growth curvature. In the case of shoots, the auxin is assumed to promote cell extension, resulting in an upward curvature. There is some evidence to support this story but there are also some aspects that have not been explained yet.

There is less evidence in favour of a similar explanation of phototropism, in which the auxin is presumed to be unequally distributed as a result of one-sided illumination.

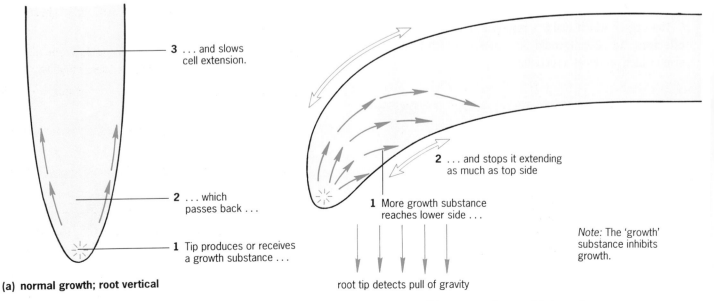

3 . . . and slows cell extension.

2 . . . which passes back . . .

1 Tip produces or receives a growth substance . . .

(a) normal growth; root vertical

2 . . . and stops it extending as much as top side

1 More growth substance reaches lower side . . .

Note: The 'growth' substance inhibits growth.

root tip detects pull of gravity

(b) response to one-sided gravity

Figure 5 Possible explanation of positive geotropism in roots

Use of plant growth substances

Chemicals can be manufactured which closely resemble natural growth substances and may be employed to control various aspects of growth and development of crop plants.

An artificial auxin sprayed onto tomato flowers will induce all of them to produce fruit, whether or not they have been pollinated. Another growth substance, sprayed onto fruit trees, prevents early fruit fall and enables all the fruit to be harvested at the same time.

The weedkiller, 2,4-D, is very similar to one of the auxins. When sprayed on a lawn, it affects the broad-leaved weeds (e.g. daisies and dandelions) but not the grasses. (It is called a 'selective weedkiller'.) Among other effects, it distorts the weeds' growth and speeds up their rate of respiration to the extent that they exhaust their food reserves and die.

Gardeners and horticulturists often dip cuttings into a powder ('rooting hormone') containing a growth substance which initiates the production of roots from the stem (p. 80).

In the production of plants by tissue culture (p. 81) the correct balance of growth substances is necessary to promote root and shoot formation from the undifferentiated tissue.

Survival value of behaviour

Appropriate responses to stimuli are essential to the survival of an organism. A seedling shoot which failed to respond positively to the stimulus of one-sided illumination would eventually be unable to photosynthesize. A male frog which did not release sperms when his partner laid her eggs would not father any offspring.

The chicks of a ground-nesting bird such as a pheasant, respond to their mother's warning cries by 'freezing', i.e. crouching and remaining motionless. They are thus less likely to be seen by a predator. A chick which does not respond in this way is likely to be caught and eaten and, therefore, will not leave any offspring. The chick's inappropriate behaviour has been 'weeded out' by natural selection.

QUESTIONS

1 A fly lands on your hand; you attempt to swat it but the fly gets away. What are the possible stimuli to which the fly responded?
2 Which of your sense organs respond to the stimuli of **a** light, **b** temperature, **c** chemicals?
3 What effectors are involved when you unexpectedly touch a rose thorn, cry out and remove your hand?
4 To what stimuli do wasps appear to be responding when **a** they pester you at a picnic, **b** they try to escape from a room?
5 Why do you think that most animals have their main sense organs on their heads?

PRACTICAL WORK

Use of hormone rooting powder

Cut ten shoots, about 15 cm long, from the new growth of an untrimmed *Lonicera* hedge in June–August. (*Lonicera nitida* is bush honeysuckle; cuttings from flowering currant and laurel also work well but take longer than two weeks.) Remove the leaves from the lower 5 cm of the stems and, with five of the cuttings, dip the exposed stem first into water and then into hormone rooting powder. Tap off the excess powder and push the stems into potting compost in a flowerpot labelled 'A'. Dip the other five stems into water only and push them into compost in a flowerpot labelled 'B'. **Wash your hands to remove any traces of the powder.**

Leave both pots in the same conditions of light and temperature and keep the compost moist for 2 weeks. After this time, remove the cuttings and wash the compost off the stem bases.

Result   Most of the cuttings will have developed roots from the part of the stem in the compost, but the cuttings treated with hormone rooting powder will have many more roots. Some of the untreated cuttings may have failed to develop any roots at all.

Interpretation   The hormone rooting powder contains a chemical similar to an auxin. This chemical has promoted the formation of roots in the stem cuttings.
Note The rooting powder may also contain a fungicide, which makes interpretation of the results less certain.

Checklist

- Living organisms respond to stimuli. They are said to be sensitive.
- Stimuli are such things as touch, heat, cold, and light acting on the organism.
- Animals have sensory organs for detecting these stimuli. Plants are sensitive but do not have specialized sense organs.
- Organs which make a response to a stimulus are called effectors.
- There has to be communication between receptors and effectors.
- In animals this communication is by nerve impulses or hormones.
- In plants the communication is by 'growth substances'.
- Appropriate responses to stimuli contribute to the success and survival of organisms.

EXAMINATION QUESTIONS

Section 6 Living Organisms

Do not write on this page. Where necessary copy drawings, tables or sentences.

1 Which TWO of the following living characteristics are likely to contribute most to the process of evolution by natural selection?

A Respiration **D** Reproduction
B Excretion **E** Feeding
C Response to stimuli **F** Movement

2 Name any four organisms belonging to the groups listed below that you would expect to find in a single habitat (e.g. a seashore, woodland, grassland, pond, etc.). Each organism should be from a different group. For each organism, describe a feature that you would use to put it in its group.
Algae; fungi; mosses; ferns; flowering plants; protozoans; annelids; arthropods; molluscs; fish; birds; mammals.
(a) Type of habitat. (1)
(b) Name of organism; group; feature. (6)
[W]

3 The figure shows a virus and a bacterium, not drawn to scale.

(a) Which is the bacterium? (1)
(b) Write down TWO features, shown in the figure, which helped you to decide which was which. (2)
(c) Describe briefly how a virus reproduces. (2)
[M]

4 (a) (i) Name the structures used by fish to absorb oxygen from pond water. (1)
(ii) Write down ONE feature of these structures which helps in gaseous exchange. (1)
(b) Many microscopic pond animals do not have special structures for gaseous exchange. Briefly explain why these animals do not need special structures but larger animals do. (2)
[L]

5 The diagram shows a section through a bird's wing.

(i) On the diagram, label the front and rear edges of the wing. Draw arrows to show the pattern of air flow over this wing during flight. (2)
(ii) How does the pattern of air flow you have drawn help to support the bird during gliding flight? (2)
(iii) During flapping flight, a bird moves its wings using muscles. This requires the use of energy. Explain where this energy comes from and how it is used in muscle contraction. (5)
[S]

6 Growth and reproduction are two important characteristics of living things.
(a) State the similarities and differences between growth and reproduction. (6)
(b) In reproduction the offspring may be genetically similar or dissimilar to the parent(s). With reference to asexual and sexual reproduction in named plants, explain how this can come about. (8)
(c) Describe how the radicle of a germinating seed grows in length. Include in your description the following terms: mitosis; cell enlargement; cell differentiation. (6)
[M]

7 The table has a number of questions which compare the sperm with the egg cell (ovum). Complete each line of the table using ONE of the following terms: sperm; egg cell.

Which is able to move?	
Which has more cytoplasm?	
Which is produced in greater numbers?	

(3)
[L]

8 (a) What type of substance is auxin? (1)
(b) (i) Explain the role of auxin in phototropism. (2)
(ii) Give ONE advantage of a phototropic response in stems. (1)
(c) A number of artificial auxins have been produced for use as weedkillers. Suggest how these auxins could have this effect. (2)
[L]

9 Name an insect you have studied which shows complete metamorphosis in its life cycle. List THREE external features of this animal which help to identify it as an insect. (3)
[L]

11 (a) What is the difference between asexual and sexual reproduction in terms of (i) number of parents involved and (ii) the variation in the offspring? (2)
(b) What are the advantages and disadvantages of the differences stated in (a) (ii)? (4)
(c) Some flowering plants use runners to reproduce asexually. Describe how this happens. (4)
[M]

Micro-organisms
and humans

False colour electronmicrograph showing the effect of an antibiotic on the bacterium *Staphylococcus aureus*. On the right are the remains of a bacterium destroyed by bursting; the bacterium on the left is, as yet, undamaged and is in the process of dividing. Magnification: x125,000.

37 BIOTECHNOLOGY

Biotechnology can be defined as the application of biological organisms, systems or processes to manufacturing and service industries.

Although biotechnology is 'hot news', we have been making use of it for hundreds of years. Wine-making, beer brewing, the baking of bread and the production of cheese all depend on fermentation processes brought about by yeasts, other fungi and bacteria, or enzymes from these organisms.

Antibiotics, such as penicillin, are produced by mould fungi or bacteria. The production of industrial chemicals such as citric acid or lactic acid needs bacteria or fungi to bring about essential chemical changes.

Sewage disposal (p. 256) depends on bacteria in the filter beds to form the basis of the food chain which purifies the effluent.

Biotechnology is not concerned solely with the use of micro-organisms. Cell cultures and enzymes also feature in modern developments. In this chapter, however, there is space to consider only a representative sample of biotechnological processes which use micro-organisms.

The processes are described only in outline and relate to the basic, traditional steps. In modern processes, genetically engineered organisms, industrially produced enzymes and a variety of different substrates, nutrients, and additives may be involved.

| *Fermentation products*

The term 'fermentation' does not apply only to alcoholic fermentation but to a wide range of reactions, brought about by enzymes or micro-organisms. On p. 21 the anaerobic respiration of glucose to alcohol or lactic acid was described as a form of fermentation.

Micro-organisms which bring about fermentation are using the chemical reaction to produce the energy which they need for their living processes. The reactions which are useful in fermentation biotechnology are mostly those that produce incompletely oxidized compounds. A reaction that goes all the way to carbon dioxide and water is not much use in this context.

The micro-organisms are encouraged to grow and multiply by providing nutrients such as glucose, with added salts and, possibly, vitamins. Oxygen or air is bubbled through the culture if the reaction is aerobic, or excluded if the process is anaerobic. An optimum pH and temperature are maintained for the species of microbe being cultured.

Yoghurt

Yoghurt and cheese have been made by herdsmen for thousands of years. Both products result from the fermentation of milk by bacteria. Initially these bacteria came from natural sources, e.g. from the animals themselves but, since there would be a mixture of different bacteria in different conditions, the products were often unpredictable.

Today, the micro-organisms and the conditions are carefully controlled so that the wanted product is assured.

Milk from cows, sheep or goats may be used to prepare yoghurt. On a commercial scale the fat and protein content of the milk is adjusted and the milk is homogenized. This is a mechanical process which breaks up the fat droplets and prevents them from separating out. The milk is then pasteurized (p. 101).

Pasteurization destroys pathogenic (potentially harmful) bacteria and other micro-organisms that might cause unwanted changes.

The pasteurized milk is then fermented by adding a 'starter culture' of bacteria, usually a mixture of *Streptococcus thermophilus* and *Lactobacillus bulgaricus*. These bacteria act on the milk sugar, **lactose**, and convert it to lactic acid which, in turn coagulates the milk protein, **casein**, to produce the thick creamy consistency of yoghurt. The fermentation works best at a temperature of 46 °C and, when complete, the yoghurt is cooled to 5 °C to stop the bacterial processes. The lactic acid gives the yoghurt its slightly sour taste.

Throughout the process, the materials and containers must be kept in a sterile condition, i.e. free from any micro-organisms that might compete with the *Streptococcus* and *Lactobacillus* and produce unwanted substances.

Cheese

The basic processes for cheese production are the same as for yoghurt. The composition of the milk is adjusted; it is homogenized and pasteurized. A variety of bacteria cultures may be used to ferment the milk, depending on the type of cheese to be produced, but they are usually species of *Streptococcus* or *Lactobacillus* and they work best at about 40 °C. The process differs from yoghurt production, at this stage, in that a mixture of enzymes, called **rennet** is added. Rennet contains the enzyme **chymosin** which coagulates milk casein and forms the semi-solid 'curds'. Chymosin can be obtained from calves' stomachs but is now mainly produced from genetically engineered yeast which carries the chymosin gene.

The liquid 'whey' is drained from the curds which are partially dried and compressed. The degree of drying and the method of ripening depends on the variety of cheese required. In principle, the bacterial enzymes act on the proteins and fats in the curd, partly digesting them to amino acids and fatty acids which give the cheese its flavour and aroma.

For varieties of 'blue cheese', *Penicillium* spores are added at the fermentation stage (p. 304).

Figure 2 Fermenting beer. The wort is being fermented in a large vat.

Figure 1 Cheese making on a small scale. The casein has formed the curds which are being scooped out from the liquid whey.

Beer

As with yoghurt and cheese, the production of beer and wine dates back thousands of years. The fermentation in this case is brought about by yeast, not bacteria.

Commercial techniques have refined the traditional methods so that the process is controlled from start to finish.

Beer is made from barley. The barley grains are soaked in water for 2 days at 15–20 °C and then allowed to germinate for 4 days. During germination, the starch-digesting enzymes in the seeds are activated. The germinating seeds are then dried at temperatures which kill the seeds but do not denature the enzymes. This process is called 'malting' and the product is 'barley malt'.

The dried grains are crushed (milled) and mixed with water at about 65 °C and extra starch from wheat or rice is added. The barley enzymes now digest the starch to maltose and glucose. The sugary solution (called 'wort') is filtered and boiled with hops which suppress the growth of bacteria and give the beer its bitter flavour.

Yeast is added to the wort. The most commonly used species of yeast are *Saccharomyces cerevisiae* and *S. carlsbergensis*. Enzymes in the yeast convert maltose to glucose. In the early stages of fermentation, yeast respires aerobically, oxidizing the glucose to carbon dioxide and water and dividing rapidly. As the oxygen supply diminishes, anaerobic respiration takes over and the glucose is oxidized only as far as alcohol and carbon dioxide (p. 21).

Fermentation takes 5–14 days, depending on the type of beer, and results in the conversion of all the sugar to alcohol, giving a 3–5 per cent solution. After this, the dead yeast, proteins and hop resins are allowed to settle out, helped by 'clarification agents'. Beer for kegs, bottles or cans is pasteurized and its carbon dioxide level is adjusted to give the necessary 'sparkle' and fizzy taste. 'Real ale' is not pasteurized; the residual living yeast continues fermenting and producing carbon dioxide in the barrel.

The flavour of the final product depends on a multitude of substances, as well as alcohol and carbon dioxide, produced in the course of fermentation.

Wine

Grape juice is the basis for wine-making, and yeast is the micro-organism that causes the fermentation.

The grapes are crushed and the extracted juice is treated with sulphur dioxide to kill bacteria and naturally occurring yeasts. The sugar content and pH are adjusted before adding a starter culture of yeast, usually a variety of *Saccharomyces cerevisiae*. The natural yeasts on the grape skins would bring about fermentation but the outcome would be unpredictable.

The juice is aerated at first; the yeast respires aerobically and its population increases rapidly. When aeration stops, the yeast ferments the grape sugar (glucose) anaerobically to alcohol and carbon dioxide. The alcohol content rises to 10–15 per cent, a concentration which eventually kills the yeast. Low temperatures prolong the fermentation and increase the alcohol content.

The wine is transferred to barrels and any particles are allowed to settle, with the aid of clearing agents. This may be repeated several times until the wine is clear, then it is bottled.

The variety of wine depends on the type of grape used and the details of fermentation and storage. Red wine comes from 'black' grapes; the skins are left in the grape juice for a time and the alcohol dissolves out the red pigment. White wines can be prepared from black or white grapes but the skins are removed after crushing. The taste and flavour ('bouquet') of wine depends on the organic acids and many aromatic organic compounds present in the grape juice and produced during fermentation.

Vinegar

Acetobacter is a genus of rod-shaped bacteria widely distributed in nature and especially in rotting plant remains. Different species of *Acetobacter* can oxidize a variety of organic compounds to produce organic acids. Some of these species are exploited in the production of vinegar from beer, cider or wine (*vin aigre* (Fr.) = sour wine).

Malt vinegar is prepared from crushed malted barley which is mixed with hot water and fermented with yeast to produce a rough kind of beer called **gyle**.

The process is much the same as in beer brewing but there is no need to filter or sterilize the liquid, which is allowed to settle for 3–4 months.

The gyle is then inoculated with *Acetobacter* and pumped into large vats which contain two layers of birch twigs. The gyle is pumped over the twigs and the *Acetobacter* grows on the large surface they provide.

With air supply and temperature carefully controlled, most of the ethanol (alcohol) in the gyle is converted aerobically to ethanoic (acetic) acid.

$$C_2H_5OH + O_2 \rightarrow CH_3COOH + H_2O + energy$$
ethanol   oxygen   ethanoic acid water

The product is allowed to mature for up to 6 months before filtering and bottling.

There are more traditional and more commercial ways of making vinegar, but this is the principle behind them all.

If wine, beer or cider is left exposed to the air for any length of time, *Acetobacter* will arrive and oxidize the alcohol to ethanoic acid making the beverage sour and undrinkable.

Bread

Yeast is the micro-organism used in bread-making but the only fermentation product needed is carbon dioxide. The carbon dioxide makes bubbles in the bread dough. These bubbles make the bread 'light' in texture.

Flour, water, salt, oil and yeast are mixed to make a dough. Yeast has no enzymes for digesting the starch in flour but the addition of water activates the amylases already present in flour and these digest some of the starch to sugar. With highly refined white flour, it may be necessary to add sugar to the dough. The yeast then ferments the sugar to alcohol and carbon dioxide.

A protein called **gluten** gives the dough a sticky, plastic texture which holds the bubbles of gas. The dough is repeatedly folded and stretched ('kneaded') either by hand, in the home, or mechanically in the bakery. The dough is then left for an hour or two at a temperature of about 27 °C while the yeast does its work. The accumulating carbon dioxide bubbles make the dough rise to about double its volume (Fig. 3). The dough may then be kneaded again or put straight into baking tins and into an oven at about 200 °C. This temperature makes the bubbles expand more, kills the yeast and evaporates the small quantities of alcohol before the dough turns into bread.

Figure 3 Carbon dioxide produced by the yeast has caused the dough to rise

Antibiotics

When micro-organisms are used for the production of antibiotics, it is not their fermentation products which are wanted, but complex organic compounds, called **antibiotics**, that they synthesize.

Most of the antibiotics we use come from bacteria or fungi which live in the soil. There is assumed to be fierce competition between the populations of rapidly dividing soil micro-organisms. The antibiotics they produce are thought to suppress the growth of some of the rival competitors.

One of the most prolific sources of antibiotics are the *Actinomycetes*. These are filamentous bacteria which resemble microscopic mould fungi. The actinomycete *Streptomyces* produces the antibiotic **streptomycin**.

Perhaps the best known antibiotic is **penicillin** which is produced by the mould fungus *Penicillium* (p. 304) and was discovered by Sir Alexander Fleming in 1928. Penicillin is still an important antibiotic but it is produced by mutant forms of a different species of *Penicillium* from that studied by Fleming. The different mutant forms of the fungus produce different types of penicillin. These are chemically altered in the laboratory to make them more effective and to 'tailor' them for use with different diseases. 'Ampicillin', 'methicillin' and 'oxacillin' are all chemically altered forms of penicillin.

Antibiotics attack bacteria in a variety of ways. Some of them disrupt the production of the cell wall and so prevent the bacteria from reproducing, or even cause them to burst open; some interfere with protein synthesis and thus arrest bacterial growth. Those that stop bacteria from reproducing are said to be **bacteriostatic**, those that kill the bacteria are **bacteriocidal**.

Animal cells do not have cell walls and the cell structures involved in protein production are different. Consequently, antibiotics do not damage human cells although they may produce some side effects such as allergic reactions.

Figure 4 Penicillin production. *Penicillium* spores are being inoculated on to maize under strictly aseptic conditions. The culture will later be transferred to a 10 000 litre fermenter.

Commercial production

Antibiotics are produced in giant fermenting tanks, up to 100 000 litres in capacity. The tanks are filled with a nutrient solution. For penicillin production, the carbohydrate source is sugar, mainly lactose or 'corn-steep liquor' a by-product of the manufacture of cornflour and maize starch; it contains amino acids as well as sugars. Mineral salts are added, the pH is adjusted to between 5 and 6, the temperature is maintained at about 26 °C, air is blown through the liquid and it is stirred. The principles of industrial fermentation are shown in Fig. 5. The nutrient liquid is seeded with a culture of the appropriate micro-organism which is allowed to grow for a day or two (Fig. 4). Sterile conditions are essential. If 'foreign' bacteria or fungi get into the system they can completely disrupt the process. As the nutrient supply diminishes, the micro-organisms begin to secrete their antibiotics into the medium.

The nutrient fluid containing the antibiotic is filtered off and the antibiotic extracted by crystallization or other methods.

Single-cell protein

The object of this exercise is to produce micro-organisms in bulk to use as human food or animal feed. Unicellular algae, fungi and bacteria have been used.

The reasoning behind such projects was expressed by an eminent biologist many years ago when he said, 'In 24 hours half a tonne of bullock will make 500 grams of protein; half a tonne of yeast will make 50 tonnes and needs only a few square yards to do it on.' Moreover, the feedstocks for micro-organisms can often be produced cheaply from the waste products of other commercial processes.

In theory, such a prodigous rate of reproduction in micro-organisms could make up for any world shortage of protein. In fact, very few projects have turned out to be commercially viable.

In countries with prolonged sunlight and high temperatures, unicellular algae can be grown in tanks of nutrients or in plastic pipes through which the nutrient circulates. The dried algae can be used as cattle feed. However, the cost of production is often greater than producing crops on the same area, but it may have applications in arid regions.

ICI set up a plant, in 1980, to produce single-cell protein (SCP) from bacteria. The bacteria were grown in a feedstock of dilute methanol (methyl alcohol), with ammonia as a source of nitrogen, in fermenters similar to the one in Fig. 5. The bacteria were filtered off and dried to produce an animal feed called 'Pruteen'. Unfortunately, a slump in the price of soya flour and skimmed milk (both used as animal feed) made the process unprofitable.

Other drawbacks to SCP are the reluctance of people to eat food made from bacteria and the fact that the products are rather tasteless.

One success story is the production of **myco-protein**. A filamentous fungus, *Fusarium graminearum* is grown in a medium containing glucose and mineral salts at 30 °C using specially designed fermenters. The process used is called **continuous culture**. The medium is circulated round the fermenter and the fungus is filtered off at a steady rate and dried. The culture medium is topped up as necessary. This is cheaper and simpler than the **batch processes** described above and avoids repeated sterilization of the fermenter vessels.

The filamentous nature of the fungus gives the product a texture similar to meat but it has a bland taste. It can be flavoured with natural products and cooked in a similar way to meat (Fig. 6).

Its attraction is that it is rich in protein, low in fat and has a good level of dietary fibre. This is a healthier combination than is found in most meat products.

It will be clear by now that 'single-cell protein' is a misnomer. *Fusarium* consists of hyphae rather than single cells and its product contains several nutrients in addition to protein.

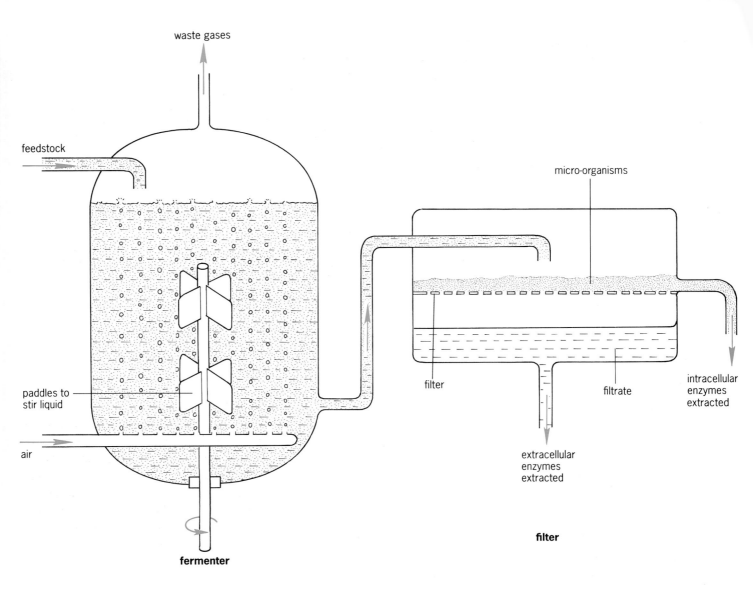

Figure 5 Principles of enzyme production from micro-organisms

Figure 6 Quorn myco-protein

Enzymes

Industrial uses of enzymes are discussed on p. 18. Many of these enzymes are produced by biotechnology using yeasts, fungi or bacteria. The micro-organisms are allowed to grow in fermenters (Figs. 5 and 8) using suitable feedstocks, e.g. starch solution. If the enzymes are extracellular (p. 15) then the liquid feedstock is filtered from the organism and the enzyme is extracted. If the enzymes are intracellular, the micro-organisms have to be filtered from the feedstock. They are then crushed and the enzymes extracted with water or other solvents.

By the techniques of genetic engineering (p. 206), new genes can be introduced into the microbes to 'improve' the action of the enzymes coded for by the genes (e.g. making the enzymes more heat stable).

Genes from entirely different organisms can be introduced to make the microbes produce enzymes which are not in their normal repertoire. For example, the mammalian chymosin gene has been introduced into yeast cells (see 'Cheese' above).

One effective way of using enzymes is by 'immobilizing' them. The enzymes or the micro-organisms which produce them are held in or on beads or membranes of an insoluble and inert substance, e.g. plastic. The beads or membranes are packed into columns and the substrate is poured over them at the optimum rate. This method has the advantage that the enzyme is not lost every time the product is extracted. Immobilized enzymes also allow the process to take place in a continuous way rather than a batch at a time.

Sludge digestion

In this process, the principal role of the micro-organisms is to react with sewage sludge or farmyard slurry and convert it to harmless or useful substances. This is of great value in reducing potential sources of pollution.

The sludge or slurry is fed into large, enclosed tanks (anaerobic digesters) and heated initially to about 25 °C. Conditions soon become anaerobic and the wide range of bacteria in the waste material digest the organic compounds to fatty acids, monosaccharides and amino acids. These are acted on by other bacteria to produce organic acids and alcohols. Finally, **methanogenic** bacteria convert these to methane (CH_4) and water.

The methane is drawn off and used as an energy source to provide the heat needed to start the reactions and there is usually a substantial surplus for other purposes such as central heating or driving the compressors for activated sludge (p. 257).

The advantages of the anaerobic bio-digester are:

1 the bulk of the sludge is greatly reduced
2 the high temperatures of anaerobic fermentation destroy potential pathogens so that the residue can be safely used as a soil conditioner or compost and the remaining liquid can be used or sold as a liquid fertilizer
3 the odours are greatly lessened (particularly valuable in intensive pig farms)
4 dumping of sludge is reduced and the potential for water pollution with farmyard slurry is avoided
5 the methane ('biogas') is a valuable, relatively non-polluting source of energy.

Biogas production is not confined to sludge and slurry. Many organic wastes, e.g. those from sugar factories, can be fermented anaerobically to produce biogas. In developing countries, biogas generators use animal dung to produce methane for whole villages.

QUESTIONS

1 Name three micro-organisms used in industrial fermentation and state their fermentation products.
2 Apart from using their fermentation products, how else are micro-organisms exploited commercially?
3 What are the general conditions that need to be controlled for successful growth of micro-organisms on a large scale? Why is it necessary to ensure sterile conditions?
4 What are 'curds and whey'?
5 Why do you think bio-digesters need anaerobic conditions?
6 Yeast can digest sugar but not starch, so how can fermentation proceed in beer brewing and bread-making where the original substrate is cereal starch?
7 Why does wine 'go off' if left exposed to the air?
8 How do modern penicillins differ from the one discovered in 1928?

PRACTICAL WORK

1 Fermentation in yeast

An experiment to investigate anaerobic respiration in yeast is described on p. 25 and can be modified to explore the effect of temperature on the rate of fermentation. The carbon dioxide can be collected in a burette or measuring cylinder as shown in Fig. 7. The volume of gas collected in a given time represents the rate of fermentation and alcohol production. The temperature of the water bath can be raised by 5 ° at a time. It is not necessary, in this experiment, to have anaerobic conditions.

Figure 7 Exploring the effect of temperature on the rate of fermentation

2 The function of yeast in bread dough *

One day before the experiment make a yeast suspension 'starter' by mixing a level teaspoonful of dried yeast with half a teaspoonful of sugar and 20 cm^3 water, in a small flask or bottle. Plug the mouth of the flask with cotton wool and leave it in a warm place for 24 hours.

Mix the following in measures of level teaspoons: plain white flour, 2; sugar, ½; yeast suspension, ½; water, 2. This should make a sticky dough.

Withdraw the plunger of a 1 cm^3 syringe to the 0.5 mark and then draw up about 0.3 cm^3 dough into the syringe. Note the level of the dough and plug the nozzle of the syringe with a piece of sharpened matchstick. Place the syringe in a water bath kept at about 35 °C and note the level of the dough at 5-minute intervals.

A control should be set up using the same mixture but made with yeast suspension that has been boiled for a minute to kill the yeast cells.

The results should indicate what part is played by yeast in making bread. The experiment also demonstrates anaerobic respiration, since there is little or no air present in the bread dough.

The rate of anaerobic respiration at different temperatures may be compared if similar syringes are placed in water baths at 15, 25 and 45 °C. The rate at which the dough rises is a measure of the rate of carbon dioxide production.

*After P.W. Freeland, *School Science Review* No. 194, 1974.

Figure 8 Commercial fermentation unit for production of genetically engineered products such as enzymes and insulin

Checklist

- Biotechnology is the application of living organisms, systems or processes in industry.

- Many biotechnological processes use micro-organisms (fungi and bacteria) to bring about the reactions.

- Most biotechnological processes are classed as 'fermentations'.

- Fermentation may be aerobic or anaerobic.

- The required product of biotechnology may be the organism itself (e.g. SCP) or one of its products (e.g. alcohol).

- Yoghurt and cheese are made from fermented milk using bacteria.

- Beer and wine are produced from fermented sugars using yeast.

- Antibiotics are produced from bacteria and fungi.

- Enzymes from micro-organisms can be produced on an industrial scale and used in other biotechnology processes.

- Sterile conditions are essential in biotechnology to avoid unwanted products from contaminating microbes.

38 DISEASE: CAUSES, TRANSMISSION AND CONTROL

Bacterial diseases
Salmonella food poisoning, gonorrhoea, syphilis.

Virus diseases
Common cold, influenza, AIDS.

Fungal parasites
Tinea.

Protozoan parasites
Malaria, amoebic dysentery.

Disease transmission
By air, water and food; vectors; contagion.

Methods of prevention
Natural barriers: prevention of transmission by air, water and food and contagion; water treatment; immunization; drug therapy.

Non-transmissible diseases

These are diseases which are not infectious. In origin they may be genetic (e.g. cystic fibrosis, p. 215), nutritional (e.g. vitamin deficiency diseases, p. 97), degenerative (e.g. arthritis, heart attacks, strokes), or they may have environmental causes (e.g. silicosis resulting from exposure to mineral dust), or no external or clearly defined caused (e.g. diabetes and many cancers).

Transmissible diseases

These are the infectious diseases which can be passed from one person to another. They are caused by viruses, bacteria, fungi or protozoa and some of them are described below.

Bacterial diseases

Most bacteria are useful or harmless as explained on pp. 237 and 300. Even the bacteria which live and reproduce on our skin or in our breathing passages or intestines are usually harmless. In fact, they may keep more harmful competitors away.

Some bacteria in the human large intestine digest vegetable fibre in our food to form fatty acids, and others produce vitamin K. It is possible that we absorb and use these substances.

However, some species of parasitic bacteria cause disease when they invade our bodies. These are called **pathogenic bacteria**.

Species of the bacterium **Streptococcus** cause sore throats, blood poisoning and scarlet fever. Species of **Clostridium** bacteria cause tetanus and botulism. **Staphylococcus** species cause tonsillitis and boils. Tuberculosis, cholera, typhoid, diphtheria, whooping cough, food poisoning, gonorrhoea and syphilis are all bacterial diseases.

The ill-effects of a bacterial disease are caused mainly by poisonous products produced either by the bacteria or by the cells which they invade. Bacterial poisons are called **toxins**. Toxins damage the cells in which the bacteria are growing. They also upset some of the systems in the body. This gives rise to a raised temperature, headache, tiredness and weakness, and sometimes to diarrhoea and vomiting.

Salmonella food poisoning

One of the commonest causes of food poisoning are the toxins produced by the bacteria *Salmonella typhimurium* or *S. enteritidis*. These bacteria live in the intestines of cattle, pigs, chickens and ducks without causing disease symptoms. Humans, however, may develop food poisoning if they drink milk or eat meat or eggs which are contaminated with *Salmonella* bacteria from the alimentary canal of an infected animal.

Intensive methods of animal rearing may contribute to a spread of infection unless care is taken to reduce the exposure of animals to infected faeces.

The symptoms of food poisoning are diarrhoea, vomiting and abdominal pain. They occur from 12 to 24 hours after eating the contaminated food. Although these symptoms are unpleasant, the disease is not usually serious and does not need treatment with drugs. Elderly people and very young children, however, may be made very ill by food poisoning.

The *Salmonella* bacteria are killed when meat is cooked or milk is pasteurized (p. 101). Infection is most likely if untreated milk is drunk, meat is not properly cooked, or cooked meat is contaminated with bacteria transferred from raw meat (Fig. 1). Frozen poultry must be thoroughly defrosted before cooking, otherwise the inside of the bird may not get hot enough during cooking to kill the *Salmonella*.

Figure 1 Transmission of *Salmonella* food poisoning

It follows that, to avoid the disease, all milk should be pasteurized and meat should be thoroughly cooked. People such as shop assistants and cooks should not handle cooked food at the same time as they handle raw meat. If they must do so, they should wash their hands thoroughly between the two activities.

The liquid which escapes when a frozen chicken is defrosted may contain *Salmonella* bacteria. The dishes and utensils used while the bird is defrosting must not be allowed to come into contact with any other food.

Uncooked meat or poultry should not be kept alongside any food which is likely to be eaten without cooking. Previously cooked meat should never be warmed up; the raised temperature accelerates the reproduction of any bacteria present. The meat should be eaten cold or cooked at a high temperature.

In the past few years there has been an increase in the outbreaks of *Salmonella* food poisoning in which the bacteria are resistant to antibiotics. Some scientists suspect that this results from the practice of feeding antibiotics to farm animals to increase their growth rate. This could allow populations of drug-resistant *Salmonellae* to develop.

In the 1970s another genus of bacteria, *Campylobacter*, was identified as a cause of food poisoning. This bacterium causes acute abdominal pains and diarrhoea for about 24 hours. The sources of infection are thought to be undercooked poultry and unpasteurized milk. In recent years, *Campylobacter* outbreaks have become more common than those due to *Salmonella*.

Gonorrhoea

This is a sexually transmitted disease. That means that it is almost always caught by having sexual intercourse with an infected person. The disease is caused by a bacterium, *Neisseria gonorrhoea*.

The first symptoms in men are pain and a discharge of pus from the urethra. In women, there may be similar symptoms, or no symptoms at all.

In men, the disease leads to a blockage of the urethra and to sterility. A woman can pass the disease to her child during birth. The bacteria in the vagina invade the baby's eyes and cause blindness.

The disease can be cured with penicillin, but some strains of *Neisseria* have become resistant to this antibiotic. There is no immunity to gonorrhoea. Having the disease once does not prevent you catching it again.

Syphilis

Syphilis is also a sexually transmitted disease. It is caused by a bacterium called *Treponema pallidum*.

In the first stage of the disease, a lump or ulcer appears on the penis or the vulva, one week to three months after being infected. The ulcer usually heals without any treatment after about six weeks. By this time the bacteria have entered the body and may affect any tissue or organ. There may be a skin rash, a high temperature and swollen lymph nodes. But the symptoms are variable and the infected person may appear to be in good health for many years. But if the disease is not treated in the early stages, the bacteria will in time cause inflammation almost anywhere in the body. They can do permanent damage to the blood vessels, heart or brain, leading to paralysis and insanity.

In a pregnant woman, the bacteria can get across the placenta and infect the fetus.

Penicillin will cure syphilis. But unless it is used in the early stages of the disease, the bacteria may do permanent damage.

QUESTIONS

1 Explain why rewarming cooked meat might lead to food poisoning.
2 In preparing for a reception, a cook defrosted a frozen chicken, placed it in the oven and then sliced some cooked ham which he put into the refrigerator. Then he washed his hands before preparing the salad.
   After the reception several guests who ate the cold chicken and ham salad suffered from food poisoning.
   a How could they have become infected, in spite of the chicken being cooked at a high temperature and the ham being refrigerated?
   b How could the outbreak have been avoided?
3 Syphilis and gonorrhoea are sexually transmitted diseases. So how is it that babies can be infected?

Viral diseases

Viruses can reproduce only inside other cells, and so all viruses are parasitic. In humans they cause diseases such as colds, influenza, herpes, mumps, measles, chickenpox, rubella, hepatitis and AIDS. Some viruses can remain dormant (inactive) in certain body cells without immediately producing symptoms of disease.

Viruses do not produce toxins. The harm they cause is probably the result of the destruction of the cells they invade. When they destroy cells in, for example, the lining of the windpipe and bronchi, bacteria can invade the damaged tissues.

After entering the body and reproducing in a small group of cells, the viruses may be carried to other organs in the circulation.

Although there are many effective drugs against bacterial infections, drugs against viral infections are only just being developed. However, some viral infections do give long-term immunity, and many virus diseases can be prevented by immunization (p. 355).

The common cold

This is caused by a **rhinovirus**. It is spread by droplet infection and contact. The symptoms of the disease develop within 12–78 hours after infection and are very familiar: dry throat, watering of the eyes, production of watery mucus from the nose and swollen (congested) nasal membranes, making it difficult to breathe through the nose.

These symptoms last for a few days. But the damage done by the virus to the nose and throat membranes often allows *Streptococcus* bacteria to invade. This **secondary bacterial infection** may give rise to a sore throat, a cough and catarrh.

Although the body develops temporary immunity to the 'cold' virus, there are many different strains of rhinovirus, and other species of virus also cause colds. Immunity to one of these strains does not extend to others. This is why you can have one cold after another.

There is no cure for a cold. Antibiotics are ineffective against rhinoviruses, but they may be prescribed if an acute or persistent secondary bacterial infection appears.

Influenza ('flu)

Influenza is caused by a virus which exists in three strains, A, B and C. The virus attacks the lining of the throat and breathing passages, giving rise to inflammation of the trachea, bronchi and bronchioles. The patient will have a raised temperature, headache, dry cough and a mild sore throat and will feel generally 'rotten'. The symptoms subside in 2–4 days, but the damage to the respiratory linings may allow *Streptococci* to invade, causing a secondary bacterial infection.

There are no specific drugs. Aspirin helps to lower the temperature, and antibiotics may be used against any secondary infection.

A bout of infection confers immunity for several years. But the 'A' strain of the virus undergoes mutations (p. 205) very readily and immunity to one form is not effective against other mutants.

There are sometimes severe epidemics of influenza. If it is known which mutant is responsible, it is possible to prepare a vaccine which gives protection for a few months to people most at risk and to doctors and nurses dealing with the epidemic.

AIDS

The initials stand for Acquired Immune Deficiency Syndrome. (A 'syndrome' is a pattern of symptoms associated with a particular disease.) The virus which causes AIDS is the human immunodeficiency virus (**HIV**). It attacks certain kinds of lymphocyte (p. 114), and thus weakens the body's immune responses.

As a result, the patient has little or no resistance to a range of diseases that would not normally invade the body. Once the symptoms appear, the outlook for at least 50 per cent of the patients is bleak. Death does not result from the virus itself, but from diseases such as pneumonia, blood disorders, skin cancer or damage to the nervous system which the body cannot resist.

After a person has been infected, years may pass before symptoms develop. So people may carry the virus yet not show any symptoms. They can still infect other people, however. It is not known for certain what proportion of HIV carriers will eventually develop AIDS: perhaps 30–50 per cent, or even more, may do so.

HIV is transmitted by direct infection of the blood. Drug users who share needles contaminated with infected blood run a high risk of the disease. It can also be transmitted sexually, both between men and women and, especially, between homosexual men who practise anal intercourse.

Haemophiliacs (p. 215) have also fallen victim to AIDS. Haemophiliacs have to inject themselves with a blood product which contains a clotting factor. Before the risks were recognized, infected carriers sometimes donated blood which was used to produce the clotting factor.

Children born to HIV carriers may also contract the disease.

There is no evidence to suggest that the disease can be passed on by droplets (p. 352), by saliva or by normal everyday contact.

The disease has appeared comparatively recently in the industrialized countries. There are no very effective drugs or vaccines against it at the moment, though intensive work is taking place on both fronts.

There is a range of blood tests designed to detect HIV infection. These tests do not detect the viruses but do indicate whether antibodies to the virus are in the blood. If HIV antibodies are present, the person is said to be **HIV positive**. The tests vary in their reliability and some are too expensive for widespread use. The American Food and Drug Administration claims a 99.8 per cent accuracy but this figure is disputed.

QUESTIONS

4 Antibiotics are ineffective against virus diseases. Why then are antibiotics sometimes given to people suffering from a virus infection such as influenza?
5 Blood donors now have their blood tested to see if it contains antibodies to AIDS. What do you think is the value of this test?
6 People living in closed, isolated communities, such as Arctic explorers, often get a cold or two at first and then have several months free from colds. However, when the supply ship arrives the colds start again. Try to explain this phenomenon.

Fungal parasites

Tinea ('ringworm')

Several species of fungus give rise to the various forms of this disease. The fungus attacks the cornified layer of the skin and produces a patch of inflamed tissue. On the skin the infected patch spreads outwards and heals in the centre, giving a ring-like appearance ('ringworm').

The different species of tinea fungi may live on the skin of humans or domestic animals, or in the soil. The region of the body affected will depend on the species of fungus.

One kind affects the scalp and causes circular bald patches. The hair usually grows again when the patient recovers from the disease.

The species of fungus which affect the feet usually cause cracks in the skin between the toes. This is known as 'athlete's foot'.

Tinea of the crutch is a fungus infection, occurring usually in males, which affects the inner part of the thighs on each side of the

scrotum. It causes a spreading, inflamed area of skin with an itching or burning sensation.

All forms of the disease are very contagious. That means, they are spread by contact with an infected person or their personal property. Tinea of the scalp is spread by using infected hairbrushes, combs or pillows. Tinea of the crutch can be caught by using towels or bedclothes contaminated by the fungus or its spores, and 'athlete's foot' by wearing infected socks or shoes, or from the floors of showers and swimming pools.

When an infection is diagnosed, the clothing, bed linen, hair brushes, combs or towels must be boiled to destroy the fungus. It is best, anyway, to avoid sharing these items as their owners may be carrying the infection without knowing or admitting it.

In young people, tinea infections often clear up without treatment. Where treatment is needed, a fungicide cream or dusting powder is applied to the affected areas of skin. Infected feet may be dipped in a solution of potassium permanganate (potassium manganate(VII)).

Protozoan parasites

There are relatively few protozoa which parasitize humans but those that do, cause serious diseases. Two of these are malaria and amoebic dysentery.

Malaria

About 280 million people suffer from malaria in over 100 countries (Fig. 2), and 2 million or more die each year from the disease.

The disease is caused by a protozoan parasite called *Plasmodium* which is transmitted from person to person by the bites of infected mosquitoes of the genus *Anopheles*. The mosquito is said to be the **vector** of the disease. When a mosquito 'bites' a human, it inserts its sharp, pointed mouth parts through the skin till they reach a capillary (p. 311). The mosquito then injects saliva to stop the blood from clotting. If the mosquito is infected, it will also inject hundreds of malarial parasites.

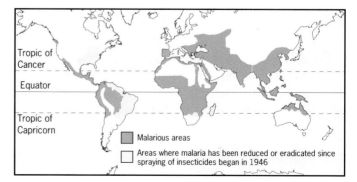

Figure 2 The world-wide distribution of malaria

The parasites reach the liver via the circulation and burrow into the liver cells and reproduce. A week or two later, the daughter cells break out of the liver cells and invade the red blood cells. Here infected mosquito injects *Plasmodium* parasites

parasite enters liver cell

enters new red cell

onset of fever

escapes

reproduces

enters red blood cell

escapes

reproduces

Figure 3 *Plasmodium*, the malarial parasite

they reproduce rapidly and then escape from the original red cells to invade others (Fig. 3). This cycle of reproduction in the red cells takes two or three days (depending on the species of *Plasmodium*). Each time the daughter *Plasmodia* are released simultaneously from thousands of red cells the patient experiences the symptoms of malaria. These are chills accompanied by violent shivering followed by a fever and profuse sweating (Fig. 4). With so many red cells being destroyed, the patient will also become anaemic (p. 96).

If a mosquito sucks blood from an infected person, it will take up the parasites in the red cells. The parasites reproduce in the mosquito and finally invade the salivary glands, ready to infect the next human.

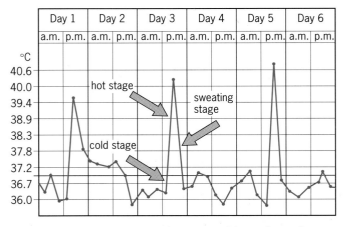

Figure 4 Temperature graph of a patient with one form of malaria

Control

There are drugs which kill the parasites in the bloodstream but they do not reach those in the liver. The parasites in the liver may emerge at any time and start the cycle again. If these drugs are taken by a healthy person before entering a malarious country, they kill any parasites as soon as they are injected. This is a protective or **prophylactic** use of the drug.

Unfortunately there are now many mutant forms of *Plasmodium* which have developed resistance to these drugs.

A great deal of work has been devoted to finding an effective vaccine, without much success until recently. Trials are currently taking place, in Colombia, of a promising vaccine which may offer at least partial protection against the disease.

The most far-reaching form of malarial control is based on the elimination of the mosquito. It is known that mosquitoes lay their eggs in stagnant water and that the larvae hatch, feed and grow in the water, but have to come to the surface to breathe air.

Spraying stagnant water with oil and insecticides suffocates or poisons the larvae and pupae. Spraying must include not only lakes and ponds but any accumulation of fresh water which mosquitoes can reach, e.g. drains, gutters, tanks, tin cans and old car tyres. By draining swamps and turning sluggish rivers into swifter streams, the breeding grounds of the mosquito are destroyed.

Spraying the walls of dwellings with chemicals like DDT was once very effective because the insecticide remained active for several months and the mosquito picked up a lethal dose merely by settling on the wall.

However, in at least 60 countries, many species of *Anopheles* have developed resistance to these insecticides and this method of control is now far less effective. The emphasis has changed back to the removal of the mosquito's breeding grounds or the destruction of the larvae and pupae.

Amoebic dysentery

Entamoeba histolytica is a species of small amoebae which normally live harmlessly in the human intestine, feeding on food particles or bacteria. In certain conditions, however, *Entamoeba* invades the lining of the intestine causing ulceration and bleeding, with pain, vomiting and diarrhoea; the symptoms of amoebic dysentery.

The diarrhoea and vomiting lead to a loss of water and salts from the body and if they persist for very long can cause **dehydration**. Dehydration, if untreated, can lead to kidney failure and death. The treatment for dehydration is to give the patient a carefully prepared mixture of water, salts and sugar. The intestine absorbs this solution more readily than water and it restores the volume and concentration of the body fluids. This simple, effective and inexpensive treatment is called **oral rehydration therapy** and has probably saved thousands of lives since it was first discovered. There are also drugs which attack the *Entamoebae*.

The faeces of infected people contain *Entamoebae* which, if they reach food or drinking water, can infect other people. The disease is prevalent in tropical, sub-tropical and, to some extent, temperate countries and is associated with low standards of hygiene and sanitation.

QUESTIONS

7 a What is the difference between HIV and AIDS?
  b What does 'HIV positive' mean?
  c Which groups of people are most at risk of HIV infection?
8 a What are the two main lines of attack on malaria?
  b What is the connection between stagnant water and malaria?
  c What are the principal 'set-backs' in the battle against malaria?

Disease transmission

Airborne, 'droplet' or aerosol infection

When we sneeze, cough, laugh, speak or just breathe out, we send a fine spray of liquid drops into the air. These droplets are so tiny that they remain floating in the air for a long time. They may be breathed in by other people or fall onto exposed food (Fig. 5). If the droplets contain viruses or bacteria, they may cause disease when they are eaten with food or inhaled.

Virus diseases like colds, 'flu, measles and chickenpox are spread in this way. So are the bacteria (*Streptococci*) that cause sore throats. When the water in the droplets evaporates, the bacteria often die as they dry out. The viruses remain infectious, however, floating in the air for a long time.

In buses, trains, cinemas and discos, the air is warm and moist, and full of floating droplets. These are places where you are likely to pick up one of these infections.

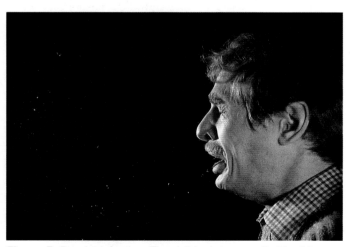

Figure 5 Droplet infection. The visible drops expelled by this sneeze will soon sink to the floor, but smaller droplets will remain suspended in the air.

Contamination of water

If disease bacteria get into water supplies used for drinking, hundreds of people can become infected. Diseases of the alimentary canal, like typhoid and cholera, are especially dangerous. Millions of bacteria infest the intestinal lining of a sick person.

Some of these bacteria will pass out with the faeces. If the faeces get into streams or rivers, the bacteria may be carried into reservoirs of water used for drinking. Even if faeces are left on the soil or buried, rain water may wash the bacteria into a nearby stream.

To prevent this method of infection, faeces must be made harmless, and drinking water purified.

Contamination of food

Contamination by people

The bacteria most likely to get into food are the ones which cause diseases of the alimentary canal such as typhoid and *Salmonella* food poisoning (p. 347). The bacteria are present in the faeces of infected people and may reach food from the unwashed hands of the sufferer.

People recovering from one of these diseases may feel quite well. But bacteria may still be present in their faeces. If they don't wash their hands thoroughly after going to the lavatory, they may have small numbers of bacteria on their fingers. If they then handle food, the bacteria may be transferred to the food. When this food is eaten by healthy people, the bacteria will multiply in their bodies and give them the disease.

People working in food shops, kitchens and food-processing factories could infect thousands of other people in this way if they were careless about their personal cleanliness.

Some forms of food poisoning result from poisons (toxins) that are produced by bacteria which get into food. Cooking kills the bacteria in the food, but does not destroy the toxins which cause the illness. Only one form of this kind of food poisoning, called **botulism**, is dangerous. It is also very rare.

Contamination by houseflies

Flies walk about on food. They place their mouth parts on it and pump saliva onto the food. Then they suck up the digested food as a liquid.

This would not matter much if flies fed only on clean food, but they also visit decaying food or human faeces. Here they may pick up bacteria on their feet or their mouth parts. They then alight on our food and the bacteria on their bodies are transferred to the food. Fig. 6 shows the many ways in which this can happen.

Vectors

A vector is an animal, often an insect, which carries pathogens from one person to another. Some species of mosquito carry the protozoa which cause malaria (p. 350), and other species carry the viruses of yellow fever. Rat fleas carry the bacteria of bubonic plague. Houseflies spread polio, and a variety of intestinal diseases, such as food poisoning.

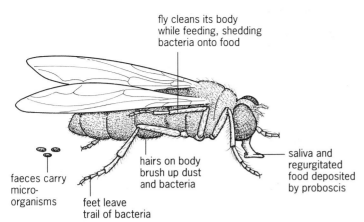

fly cleans its body while feeding, shedding bacteria onto food

saliva and regurgitated food deposited by proboscis

hairs on body brush up dust and bacteria

feet leave trail of bacteria

faeces carry micro-organisms

Figure 6 Transmission of bacteria by houseflies

Contagion

A contagious disease can be spread by contact with an infected person, or with their clothing, bed linen or towels. The fungus disease tinea (described on p. 350) is very contagious.

Colds may also be spread by contact. The viruses are transferred from the hands of an infected person to the hands of healthy people, who then infect themselves by touching their eyes or nose.

Sexually transmitted diseases

Syphilis and gonorrhoea are spread almost exclusively by sexual intercourse with an infected person. AIDS too is spread mainly by sexual contact, though drug addicts who share contaminated needles are also at risk.

There is no easy way of recognizing an infected person. People who have many sexual partners, such as prostitutes, are the most likely to be infected.

QUESTIONS

9 Why should people who sell, handle and cook food be particularly careful about their personal hygiene?

10 Coughing or sneezing without covering the mouth and nose with a handkerchief is thought to be inconsiderate behaviour. Why is this?

11 Inhaling cigarette smoke can stop the action of cilia in the trachea and bronchi for about 20 minutes. Why should this increase a smoker's chance of catching a respiratory infection?

12 Although wasps often walk on our food, they are not suspected of spreading disease. Why do you think they are less harmful in this respect than houseflies?

Methods of preventing spread of disease

Natural barriers

Although many bacteria live on the surface of the skin, the outer layer of the epidermis (p. 145) seems to act as a barrier which stops them getting into the body. But if the skin is cut or damaged, the bacteria may get into the deeper tissues and cause infection.

The sweat glands and sebaceous glands may produce substances which kill bacteria and keep down their numbers on the skin. Tears contain an enzyme called **lysozyme**. This dissolves the cell walls of some bacteria and so protects the eyes from infection.

The acid conditions in the stomach destroy most of the bacteria which may be taken in with food. The moist lining of the nasal passages traps many bacteria. So does the mucus produced by the lining of the trachea and bronchi. The ciliated cells of these organs carry the trapped bacteria away from the lungs.

When bacteria get through these barriers, the body has two more lines of defence—the white cells and the antibodies. The way these work is described on p. 123.

Airborne infection

There is no single way to prevent the spread of airborne pathogens, apart from programmes of immunization against individual diseases.

Children with infectious diseases such as rubella or chickenpox are usually kept away from school during their infectious period. However, it is probably better to catch these 'childhood' diseases when young and so develop immunity, because older people are more severely affected.

You might perhaps reduce your chances of catching a cold or 'flu by avoiding crowded, humid, poorly ventilated places. But you might not enjoy living in permanent isolation.

Similarly, you can hardly stay at home every time you have a cold. But you might infect fewer people if you cough and sneeze into a handkerchief, rather than spraying droplets over a wide area.

In an operating theatre, the surgeon and staff wear sterile gauze masks to trap any droplets that might otherwise infect the patient (Fig. 7).

Figure 7 Prevention of infection in an operating theatre. Staff wear sterile masks and clothing in order to protect the patient from bacteria from their own bodies.

Waterborne infection

On a small scale, simply boiling the water used for drinking will destroy any pathogens. On a large scale, water supplies are protected by (a) ensuring that untreated human sewage cannot reach them and (b) treating the water to make it safe.

Sewage treatment

This has been described on pp. 256–258.

Water treatment

Rain water may be acid, but it is unlikely to contain pathogens. Water from fast flowing rivers which come from deserted hills is not likely to be contaminated. Underground water that has filtered through permeable rock should contain few bacteria. Slow-moving rivers that pass near human settlements are likely to contain some disease bacteria, however.

The treatment used to purify water depends on how clean the source is. Usually it is similar to the process shown in Fig. 8.

Water is pumped from a river or an underground source. River water is passed through a grid to remove floating objects, such as twigs or bottles. The water is allowed to remain for some time in a settling tank where small suspended particles settle out.

The next step is to remove bacteria from the water by passing it through a filter. There are various types of filter. In all of them the water passes through a layer of jelly-like material supported on a bed of sand (Fig. 9). In most cases the gelatinous layer is formed by bacteria and protozoa growing on the sand. The filters eventually become clogged. So from time to time they are cleaned and the jelly layer is re-formed.

Chlorine gas is added to the filtered water and remains in contact with it for long enough to kill any bacteria which have passed through the filter. How much chlorine is added and the length of the contact time both depend on how contaminated the water source is likely to be. Most of the chlorine disappears before the water reaches the consumers.

Ozone is to largely replace chlorine in the next few years.

The purified water is pumped to a high-level reservoir or water tower. These are enclosed to ensure that no pathogens can get into the water. The height of the reservoir provides the pressure needed to deliver the water to the consumer.

Figure 9 Hampton water purification works. The main building contains the pumping equipment and the chlorination plant. The 'ponds' in the foreground are slow sand filters, one of which is drained for cleaning. Behind the pumping station are more filters (left) and two reservoirs (right).

Food-borne infection

People

People who handle and prepare food need to be extremely careful about their personal hygiene. It is essential that they wash their hands before touching food, particularly after they have visited the lavatory (Fig. 10). Hand-washing is also important after handling raw meat, particularly poultry (see p. 348).

Some people carry intestinal pathogens without showing any symptoms of disease. These people are called 'carriers'. Once identified, they should not be allowed to work in canteens or food-processing factories.

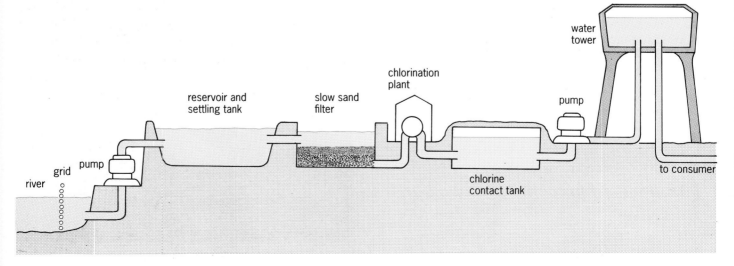

Figure 8 The principles of water purification

Figure 10 Hygienic handling of food. The shop assistant avoids handling meat with the fingers by always interposing a sheet of paper.

Houseflies

Houseflies must be prevented from carrying pathogens to food. There are four main ways to do this:

1 Keep all unwrapped food in fly-proof containers such as refrigerators or larders (Fig. 11).
2 Enclose all food waste in fly-proof dustbins so that the flies cannot pick up bacteria.
3 Never leave human faeces where flies can reach them (if faeces cannot be flushed into the sewage system, they must be buried).
4 Destroy houseflies, wherever possible, in the places where they breed, such as rubbish tips and manure heaps.

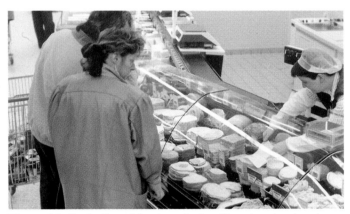

Figure 11 Protection of food on display. The glass barrier stops customers from touching the food and helps stop droplets from coughs and sneezes falling on the food.

Processing

Cooking destroys any bacteria present in food. Refrigeration and freezing slow down or prevent bacterial reproduction. Dehydration, pickling, salting and canning are methods of preserving food. These processes destroy the bacteria which might cause disease. They also stop the food from going bad (see p. 100).

Contagion

Prevention of the spread of tinea is described on p. 350. The methods listed there apply equally well to any contagious disease other than those transmitted sexually.

Sexual transmission

The best way to avoid sexually transmitted diseases is to avoid having sexual intercourse with an infected person. However, the symptoms of the disease are often not obvious and it is difficult to recognize an infected individual. So the disease is avoided by not having sexual intercourse with a person who *might* have the disease. Such persons are (a) prostitutes who offer sexual intercourse for money, (b) people who are known to have had sexual relationships with many others ('slept around'), (c) casual acquaintances whose general background and past sexual activities are not known.

These are good reasons, among many others for being faithful to one partner.

The risk of catching a sexually transmitted disease can be greatly reduced if the man uses a condom (p. 160). This acts as a barrier to bacteria or viruses.

If a person suspects that he or she has caught a sexually transmitted disease, treatment must be sought at once. Information about treatment can be obtained by phoning one of the numbers listed under 'Venereal Disease' or 'Health Information Service' in the telephone directory. Treatment is always confidential. The patients must, however, ensure that anyone they have had sexual contact with also gets treatment. There is no point in one partner being cured if the other is still infected.

Immunization

In the course of an infectious disease, the patient makes antibodies to the pathogen. The antibodies to some diseases remain in the blood for a long time. Others can be produced again very rapidly. So a pathogen which gets into the body for the second time is quickly destroyed. The person is said to be **immune** to that disease. Immunity may last for a few months or for many years, depending on the disease (see also p. 124).

Not all diseases produce immunity. There seems to be no immunity to syphilis or gonorrhoea.

You can acquire immunity to some diseases without actually catching them. One way is to be injected with a harmless form of the pathogen or its toxin (p. 347). This makes your body produce the antibody, without suffering the symptoms of the disease. Then if the real pathogens enter your body, you have the antibodies ready to attack them. You have been **immunized** against that disease.

The substance with which you are injected is called a **vaccine**. The vaccine against whooping cough consists of the killed bacteria. The dead bacteria cannot cause the disease, but they do stimulate the antibody reaction. The vaccines for diphtheria and tetanus are the toxins extracted from the bacteria and made harmless by heat or chemicals. The BCG vaccine against tuberculosis consists of a harmless form of bacteria closely related to the TB bacteria. The polio vaccine is a harmless form of the virus and can be taken by mouth.

You were probably immunized against diphtheria, polio, tetanus, whooping cough and measles during your first five years, and against tuberculosis between 10 and 13 years. Girls are usually immunized against rubella (Fig. 12). Babies may receive a

Figure 12 Immunization. The girl is being immunized against rubella (German measles).

combined inoculation against measles, mumps and rubella (MMR) at 18 months. This replaces the measles vaccination at 15 months and the later rubella inoculation.

There is a small risk of serious side effects from vaccines, just as there is with all medicines. These risks are always far lower than the risk of catching the disease itself. For example, the measles vaccine carries a risk of 1 in 87 000 of causing encephalitis (inflammation of the brain). This is much less than the risk of getting encephalitis as a result of catching measles. Also, the vaccines themselves are becoming much safer, and the risk of side effects is now almost nil.

Routine immunization not only protects the individual but also prevents the spread of infectious disease. Diseases like diphtheria and whooping cough were once common, and are now quite rare. This is the result of improved social conditions and routine immunization. Smallpox was completely wiped out throughout the world by a World Health Organization programme of immunization between 1959 and 1980.

Drug therapy

Drugs cure diseases rather than prevent them. But if a drug cures a disease before the infectious period is over, it might help to reduce the spread of that disease.

Although drugs play a valuable part in keeping us healthy, many people have come to rely on them for trivial ailments. Many diseases are self-limiting—the patient will get better quite quickly without using drugs. However, when some people consult a doctor, they may feel cheated if medicine of some kind or other is not prescribed. In addition, high-pressure advertising gives the idea that headaches, coughs and colds, indigestion and other minor complaints can be relieved by taking patent medicines. In some cases this may be true. But many of these conditions will get better anyway.

There are fewer effective drugs against viruses than against bacteria. This is because the virus is so closely involved in its host cell's physiology that any chemical that harms the virus will also harm the cell.

Antibiotics

The ideal drug for curing disease would be a chemical that destroyed the pathogen without harming the tissues of the host. In practice, modern antibiotics such as penicillin come pretty close to this ideal for bacterial infections. Antibiotics cause the bacterial cell wall to break down by interfering with the chemical processes which maintain it. Since these processes do not occur in animal cells, the host tissues are unaffected. Nevertheless, many antibiotics cause some side effects, such as allergic reactions in some people.

Not all bacteria are killed by antibiotics. Some bacteria have a nasty habit of mutating to forms which are resistant to these drugs. For this reason it is important not to use antibiotics in a diluted form, for too short a period or for trivial complaints. These practices lead to a build-up of a resistant population of bacteria. The drug resistance can be passed from harmless bacteria to pathogens.

QUESTIONS

13 How might a harmful bacterium be destroyed or removed by the body if it arrived
  a on the cornea,
  b on the hand,
  c in a bronchus,
  d in the stomach?
14 After a disaster such as an earthquake, the survivors are urged to boil all drinking water. Why do you think this is so?
15 Revise pp. 123–124. Now explain why immunization against diphtheria does not protect you against polio as well.
16 Even if there have been no cases of diphtheria in a country for many years, children may still be immunized against it. What do you think is the point of this?
17 Why, do you think, has it not been possible to produce a vaccine against the common cold?

Checklist

- Transmissible diseases are infections caused by viruses, bacteria, fungi or protozoa.

- Non-transmissible diseases may be inherited, or acquired during the lifetime.

- Infectious diseases may be transmitted by air, water, food or contact.

- Sexually transmitted diseases are caught during sexual intercourse with an infected person.

- Waterborne diseases are controlled by sewage treatment and water purification.

- Food-borne diseases are controlled by (a) hygienic handling, (b) wrapping or covering, (c) preservation and processing, (d) control of houseflies.

- Sexually transmitted diseases can be prevented by avoiding sexual contact with infected people.

- A vaccine stimulates the blood system to produce antibodies against a disease, without causing the disease itself.

- The presence of antibodies in the blood, or the ability to produce them rapidly, gives immunity to a disease.

- Systematic immunization can protect whole populations.

- Antibiotics are very effective against disease bacteria. But bacteria can become resistant if the antibiotics are used unwisely.

EXAMINATION QUESTIONS

Section 7 Micro-organisms and humans

Do not write on this page. Where necessary copy drawings, tables or sentences.

1 Yeast is used in bread-making. This is because the yeast:
A Produces carbon dioxide   **C** Digests starch
B Produces alcohol   **D** Releases enzymes

2 (a) What does the term biotechnology mean?

(b) Application	Organism
Baking	Saccharomyces cerevisiae
Lager and beer production	Saccharomyces carlsbergensis
Yoghurt production	Streptococcus thermophilus
Vinegar production	Gluconbacter suboxidans

(i) What word is used to describe the type of organisms shown in the table?   (ii) Which organism is used in the production of yoghurt?
(c) Describe THREE methods of preserving food. For each method say what is done to the food to preserve it and how the treatment works. (9)
[M]

3 (a) Complete the table below for the human disorders athlete's foot, syphilis, and influenza, by naming the type of pathogen, ONE major symptom and ONE control method. Parts of the table have been completed for you.

Human disorder	Type of pathogen	ONE major symptom	ONE control method
Athlete's foot		Itchy skin between toes	
Syphilis	Bacterium		
Influenza			None (prevention by vaccination)

(b) Suggest why a control method that works against one of these pathogens may not work against the other two. (8)
[L]

4 (a) Describe how malaria is spread by female mosquitoes. (2)
(b) Attempts to control malaria include controlling mosquitoes and attacking the malarial parasite once it has infected the human host. Describe FOUR methods of control. (8)
[M]

5 An island has a large population of mice. A disease which can kill mice spreads to the island and infects the mice. Write an account of how the mouse population is likely to be affected by the disease over the next few years. Your account should refer to: how the disease may spread; how the organisms that cause the disease are killed inside the bodies of the mice; how resistance to the disease might affect the size of the mouse population over the next few years. Marks will be given both for showing knowledge and understanding and for the way your account is organised and expressed. (8)
[N]

6 We can be immunised against diseases such as diphtheria by having a small quantity of weakened toxins injected into our blood. These toxins are produced by the bacteria.
(a) Why do we not become ill with the disease after the injection? (1)
(b) Explain what happens in our bodies to make us immune to diphtheria after an injection of the weakened toxin. (2)
[S]

7 Listed below are five human diseases.
AIDS   CYSTIC FIBROSIS   GERMAN MEASLES
TETANUS   WHOOPING COUGH
(a) Underline the one inherited disease. (1)
(b) Explain how the disease is inherited. (2)
[NI]

8 The Brown family suffered from food poisoning after eating chicken. Health inspectors identified *Salmonella* bacteria in the remains of the chicken.
(a) The chicken was well cooked which killed the bacteria. Why did the family suffer from food poisoning? (2)
(b) The health inspectors identified *Salmonella* bacteria at the farm which produced the chicken.
(i) Where are *Salmonella* bacteria found inside a chicken? (1)
(ii) How do *Salmonella* bacteria leave one chicken and spread to others? (1)
(c) Mr. Brown had left the chicken in his car on a hot, sunny day before taking it home. How did this help to cause food poisoning? (2)
[L]

9 Antiseptics and antibiotics are both used to kill bacteria.
(a) Use the word ANTISEPTIC or ANTIBIOTIC to complete the following sentence.
Phenol is an .... penicillin is an .... (2)
(b) What substance, ANTISEPTIC or ANTIBIOTIC is used to kill harmful bacteria inside the human body? (1)
(c) Describe TWO ways white blood cells kill harmful bacteria in the human body? (2)
[NI]

APPENDIX 1: REAGENTS

(Only teachers or technicians should prepare these.)

Acetic orcein   It is simplest to buy the concentrated solution (3.3% orcein in glacial acetic acid). Dilute 10 cm^3 with 12 cm^3 water just before use. The diluted stain does not keep.

Benedict's solution   Dissolve 170 g sodium citrate and 100 g sodium carbonate in 800 cm^3 distilled water. Add a solution of copper sulphate made from 17 g copper sulphate in 200 cm^3 distilled water.

Bicarbonate indicator   (*See* Hydrogencarbonate indicator.)

Biuret test   10% sodium hydroxide; 1% copper sulphate.

Cobalt chloride paper   Dissolve 1 g cobalt chloride in 20 cm^3 distilled water. Soak filter paper in the solution and allow to dry.

Ethanoic (acetic) acid (M/10)   Place 6 cm^3 glacial ethanoic (acetic) acid in a graduated flask and make up the volume to 1 litre with distilled water.

Formalin   Dilute 1 part of 40% formaldehyde with 39 parts of water.

Glucose phosphate   Make a 5% solution in distilled water. (Glucose phosphate costs over £2 per gram. Make only enough to meet your needs.)

Hydrochloric acid (M/10)   Dilute 10 cm^3 concentrated acid with 990 cm^3 distilled water.

Hydrogencarbonate indicator (bicarbonate indicator)   Dissolve 0.2 g thymol blue and 0.1 g cresol red powders in 20 cm^3 ethanol. Dissolve 0.84 g 'Analar' sodium hydrogen-carbonate (sodium bicarbonate) in 900 cm^3 distilled water. Add the alcoholic solution to the hydrogencarbonate and make the volume up to 1 litre with distilled water. Just before use, dilute the appropriate amount of this solution 10 times, i.e. add 9 times its own volume of distilled water.

   To bring the solution into equilibrium with atmospheric air, bubble air from outside the laboratory through the diluted indicator using a filter pump or aquarium pump. After 10 minutes, the dye should be red.

Hydrogen peroxide   Use a 20-volume solution, from suppliers or pharmacists.

Hormone rooting powder   Available from hardware shops or garden centres for both woody and soft-stemmed varieties.

Iodine solution   Grind 1 g iodine with 1 g potassium iodide in a mortar with distilled water. Make up to 100 cm^3 and dilute 5 cm^3 of this solution with 100 cm^3 water for experiments.

Lime water   Shake distilled water with an excess of calcium hydroxide and allow the lime to settle. Decant off the clear liquid. Before use, test the liquid by bubbling exhaled air through it.

Manometer liquid   (Experiment 1, p.23) Try water with a few drops of liquid detergent plus a cooking dye such as cochineal substitute.

Methylene blue   Dissolve 0.5 g methylene blue in 30 cm^3 ethanol and dilute with 100 cm^3 distilled water.

PIDCP (Phenol-indo-2,6-dichlorophenol)   Make a 0.1% solution in distilled water.

Pyrogallic acid   Dissolve 10 g pyrogallic acid in 100 cm^3 2N sodium hydroxide solution just before use. Alternatively, place 1 g pyrogallic acid in each flask and add 10 cm^3 2N sodium hydroxide.

Soda-lime   Use the self-indicating form which changes colour when it loses its activity.

Sodium carbonate (M/20)   Dissolve 5.3 g anhydrous sodium carbonate in 1 litre distilled water.

Sodium hydrogencarbonate   Use a 10% solution.

Sugar solution (osmosis experiments)   Dissolve sucrose in its own weight of tap water.

Starch solution (1%)   Shake 1 g starch powder with 100 g water and heat the mixture gently, with stirring, until the liquid just starts to boil.

Water cultures   2 g calcium nitrate, 0.5 g each potassium nitrate, magnesium sulphate and potassium phosphate (KH_2PO_4) in 2 litres distilled water. Add a few drops of iron (II) chloride solution.

   No nitrate; use potassium and calcium chlorides instead of the nitrates.

   No calcium; use potassium nitrate instead of calcium nitrate.

   No phosphate; use potassium sulphate instead of the phosphate.

Water Unless distilled water is specified in this list of reagents, tap water is adequate.

APPENDIX 2: FURTHER EXAMINATION QUESTIONS

Do not write on the following pages. Where necessary copy drawings, tables or sentences.

1 **(a)** The diagram shows a root hair cell from near the tip of a young root.

(i) This cell needs oxygen. Name the process by which oxygen enters the cell from the air in the soil. (1)
(ii) In early spring, the soil is very cold. What effect will this have on the uptake of oxygen? (1)
(b) (i) Describe the process by which water enters the root hair cells. (3)
(ii) A seaside garden is flooded by the sea in an exceptionally severe storm. Several of the plants wilt and die. Explain why flooding with sea water caused the plants to wilt. (2)
[N]

2 The drawing shows a germinating bean seed.

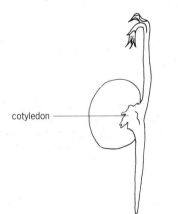

cotyledon

(a) During germination, enzymes are released in the cotyledon to digest stored food.
(i) Name an enzyme which will digest:
(A) stored protein (B) stored starch. (2)
(ii) The enzyme which digests protein cannot digest starch. Why? (1)
(iii) The beans did not germinate when planted in acidic compost. What effect would acid have had on the bean's enzymes? (2)
(b) Humans eat beans. Of what use is the starch stored in a bean to the human body? (2)
[S]

3 **(a)** Describe with full experimental details, how you would compare the growth rates of seedlings of TWO different varieties of the same species of plant. (8)
(b) Man is constantly exploring ways of producing more food of the right type and quality. Artificial selection and biotechnology can each play a part in helping to solve this problem.
Explain what is meant by: (i) artificial selection,
(ii) biotechnology.
In each case, describe ONE method by which the production of food is made more effective. (12)
[M]

4 **(a)** Plants and animals need support to provide shape and form so that they can function effectively and efficiently.
(i) Compare how support is provided during the life of a woody plant with that provided during the life of a mammal. (8)
(ii) Describe how muscles, tendons and bones are involved in the raising of the forearm at the human elbow joint. (5)
(b) All viruses are parasites.
(i) State the meaning of the term parasite.
(ii) State TWO ways in which HIV (human immuno-deficiency virus) is different from each of the following:
(A) a bacterium, (B) a fungus. (4)
(iii) Describe the biological basis of ONE method of preventing the spread of HIV. (2)
[M]

5 **(a)** (i) Explain what is meant by a tissue.
(ii) Why is human blood regarded as a tissue? (6)
(b) Organs are made of tissues which perform certain specialised functions. Describe and explain how this statement is true for each of the following organs:
(i) the leaf of a flowering plant; (ii) the human eye.
(7 each)
[M]

6 What do you understand by the following terms? They are all linked to the nitrogen cycle.
(a) Nitrification (2)
(b) Nitrogen fixation (2)
(c) Leaching (1)
[L]

7 The complex organic compounds found in human food are built up from simpler sub-units. Digestion of food releases these simpler sub-units.
(a) Complete the following table to show which simple sub-units are present in which complex food compounds. (2)

Complex food	Simple sub-unit
Starch	Glucose
Protein	
	Fatty acids and glycerol

(b) (i) Describe how starch is changed into glucose in the human body. (2)
(ii) Why must starch be changed to glucose before it can be used in the human body? (3)
[S]

8 (a) Explain (i) the importance of an increased pulse rate during exercise, and, (ii) why it takes several minutes for the pulse rate to return to normal after exercise. (6)
(b) If an athlete is a regular smoker, explain how this might affect both his pulse and his breathing rate after exercise. (6)
[M]

9 (a) Name two substances in cigarette smoke which can harm a person's respiratory system. (2)
(b) People who have regularly smoked cigarettes for several years may become short of breath when they take exercise. They may also have a persistent cough. Explain what the cigarette smoke may have done to their lungs and respiratory tract to bring about these two effects. (5)
[S]

10 (a) Describe how a molecule of oxygen passes from the air inside an alveolus to a LIVER cell. Include in your account details of the route along which the molecule will pass. (12)
(b) Smoking is known to reduce the supply of oxygen to organs such as the liver. Explain:
(i) the direct and (ii) the indirect effects that smoking can have on the uptake and transport of oxygen in the body. (8)
[M]

11 The diagram shows a section through a human eye with rays of light from a nearby object.

(a) Name the parts labelled A and B. (2)
(b) Give the function of the parts C and D. (2)
(c) Name ONE type of cell found in the retina. (1)
(d) On the diagram, extend the rays of light to show how they would be focused on the retina. (3)
(e) How would the following structures of the eye change if a person moved from a dark room into the bright light?
(i) Pupil
(ii) Eyelid (2)
[NI]

12 (a) Explain the functions of the parts of the human male reproductive system. (7)

(b) Give an account of the reproductive processes which bring about the formation of a mature fetus. (8)
[L]

13 (a) Describe how a one-way flow of blood is maintained through the circulatory system of a mammal. (5)
(b) (i) Describe how oxygen from the air reaches a muscle cell. (4)
(ii) Suggest ways in which exercise increases the rate at which oxygen is supplied to the muscles. (3)
(c) Describe briefly how osmosis is involved with the opening and closing of stomata of a green plant. (5)
[M]

14 (a) (i) Name the process occurring in the diagram below. (1)

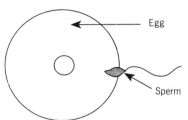

(ii) If the sperm cell carried an X chromosome, what sex would the developing fetus be? (1)
(iii) Give two reasons for the difference in size between the egg and the sperm. (2)
(b) (i) Name a hormone that controls the development of female secondary sexual characteristics. (1)
(ii) Give one example of a human female secondary sexual characteristic. (1)
[NI]

15 (a) (i) What does the term chromosome mean? (2)
(ii) How many chromosomes are there in a human sperm and a fertilised human egg? (2)
(iii) What is the name of the cell division that produces sperms and eggs? (1)
(b) Describe how organisms may have evolved by a process of natural selection. (5)
[M]

16 Very few woodlice are active on a very cold winter day. On the same day, waterlice can be seen moving about freely on the bottom of a shallow pond even when the surface is frozen. Describe THREE features of water which allow this difference to occur. (3)
[L]

17 In the Philippines many people are employed on coconut plantations. A beetle was discovered to be a serious pest to the coconuts. To protect the crop, genetic engineers have produced a parasitic fungus which grows and feeds on the beetle.
(a) What do you understand by the term parasite? (1)
(b) What structure does the fungus produce to infect the beetle? (1)
(c) State two important factors which should be considered before releasing the fungus into the environment. (2)
(d) State two advantages of using the fungus compared to using an insecticide. (2)
[W]

18 The figure is a graph of human death rate and birth rate between 1940 and 1970.

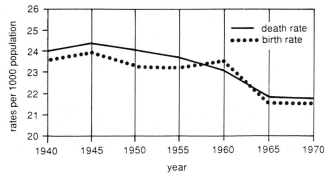

(a) During which TWO periods of time, between 1940 and 1970, was the death rate greater than the birth rate? (2)
(b) Between 1965 and 1970 what were the trends in human death rate and birth rate and how did these affect the population size? (2)
[M]

19 In Atlanta, USA an oral vaccine has been developed which increases the body's natural immunity to tooth decay. We get fewer cavities in our teeth as we get older. When bacteria which cause tooth decay are swallowed, antibodies are produced.
(a) Using words from the statement above, fill in the blanks.
(i) A small amount of .... is given to a person by mouth.
(ii) After a few days the person produces ....
(iii) These act on the .... which cause tooth decay. (3)
(b) Suggest two other reasons why we may have fewer tooth cavities as we get older. (2)
(c) Explain how our natural immunity to tooth decay increases as we get older. (3)
[W]

20 (a) An adult blowfly lands on the surface of an open meat sandwich on a hot summer's day and lays eggs.
(i) Name ONE type of micro-organism which could be transferred from the blowfly to the meat. (1)
(ii) Explain why the growth of some micro-organisms on the meat might have been reduced if a large growth of penicillium mould was present on the bread on which the meat was placed. (4)
(iii) Describe how the blowfly eggs will eventually develop into adult blowflies. (4)
(b) (i) Name TWO factors which can limit the size of a human population. (2)
(ii) What would happen to the size of a human population if these factors were removed, and suggest two social problems that could arise if such a change occurred. (3)
(c) (i) Name TWO organisms which take part in a predator-prey relationship. (2)
(ii) What effect does an increase in the prey population have on the predator population? (1)
[M]

21 A bacterium reproduces by simple division into two. This process, given ideal conditions, can result in billions of bacteria from one bacterium in 24 hours.

(a) (i) Name one human disease caused by bacteria. (1)
(ii) State two ways in which (A) the human body provides ideal conditions for the growth of bacteria which cause this disease, (2)
(B) human blood is able to prevent a continued high reproduction rate of these bacteria. (2)
(b) When an agar plate is exposed to the air and incubated for 24 hours, colonies of bacteria become visible to the naked eye.
(i) Suggest a control experiment that could be used to confirm that the bacteria had come from the air and not the agar. (1)
(ii) Explain how the colonies become visible on the plate. (2)
[W]

22 In West Wales a scientist has set up a leech farm where he breeds thousands of leeches. Enzymes are obtained from leeches. These are used to break down blood clots in humans.
(a) What do you understand by the term enzyme? (2)
(b) Explain how the enzyme might be used in the treatment of people with a type of heart problem. (1)
(c) Leeches belong to the same group as earthworms. State two features of this group. (2)
[W]

23 (a) (i) Describe the structure of a spermatozoon. (2)
(ii) Describe how the parts of the male reproductive system bring about the production of spermatozoa and their release into the body of a female. (8)
(iii) Without the secretions of the prostate gland, sperm passed to the female will be inactive. Suggest what might be present in the secretion from the prostate gland. (2)
(b) A number of diseases can be transmitted by sexual contact.
(i) Name ONE such disease and explain the physical effects and social problems associated with this named disease. (6)
(ii) Suggest TWO ways in which the risk of transmission of such diseases can be reduced. (2)
[M]

24 The graph shows the changes in the amount of chromosome material in the nucleus of dividing cells of a broad bean root over a period of 24 hours.

(a) (i) What is the name given to the process of nuclear division in the root cells? (1)
(ii) In which region of the root would the cell division take place? (1)
(b) (i) What is happening at X? (1)
(ii) Explain what is happening at Y when the amount of chromosome material is halved. (1)
[L]

25 (a) (i) Describe in outline the process of mitosis. (6)
(ii) State TWO ways in which meiosis differs from mitosis and explain the significance of each difference. (4)
(iii) State where meiosis takes place in males and females. (2)
(b) The effect of a mutation occurring during meiosis can have a serious effect on a future child.
(i) What is a mutation? (2)
(ii) Name ONE condition caused by a chromosome mutation occurring during meiosis. Explain how this type of mutation could have occurred and given rise to a child showing the condition. (6)
[M]

26 Sickle cell anaemia is a disease in which people produce abnormal haemoglobin in their red blood cells. Normal haemoglobin is produced by the allele H. Abnormal haemoglobin is produced by the allele S.
(a) Complete the table below to show the cross between a parent with the genotype HH and a parent who has one allele of each type, HS.

Parent genotypes	HH x HS
Gametes	
Offspring genotypes	(2)

(b) (i) Complete the table below to show a cross between two parents who both have one allele of each type.

Parent genotypes	
Gametes	
Offspring genotypes	(3)

(ii) What term is used to describe the genotype of a person who has both the H and the S allele? (1)
(c) How was the S allele first formed? (2)
[L]

27 Over the centuries people have improved the quality of crop plants and farm animals by careful breeding programmes and artificial selection. Use your knowledge of variation and heredity to explain as fully as you can,
(a) how these methods have resulted in crops and animals which are much better than their 'wild' ancestors, and
(b) why it may be beneficial to preserve examples of the original 'wild' species in their natural habitats.
Marks will be given both for showing knowledge and understanding and for the way in which your account is organised and expressed. (10)
[N]

28 When the power station at Chernobyl leaked many people were exposed to high levels of radioactivity.
(a) Explain why the disaster at Chernobyl was so dangerous for the surrounding people. (2)
(b) What affect could this have on the children born to people exposed to the disaster. (1)
[NI]

APPENDIX 3: SOME LANDMARKS IN THE HISTORY OF MEDICINE

Edward Jenner and vaccination

The history of immunisation centres on the disease **smallpox**, which is caused by a virus. Only a few years ago it was a serious, world-wide disease causing hundreds of thousands of deaths.

It had long been noticed that people who had recovered from smallpox, never caught the disease again. In the late 1600s, this observation was exploited in countries such as Greece, Turkey, China and India. Fluid from the blisters, which characterised the disease, was introduced into healthy people through cuts in the skin. The patient suffered a mild form of smallpox but was, thereafter, immune to the disease. It was a risky practice however, and some people developed smallpox and died as a result of the vaccination.

In the 1750s, a Suffolk surgeon, Robert Sutton, refined the technique with considerable success. **Edward Jenner** (1749-1823) is usually given the credit for smallpox vaccination. While using Sutton's technique he noticed that milkmaids, who had caught 'cowpox' from infected cows, did not develop the mild symptoms of illness after vaccination.

In 1796, Jenner conducted a crucial, if somewhat risky, experiment. He took fluid from a blister on a milkmaid's hand and injected it into a young boy. Two months later, he inoculated the boy with smallpox and demonstrated that the boy was immune. After publication of the results, the practice spread widely throughout Europe, reducing deaths from smallpox by about two thirds.

Jenner called his technique 'vaccination' to distinguish it from inoculation with smallpox. 'Vacca' is latin for 'cow' and 'vaccinia' is the medical name for cowpox. We now know that viruses and bacteria often lose much of their virulence if they are allowed to pass through different animals or are cultured in a particular way. Such non-virulent microbes are said to be **attenuated**. Jenner and his contemporaries, of course, knew nothing about viruses or attenuation but their shrewd observations, logical deductions and bold experiments led to a massive reduction in suffering.

In 1967, the World Health Organisation embarked on a programme to eradicate smallpox from the whole world. The strategy was to trace all cases of smallpox and isolate the patients so that they could not pass on the disease. Everyone

Figure 1 Edward Jenner

at risk was then vaccinated. By 1987 the disease had been eradicated.

Joseph Lister and antiseptic surgery

With the discovery of anaesthetics in the mid 1800s surgery became less traumatic, but there was still a high death rate resulting from the wounds becoming infected.

Joseph Lister (1827-1912) was a surgeon practising in Glasgow. In 1865 he read of Louis Pasteur's successful experiments which had led to the **germ theory** of disease. Lister reasoned that if the disease-causing microbes could be killed, the surgical wounds would heal without becoming infected.

Consequently he sterilised his instruments with carbolic acid and had an assistant spraying a mist of carbolic acid over the area of the surgery. This reduced deaths following amputation from 43% to 15%.

Carbolic acid (phenol) certainly killed the pathogens but was a very strong antiseptic, unpleasant to inhale and it interfered with the natural healing process. Subsequently, Lister used milder antiseptics and sterilised his instruments and dressings by heating them.

The sterilisation process was eventually extended to include the surgeon's clothing and the operating theatre itself, giving rise to **aseptic**, rather than **antiseptic** surgery.

Alexander Fleming and penicillin

Before 1934 there were few effective drugs. Some herbal preparations may have been useful; after all, many of our present day drugs are derived from or based on plant products. Quinine, for example, was used for the treatment of malaria and was extracted from a specific kind of tree bark.

In 1935 a group of chemicals called **sulphanilamides** were found to be effective against some bacterial diseases such as blood poisoning, pneumonia and septic wounds.

Alexander Fleming had discovered penicillin in 1928, seven years before the use of sulphanilamides, but he had been unable to purify it and test it on humans. Fleming was a bateriologist working at St. Mary's hospital in London. In 1928 he was studying different strains of *Staphylococcus* bacteria. He had made some cultures on agar plates and left them on the laboratory bench during a four week holiday. When he returned he noticed that one of the plates had been contaminated by a mould fungus and that round the margins of the mould there was a clear zone with no bacteria growing.

Fleming reasoned that a substance had diffused out of the mould colony and killed the bacteria. The mould was identified as *Penicillium notatum* and the supposed anti-bacterial chemical was called penicillin. Fleming went on to culture the *Penicillium* on a liquid meat broth medium and showed that the broth contained penicillin which suppressed the growth of a wide range of bacteria. Two research assistants at St. Mary's tried to obtain a pure sample of penicillin, free from all the other substances in the broth. Although they succeeded, the procedure was cumbersome and the product was unstable. By this time, Fleming seemed to have lost interest and to assume that penicillin would be too difficult to extract and too unstable to be of medical value.

In 1939, **Howard Florey** (a pathologist) and **Boris Chain** (a biochemist) working at Oxford, succeeded in preparing reasonably pure penicillin and making it stable. Techniques of extraction had improved dramatically in 10 years and, in particular, freeze-drying enabled a stable water-soluble powder form of penicillin to be produced.

World War II was an urgent incentive for the production of penicillin in large quantities and this undoubtedly saved many lives that would otherwise have been lost as a result of infected wounds.

Once Boris Chain had worked out the molecular structure of penicillin, it became possible to modify it chemically and produce other forms of penicillin that attacked a different range of bacteria or had different properties. For example, ampicillin is a modified penicillin which can be taken by mouth rather than by injection.

Because penicillin was the product of a mould, chemists searched for other moulds, particularly those present in the soil, which might produce antibiotics. A large number of these were discovered, including streptomycin (for tuberculosis), chloramphenicol (for typhoid), aureomycin and terramycin (broad spectrum antibiotics which attack a wide range of bacteria.) The ideal drug is one which kills or suppresses the growth of harmful cells, such as bacteria or cancer cells, without damaging the body cells. Scientists have been trying for years to find a 'magic bullet' which 'homes in' exclusively on its target cells. For bacterial diseases, antibiotics come pretty close to the ideal, though the bacteria do seem able to develop resistant forms after a few years.

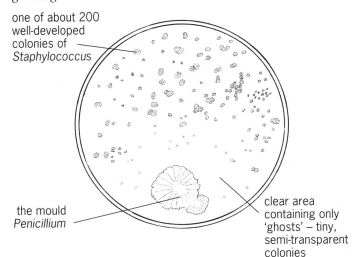

one of about 200 well-developed colonies of *Staphylococcus*

the mould *Penicillium*

clear area containing only 'ghosts' – tiny, semi-transparent colonies

Figure 2 Appearance of the *Staphylococcus* colonies on Fleming's petri dish

GLOSSARY

(A) Scientific Terms

Acid A sharp-tasting chemical, often a liquid. Some acids can dissolve metals and turn them into soluble salts. Nitric acid acts on copper and forms copper nitrate, which dissolves to give a blue-coloured solution. Plant and animal acids (amino acids, fatty acids) are weaker and do not dissolve metals. Amino acids and fatty acids are organic acids. Hydrochloric, sulphuric and nitric acids are called mineral acids or inorganic acids.

Agar A clear jelly extracted from one kind of seaweed. On its own it will not support the growth of bacteria or fungi, but will do so if food substances (e.g. potato juice or Bovril) are dissolved in it. Agar with different kinds of food dissolved in it is used to grow different kinds of micro-organism.

Alcohol Usually a liquid. There are many kinds of alcohol but the commonest is ethanol (or ethyl alcohol) which occurs in wines, spirits, beer, etc. It is produced by fermentation of sugar. Ethanol vaporises quickly and easily catches fire.

Alkali The opposite of an acid. An alkali can neutralise an acid and so remove its acid properties. Sodium hydroxide (NaOH) is an alkali. It neutralises hydrochloric acid (HCl) to form a salt, sodium chloride.

$$Na\boxed{OH + H}Cl \longrightarrow \underset{\text{salt}}{NaCl} \quad + \quad \underset{\text{water}}{H_2O}$$

Atom The smallest possible particle of an element. Even a microscopic piece of iron would be made up of millions of iron atoms. When we write formulae, the letters represent atoms. So H_2O for water means an atom of oxygen joined to two atoms of hydrogen.

Carbon A black, solid non-metal which occurs as charcoal or soot, for example. Its atoms are able to combine together to make ring or chain molecules (see p.12). These molecules make up most of the chemicals of living organisms (see 'Organic'). One of the simplest compounds of carbon is carbon dioxide (CO_2).

Carbon dioxide A gas which forms 0.03 per cent (by volume) of the air. It is produced when carbon-containing substances burn ($C + O_2 \longrightarrow CO_2$). It is also produced by the respiration of plants and animals. It is taken up by green plants to make food during photosynthesis.

Catalyst A substance which makes a chemical reaction go faster but does not get used up in the reaction. Platinum is a catalyst which speeds up the rate at which nitrogen and hydrogen combine to form ammonia, but does not get used up. Enzymes are catalysts for chemical reactions inside living cells.

Caustic A caustic substance can damage the skin and clothing and therefore should be handled with great care.

Compound Two or more elements joined together form a compound. Carbon dioxide, CO_2, is a compound of carbon and oxygen. Potassium nitrate, KNO_3, is a compound of potassium, nitrogen and oxygen.

Cubic centimetre (cm^3) This is a unit of volume. A tea-cup holds about $200cm^3$ liquid. One thousand cubic centimetres are called a cubic decimeter (dm^3) but this volume is also called a litre. Some measuring instruments are marked in millilitres (ml). A millilitre is a thousandth of a litre and therefore the same volume as a cubic centimetre. So $1\ cm^3 = 1$ ml.

Density This is the weight (mass) of a given volume of a substance. Usually it is the weight in grams of one cubic centimetre of the substance, e.g. $1\ cm^3$ lead weighs 11 grams, so its density is 11 grams per cm^3. The density of water at $4^{\circ}C$ is 1 g per cm^3.

Diffusion The random movement of molecules by which gases or dissolved substances move from a region of high concentration to a region of low concentration.

Dissolve A substance which mixes with a liquid and seems to 'disappear' in the liquid is said to dissolve. Sugar dissolves in water to make a solution.

Element An element is a substance which cannot be broken down into anything else. Sulphur is a non-metallic element. Iron is a metallic element. Oxygen and nitrogen are gaseous elements. Water (H_2O) is not an element because it can be broken down into hydrogen and oxygen.

Energy This can be heat, movement, light, electricity, etc. Anything which can be harnessed to do some kind of work is energy. Food consists of substances containing chemical energy. When food is turned into carbon dioxide and water by respiration, energy is released to do work such as making muscles contract.

Filtrate The clear solution which passes through a filter; e.g. if a mixture of copper sulphate solution and sand is filtered, the blue copper sulphate solution which passes through the filter paper is called the filtrate.

Formula A way of showing the chemical composition of a substance. Letters are chosen to represent elements, and numbers show how many atoms of each element are present. The letter for carbon is C and for oxygen is O. A molecule of carbon dioxide is one atom of carbon joined to two atoms of oxygen and the formula is CO_2. There are more elements than letters in the alphabet, so some of the elements have two letters, e.g. Mg for magnesium. Other elements have letters standing for the latin name, e.g. sodium is Na (= natrium).

Gram (g) A unit of weight in the metric system.
  A penny weighs 3.5 grams.
  A pack of butter is 225 grams.
  1000 grams is a kilogram (kg).
  One thousandth of a gram is a milligram (mg).

Hydrogen Hydrogen is a gas which burns very readily. It is present in only tiny amounts in the air but forms part of many compounds such as water (H_2O), and organic compounds like carbohydrates (e.g. $C_6H_{12}O_6$ glucose) and fats.

Inorganic Substances like iron, salt, oxygen and carbon dioxide are inorganic. They do not have to come from a living organism. Salt is in the sea, iron is part of a mineral in the ground, oxygen is in the air. Inorganic substances can be made by industrial processes or extracted from minerals.

Insoluble An insoluble substance is one which will not dissolve. Sugar is soluble in water but insoluble in petrol.

Ion An atom or small group of atoms carrying an electric charge and having different properties from the uncharged atoms. A solution of sodium nitrate in water will consist of ions of positively charged sodium (Na^+) and negatively charged nitrate (NO_3^-).

Joule Just as a centimetre is a unit of length, a joule (J) is a unit of energy. For example 4.2 joules of heat energy will raise the temperature of 1 gram of water by 1^0C. A kilojoule (kJ) is 1000 joules. The older unit of energy, the calorie, is often still used to express the energy value of food. 1 calorie = 4.2 joules.

Lime water A weak solution of lime (calcium hydroxide) in water. When carbon dioxide bubbles through this solution, it reacts with the calcium hydroxide to form calcium carbonate (chalk) which is insoluble and forms a cloudy suspension. This makes lime water a good test for carbon dioxide.

$$Ca(OH)_2 + CO_2 \longrightarrow CaCO_3 + H_2O$$

Manometer An instrument which measures pressure by the displacement of a liquid in a U-tube.

Mass This is the amount of matter in an object. The more mass an object has, the more it weighs, so mass can be measured by weighing something. However, if the force of gravity becomes less, as on the Moon, the same object will weigh less even though the amount of matter in it (its mass) has not changed. So mass and weight are related, but are not the same.

Molecule The smallest amount of a substance which you can have. For example, the water molecule is H_2O, that is, two atoms of hydrogen joined to one atom of oxygen. A drop of water consists of countless millions of molecules of H_2O moving about in all directions and with a lot of space between them.

Organic This usually refers to a substance produced by a living organism. Organic chemicals are things like carbohydrates, protein and fat. They have very large molecules and are often insoluble in water. Inorganic chemicals are usually simple substances like sodium chloride (salt) or carbon dioxide (CO_2).

Oxygen Oxygen is a gas which makes up about 20 per cent (by volume) of the air. It combines with other substances and oxidizes them, sometimes producing heat and light energy. In plants and animals it combines with food to release energy.

Permeable Allows liquids or gases to pass through. A cotton shirt is permeable to rain but a PVC mackintosh is impermeable. Plant cell walls are permeable to water and dissolved substances.

pH This is a measure of how acid or how alkaline a substance is. A pH of 7 is neutral. A pH in the range 8–11 is alkaline; pH's in the 6–2 range are acid; pH 6 is slightly acid; pH2 is very acid.

PIDCP The initials of an organic chemical called phenol-indo-dichlorophenol. It changes from blue to colourless in the presence of certain chemicals, including Vitamin C.

Pigment A chemical which has a colour. Haemoglobin in blood is a red pigment; chlorophyll in leaves is a green pigment. A black pigment called melanin may give a dark colour to human skin, hair and eyes.

Reaction (chemical) A change which takes place when certain chemicals meet or are acted on by heat or light. The change results in the production of new substances. When paper burns, a reaction is taking place between the paper and oxygen in the air.

Salt A salt is a compound formed from an acid and a metal. Salts have double-barrelled names like sodium chloride (NaCl) and potassium nitrate (KNO_3). The first name is usually a metal and the second name is the acid. Potassium (K) is a metal, and the nitrate (NO_3) comes from nitric acid (HNO_3).

Sodium hydrogencarbonate At one time this was called sodium bicarbonate. It is a salt which is used to make carbon dioxide in experiments. Its formula is $NaHCO_3$.

Sodium hydroxide (NaOH) An alkali with caustic properties, i.e. its solution will dissolve flesh, wood and fabrics.

Soluble A soluble substance is one which will dissolve in a liquid. Sugar is soluble in water.

Solution When something like sugar or salt dissolves in water it forms a solution. The molecules of the solid become evenly spread through the liquid.

Volume The amount of space something takes up, or the amount of space inside it. A milk bottle has an internal volume of one pint. Your lungs have a volume of about 5 litres; they can hold up to 5 litres of air. This cube has a volume of 8 cubic centimetres (8 cm^3).

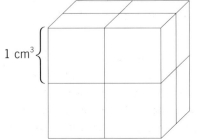

1 cm^3

(B) Biological Terms

(References in brackets are to pages.)

Abdomen (291) The part of the body below the diaphragm which contains stomach, kidneys, liver, etc.; in insects it refers to the third region of the body.

Accommodation (174) Changing the shape (and focal length) of the eye lens to focus on near or distant objects.

Active transport (30) The transport of a substance across a cell membrane with expenditure of energy, often against a concentration gradient.

Adaptation (43) The development, during evolution of an organism, of structures or processes which make it more efficient in its environment.

Alleles (212) Alternative forms of a gene, occupying the same place on a chromosome and affecting the same characteristics but in different ways.

Anabolism (22) The building up of complex substances from simpler ones.

Antibiotic (342) An anti-bacterial drug derived from a fungus or bacterium.

Aseptic technique (305) Method of handling materials or apparatus so that unwanted micro-organisms are excluded.

Asexual reproduction (321) Reproduction without the involvement of gametes.

Assimilation (110) Absorption of substances which are then built into other compounds in the organism.

Autotroph (309) An organism which can build up its organic materials from inorganic substances.

Auxin (336) A chemical which affects the rate of growth in plants.

Basal metabolism (22) The minimum rate of chemical activity needed to keep an organism alive.

Biodegradable (258) Able to be broken down to simple inorganic substances by the action of bacteria and fungi.

Biomass (233) The weight (mass) of all the organisms in a population, community or habitat.

Biosphere (266) The part of the Earth which contains living organisms.

Biotechnology (340) The use of living organisms or biological processes for industrial, agricultural or medical processes.

Biotic factors (277) The environmental effects of other living things on an organism.

Cardiac To do with the heart.

Catabolism (22) The breakdown of complex substances to simpler substances in the cell, with a release of energy.

Clone (206) A population of organisms derived by asexual reproduction from a single individual.

Control (24) An experiment which is set up to ensure that only the condition being investigated has affected the results.

Co-ordination (179) The process which makes the different systems in an organism work effectively together.

Cortex (140) An outer layer.

Denature (12) Destroy the structure of a protein by means of heat or chemicals.

Detoxication (113) The process by which the liver makes poisonous chemicals harmless.

Dialysis (31) The separation of small molecules from large molecules in solution by a selectively permeable membrane.

Ecosystem (265) A community of inter-dependent organisms and the environment in which they live.

Emulsify (108) Break-up of oil or fat into tiny droplets which remain suspended in water as an emulsion.

Eutrophic (246) An aquatic environment well supplied with nutrients for plant growth.

Fermentation (21) A form of anaerobic respiration in which carbohydrate is broken down to carbon dioxide and, in some cases, alcohol.

Gastric (108) To do with the stomach.

Gene (202) A sequence of chemicals in a chromosome which controls the development of a particular characteristic in an organism.

Genetic code (203) The sequence of bases in a molecule of DNA which specifies the order of amino acids in protein.

Genetic engineering (206) Altering the genetic constitution of an organism by introducing new DNA into its chromosomes.

Genotype (211) The combination of genes present in an organism.

Genus (285) One of the categories in classification; a group of closely related species.

Gestation (156) The period of growth and development of a fetus in the uterus of a mammal.

Hepatic (120) To do with the liver.

Heterotroph (309) An organism which feeds by taking in organic substances made by other organisms.

Heterozygous (211) Carrying a pair of contrasted genes for any one heritable characteristic; will not breed true for this characteristic.

Homeostasis (112) Keeping the composition of the body fluids within narrow limits.

Homoiothermic (149) Animals whose body temperature is maintained at a constant level, usually above that of their surroundings. Sometimes called 'warm-blooded'.

Homologous chromosomes (200) A pair of corresponding chromosomes of the same shape and size; one from each parent.

Homozygous (211) Possessing a pair of identical genes controlling the same characteristic; will breed true for this characteristic.

Hypothesis (26) A provisional explanation for an observation; it can be tested by experiments.

Immunity (125) Ability of an organism to resist infection, usually because it carries antibodies in its blood.

Implantation (154) The process in which an embryo becomes attached to the lining of the uterus.

Incubate (306) Maintain at a raised temperature, e.g. birds' eggs or bacteria cultures.

Inflorescence (72) A group of flowers on the same stalk.

Inhibit (188) Slow down a process, or prevent its happening.

Inoculation (126) Deliberate infection with a mild form of disease to stimulate the formation of antibodies. (In the case of culture methods for bacteria or fungi, 'inoculation' means introducing the organism to the culture medium.)

Laparoscopy (160) A method of examining the inside of the abdomen by inserting an optical instrument (an endoscope) through the abdominal wall.

Laparotomy (160) Any operation which involves opening or entering the abdomen by surgery.

Limiting factor (43) A condition which limits the rate of a process, e.g. shortage of light for photosynthesis.

Metabolism (22) All the chemical changes going on in the cells of an organism which keep it alive.

Metamorphosis (326) The relatively sudden change by which the larval form of an insect or amphibian becomes an adult.

Monoculture (243) Growing a single species of crop plant, usually in the same ground for successive years.

Mutation (205) A spontaneous change in a gene or chromosome, which may affect the appearance or physiology of an organism.

Parasite (309) An organism living in or on another organism (the host). The parasite derives its food from the host.

Pathogen (263) A parasite which causes disease or harms its host in other ways.

Phenotype (211) The observable characteristics of an organism which are genetically controlled.

Plankton (233) The community of small plants and animals floating in the surface waters of an aquatic environment.

Plasmid (206) Small circle of DNA in a bacterium. Often carries genes for antibiotic resistance. Used in genetic engineering to introduce foreign DNA.

Poikilothermic (149) Having a body temperature which fluctuates with that of the environment. Sometimes referred to (inaccurately) as 'cold-blooded'.

Predator (113) An animal which kills and eats other animals.

Protoctista (228) Single-celled organisms which have a proper nucleus.

Protophyta (284) Those protoctista which have chlorophyll and make their food by photosynthesis.

Protozoa (257) Those protoctista which take in solid food and digest it.

Puberty (158) The period of growth during which humans become sexually mature.

Receptor (171) A sense organ which detects a stimulus.

Recessive (210) A gene which, in the presence of its contrasting allele, is not expressed in the phenotype.

Recombinant (206) Refers to the insertion, by genetic engineering, of foreign DNA into a genotype.

Recycling (235, 259) As a biological term this means the return of matter to the soil, air or water, and its re-use by other organisms. In daily life, it means the re-use of manufactured materials such as paper, glass and metals.

Renal (120) To do with the kidneys.

Replication (204) Production of a duplicate set of chromosomes prior to cell division.

Section (2) A thin slice of tissue which can be examined under the microscope.

Sensitivity (335) The ability to detect and respond to a stimulus.

Sphincter (107) A band of circular muscle which can contract to constrict or close a tubular organ.

Spore (295) A cell or small group of cells which can grow into a new organism.

Stimulus (88) An event in the surroundings or in the internal anatomy of an organism, which provokes a response.

Stoma (54) A structure, in the epidermis of a plant, which consists of a pore enclosed by two guard cells. It permits gaseous exchange with the atmosphere.

Symbiosis (313) A very close association between two unrelated organisms. Each organism derives some benefit from the association. May also be used to describe any close association, whether beneficial or harmful.

Toxin (125) A poisonous protein produced by pathogenic bacteria.

Toxoid (126) A toxin which has been treated to make it harmless, but can still cause the body to make antibodies.

Transgenic (207) An organism carrying DNA derived from an unrelated organism.

Trophic level (233) An organism's position in a food chain, e.g. primary or secondary consumer.

Turgor (32) The pressure built up in a plant cell as a result of taking in water by osmosis.

Vascular To do with vessels: blood vessels or xylem and phloem.

(C) Some Derivations

A great many terms used in Biology are derived from Greek or Latin words. In some cases, knowing the general meaning of a prefix or suffix helps to understand or recognize the term. Unless indicated as Latin (L), all the words below are Greek.

autos = self, e.g. autotroph
bios = life, e.g. biology, biomass
bis (L) = twice, e.g. binary fission, bicuspid valve
chloros = pale green, e.g. chlorophyll
chroma = colour, e.g. chromosome (takes up coloured stains)
dia = across, e.g. diaphragm, dialysis
dis = twice, e.g. dipeptide, diploid
ektos = outside, e.g. ectoparasite
epi = upon (above), e.g. epidermis, epicotyl
exo = outside, e.g. exoskeleton, exocytosis
haima = blood, e.g. haemoglobin, haemophilia
heteros = other (i.e. different), e.g. heterozygous
homos = same, e.g. homozygous, homologous
hypo = under, e.g. hypothermia, hypocotyl
inter (L) =between, e.g. inter-cellular, intercostal
intra (L) = within, e.g. intra-cellular, intra-uterine
kytos = vessel (a cell), e.g. cytoplasm, leucocyte
lipos = fat, e.g. lipid, lipase
lysis = dissolution, e.g. lysozyme, dialysis
mesos = middle, e.g. mesophyll, mesenteric
meta = after (change), e.g. metamorphosis
mikros = little, e.g. microvilli, micropyle
morphe = form (shape), e.g. metamorphosis, morphology
phagein = to eat, e.g. phagocyte, oesophagus
phyllon = leaf, e.g. mesophyll, chlorophyll
phyton = plant, e.g. phytoplankton, saprophyte
polys = many, e.g. polypeptide, polysaccharide
protos = first formed, e.g. protista, protoplasm
rhiza = root, e.g. rhizoid, rhizome
semi(L) = half, e.g. semi-lunar valve, semicircular canal
sub (L) = under, e.g. subsoil, subclavian
sym-, syn- = together, e.g. symbiosis, synapse
treis = three, e.g. tripeptide, tricuspid valve
trophe = food, e.g. autotroph, trophic level
unus (L) = one, e.g. unicellular, unisexual
vas (L) = vessel, e.g. vascular bundle, vaso-dilation
zoion = animal, e.g. zooplankton, zoology
zygon = yolk, e.g. zygote, homozygous

INDEX